Wavelet Transforms and Their Applications

Lokenath Debnath

Wavelet Transforms and Their Applications

With 69 Figures

Birkhäuser
Boston • Basel • Berlin

Lokenath Debnath
Department of Mathematics
University of Texas—Pan American
Edinburg, TX 78539-2999
USA

Library of Congress Cataloging-in-Publication Data
Debnath, Lokenath.
Wavelet transforms and their applications / Lokenath Debnath.
p. cm.
Includes bibliographical references and index.
ISBN 0-8176-4204-8 (alk. paper)
1. Wavelets (Mathematics) 2. Signal processing—Mathematics. I. Title.
QA403.3 .D43 2001
621.382'2—dc21 2001035266

Printed on acid-free paper.

ISBN 0-8176-4204-8
ISBN 3-7643-4204-8 SPIN 10773883

Production managed by Louise Farkas; manufacturing supervised by Jerome Basma.
Typeset by the author.
Printed and bound by Edwards Brothers, Inc., Ann Arbor, MI.
Printed in the United States of America.

9 8 7 6 5 4 3 2 1

Birkhäuser Boston Basel Berlin
A member of BertelsmannSpringer Science+Business Media GmbH

Contents

Preface

Overview

Historically, the concept of "ondelettes" or "wavelets" originated from the study of time-frequency signal analysis, wave propagation, and sampling theory. One of the main reasons for the discovery of wavelets and wavelet transforms is that the Fourier transform analysis does not contain the local information of signals. So the Fourier transform cannot be used for analyzing signals in a joint time and frequency domain. In 1982, Jean Morlet, in collaboration with a group of French engineers, first introduced the idea of wavelets as a family of functions constructed by using translation and dilation of a single function, called the *mother wavelet*, for the analysis of nonstationary signals. However, this new concept can be viewed as the synthesis of various ideas originating from different disciplines including mathematics (Calderón-Zygmund operators and Littlewood-Paley theory), physics (coherent states in quantum mechanics and the renormalization group), and engineering (quadratic mirror filters, sideband coding in signal processing, and pyramidal algorithms in image processing).

Wavelet analysis is an exciting new method for solving difficult problems in mathematics, physics, and engineering, with modern applications as diverse as wave propagation, data compression, image processing, pattern recognition, computer graphics, the detection of aircraft and submarines, and improvement in CAT scans and other medical image technology. Wavelets allow complex information such as music, speech, images, and patterns to be decomposed into elementary forms, called the *fundamental building blocks*, at different positions and scales and subsequently reconstructed with high precision. With ever greater demand for mathematical tools to provide both theory and applications for science and engineering, the utility and interest of wavelet analysis seem more clearly established than ever. Keeping these things in mind, our main goal in this modest book has been to provide both a systematic exposition of the basic ideas and results of wavelet transforms and some applications in time-frequency signal analysis and turbulence.

Audience and Organization

This book is appropriate for a one-semester course in wavelet transforms with applications. There are two basic prerequisites for this course: Fourier transforms and Hilbert spaces and orthonormal systems. The book is also intended to serve as a ready reference for the reader interested in advanced study and research in various areas of mathematics, physics, and engineering to which wavelet analysis can be applied with advantage. While teaching courses on integral transforms and wavelet transforms, the author has had difficulty choosing textbooks to accompany lectures on wavelet transforms at the senior undergraduate and/or graduate levels. Parts of this book have also been used to accompany lectures on special topics in wavelet transform analysis at U.S. and Indian universities. I believe that wavelet transforms can be best approached through a sound knowledge of Fourier transforms and some elementary ideas of Hilbert spaces and orthonormal systems. In order to make the book self-contained, Chapters 2 and 3 deal with Hilbert spaces and orthonormal systems and Fourier transforms with examples of applications. It is not essential for the reader to know everything about these topics, but limited knowledge of at least some of them would be sufficient. There is plenty of material in this book for a one-semester graduate-level course for mathematics, science, and engineering students. Many examples of applications to problems in time-frequency signal analysis and turbulence are included.

The first chapter gives a brief historical introduction and basic ideas of Fourier series and Fourier transforms, Gabor transforms, and the Wigner-Ville distribution with time-frequency signal analysis, wavelet transforms, wavelet bases, and multiresolution analysis. Some applications of wavelet transforms are also mentioned.

Chapter 2 deals with Hilbert spaces and orthonormal systems. Special attention is given to the theory of linear operators on Hilbert spaces, with some emphasis on different kinds of operators and their basic properties. The fundamental ideas and results are discussed, with special attention given to orthonormal systems, linear functionals, and the Riesz representation theorem.

The third chapter is devoted to the theory of Fourier transforms and their applications to signal processing, differential and integral equations, and mathematical statistics. Several important results including the approximate identity theorem, convolution theorem, various summability kernels, general Parseval relation, and Plancherel's theorem are discussed in some detail. Included are Poisson's summation formula, Gibbs's phenomenon, the Shannon sampling theorem, and Heisenberg's uncertainty principle.

Chapter 4 is concerned with classification of signals, joint time-frequency analysis of signals, and the Gabor transform and its basic properties, including the inversion formula. Special attention is given to frames and frame operators, the discrete Gabor transform, and the Gabor representation problem. Included are the Zak transform, its basic properties, including the Balian-Low theorem, and applications for studying the orthogonality and completeness of Gabor frames in the critical case.

The Wigner-Ville distribution and time-frequency signal analysis are the main topics of Chapter 5. The basic structures and properties of the Wigner-Ville distribution and the ambiguity function are discussed in some detail. Special attention is paid to fairly exact mathematical treatment with examples and applications in the time-frequency signal analysis. The relationship between the Wigner-Ville distribution and ambiguity functions is examined with radar signal analysis. Recent generalizations of the Wigner-Ville distribution are briefly described.

Chapter 6 is devoted to wavelets and wavelet transforms with examples. The basic ideas and properties of wavelet transforms are discussed with special emphasis given to the use of different wavelets for resolution and synthesis of signals. This is followed by the definition and properties of discrete wavelet transforms.

In Chapter 7, the idea of multiresolution analysis with examples and construction of wavelets is described in some detail. This chapter includes properties of scaling functions and orthonormal wavelet bases and construction of orthonormal wavelets. Also included are treatments of Daubechies' wavelet and algorithms, discrete wavelet transforms, and Mallat's pyramid algorithm.

Chapter 8 deals with Newland's harmonic wavelets and their basic properties. Special attention is given to properties of harmonic scaling functions, wavelet expansions, and Parseval's formula for harmonic wavelets.

The final chapter is devoted to a brief discussion of the Fourier transform analysis and the wavelet transform analysis of turbulence based on the Navier-

Stokes equations and the equation of continuity. Included are fractals, multifractals, and singularities in turbulence. This is followed by Farge's and Meneveau's wavelet transform analyses of turbulence in some detail. Special attention is given to the adaptive wavelet method for computation and analysis of turbulent flows.

Salient Features

The book contains a large number of worked examples, examples of applications, and exercises which are either directly associated with applications or phrased in terms of mathematical, physical, and engineering contexts in which theory arises. It is hoped that they will serve as useful self-tests for understanding of the theory and mastery of wavelets, wavelet transforms, and other related topics covered in this book. A wide variety of examples, applications, and exercises should provide something of interest for everyone. The exercises truly complement the text and range from elementary to the challenging.

This book is designed as a new source for modern topics dealing with wavelets, wavelet transforms, Gabor transforms, the Wigner-Ville distribution, multiresolution analysis, and harmonic wavelets and their applications for future development of this important and useful subject. Its main features are listed below:

1. A detailed and clear explanation of every concept and method which is introduced, accompanied by carefully selected worked examples, with special emphasis being given to those topics in which students experience difficulty.
2. Special emphasis is given to the joint time-frequency signal analysis and the ambiguity functions for the mathematical analysis of sonar and radar systems.
3. Sufficient flexibility in the book's organization so as to enable instructors to select chapters appropriate to courses of different lengths, emphases, and levels of difficulty.
4. A wide spectrum of exercises has been carefully chosen and included at the end of each chapter so that the reader may develop both manipulative skills in the theory and applications of wavelet analysis and a deeper insight into this most modern subject. Answers and hints for selected exercises are provided at the end of the book for additional help to students.

5. The book provides important information that puts the reader at the forefront of current research. An updated Bibliography is included to stimulate new interest in future study and research.

Acknowledgments

In preparing the book, the author has been encouraged by and has benefited from the helpful comments and criticism of a number of faculty and postdoctoral and doctoral students of several universities in the United States, Canada, and India. The author expresses his grateful thanks go to these individuals for their interest in the book. My special thanks go to Jackie Callahan and Ronee Trantham who typed a manuscript with many diagrams and cheerfully put up with constant changes and revisions. In spite of the best efforts of everyone involved, some typographical errors doubtlessly remain. I do hope that these are both few and obvious and will cause minimal confusion. Finally, the author wishes to express his special thanks to Lauren Schultz, associate editor, Wayne Yuhasz, executive editor, and the staff of Birkhäuser for their help and cooperation. I am deeply indebted to my wife, Sadhana, for her understanding and tolerance while the book was being written.

Edinburg, Texas *Lokenath Debnath*

Chapter 1

Brief Historical Introduction

"If you wish to foresee the future of mathematics our proper course is to study the history and present condition of the science."

Henri Poincaré

1.1 Fourier Series and Fourier Transforms

Historically, Joseph Fourier (1770-1830) first introduced the remarkable idea of expansion of a function in terms of trigonometric series without giving any attention to rigorous mathematical analysis. The integral formulas for the coefficients of the Fourier expansion were already known to Leonardo Euler (1707-1783) and others. In fact, Fourier developed his new idea for finding the solution of heat (or Fourier) equation in terms of Fourier series so that the Fourier series can be used as a practical tool for determining the Fourier series solution of partial differential equations under prescribed boundary conditions. Thus, the Fourier series of a function $f(x)$ defined on the interval $(-\ell,\ell)$ is given by

$$f(x)=\sum_{n=-\infty}^{\infty} c_n \exp\left(\frac{in\pi x}{\ell}\right), \tag{1.1.1}$$

where the Fourier coefficients are

$$c_n=\frac{1}{2\ell}\int_{-\ell}^{\ell} f(t)\exp\left(-\frac{in\pi t}{\ell}\right) dt. \tag{1.1.2}$$

In order to obtain a representation for a non-periodic function defined for all real x, it seems desirable to take the limit as $\ell\to\infty$, that leads to the formulation of the famous Fourier integral theorem:

$$f(x)=\frac{1}{2\pi}\int_{-\infty}^{\infty}e^{i\omega x}\,d\omega\int_{-\infty}^{\infty}e^{-i\omega t}\,f(t)\,dt. \tag{1.1.3}$$

Mathematically, this is a continuous version of the completeness property of Fourier series. Physically, this form (1.1.3) can be resolved into an infinite number of harmonic components with continuously varying frequency $\left(\frac{\omega}{2\pi}\right)$ and amplitude,

$$\frac{1}{2\pi}\int_{-\infty}^{\infty}f(t)\,e^{-i\omega t}\,dt, \tag{1.1.4}$$

whereas the ordinary Fourier series represents a resolution of a given function into an infinite but discrete set of harmonic components. The most significant method of solving partial differential equations in closed form, which arose from the work of P.S. Laplace (1749-1827), was the Fourier integral. The idea is due to Fourier, A.L. Cauchy (1789-1857), and S.D. Poisson (1781-1840). It seems impossible to assign priority for this major discovery, because all three presented papers to the Academy of Sciences of Paris simultaneously. They also replaced the Fourier series representation of a solution of partial differential equations of mathematical physics by an integral representation and thereby initiated the study of Fourier integrals. At any rate, the Fourier series and Fourier integrals, and their applications were the major topics of Fourier's famous treatise entitled *Théore Analytique de la Chaleur* (*The Analytical Theory of Heat*) published in 1822.

In spite of the success and impact of Fourier series solutions of partial differential equations, one of the major efforts, from a mathematical point of view, was to study the problem of convergence of Fourier series. In his seminal paper of 1829, P.G.L. Dirichlet (1805-1859) proved a fundamental theorem of pointwise convergence of Fourier series for a large class of functions. His work has served as the basis for all subsequent developments of the theory of Fourier series which was profoundly a difficult subject. G.F.B. Riemann (1826-1866) studied under Dirichlet in Berlin and acquired an interest in Fourier series. In 1854, he proved necessary and sufficient conditions which would give convergence of a Fourier series of a function. Once Riemann declared that Fourier was the first who understood the nature of trigonometric series in an exact and complete manner. Later on, it was recognized that the Fourier series of a continuous function may diverge on an arbitrary set of measure zero. In 1926, A.N. Kolmogorov proved that there exists a Lebesgue integrable function whose

Fourier series diverges everywhere. The fundamental question of convergence of Fourier series was resolved by L. Carleson in 1966 who proved that the Fourier series of a continuous function converges almost everywhere.

In view of the abundant development and manifold applications of the Fourier series and integrals, the fundamental problem of series expansion of an arbitrary function in terms of a given set of functions has inspired a great deal of modern mathematics.

The Fourier transform originated from the Fourier integral theorem that was stated in Fourier's treatise entitled *La Théorie Analytique de la Chaleur,* and its deep significance has subsequently been recognized by mathematicians and physicists. It is generally believed that the theory of Fourier series and Fourier transforms is one of the most remarkable discoveries in the mathematical sciences and has widespread applications in mathematics, physics, and engineering. Both Fourier series and Fourier transforms are related in many important ways. Many applications, including the analysis of stationary signals and real-time signal processing, make an effective use of the Fourier transform in time and frequency domains. The Fourier transform of a signal or function $f(t)$ is defined by

$$\mathscr{F}\{f(t)\} = \hat{f}(\omega) = \int_{-\infty}^{\infty} \exp(-i\omega t) f(t)\, dt = \left(f, e^{i\omega t}\right), \tag{1.1.5}$$

where $\hat{f}(\omega)$ is a function of frequency ω and $\left(f, e^{i\omega t}\right)$ is the inner product in a Hilbert space. Thus, the transform of a signal decomposes it into a sine wave of different frequencies and phases, and it is often called the *Fourier spectrum.*

The remarkable success of the Fourier transform analysis is due to the fact that, under certain conditions, the signal $f(t)$ can be reconstructed by the Fourier inversion formula

$$f(t) = \mathscr{F}^{-1}\left\{\hat{f}(\omega)\right\} = \frac{1}{2\pi} \int_{-\infty}^{\infty} \exp(i\omega t)\, \hat{f}(\omega)\, d\omega = \frac{1}{2\pi}\left(\hat{f}, e^{-i\omega t}\right). \tag{1.1.6}$$

Thus, the Fourier transform theory has been very useful for analyzing harmonic signals or signals for which there is no need for local information.

On the other hand, Fourier transform analysis has also been very useful in many other areas, including quantum mechanics, wave motion, and turbulence. In these areas, the Fourier transform $\hat{f}(k)$ of a function $f(x)$ is defined in the space and wavenumber domains, where x represents the space variable and k is the wavenumber. One of the important features is that the trigonometric kernel $\exp(-ikx)$ in the Fourier transform oscillates indefinitely, and hence, the localized information contained in the signal $f(x)$ in the x-space is widely distributed among $\hat{f}(k)$ in the Fourier transform space. Although $\hat{f}(k)$ does not lose any information of the signal $f(x)$, it spreads out in the k-space. If there are computational or observational errors involved in the signal $f(x)$, it is almost impossible to study its properties from those of $\hat{f}(k)$.

In spite of some remarkable successes, Fourier transform analysis seems to be inadequate for studying the above physical problems for at least two reasons. First, the Fourier transform of a signal does not contain any local information in the sense that it does not reflect the change of wavenumber with space or of frequency with time. Second, the Fourier transform method enables us to investigate problems either in the time (space) domain or in the frequency (wavenumber) domain, but not simultaneously in both domains. These are probably the major weaknesses of the Fourier transform analysis. It is often necessary to define a single transform of time and frequency (or space and wavenumber) that can be used to describe the energy density of a signal simultaneously in both time and frequency domains. Such a single transform would give complete time and frequency (or space and wavenumber) information of a signal.

1.2 Gabor Transforms

In quantum mechanics, the Heisenberg uncertainty principle states that the position and momentum of a particle described by a wave function $\psi \in L^2(\mathbb{R})$ cannot be simultaneously and arbitrarily small. Motivated by this principle in 1946, Dennis Gabor, a Hungarian-British physicist and engineer who won the 1971 Nobel Prize in physics, first recognized the great importance of localized time and frequency concentrations in signal processing. He then introduced the windowed Fourier transform to measure localized frequency components of sound waves. According to the Heisenberg uncertainty principle, the energy spread of a signal and its Fourier transform cannot be simultaneously and arbitrarily small. Gabor first identified a signal with a family of waveforms which are well-concentrated in time and in frequency. He called these elementary waveforms as the *time-frequency atoms* that have a minimal spread in a time-frequency plane.

In fact, Gabor formulated a fundamental method for decomposition of signals in terms of elementary signals (or atomic waveforms). His pioneering approach has now become one of the standard models for time-frequency signal analysis.

In order to incorporate both time and frequency localization properties in one single transform function, Gabor first introduced the *windowed Fourier transform* (or the *Gabor transform*) by using a Gaussian distribution function as a window function. His major idea was to use a time-localization window function $g_a(t-b)$ for extracting local information from the Fourier transform of a signal, where the parameter a measures the width of the window, and the parameter b is used to translate the window in order to cover the whole time domain. The idea is to use this window function in order to localize the Fourier transform, then shift the window to another position, and so on. This remarkable property of the Gabor transform provides the local aspect of the Fourier transform with time resolution equal to the size of the window. In fact, Gabor (1946) used $g_{t,\omega}(\tau)=\bar{g}(\tau-t)\exp(i\omega\tau)$ as the window function by translating and modulating a function g, where $g(\tau)=\pi^{-\frac{1}{4}}\exp\left(-2^{-1}\tau^2\right)$, which is the so-called *canonical coherent states in quantum physics*. The *Gabor transform* (*windowed Fourier transform*) of f with respect to g, denoted by $\tilde{f}_g(t,\omega)$, is defined as

$$\mathcal{G}[f](t,\omega)=\tilde{f}_g(t,\omega)=\int_{-\infty}^{\infty} f(\tau)g(\tau-t)\,e^{-i\omega\tau}d\tau=\left(f,\bar{g}_{t,\omega}\right), \tag{1.2.1}$$

where $f,g\in L^2(\mathbb{R})$ with the inner product (f,g). In practical applications, f and g represent signals with finite energy. In quantum mechanics, $\tilde{f}_g(t,\omega)$ is referred to as the *canonical coherent state representation* of f. The term "coherent states" was first used by Glauber (1964) in quantum optics. The inversion formula for the Gabor transform is given by

$$\mathcal{G}^{-1}\left[\tilde{f}_g(t,\omega)\right]=f(\tau)=\frac{1}{2\pi}\frac{1}{\|g\|^2}\int_{-\infty}^{\infty}\int_{-\infty}^{\infty}\tilde{f}_g(t,\omega)\bar{g}(\tau-t)e^{i\omega t}dt\,d\omega. \tag{1.2.2}$$

In terms of the sampling points defined by $t=mt_0$ and $\omega=n\omega_0$, where m and n are integers and ω_0 and t_0 are positive quantities, the discrete Gabor functions are defined by $g_{m,n}(t)=\bar{g}(t-mt_0)\exp(-in\omega_0 t)$. These functions are called the

Weyl-Heisenberg coherent states, which arise from translations and modulations of the Gabor window function. From a physical point of view, these coherent states are of special interest. They have led to several important applications in quantum mechanics. Subsequently, various other functions have been used as window functions instead of the Gaussian function that was originally introduced by Gabor. The discrete Gabor transform is defined by

$$\tilde{f}(m,n) = \int_{-\infty}^{\infty} f(t)\, \overline{g}_{m,n}(t)\, dt = \left(f, g_{m,n}\right). \tag{1.2.3}$$

The double series $\sum_{m,n=-\infty}^{\infty} \tilde{f}(m,n)\, g_{m,n}(t)$ is called the *Gabor series* of $f(t)$.

In many applications, it is more convenient, at least from a numerical point of view, to deal with discrete transforms rather than continuous ones.The discrete Gabor transform is defined by

$$\tilde{f}\left(mt_0, n\omega_0\right) = \frac{1}{\sqrt{2\pi}} \int_{-\infty}^{\infty} f(\tau)\, g_{m,n}(\tau)\, d\tau = \frac{1}{\sqrt{2\pi}} \left(f, \overline{g}_{m,n}\right). \tag{1.2.4}$$

If the functions $\left\{g_{m,n}(t)\right\}$ form an orthonormal basis or, more generally, if they form a frame on $L^2(\mathbb{R})$, then $f \in L^2(\mathbb{R})$ can be reconstructed by the formula

$$f(t) = \sum_{m,n=-\infty}^{\infty} \left(f, g_{m,n}\right) g_{m,n}^{*}(t), \tag{1.2.5}$$

where $\left\{g_{m,n}^{*}(t)\right\}$ is the dual frame of $\left\{g_{m,n}(t)\right\}$. The discrete Gabor transform deals with a discrete set of coefficients which allows efficient numerical computation of those coefficients. However, Henrique Malvar (1990a,b) recognized some serious algorithmic difficulties in the Gabor wavelet analysis. He resolved these difficulties by introducing new wavelets which are now known as the *Malvar wavelets* and fall within the general framework of the window Fourier analysis. From an algorithmic point of view, the Malvar wavelets are much more effective and superior to Gabor wavelets and other wavelets.

1.3 The Wigner-Ville Distribution and Time-Frequency Signal Analysis

In a remarkable paper, Eugene Paul Wigner (1932), the 1963 Nobel Prize Winner in Physics, first introduced a new function $W_\psi(x,p)$ of two independent variables from the wave function ψ in the context of quantum mechanics defined by

$$W_\psi(x,p) = \frac{1}{h}\int_{-\infty}^{\infty} \psi\left(x+\frac{1}{2}t\right)\overline{\psi}\left(x-\frac{1}{2}t\right)\exp\left(\frac{ipt}{\hbar}\right)dt, \qquad (1.3.1)$$

where ψ satisfies the one-dimensional Schrödinger equation, the variables x and p represent the quantum-mechanical position and momentum respectively, and $h = 2\pi\hbar$ is the Planck constant. The Wigner function $W_\psi(x,p)$ has many remarkable properties which include the space and momentum marginal integrals

$$\frac{1}{2\pi}\int_{-\infty}^{\infty} W_\psi(x,p)\,dp = |\psi(x)|^2, \quad \int_{-\infty}^{\infty} W_\psi(x,p)\,dx = |\hat{\psi}(p)|^2. \qquad (1.3.2\text{a,b})$$

These integrals represent the usual position and momentum energy densities. Moreover, the integral of the Wigner function over the whole (x,p) space is

$$\frac{1}{2\pi}\int_{-\infty}^{\infty}\int_{-\infty}^{\infty} W_\psi(x,p)\,dx\,dp = \frac{1}{2\pi}\int_{-\infty}^{\infty} |\hat{\psi}(p)|^2\,dp = \int_{-\infty}^{\infty} |\psi(x)|^2\,dx. \qquad (1.3.3)$$

This can be interpreted as the total energy over the whole position-momentum plane (x,p).

As is well-known, the Fourier transform analysis is a very effective tool for studying stationary (time-independent) signals (or waveforms). However, signals (or waveforms) are, in general, nonstationary. Such signals or waveforms cannot be analyzed completely by the Fourier analysis. Therefore, a complete analysis of nonstationary signals (or waveforms) requires both time-frequency (or space-wavenumber) representations of signals. In 1948, Ville proposed the Wigner distribution of a function or signal $f(t)$ in the form

$$W_f(t,\omega) = \int_{-\infty}^{\infty} f\left(t+\frac{\tau}{2}\right)\bar{f}\left(t-\frac{\tau}{2}\right)e^{-i\omega\tau}\,d\tau \qquad (1.3.4)$$

for analysis of the time-frequency structures of nonstationary signals, where $\bar{f}(z)$ is the complex conjugate of $f(z)$. Subsequently, this time-frequency representation (1.3.4) of a signal f is known as the *Wigner-Ville distribution* (WVD) which is one of the fundamental methods that have been developed over the years for the time-frequency signal analysis. An extensive study of this distribution was made by Claasen and Mecklenbräuker (1980) in the context of the time-frequency signal analysis. Besides other linear time-frequency representations, such as the short-time Fourier transform or the Gabor transform, and the Wigner-Ville distribution plays a central role in the field of bilinear/quadratic time-frequency representations. In view of its remarkable mathematical structures and properties, the Wigner-Ville distribution is now well-recognized as an effective method for the time-frequency (or space-wavenumber) analysis of nonstationary signals (or waveforms), and nonstationary random processes. In recent years, this distribution has served as a useful analysis tool in many fields as diverse as quantum mechanics, optics, acoustics, communications, biomedical engineering, signal processing, and image processing. It has also been used as a method for analyzing seismic data, and the phase distortion involved in a wide variety of audio engineering problems. In addition, it has been suggested as a method for investigating many important topics including instantaneous frequency estimation, spectral analysis of nonstationary random signals, detection and classification of signals, algorithms for computer implementation, speech signals, and pattern recognition.

In sonar and radar systems, a real signal is transmitted and its echo is processed in order to find out the position and velocity of a target. In many situations, the received signal is different from the original one only by a time translation and the Doppler frequency shift. In the context of the mathematical analysis of radar information, Woodward (1953) reformulated the theory of the Wigner-Ville distribution. He introduced a new function $A_f(t,\omega)$ of two independent variables t,ω from a radar signal f in the form

$$A_f(t,\omega)=\int_{-\infty}^{\infty} f\left(\tau+\frac{t}{2}\right)\bar{f}\left(\tau-\frac{t}{2}\right)e^{-i\omega\tau}\,d\tau. \tag{1.3.5}$$

This function is now known as the *Woodward ambiguity function* and plays a central role in radar signal analysis and radar design. The ambiguity function has been widely used for describing the correlation between a radar signal and its Doppler-shifted and time-translated version. It was also shown that the ambiguity function exhibits the measurement between ambiguity and target resolution, and for this reason it is also known as the *radar ambiguity function.* In analogy with the Heisenberg uncertainty principle in quantum mechanics, Woodward also formulated a *radar uncertainty principle*, which says that the range and velocity (range rate) cannot be measured exactly and simultaneously. With the activity surrounding the radar uncertainty principle, the representation theory of the Heisenberg group and ambiguity functions as special functions on the Heisenberg group led to a series of many important results. Subsequently, considerable attention has been given to the study of radar ambiguity functions in harmonic analysis and group theory by several authors, including Wilcox (1960), Schempp (1984), and Auslander and Tolimieri (1985).

From theoretical and application points of view, the Wigner-Ville distribution plays a central role and has several important and remarkable structures and properties. First, it provides a high-resolution representation in time and in frequency for some nonstationary signals. Second, it has the special property of satisfying the time and frequency marginals in terms of the instantaneous power in time and energy spectrum in frequency. Third, the first conditional moment of frequency at a given time is the derivative of the phase of the signal at that time. The derivative of the phase divided by 2π gives the *instantaneous frequency* which is uniquely related to the signal. Moreover, the second conditional moment of frequency of a signal does not have any physical interpretation. In spite of these remarkable features, its energy distribution is *not* nonnegative and it often possesses severe cross-terms, or interference terms between different time-frequency regions, leading to undesirable properties.

In order to overcome some of the inherent weaknesses of the Wigner-Ville distribution, there has been considerable recent interest in more general time-frequency distributions as a mathematical method for time-frequency signal analysis. Often, the Wigner-Ville distribution has been modified by smoothing in one or two dimensions, or by other signal processing. In 1966, Cohen introduced a general class of bilinear shift-invariant, quadratic time-frequency distributions in the form

$$C_f(t,\nu)=\int_{-\infty}^{\infty}\int_{-\infty}^{\infty}\int_{-\infty}^{\infty}\exp\left[-2\pi i\left(\nu\tau+st-rs\right)\right]g(s,\tau)$$

$$\times f\left(r+\frac{\tau}{2}\right)\bar{f}\left(r-\frac{\tau}{2}\right)d\tau\,dr\,ds, \tag{1.3.6}$$

where the given kernel $g(s,\tau)$ generates different distributions which include windowed Wigner-Ville, Choi-Williams, spectrogram, Rihaczek, Born-Jordan, and Page distributions.

In modern time-frequency signal analysis, several alternative forms of the Cohen distribution seem to be convenient and useful. A function u is introduced in terms of the given kernel $g(s,\tau)$ by

$$u(r,\tau) = \int_{-\infty}^{\infty} g(s,\tau)\, \exp(2\pi i s r)\, ds \tag{1.3.7}$$

so that the Cohen distribution takes the general form

$$C_f(t,\nu) = \int_{-\infty}^{\infty}\int_{-\infty}^{\infty} u(r-t,\tau)\, f\left(r+\frac{\tau}{2}\right) \bar{f}\left(r-\frac{\tau}{2}\right) \exp(-2\pi i \nu \tau)\, d\tau\, dr. \tag{1.3.8}$$

The general Cohen distribution can also be written in terms of an ambiguity function as

$$C_f(t,\nu) = \int_{-\infty}^{\infty}\int_{-\infty}^{\infty} A(s,\tau) \exp\left[-2\pi i(st+\nu\tau)\right] ds\, d\tau, \tag{1.3.9}$$

where $A(s,\tau)$ is the general ambiguity function of f and g defined by

$$A(s,\tau) = g(s,\tau) \int_{-\infty}^{\infty} f\left(r+\frac{\tau}{2}\right) \bar{f}\left(r-\frac{\tau}{2}\right) \exp(2\pi i r s)\, dr. \tag{1.3.10}$$

As a natural generalization of the Wigner-Ville distribution, another family of bilinear time-frequency representations was introduced by Rihaczek in 1968. This is called the *generalized Wigner-Ville* (GWV) distribution or more appropriately, the *Wigner-Ville-Rihaczek* (WVR) distribution which is defined for two signals f and g by

$$R_{f,g}^{\alpha}(t,\omega) = \int_{-\infty}^{\infty} f\left(t+\left(\frac{1}{2}-\alpha\right)\tau\right) \bar{g}\left(t-\left(\frac{1}{2}+\alpha\right)\tau\right) e^{-i\omega\tau} d\tau, \tag{1.3.11}$$

where α is a real constant parameter. In particular, when $\alpha = 0$, (1.3.11) reduces to the Wigner-Ville distribution, and when $\alpha = 2^{-1}$, (1.3.11) represents the *Wigner-Rihaczek* distribution in the form

$$R_{f,g}^{\frac{1}{2}}(t,\omega) = f(t) \int_{-\infty}^{\infty} \bar{g}(t-\tau)\, e^{-i\omega\tau} d\tau = f(t)\, e^{-i\omega t}\, \hat{\bar{g}}(\omega). \tag{1.3.12}$$

The main feature of these distributions is their time- and frequency-shift invariance. However, for some problems where the scaling of signals is important, it is necessary to consider distributions which are invariant to translations and compressions of time, that is, $t \to at+b$ (affine transformations). Bertrand and Bertrand (1992) obtained another general class of distributions which are called *affine time-frequency distributions* because they are invariant to affine transformations. Furthermore, extended forms of the various affine distributions are also introduced to obtain representations of complex signals on the whole time-frequency plane. The use of the real signal in these forms shows the effect of producing symmetry of the result obtained with the analytic signal. In any case, the construction based on the affine group, which is basic in signal analysis, ensures that no spurious interference will ever occur between positive and negative frequencies. Special attention has also been given to the computational aspects of broadband functionals containing stretched forms of the signal such as affine distributions, wavelet coefficients, and broadband ambiguity functions. Different methods based on group theory have also been developed to derive explicit representations of joint time-frequency distributions adapted to the analysis of wideband signals.

Although signal analysis orginated more than fifty years ago, there has been major development of the time-frequency distributions approach in the basic idea of the method to develop a joint function of time and frequency, known as a time-frequency distribution, that can describe the energy density of a signal simultaneously in both time and frequency domains. In principle, the joint time-frequency distributions characterize phenomena in the two-dimensional time-frequency plane. Basically, there are two kinds of time-frequency representations. One is the quadratic method describing the time-frequency distributions, and the other is the linear approach including the Gabor transform and the wavelet transform. Thus, the field of time-frequency analysis has evolved into a widely recognized applied discipline of signal processing over the last two decades. Based on studies of its mathematical structures and properties by many authors including de Bruijn (1967, 1973), Claasen and Mecklenbräuker (1980), Boashash (1992), Mecklenbräuker and Hlawatsch (1997), the Wigner-Ville distribution and its various generalizations with applications were brought to the attention of larger mathematical, scientific, and engineering communities. By any assessment, the Wigner-Ville distribution has served as the fundamental basis for all subsequent classical and modern developments of time-frequency signal analysis and signal processing.

1.4 Wavelet Transforms

Historically, the concept of "ondelettes" or "wavelets" started to appear more frequently only in the early 1980's. This new concept can be viewed as a synthesis of various ideas originating from different disciplines including mathematics (Calderón-Zygmund operators and Littlewood-Paley theory), physics (the coherent states formalism in quantum mechanics and the renormalization group), and engineering (quadratic mirror filters, sideband coding in signal processing, and pyramidal algorithms in image processing). In 1982, Jean Morlet, a French geophysical engineer, discovered the idea of the wavelet transform, providing a new mathematical tool for seismic wave analysis. In Morlet's analysis, signals consist of different features in time and frequency, but their high-frequency components would have a shorter time duration than their low-frequency components. In order to achieve good time resolution for the high-frequency transients and good frequency resolution for the low-frequency components, Morlet (1982a,b) first introduced the idea of wavelets as a family of functions constructed from translations and dilations of a single function called the "*mother wavelet*" $\psi(t)$. They are defined by

$$\psi_{a,b}(t) = \frac{1}{\sqrt{|a|}}\,\psi\left(\frac{t-b}{a}\right), \qquad a,b \in \mathbb{R},\ a \neq 0, \tag{1.4.1}$$

where a is called a *scaling parameter* which measures the degree of compression or scale, and b a *translation parameter* which determines the time location of the wavelet. If $|a|<1$, the wavelet (1.4.1) is the compressed version (smaller support in time-domain) of the mother wavelet and corresponds mainly to higher frequencies. On the other hand, when $|a|>1$, $\psi_{a,b}(t)$ has a larger time-width than $\psi(t)$ and corresponds to lower frequencies. Thus, wavelets have time-widths adapted to their frequencies. This is the main reason for the success of the Morlet wavelets in signal processing and time-frequency signal analysis. It may be noted that the resolution of wavelets at different scales varies in the time and frequency domains as governed by the Heisenberg uncertainty principle. At large scale, the solution is coarse in the time domain and fine in the frequency domain. As the scale a decreases, the resolution in the time domain becomes finer while that in the frequency domain becomes coarser.

Morlet first developed a new time-frequency signal analysis using what he called "wavelets of constant shape" in order to contrast them with the analyzing functions in the short-time Fourier transform which do not have a constant shape. It was Alex Grossmann, a French theoretical physicist, who quickly recognized the importance of the Morlet wavelet transforms which are somewhat similar to the formalism for coherent states in quantum mechanics, and developed an exact inversion formula for this wavelet transform. Unlike the Weyl-Heisenberg coherent states, these coherent states arise from translations and dilations of a single function. They are often called *affine coherent states* because they are associated with an affine group (or "$ax+b$" group). From a group-theoretic point of view, the wavelets $\psi_{a,b}(x)$ are in fact the result of the action of the operators $U(a,b)$ on the function ψ so that

$$[U(a,b)\psi](x) = \frac{1}{\sqrt{|a|}}\,\psi\left(\frac{x-b}{a}\right). \tag{1.4.2}$$

These operators are all unitary on the Hilbert space $L^2(\mathbb{R})$ and constitute a representation of the "$ax+b$" group:

$$U(a,b)\,U(c,d) = U(ac, b+ad). \tag{1.4.3}$$

This group representation is *irreducible*, that is, for any non-zero $f \in L^2(\mathbb{R})$, there exists no nontrivial g orthogonal to all the $U(a,b)f$. In other words, $U(a,b)f$ span the entire space. The coherent states for the affine $(ax+b)$-group, which are now known as *wavelets*, were first formulated by Aslaksen and Klauder (1968, 1969) in the context of more general representations of groups. The success of Morlet's numerical algorithms prompted Grossmann to make a more extensive study of the Morlet wavelet transform which led to the recognition that wavelets $\psi_{a,b}(t)$ correspond to a square integrable representation of the affine group. Grossmann was concerned with the wavelet transform of $f \in L^2(\mathbb{R})$ defined by

$$\mathcal{W}_\psi[f](a,b) = (f, \psi_{a,b}) = \frac{1}{\sqrt{|a|}} \int_{-\infty}^{\infty} f(t)\,\overline{\psi\left(\frac{t-b}{a}\right)}\,dt, \tag{1.4.4}$$

where $\psi_{a,b}(t)$ plays the same role as the kernel $\exp(i\omega t)$ in the Fourier transform. Like the Fourier transformation, the continuous wavelet transformation $\mathcal{W}_\psi$ is linear. However, unlike the Fourier transform, the continuous wavelet transform is not a single transform, but any transform obtained in this way. The inverse wavelet transform can be defined so that f can be reconstructed by means of the formula

$$f(t) = C_\psi^{-1} \int_{-\infty}^{\infty}\int_{-\infty}^{\infty} \mathcal{W}_\psi[f](a,b)\,\psi_{a,b}(t)\left(a^{-2}da\right)db, \tag{1.4.5}$$

provided C_ψ satisfies the so called *admissibility condition*

$$C_\psi = 2\pi \int_{-\infty}^{\infty} \frac{|\hat{\psi}(\omega)|^2}{|\omega|}\,d\omega < \infty, \tag{1.4.6}$$

where $\hat{\psi}(\omega)$ is the Fourier transform of the mother wavelet $\psi(t)$.

Grossmann's ingenious work also revealed that certain algorithms that decompose a signal on the whole family of scales, can be utilized as an efficient tool for multiscale analysis. In practical applications involving fast numerical algorithms, the continuous wavelet can be computed at discrete grid points. To do this, a general wavelet ψ can be defined by replacing a with $a_0^m\ (a_0 \neq 0,1)$, b with $nb_0a_0^m\ (b_0 \neq 0)$, where m and n are integers, and making

$$\psi_{m,n}(t) = a_0^{-m/2}\psi\left(a_0^{-m}t - nb_0\right). \tag{1.4.7}$$

The discrete wavelet transform of f is defined as the doubly indexed sequence

$$\tilde{f}(m,n) = \mathcal{W}[f](m,n) = \left(f, \psi_{m,n}\right) = \int_{-\infty}^{\infty} f(t)\,\overline{\psi}_{m,n}(t)\,dt, \tag{1.4.8}$$

where $\psi_{m,n}(t)$ is given by (1.4.7). The double series

$$\sum_{m,n=-\infty}^{\infty} \tilde{f}(m,n)\,\psi_{m,n}(t) \tag{1.4.9}$$

is called the *wavelet series* of f, and the functions $\{\psi_{m,n}(t)\}$ are called the *discrete wavelets*, or simply *wavelets*. However, there is no guarantee that the original function f can be reconstructed from its discrete wavelet coefficients in general. The reconstruction of f is still possible if the discrete lattice has a very fine mesh. For very coarse meshes, the coefficients may not contain sufficient information for determination of f from these coefficients. However, for certain values of the lattice parameter (m,n), a numerically stable reconstruction formula can be obtained. This leads to the concept of a "frame" rather than bases. The notion of the frame was introduced by Duffin and Schaeffer (1952) for the study of a class of nonharmonic Fourier series to which Paley and Wiener made fundamental contributions. They discussed related problems of nonuniform sampling for band-limited functions.

In general, the function f belonging to the Hilbert space, $L^2(\mathbb{R})$ (see Debnath and Mikusinski, 1999), can be completely determined by its discrete wavelet transform (wavelet coefficients) if the wavelets form a complete system in $L^2(\mathbb{R})$. In other words, if the wavelets form an orthonormal basis or a frame of $L^2(\mathbb{R})$, then they are complete. And f can be reconstructed from its discrete wavelet transform $\{\tilde{f}(m,n) = (f, \psi_{m,n})\}$ by means of the formula

$$f(x) = \sum_{m,n=-\infty}^{\infty} (f, \psi_{m,n}) \psi_{m,n}(x), \tag{1.4.10}$$

provided the wavelets form an orthonormal basis.

On the other hand, the function f can be determined by the formula

$$f(x) = \sum_{m,n=-\infty}^{\infty} (f, \psi_{m,n}) \tilde{\psi}_{m,n}(x) \tag{1.4.11}$$

provided the wavelets form a frame and $\{\tilde{\psi}_{m,n}(x)\}$ is the dual frame.

For some very special choices of ψ and a_0, b_0, the $\psi_{m,n}$ constitute an orthonormal basis for $L^2(\mathbb{R})$. In fact, if $a_0 = 2$ and $b_0 = 1$, then there exists a function ψ with good time-frequency localization properties such that

$$\psi_{m,n}(x) = 2^{-m/2} \psi(2^{-m} x - n) \tag{1.4.12}$$

form an orthonormal basis for $L^2(\mathbb{R})$. These $\{\psi_{m,n}(x)\}$ are known as the *Littlewood-Paley wavelets*. This gives the following representation of f

$$f(x) = \sum_{m,n} (f, \psi_{m,n}) \psi_{m,n}(x) \tag{1.4.13}$$

which has a good space-frequency localization. The classic example of a wavelet ψ for which the $\psi_{m,n}$ defined by (1.4.12) constitute an orthonormal basis for $L^2(\mathbb{R})$ is the Haar wavelet

$$\psi(x) = \begin{cases} 1, & 0 \le x < \frac{1}{2} \\ -1, & \frac{1}{2} \le x < 1 \\ 0, & \text{otherwise} \end{cases}. \tag{1.4.14}$$

Historically, the first orthonormal wavelet basis is the Haar basis, which was discovered long before the wavelet was introduced. It may be observed that the Haar wavelet ψ does not have good time-frequency localization and that its

Fourier transform $\hat{\psi}(k)$ decays like $|k|^{-1}$ as $k \to \infty$. The joint venture of Morlet and Grossmann led to a detailed mathematical study of the wavelet transforms and their applications. It became clear from their work that, analogous to the Fourier expansion of functions, the wavelet transform analysis provides a new method for decomposing a function (or a signal).

In 1985, Yves Meyer, a French pure mathematician, recognized the deep connection between the Calderón formula in harmonic analysis and the new algorithm discovered by Grossmann and Morlet (1984). He also constructed an orthonormal basis, for the Hilbert space $L^2(\mathbb{R})$, of wavelets $\psi_{m,n}$ defined by (1.4.12) based on the mother wavelet ψ with compact support and C^∞ Fourier transform $\hat{\psi}$. This basis turned out to be an unconditional basis for all L^p spaces $(1 < p < \infty)$, Sobolev spaces, and other spaces. Furthermore, in a Hilbert space, a normalized basis turns out to be an unconditional basis if and only if it is also a frame. Such a basis is called the *Riesz basis*. However, if $\{\psi_n\}$ is an orthonormal basis, then $e_n = (1+n^2)^{-\frac{1}{2}}(n\psi_1 + \psi_n)$ is an example of a basis of normalized vectors that is not a Riesz basis. Using the knowledge of the Calderón-Zygmund operators and the Littlewood-Paley theory, in 1985-86 Meyer (1990) successfully gave a mathematical foundation of the wavelet theory. The Meyer basis has become a more powerful tool than the Haar basis.

Even though the mother wavelet in the Meyer basis decays faster than any inverse polynomials, the constants involved are very large so that it is not very well-localized. Lemarié and Meyer (1986) extended the Meyer orthonormal basis to more than one dimension. One of the new orthonormal wavelet bases for $L^2(\mathbb{R})$ with localization properties in both time and frequency was first constructed by Strömberg in 1982. His wavelets are in C^n, where n is arbitrary but finite and decays exponentially. He also proved that the orthonormal wavelet basis defined by (1.4.12) is, in fact, an unconditional basis for the Hardy space $\mathcal{H}^1(\mathbb{R})$ which consists of real-valued functions $u(x)$ if and only if $u(x)$ and its Hilbert transform $\hat{u}(\kappa)$ belong to $L^1(\mathbb{R})$. In fact, $\mathcal{H}^1(\mathbb{R})$ is the real version of the holomorphic Hardy space $\mathcal{H}^1(\mathbb{R})$ whose elements are $u(x) + i\,v(x)$, where $u(x)$ and $v(x)$ are real-valued functions. A function $f(z)$, where $z = x + i\,y$, belongs to the Hardy space $H^p(\mathbb{R})$, $0 \le p \le \infty$, if it is holomorphic in the upper half $(y > 0)$ of the complex plane and if

$$\| f \|_p = \sup_{y>0} \left[\int_{-\infty}^{\infty} | f(z) |^p \, dx \right]^{\frac{1}{p}} < \infty. \tag{1.4.15}$$

If this condition is satisfied, the upper bound, taken over $y > 0$, is also the limit as $y \to 0$. Moreover, $f(z)$ converges to a function $f(x)$ as $y \to 0+$, where convergence is in the sense of the L^p-norm. The space $H^p(\mathbb{R})$ can thus be identified with a closed subspace of $L^p(\mathbb{R})$. The Hardy space $H^2(\mathbb{R})$ plays a major role in signal processing. The real part of an analytic signal $F(t) = f(t) + i\,g(t)$, $t \in \mathbb{R}$, represents a real signal $f(t)$ with finite energy given by

$$\| f \| = \left[\int_{-\infty}^{\infty} | f(t) |^2 \, dt \right]^{\frac{1}{2}}. \tag{1.4.16}$$

If F has finite energy, then $F \in H^2(\mathbb{R})$.

For more information about the history of wavelets, the reader is referred to Debnath (1998c).

1.5 Wavelet Bases and Multiresolution Analysis

In late 1986, Meyer and Mallat recognized that construction of different wavelet bases can be realized by the so-called *multiresolution analysis*. This is essentially a framework in which functions $f \in L^2(\mathbb{R}^d)$ can be treated as a limit of successive approximations $f = \lim_{m \to -\infty} P_m f$, where the different $P_m f$ for $m \in \mathbb{Z}$ correspond to smoothed versions of f with a smoothing-out action radius of the order 2^m. The wavelet coefficients $(\psi_{m,m}, f)$ for a fixed m then correspond to the difference between the two successive approximations $P_{m-1} f$ and $P_m f$. In the late 1980's, efforts for construction of orthonormal wavelet bases continued rapidly. Battle (1987) and Lemarié (1988, 1989) independently constructed spline orthonormal wavelet bases with exponential decay properties. At the same time, Tchamitchian (1987) gave a first example of biorthogonal wavelet bases. These different orthonormal wavelet bases have been found to be very useful in applications to signal processing, image processing, computer vision, and quantum field theory.

The construction of a "painless" nonorthogonal wavelet expansion by Daubechies, Grossmann, and Meyer (1986) can be considered one of the major achievements in wavelet analysis. During 1985-86, further work of Meyer and Lemarié on the first construction of a smooth orthonormal wavelet basis on $\mathbb{R}$ and then on $\mathbb{R}^n$ marked the beginning of their famous contributions to the wavelet theory. Many experts realized the importance of the existence of an orthonormal basis with good time-frequency localization. Particularly, Stéphane Mallat recognized that some quadratic mirror filters (QMF) play an important role in the construction of orthogonal wavelet bases generalizing the classic Haar system. Lemarié and Meyer (1986) and Mallat (1988, 1989a,b) discovered that orthonormal wavelet bases of completely supported wavelets could be constructed systematically from a general formalism. Their collaboration culminated with a major discovery by Mallat (1989a,b) of a new formalism, the so-called *multiresolution analysis*. The concept of multiresolution analysis provided a major role in Mallat's algorithm for the decomposition and reconstruction of an image in his work. The fundamental idea of multiresolution analysis is to represent a function as a limit of successive approximations, each of which is a "smoother" version of the original function. The successive approximations correspond to different resolutions, which leads to the name multiresolution analysis as a formal approach to constructing orthogonal wavelet bases using a definite set of rules and procedures. It also provides the existence of so-called *scaling functions* and *scaling filters* which are then used for construction of wavelets and fast numerical algorithms. In applications, it is an effective mathematical framework for hierarchical decomposition of a signal or an image into components of different scales represented by a sequence of function spaces on $\mathbb{R}$. Indeed, Mallat developed a very effective numerical algorithm for multiresolution analysis using wavelets. It was also Mallat who constructed the wavelet decomposition and reconstruction algorithms using the multiresolution analysis. This brilliant work of Mallat has been the major source of many recent new developments in wavelet theory. According to Daubechies (1992), "...The history of the formulation of multiresolution analysis is a beautiful example of applications stimulating theoretical development." While reviewing two books on wavelets in 1993, Meyer made the following statement on wavelets: "Wavelets are without doubt an exciting and intuitive concept. The concept brings with it a new way of thinking, which is absolutely essential and was entirely missing in previously existing algorithms."

Inspired by the work of Meyer and stimulated by the exciting developments in wavelets, Ingrid Daubechies (1988a,b, 1990) made a new remarkable contribution to wavelet theory and its applications. The combined influence of Mallat's work and Burt and Adelson's (1983a,b) pyramid algorithm used in image analysis led to her major construction of an orthonormal wavelet basis of compact support. Her 1988b paper, dealing with the construction of the first orthonormal basis of continuous, compactly supported wavelets for $L^2(\mathbb{R})$ with some degree of smoothness, produced a tremendous positive impact on the study of wavelets and their diverse applications. Her discovery of an orthonormal basis for $L^2(\mathbb{R})$ of the form $2^{m/2}\psi_r(2^m t - n)$, $m, n \in \mathbb{Z}$, with the support of ψ_r in the interval $[0, 2r+1]$, created a lot of excitement in the study of wavelets. If $r = 0$, Daubechies' result reduces to the Haar system. This work explained the significant connection between the continuous wavelet on $\mathbb{R}$ and the discrete wavelets on $\mathbb{Z}$ and $\mathbb{Z}_N$, where the latter have become extremely useful for digital signal analysis. Although the concept of frame was introduced by others, Daubechies et al. (1986) successfully computed numerical estimates for the frame bounds for a wide variety of wavelets. In spite of the tremendous success, it is not easy to construct wavelets that are symmetric, orthogonal and compactly supported. In order to handle this problem, Cohen et al. (1992) investigated biorthogonal wavelets in some detail. They have shown that these wavelets have analytic representations with compact support. The dual wavelets do not have analytic representations, but they do have compact support.

In recent years, another class of wavelets, *semiorthogonal wavelets*, have received some attention. These represent a class of wavelets which are orthogonal at different scales and, for wavelets with nonoverlapping support, at the same scale. Chui and Wang (1991, 1992) and Micchelli (1991) independently studied semiorthogonal wavelets. The former authors constructed B-spline wavelets using linear splines. Then, they used the B-spline wavelets without orthogonalization to construct the semiorthogonal B-spline wavelets. On the other hand, Battle (1987) orthogonalized the B-spline and used these scaling functions to construct orthogonal wavelets. Thus, the difference between Chui and Wang's and Battle's constructions lies in the orthogonal property of the scaling function.

1.6 Applications of Wavelet Transforms

Both Weierstrass and Riemann constructed famous examples of everywhere continuous and nowhere differentiable functions. So the history of such functions is very old. More recently, Holschneider (1988) and Holschneider and Tchamitchian (1991a,b) have successfully used wavelet analysis to prove non-differentiability of both Weierstrass' and Riemann's functions.

On the other hand, Beylkin, Coifman, and Rokhlin (1991) and Beylkin (1992) have successfully applied multiresolution analysis generated by a completely orthogonal scaling function to study a wide variety of integral operators on $L^2(\mathbb{R})$ by a matrix in a wavelet basis. This work culminated with the remarkable discovery of new algorithms in numerical analysis. Consequently, some significant progress has been made in boundary element methods, finite element methods, and numerical solutions of partial differential equations using wavelets. As a natural extension of the wavelet analysis, Coifman et al. (1989, 1992a,b) in collaboration with Meyer and Wickerhauser discovered wavelet packets to design efficient schemes for the representation and compression of acoustic signals and images. Coifman et al. (1989, 1992a,b) also introduced the local sine and cosine transforms and studied their properties. This led them to the construction of a library of orthogonal bases by extending the method of multiresolution decomposition and using the quadratic mirror filters. Coifman et al. (1989) gave elementary proofs of the L^2 boundedness of the Cauchy integral on Lipschitz curves. Recently, there have also been significant applications of wavelet analysis to a variety of problems in diverse fields including mathematics, physics, medicine, computer science, and engineering.

In recent years, there have been many developments and new applications of wavelet analysis for describing complex algebraic functions and analyzing empirical continuous data obtained from many kinds of signals at different scales of resolution. The most widespread application of the wavelet transform so far has been for data compression. This is associated with the fact that the discrete Fourier transform is closely related to subband decomposition. We close this historical introduction by citing some of these applications which include addressing problems in signal processing, computer vision, seismology, turbulence, computer graphics, image processing, structure of galaxies in the universe, digital communication, pattern recognition, approximation theory, quantum optics, biomedical engineering, sampling theory, matrix theory, operator theory, differential equations, numerical analysis, statistics and multiscale segmentation of well logs, natural scenes, and mammalian visual systems. Wavelets allow complex information such as music, speech, images, and patterns to be decomposed into elementary forms, called *simple building blocks*, at different positions and scales. These building blocks represent a family of wavelets that are generated from a single function called "*mother wavelet*" by translation and dilation operations. The information is subsequently reconstructed with high precision. In order to describe the present state of wavelet research, Meyer (1993a) wrote as follows:

> "Today the boundaries between mathematics and signal and image processing have faded, and mathematics has benefitted from the rediscovery of wavelets by experts from other disciplines. The detour through signal and image processing was the most direct path leading from the Haar basis to Daubechies's wavelets."

Chapter 2

Hilbert Spaces and Orthonormal Systems

"The organic unity of mathematics is inherent in the nature of this science, for mathematics is the foundation of all exact knowledge of natural phenomena."

David Hilbert

"Hilbert spaces constitute at present the most important examples of Banach spaces, not only because they are the most natural and closest generalization, in the realm of 'infinite dimensions', of our classical euclidian geometry, but chiefly for the fact they have been, up to now, the most useful spaces in the applications to functional analysis."

Jean Dieudonné

2.1 Introduction

Historically, the theory of Hilbert spaces originated from David Hilbert's (1862-1943) work on quadratic forms in infinitely many variables with their applications to integral equations. During the period of 1904-1910, Hilbert published a series of six papers, subsequently collected in his classic book *Grundzüge einer allemeinen Theorie der linearen Integralgleichungen* published in 1912. It contained many general ideas including Hilbert spaces $\left(\ell^2 \text{ and } L^2\right)$, the compact operators, and orthogonality, and had a tremendous influence on mathematical analysis and its applications. After many years, John von Neumann (1903-1957) first formulated an axiomatic approach to Hilbert space and developed the modern theory of operators on Hilbert spaces. His remarkable contribution to this area has provided the mathematical foundation of quantum mechanics. Von Neumann's work has also provided an almost definite physical interpretation of quantum mechanics in terms of abstract relations in an infinite dimensional Hilbert space. It was shown that observables of a physical system can be represented by linear symmetric operators in a Hilbert space, and the eigenvalues and eigenfunctions of the particular operator that represents energy are energy levels of an electron in an atom and corresponding stationary states of the system. The differences in two eigenvalues represent the frequencies of the emitted quantum of light and thus define the radiation spectrum of the substance.

The theory of Hilbert spaces plays an important role in the development of wavelet transform analysis. Although a full understanding of the theory of Hilbert spaces is not necessary in later chapters, some familiarity with the basic ideas and results is essential.

One of the nice features of normed spaces is that their geometry is very much similar to the familiar two- and three-dimensional Euclidean geometry. Inner product spaces and Hilbert spaces are even nicer because their geometry is even closer to Euclidean geometry. In fact, the geometry of Hilbert spaces is more or less a generalization of Euclidean geometry to infinite dimensional spaces. The main reason for this simplicity is that the concept of orthogonality can be introduced in any inner product space so that the familiar Pythagorean formula holds. Thus, the structure of Hilbert spaces is more simple and beautiful, and hence, a large number of problems in mathematics, science, and engineering can be successfully treated with geometric methods in Hilbert spaces.

This chapter deals with normed spaces, the L^p spaces, generalized functions (distributions), inner product spaces (also called pre-Hilbert spaces), and Hilbert spaces. The fundamental ideas and results are discussed with special attention given to orthonormal systems, linear functionals, and the Riesz representation theorem. The generalized functions and the above spaces are illustrated by various examples. Separable Hilbert spaces are discussed in Section 2.14. Linear operators on a Hilbert space are widely used to represent physical quantities in applied mathematics and physics. In signal processing and wavelet analysis, almost all algorithms are essentially based on linear operators. The most important operators include differential, integral, and matrix operators. In Section 2.15, special attention is given to different kinds of operators and their basic properties. The eigenvalues and eigenvectors are discussed in Section 2.16. Included are several spectral theorems for self-adjoint compact operators and other related results.

2.2 Normed Spaces

The reader is presumed to have a working knowledge of the real number system and its basic properties. The set of natural numbers (positive integers) is denoted by $\mathbb{N}$, and the set of integers (positive, negative, and zero) is denoted by $\mathbb{Z}$, and the set of rational numbers by $\mathbb{Q}$. We use $\mathbb{R}$ and $\mathbb{C}$ to denote the set of real numbers and the set of complex numbers respectively. Elements of $\mathbb{R}$ and $\mathbb{C}$ are called *scalars*. Both $\mathbb{R}$ and $\mathbb{C}$ form a scalar field.

We also assume that the reader is familiar with the concept of a linear space or vector space which is an example of mathematical systems that have algebraic structure only. The important examples of linear spaces in mathematics have the real or complex numbers as the scalar field. The simplest example of a real vector space is the set $\mathbb{R}$ of real numbers. Similarly, the set $\mathbb{C}$ of complex numbers is a vector space over the complex numbers.

The concept of *norm* in a vector space is an abstract generalization of the length of a vector in $\mathbb{R}^3$. It is defined axiomatically, that is, any real-valued function satisfying certain conditions is called a norm.

Definition 2.2.1 (Norm). A real-valued function $\|x\|$ defined on a vector space X, where $x \in X$, is called a *norm* on X if the following conditions hold:

(a) $\|x\| = 0$ if and only if $x = 0$,

(b) $\|ax\| = |a|\,\|x\|$ for every $a \in \mathbb{R}$ and $x \in X$,

(c) $\|x + y\| \le \|x\| + \|y\|$ for all $x, y \in X$.

Condition (c) is usually called the *triangle inequality*. Since

$$0 = \|0\| = \|x - x\| \le \|x\| + \|-x\| = 2\,\|x\|,$$

it follows that $\|x\| \ge 0$ for every $x \in X$.

Definition 2.2.2 (Normed Space). A *normed space* is a vector space X with a given norm.

So, a normed space is a pair $(X, \|\cdot\|)$, where X is a vector space and $\|\cdot\|$ is a norm defined on X. Of course, it is possible to define different norms on the same vector space.

Example 2.2.1 (a) $\mathbb{R}$ is a real normed space with the norm defined by the absolute values, $\|x\| = |x|$.

(b) $\mathbb{C}$ becomes a complex normed space with the norm defined by the modulus, $\|z\| = |z|$.

Example 2.2.2 (a) $\mathbb{R}^N = \{(x_1, x_2, \dots, x_N) : x_1, x_2, \dots, x_N \in \mathbb{R}\}$ is a vector space with a norm defined by

$$\|x\| = \sqrt{(x_1^2 + x_2^2 + \cdots + x_N^2)}, \tag{2.2.1}$$

where $x = (x_1, x_2, \dots, x_N) \in \mathbb{R}^N$. This norm is often called the *Euclidean norm.*

(b) $\mathbb{C}^N = \{(z_1, z_2, \cdots, z_N) : z_1, z_2, \cdots, z_N \in \mathbb{C}\}$ is a normed space with a norm

$$\|z\| = \sqrt{|z_1|^2 + \cdots + |z_N|^2}, \tag{2.2.2}$$

where $z = (z_1, \cdots, z_N) \in \mathbb{C}^N$.

Example 2.2.3 The sequence space $\ell^p \, (1 \le p < \infty)$ is the set of all sequences $x = \{x_n\}_{n=1}^{\infty}$ of real (complex) numbers such that $\sum_{n=1}^{\infty} |x_n|^p < \infty$ and equipped with the norm

$$\|x\|_p = \left(\sum_{n=1}^{\infty} |x_n|^p\right)^{1/p}. \tag{2.2.3}$$

This space is a normed space.

Example 2.2.4 The vector space $C([a,b])$ of continuous functions on the interval $[a,b]$ is a normed space with a norm defined by

$$\|f\| = \left(\int_a^b |f(x)|^2 \, dx \right)^{\frac{1}{2}}, \tag{2.2.4}$$

or, with a norm defined by

$$\|f\| = \sup_{a \le x \le b} |f(x)|. \tag{2.2.5}$$

Remark. Every normed space $(X, \|\cdot\|)$ is a metric space (X, d), where the norm induces a metric d defined by

$$d(x, y) = \|x - y\|.$$

But the converse is not necessarily true. In other words, a metric space (X, d) is not necessarily a normed space. This is because of the fact that the metric is not induced by a norm, as seen from the following example.

Example 2.2.5 We denote by s the set of all sequences of real numbers with the metric

$$d(x, y) = \sum_{n=1}^{\infty} \frac{|x_n - y_n|}{2^n \left(1 + |x_n - y_n|\right)}. \tag{2.2.6}$$

This is a metric space, but the metric is not generated by a norm, so the space is not a normed space.

Definition 2.2.3 (Banach Space). A normed space X is called *complete* if every Cauchy sequence in X converges to an element of X. A complete normed space is called a *Banach space*.

Example 2.2.6 The normed spaces $\mathbb{R}^N$ and $\mathbb{C}^N$ with the usual norm as given in Examples 2.2.2(a) and 2.2.2(b) are Banach spaces.

Example 2.2.7 The space of continuous functions $C([a, b])$ with the norm defined (2.2.4) is not a complete normed space. Thus, it is not a Banach space.

Example 2.2.8 The sequence space ℓ^p as given in Example 2.2.3 is a Banach space for $p \ge 1$.

Example **2.2.9** The set of all bounded real-valued functions $M([a,b])$ on the closed interval $[a,b]$ with the norm (2.2.5) is a complete normed (Banach) space.

This is left for the reader as an exercise.

The following are some important subspaces of $M([a,b])$:

(a) $C([a,b])$ is the space of continuous functions on the closed interval $[a,b]$,

(b) $D([a,b])$ is the space of differentiable functions on $[a,b]$,

(c) $P([a,b])$ is the space of polynomials on $[a,b]$,

(d) $R([a,b])$ is the space of Riemann integrable functions on $[a,b]$.

Each of these spaces are normed spaces with the norm (2.2.5).

Example 2.2.10 The space of continuously differentiable functions $C' = C'([a,b])$ with the norm

$$\|f\| = \max_{a \le x \le b} |f(x)| + \max_{a \le x \le b} |f'(x)| \tag{2.2.7}$$

is a complete normed space.

It is easy to check that this space is complete.

2.3 The L^p Spaces

If $p \ge 1$ is any real number, the vector space of all complex-valued Lebesgue integrable functions f defined on $\mathbb{R}$ is denoted by $L^p(\mathbb{R})$ with a norm

$$\|f\|_p = \left[\int_{-\infty}^{\infty} |f(x)|^p \, dx\right]^{\frac{1}{p}} < \infty. \tag{2.3.1}$$

The number $\|f\|_p$ is called the L^p*-norm*. This function space $L^p(\mathbb{R})$ is a Banach space. Since we do not require any knowledge of the Banach space for an understanding of wavelets in this introductory book, the reader needs to know some elementary properties of the L^p-norms.

The L^p spaces for the cases $p = 1$, $p = 2$, $0 < p < 1$, and $1 < p < \infty$ are different in structure, importance, and technique, and these spaces play a very special role in many mathematical investigations.

In particular, $L^1(\mathbb{R})$ is the space of all Lebesgue integrable functions defined on $\mathbb{R}$ with the L^1-*norm* given by

$$\|f\| = \int_{-\infty}^{\infty} |f(x)|\, dx < \infty. \tag{2.3.2}$$

Definition 2.3.1 (Convergence in Norm). A sequence of functions $f_1, f_2, \ldots f_n \cdots \in L^1(\mathbb{R})$ is said to converge to a function $f \in L^1(\mathbb{R})$ in norm if $\|f_n - f\|_1 \to 0$ as $n \to \infty$.

So, the convergence in norm is denoted by $f_n \to f$ i.n. This is the usual convergence in a normed space.

Usually, the symbol $\int_{-\infty}^{\infty} f(x)\, dx$ or $\int_{\mathbb{R}} f(x)\, dx$ is used to represent the integral over the entire real line. In applications, we often need to integrate functions over bounded intervals on $\mathbb{R}$. This concept can easily be defined using the integral $\int_{\mathbb{R}} f(x)\, dx$.

Definition 2.3.2 (Integral Over an Interval). The integral of a function *f* over an interval $[a, b]$ is denoted by

$$\int_a^b f(x)\, dx$$

and defined by

$$\int_a^b f(x) \cdot \chi_{[a,b]}(x)\, dx, \tag{2.3.3}$$

where $\chi_{[a,b]}$ denotes the *characteristic* function of $[a, b]$ defined by

$$\chi_{[a,b]}(x) = \begin{Bmatrix} 1, & a \le x \le b, \\ 0, & \text{otherwise} \end{Bmatrix} \tag{2.3.4}$$

and $f\,\chi_{[a,b]}$ is the product of two functions.

In other words, $\int_a^b f(x)\,dx$ is the integral of the function equal to f on $[a,b]$ and zero otherwise.

Theorem 2.3.1 If $f \in L^1(\mathbb{R})$, then the integral $\int_a^b f(x)\,dx$ exists for every interval $[a,b]$.

The proof is left to the reader as an exercise.

The converse of this theorem is not necessarily true. For example, for the constant function $f=1$, the integral $\int_a^b f(x)\,dx$ exists for every $-\infty < a < x < b < \infty$, although $f \notin L^1(\mathbb{R})$. This suggests the following definition.

Definition 2.3.3 (Locally Integrable Functions). A function f defined on $\mathbb{R}$ is called *locally integrable* if, for every $-\infty < a < x < b < \infty$, the integral $\int_a^b f(x)\,dx$ exists.

Although this definition requires integrability of f over every bounded interval, it is sufficient to check that the integral $\int_{-n}^{n} f(x)\,dx$ exists for every positive integer n. The proof of this simple fact is left as an exercise.

Note that Theorem 2.3.1 implies that $L^1(\mathbb{R})$ is a subspace of the space of locally integrable functions.

Theorem 2.3.2 The locally integrable functions form a vector space. The absolute value of a locally integrable function is locally integrable. The product of a locally integrable function and a bounded locally integrable function is a locally integrable function.

For a proof of this theorem, the reader is referred to Debnath and Mikusinski (1999).

Theorem 2.3.3 If f is a locally integrable function such that $|f| \le g$ for some $g \in L^1(\mathbb{R})$, then $f \in L^1(\mathbb{R})$.

Proof. Let $f_n = f\,\chi_{[a,b]}$ for $n = 1,2,3,\ldots$. Then, the sequence of functions $\{f_n\}$ converges to f everywhere and $|f_n| \le g$ for every $n = 1,2,\ldots$. Thus, by the Lebesgue dominated convergence theorem, $f \in L^1(\mathbb{R})$.

The function space $L^2(\mathbb{R})$ is the space of all complex-valued Lebesgue integrable functions defined on $\mathbb{R}$ with the L^2-*norm* defined by

$$\|f\|_2 = \left[\int_{-\infty}^{\infty} |f(x)|^2\, dx\right]^{\frac{1}{2}} < \infty. \tag{2.3.5}$$

Elements of $L^2(\mathbb{R})$ will be called *square integrable functions*. Many functions in physics and engineering, such as wave amplitude in classical or quantum mechanics, are square integrable, and the class of L^2 functions is of fundamental importance.

The space $L^2[a,b]$ is the space of square integrable functions over $[a,b]$ such that $\int_a^b |f(x)|^2\, dx$ exists. Thus, the function $x^{-\frac{1}{3}} \in L^2[a,b]$, but $x^{-\frac{2}{3}} \notin L^2[a,b]$.

Remark. The fact that a function belongs to L^p for one particular value of p does not imply that it will belong to L^p for some other value of p.

Example 2.3.1 The function $|x|^{-\frac{1}{2}} e^{-|x|} \in L^1(\mathbb{R})$, but it does not belong to $L^2(\mathbb{R})$. On the other hand, $(1+|x|)^{-1} \in L^2(\mathbb{R})$, but it does not belong to $L^1(\mathbb{R})$.

Example 2.3.2 Functions $x^n e^{-|x|}$ and $(1+x^2)^{-1} \in L^1(\mathbb{R})$ for any integer n.

We add a comment here on the integrability and the local integrability. The condition of integrability is more stringent than local integrability. For example, the functions equal almost everywhere to $|x|^{-\frac{1}{2}}$ and $(1+x^2)^{-1}$, respectively, are both locally integrable, but only the latter one belongs to $L^1(\mathbb{R})$ because $|x|^{-\frac{1}{2}}$

decays very slowly as $|x| \to \infty$. The additional constraint imposed by integrability over that imposed by local integrability is associated with the nature of the function as $|x| \to \infty$. However, a function $f \in L^1(\mathbb{R})$ does not necessarily decay to zero at infinity. For example, for the function f whose graph consists of an infinite set of rectangular pulses with centers at $x = \pm 1, \pm 2, \cdots, \pm n, \cdots$, the pulse at $x = \pm n$ with height n and width n^{-3}, we obtain that $f \in L^1(\mathbb{R})$, but it does not tend to zero as $|x| \to \infty$.

We make another comment on functions in L^p spaces. If a function f belongs to $L^p(a,b)$ for some value of $p \ge 1$, then it also belongs to $L^q(a,b)$ for all q such that $1 \le q \le p$. In other words, raising a function to some power $p > 1$ makes the infinite singularities get 'worse' as p is increased. On the other hand, if a function is bounded in $\mathbb{R}$ and belongs to L^p for some $p \ge 1$, then it does belong to L^q for all $q \ge p$. In other words, raising a bounded function to some power p makes its nature at infinity get "better" as far as integrability is concerned. For example, the function $f = (1+|x|)^{-1} \in L^{1.1}(\mathbb{R})$ and is also bounded on $\mathbb{R}$, and also square integrable. However, if the condition of boundedness is relaxed, this result does not hold, even if the function is still locally bounded, that is, it is bounded on every finite interval on $\mathbb{R}$.

Definition 2.3.4 (Convolution). The convolution of two functions $f, g \in L^1(\mathbb{R})$ is defined by

$$(f * g)(x) = \int_{-\infty}^{\infty} f(x-y)\, g(y)\, dy \tag{2.3.6}$$

which exists for all $x \in \mathbb{R}$ or at least almost everywhere. Then, it defines a function which is called the *convolution* of f and g and is denoted by $f * g$.

We next discuss some basic properties of the convolution.

Theorem 2.3.4 If $f, g \in L^1(\mathbb{R})$, then the function $f(x-y)\, g(y)$ is integrable for almost all $x \in \mathbb{R}$. Furthermore, the convolution

$$(f * g)(x) = \int_{-\infty}^{\infty} f(x-y)\, g(y)\, dy \tag{2.3.7}$$

is an integrable function and $(f*g) \in L^1(\mathbb{R})$ and the following inequality holds:

$$\|f*g\|_1 \le \|f\|_1 \|g\|_1. \tag{2.3.8}$$

Proof. We refer to Debnath and Mikusinski (1999) for the proof of the first part of the theorem, that is, $(f*g) \in L^1(\mathbb{R})$.

To prove inequality (2.3.8), we proceed as follows:

$$\begin{aligned}
\|f*g\|_1 &= \int_{-\infty}^{\infty} |f*g|\, dx = \int_{-\infty}^{\infty} \left| \int_{-\infty}^{\infty} f(x-y)\, g(y)\, dy \right| dx \\
&\le \int_{-\infty}^{\infty}\int_{-\infty}^{\infty} |f(x-y)||g(y)\, dy|\, dx \\
&= \int_{-\infty}^{\infty}\int_{-\infty}^{\infty} |f(x-y)||g(y)|\, dx\, dy, \quad \text{by Fubini's Theorem} \\
&= \int_{-\infty}^{\infty} |f(x-y)|\, dx \int_{-\infty}^{\infty} |g(y)|\, dy \\
&= \int_{-\infty}^{\infty} |f(x)|\, dx \int_{-\infty}^{\infty} |g(y)|\, dy = \|f\|_1 \|g\|_1.
\end{aligned}$$

Thus, the proof is complete.

Theorem 2.3.5 If $f, g \in L^1(\mathbb{R})$, then the convolution is commutative, that is,

$$(f*g)(x) = (g*f)(x). \tag{2.3.9}$$

The proof follows easily by the change of variables.

Theorem 2.3.6 If $f, g, h \in L^1(\mathbb{R})$, then the following properties hold:

(a) $(f*g)*h = f*(g*h)$ (associative), (2.3.10)

(b) $(f+g)*h = f*h + g*h$ (distributive). (2.3.11)

We use Fubini's theorem to prove that the convolution is associative. We have

$$(f*g)*h=(g*f)*h(x)=\int_{-\infty}^{\infty}\left[\int_{-\infty}^{\infty}g(x-z-y)\,f(y)\,dy\right]h(z)\,dz$$

$$=\int_{-\infty}^{\infty}\int_{-\infty}^{\infty}f(y)\,g(x-y-z)\,h(z)\,dz\,dy$$

$$=f*(g*h)(x).$$

The proof of part (b) is left to the reader as an exercise.

Remark. The properties of convolution just described above shows that the $L^1(\mathbb{R})$ is a commutative Banach algebra under ordinary addition, multiplication defined by convolution, and $\|\cdot\|_1$ as norm. This Banach algebra is also referred to as the L^1-algebra on $\mathbb{R}$.

Theorem 2.3.7 If f is an integrable function and g is a bounded locally integrable function, then the convolution $f*g$ is a continuous function.

Proof. First, note that since $|f(x-y)\,g(y)|\le M|f(x-y)|$ for some constant M and every x, the integral $\int_{-\infty}^{\infty}f(x-y)\,g(y)\,dy$ is defined at every $x\in\mathbb{R}$ by Theorem 2.3.3. Next, we show that $f*g$ is a continuous function.

For any $x,h\in\mathbb{R}$, we have

$$|(f*g)(x+h)-(f*g)(x)|=\left|\int_{-\infty}^{\infty}f(x+h-y)\,g(y)\,dy-\int f(x-y)\,g(y)\,dy\right|$$

$$=\left|\int_{-\infty}^{\infty}[f(x+h-y)-f(x-y)]\,g(y)\,dy\right|$$

$$\le\int_{-\infty}^{\infty}|f(x+h-y)-f(x-y)||g(y)|\,dy$$

$$\le M\int_{-\infty}^{\infty}|f(0+h-y)-f(0-y)|\,dy,$$

which tends to zero as $h\to 0$ since

$$\lim_{h\to 0} \int_{-\infty}^{\infty} |f(h-y) - f(-y)|\, dy = 0.$$

Thus, the proof is complete.

2.4 Generalized Functions with Examples

The Dirac delta function $\delta(x)$ is the best known of a class of entities called *generalized functions*. The generalized functions are the natural mathematical quantities which are used to describe many abstract notions which occur in the physical sciences. The impulsive force, the point mass, the point charge, the point dipole, and the frequency response of a harmonic oscillator in a nondissipating medium are all aptly represented by generalized functions. The generalized functions play an important role in the Fourier transform analysis, and they can resolve the inherent difficulties that occur in classical mathematical analysis. For example, every locally integrable function (and indeed every generalized function) can be considered as the integral of some generalized function and thus becomes infinitely differentiable in the new sense. Many sequences of functions which do not converge in the ordinary sense to a limit function can be found to converge to a generalized function. Thus, in many ways the idea of generalized functions not only simplifies the rules of mathematical analysis but also becomes very useful in the physical sciences.

In order to give a sound mathematical formulation of quantum mechanics, Dirac in 1920 introduced the delta function $\delta(x)$ having the following properties:

$$\left.\begin{aligned} &\delta(x) = 0, \qquad x \neq 0, \\ &\int_{-\infty}^{\infty} \delta(x)\, dx = 1 \end{aligned}\right\}. \tag{2.4.1}$$

These properties cannot be satisfied by any ordinary function in classical mathematics. Hence, the delta function is not really a function in the classical sense. However, it can be regarded as the limit of a sequence of ordinary functions. A good example of such a sequence $\delta_n(x)$ is a sequence of Gaussian functions given by

$$\delta_n(x) = \sqrt{\frac{n}{\pi}} \exp\left(-nx^2\right). \tag{2.4.2}$$

Clearly, $\delta_n(x) \to 0$ as $n \to \infty$ for any $x \neq 0$ and $\delta_n(x) \to \infty$ as $n \to \infty$, as shown in Figure 2.1. Also, for all $n = 1, 2, 3, \ldots$,

$$\int_{-\infty}^{\infty} \delta_n(x)\, dx = 1$$

and

$$\lim_{n\to\infty} \int_{-\infty}^{\infty} \delta_n(x)\, dx = \int_{-\infty}^{\infty} \lim_{n\to\infty} \delta_n(x)\, dx = \int_{-\infty}^{\infty} \delta(x)\, dx = 1. \tag{2.4.3}$$

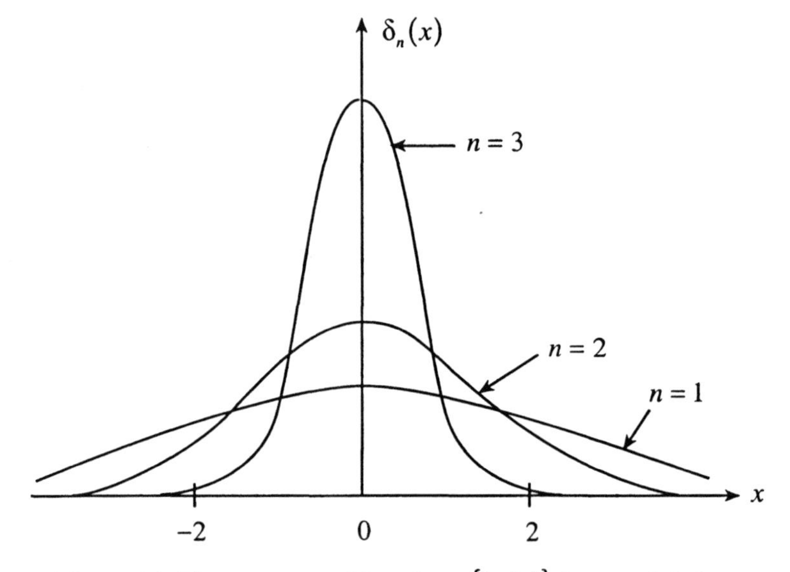

Figure 2.1. The sequence of functions $\{\delta_n(x)\}$ for $n = 1, 2, 3, \ldots$.

Thus, the Dirac delta function can be regarded as the limit of sequence $\delta_n(x)$ of ordinary functions, and we write

$$\delta(x) = \lim_{n\to\infty} \delta_n(x) = \lim_{n\to\infty} \sqrt{\frac{n}{\pi}} \exp\left(-nx^2\right). \tag{2.4.4}$$

This approach of defining new entities, such as $\delta(x)$, which do not exist as ordinary functions becomes meaningful mathematically and useful from a physical point of view.

Another alternative definition is based on the idea that if a function f is continuous at $x = a$, then $\delta(x)$ is defined by its fundamental property

$$\int_{-\infty}^{\infty} f(x)\,\delta(x-a)\,dx = f(a). \tag{2.4.5}$$

Or, equivalently,

$$\int_{-\infty}^{\infty} f(x)\,\delta(x)\,dx = f(0). \tag{2.4.6}$$

This is a rather more formal approach pioneered by Laurent Schwartz in the late 1940s. Thus, the concept of the delta function is clear and simple in modern mathematics. It has become very useful in science and engineering. Physically, the delta function represents a point mass, that is, a particle of unit mass is located at the origin. This means that a point particle can be regarded as the limit of a sequence of continuous mass distribution. The Dirac delta function is also interpreted as a probability measure in terms of the formula (2.4.5).

Definition 2.4.1 (Support of a Function). The support of a function $f : \mathbb{R} \to \mathbb{C}$ is $\{x : f(x) \neq 0\}$ and denoted by $\operatorname{supp}(f)$. A function has *bounded support* if there are two real numbers *a, b* such that $\operatorname{supp}(f) \subset (a,b)$. By a *compact support*, we mean a closed and bounded support.

Definition 2.4.2 (Smooth or Infinitely Differentiable Function). A function $f : \mathbb{R} \to \mathbb{C}$ is called *smooth* or *infinitely differentiable* if its derivatives of all orders exist and are continuous.

A function $f : \mathbb{R} \to \mathbb{C}$ is said to be *n-times continuously differentiable* if its first n derivatives exist and are continuous.

Definition 2.4.3 (Test Functions). A *test function* is an infinitely differentiable function on $\mathbb{R}$ whose support is compact. The space of all test functions is denoted by $\mathscr{D}(\mathbb{R})$ or simply by $\mathscr{D}$.

The graph of a 'typical' test function is shown in Figure 2.2.

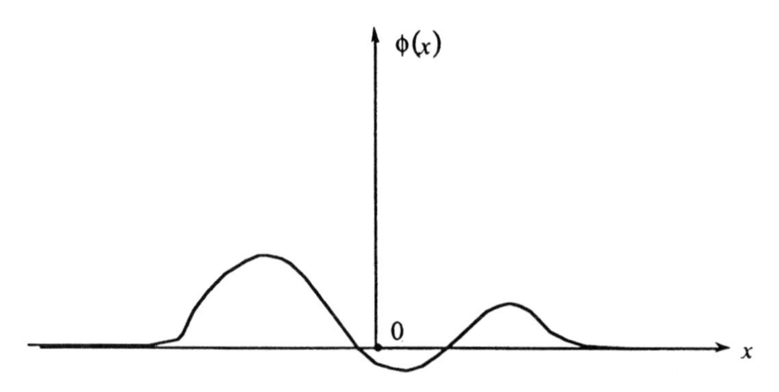

Figure 2.2. A typical test function.

Since smooth (infinitely differentiable) functions are continuous and the support of a continuous function is always closed, test functions can be equivalently defined as follows: ϕ is a test function if it is a smooth function vanishing outside a bounded set.

Example 2.4.1 A function ϕ defined by

$$\phi(x) = \begin{Bmatrix} \exp\left[\left(x^2 - a^2\right)^{-1}\right], & \text{for } |x| < a, \\ 0, & \text{otherwise} \end{Bmatrix} \tag{2.4.7}$$

is a test function with support $(-a, a)$.

Using this test function, we can easily generate a number of examples. The following are test functions:

$\phi(ax + b)$, a, b are constants and $a \neq 0$,

$f(x)\phi(x)$, f is an arbitrary smooth function,

$\phi^{(n)}(x)$, n is a positive integer.

Definition 2.4.4 (Convergence of Test Functions). Suppose $\{\phi_n\}$ is a sequence of test functions and ϕ is another test function. We say that the sequence $\{\phi_n\}$ converges to ϕ in $\mathfrak{D}$, denoted by $\phi_n \xrightarrow{\mathfrak{D}} \phi$, if the following two conditions are satisfied:

(a) $\phi_1, \phi_2, \cdots, \phi_n, \cdots$ and ϕ vanish outside some bounded interval $[a,b] \subset \mathbb{R}$,

(b) for each k, $\phi_n \to \phi(x)$ as $n \to \infty$ uniformly for some $x \in [a,b]$, where $\phi^{(k)}(x)$ denotes the kth derivative of ϕ.

Definition 2.4.5 (Generalized Function or Distribution). A continuous linear functional F on $\mathfrak{D}$ is called a *generalized function* or *distribution*. In other words, a mapping $F : \mathfrak{D} \to \mathbb{C}$ is called a *generalized function* or *distribution* if

(a) $F(a\phi + b\psi) = a\,F(\phi) + b\,F(\psi)$ for every $a, b, \in \mathbb{C}$ and $\phi, \psi \in \mathfrak{D}(\mathbb{R})$,

(b) $F(\phi_n) \to F(\phi)$ (in $\mathbb{C}$) whenever $\phi_n \to \phi$ in $\mathfrak{D}$.

The space of all generalized functions is denoted by $\mathfrak{D}'(\mathbb{R})$ or simply by $\mathfrak{D}'$. It is convenient to write (F, ϕ) instead of $F(\phi)$.

Distributions generalize the concept of a function. Formally, a function on $\mathbb{R}$ is not a distribution because its domain is not $\mathfrak{D}$. However, every locally integrable function f on $\mathbb{R}$ can be identified with a distribution F defined by

$$(F, \phi) = \int_{\mathbb{R}} f(x)\,\phi(x)\,dx. \tag{2.4.8}$$

The distribution F is said to be *generated* by the function f.

Definition 2.4.6 (Regular and Singular Distributions). A distribution $F \in \mathfrak{D}'$ is called a *regular distribution* if there exists a locally integrable function f such that

$$(F, \phi) = \int_{\mathbb{R}} f(x)\,\phi(x)\,dx \tag{2.4.9}$$

for every $\phi \in \mathfrak{D}$. A distribution that is not regular is called a *singular distribution*.

The fact that (2.4.9) defines a distribution is because of the following results.

First, the product $f\phi$ is integrable because it vanishes outside a compact support $[a,b]$. In other words,

$$(F,\phi) = \int_{-\infty}^{\infty} f(x)\,\phi(x)\,dx = \int_a^b f(x)\,\phi(x)\,dx$$

exists. Hence, F is a linear functional on $\mathcal{D}$. Also,

$$\begin{aligned}\left|(F,\phi_n)-(F,\phi)\right| &= \left|\int_a^b [\phi_n(x)-\phi(x)]\,f(x)\,dx\right| \\ &\le \int_a^b |\phi_n(x)-\phi(x)||f(x)|\,dx \\ &\le \max|\phi_n(x)-\phi(x)| \int_a^b |f(x)|\,dx \to 0 \quad \text{as } n\to\infty,\end{aligned}$$

because $\phi_n \to \phi$ uniformly. Hence,

$$(F,\phi_n) \to (F,\phi) \qquad \text{as } n\to\infty.$$

This means that F is a continuous linear functional, that is, F is a distribution.

Thus, the class of generalized functions contains elements which corresponds to ordinary functions as well as singular distributions. We now give an interpretation of (F,ϕ).

The integral $\int_{\mathbb{R}} f(x)\,\phi(x)\,dx$ in (2.4.9) can be interpreted, at least for some test function ϕ, as the average value of f with respect to probability whose density function is ϕ. Thus, (F,ϕ) can be regarded as an average value of F and of distributions as entities that have average values in neighborhoods of every point. However, in general, distributions may not have values at points. This interpretation is very natural from a physical point of view. In fact, when a quantity is measured, the result is not the exact value at a single point.

Example 2.4.2 If Ω is an open set in $\mathbb{R}$, then the functional F defined by

$$(F,\phi) = \int_{\Omega} \phi(x)\,dx \tag{2.4.10}$$

is a distribution. Note that it is a regular distribution since

$$(F,\phi) = \int_{-\infty}^{\infty} \phi(x)\,\chi_{\Omega}(x)\,dx, \tag{2.4.11}$$

where χ_{Ω} is the characteristic function of the set Ω.

In particular, if $\Omega = (0,\infty)$, we obtain a distribution

$$(H,\phi) = \int_0^\infty \phi(x)\, dx \tag{2.4.12}$$

which is called the *Heaviside function*. The symbol H is used to denote this distribution as well as the characteristic function of $\Omega = (0,\infty)$.

***Example 2.4.3* (*Dirac Distribution*).** One of the most important examples of generalized functions is the so-called *Dirac delta function* or, more precisely, the *Dirac distribution*. It is denoted by δ and defined by

$$(\delta,\phi) = \int_{-\infty}^{\infty} \phi(x)\, \delta(x)\, dx = \phi(0). \tag{2.4.13}$$

The linearity of δ is obvious. To prove the continuity, note that $\phi_n \to \phi$ in $\mathscr{D}$ implies that $\phi_n \to \phi$ uniformly on $\mathbb{R}$ and hence $\phi_n(x) \to \phi(x)$ for every $x \in \mathbb{R}$. This implies that the Dirac delta function is a singular distribution.

Example 2.4.4

(a)
$$(\delta(x-a),\phi) = (\delta(x),\phi(x+a)) = \phi(a). \tag{2.4.14}$$

(b)
$$(\delta(ax),\phi) = \frac{1}{|a|}\,\phi(0). \tag{2.4.15}$$

We have

$$(\delta(x-a),\phi) = \int_{-\infty}^{\infty} \delta(x-a)\, \phi(x)\, dx$$
$$= \int_{-\infty}^{\infty} \delta(y)\, \phi(y+a)\, dy = \phi(a).$$

This is called the *shifting property* of the delta function.

Similarly,

$$(\delta(ax),\, \phi) = \int_{-\infty}^{\infty} \delta(ax)\, \phi(x)\, dx = \int_{-\infty}^{\infty} \delta(y)\, \phi\left(\frac{y}{a}\right) \frac{dy}{a} = \frac{1}{a}\,\phi(0).$$

Hence, for $a \neq 0$,

$$\delta(ax) = \frac{1}{|a|}\,\phi(0). \tag{2.4.16}$$

The success of the theory of distributions is essentially due to the fact that most concepts of ordinary calculus can be defined for distributions. While adopting definitions and rules for distributions, we expect that new definitions and rules will agree with classical ones when applied to regular distributions. When looking for an extension of some operation A, which is defined for ordinary functions, we consider regular distributions defined by (2.4.9). Since we expect AF to be the same as Af, it is natural to define

$$(AF,\phi) = \int_{\mathbb{R}} Af(x)\,\phi(x)\,dx.$$

If there exists a continuous operation A^* which maps $\mathcal{D}$ into $\mathcal{D}$ such that

$$\int Af(x)\,\phi(x)\,dx = \int f(x)\,A^*\phi(x)\,dx,$$

then it makes sense to introduce, for an arbitrary distribution F,

$$(AF,\phi) = (F, A^*\phi).$$

If this idea is used to give a natural definition of a derivative of a distribution, it suffices to observe

$$\int_{\mathbb{R}} \left\{\frac{\partial}{\partial x} f(x)\right\} \phi(x)\,dx = -\int_{\mathbb{R}} f(x)\,\frac{\partial}{\partial x}\,\phi(x)\,dx.$$

Definition 2.4.7 (Derivatives of a Distribution). The derivative of a distribution F is a distribution F' defined by

$$\left(\frac{dF}{dx}, \phi\right) = -\left(F, \frac{d\phi}{dx}\right). \tag{2.4.17}$$

This result follows by integrating by parts. In fact, we find

$$\left(\frac{dF}{dx}, \phi\right) = \int_{-\infty}^{\infty} \frac{dF}{dx}\,\phi(x)\,dx = \left[F(x)\,\phi(x)\right]_{-\infty}^{\infty} - \int_{-\infty}^{\infty} F(x)\,\phi'(x)\,dx = -\left(F, \phi'(x)\right),$$

where the first term vanishes because ϕ vanishes at infinity.

More generally,

$$\left(F^{(k)}, \phi\right) = (-1)^k \left(F, \phi^{(k)}\right), \tag{2.4.18}$$

where $F^{(k)}(x)$ is the kth derivative of distribution F.

Thus, the extension of the idea of a function to that of a distribution has a major success in the sense that every distribution has derivatives of all orders which are again distributions.

Example 2.4.5 (Derivative of the Heaviside Function).

(a)
$$H'(x)=\delta(x). \tag{2.4.19}$$

We have

$$(H',\phi)=\int_0^\infty H'(x)\,\phi(x)\,dx=\left[H(x)\,\phi(x)\right]_0^\infty-\int_0^\infty H(x)\,\phi'(x)\,dx$$

$$=-\int_0^\infty \phi'(x)\,dx=\phi(0)=(\delta,\phi), \qquad \text{since } \phi \text{ vanishes at infinity.}$$

This proves the result.

(b) ***(Derivatives of the Dirac Delta Function).***

$$(\delta',\phi)=-(\delta,\phi')=-\phi'(0), \tag{2.4.20}$$

$$\left(\delta^{(n)},\phi\right)=(-1)^n\,\phi^{(n)}(0). \tag{2.4.21}$$

We have

$$(\delta',\phi)=\int_{-\infty}^\infty \delta'(x)\,\phi(x)\,dx=\left[\delta(x)\,\phi(x)\right]_{-\infty}^\infty-\int_{-\infty}^\infty \delta(x)\,\phi'(x)\,dx=-\phi'(0),$$

since ϕ vanishes at infinity.

Result (2.4.21) follows from a similar argument.

Example 2.4.6 If h is a smooth function and F is a distribution, then the derivative of the product (hF) is given by

$$(hF)'=hF'+h'F. \tag{2.4.22}$$

We have, for any $\phi\in\mathscr{D}$,

$$\begin{aligned}\left((hF)',\phi\right) &= -(hF,\phi') \\ &= -(F,h\phi') \\ &= -\left(F,(h\phi)'-h'\phi\right) \\ &= (F',h\phi)+(F,h'\phi) \\ &= (hF',\phi)+(h'F,\phi) \\ &= (hF'+h'F,\phi).\end{aligned}$$

This proves the result.

Example 2.4.7 The function $|x|$ is locally integrable and differentiable for all $x \neq 0$ but certainly not differentiable at $x = 0$. The generalized derivative can be calculated as follows.

For any test function ϕ, we have

$$\begin{aligned}\left(|x|',\phi\right) &= -(|x|,\phi') \\ &= -\int_{-\infty}^{\infty}|x|\,\phi'(x)\,dx = \int_{-\infty}^{0}x\,\phi'(x)\,dx - \int_{0}^{\infty}x\,\phi'(x)\,dx\end{aligned}$$

which is, integrating by parts and using the fact that ϕ vanishes at infinity,

$$= -\int_{-\infty}^{0}\phi(x)\,dx + \int_{0}^{\infty}\phi(x)\,dx. \tag{2.4.23}$$

Thus, we can write (2.4.23) in the form

$$\left(|x|',\phi\right) = \int_{-\infty}^{\infty}\operatorname{sgn}(x)\,\phi(x)\,dx = (\operatorname{sgn},\phi) \qquad \text{for all } \phi \in \mathfrak{D}.$$

Therefore,

$$|x|' = \operatorname{sgn}(x), \tag{2.4.24}$$

where $\operatorname{sgn}(x)$ is called the *sign function*, defined by

$$\operatorname{sgn}(x) = \begin{Bmatrix} 1, & x > 0 \\ -1, & x < 0 \end{Bmatrix}. \tag{2.4.25}$$

Obviously,

$$H(x) = \frac{1}{2}(1 + \operatorname{sgn} x). \tag{2.4.26}$$

Or, equivalently,

$$\operatorname{sgn} x = 2H(x) - 1. \tag{2.4.27}$$

Thus,

$$\frac{d}{dx}(\operatorname{sgn} x) = 2H'(x) = 2\delta(x). \tag{2.4.28}$$

Definition 2.4.8 (Antiderivative of a Distribution). If F is a distribution on $\mathbb{R}$ and $F \in \mathscr{D}'(\mathbb{R})$, a distribution G on $\mathbb{R}$ is called an *antiderivative* of F if $G' = F$.

Theorem 2.4.1 Every distribution has an antiderivative.

Proof. Suppose $\phi_0 \in \mathscr{D}(\mathbb{R})$ is a fixed test function such that

$$\int_{-\infty}^{\infty} \phi_0(x)\, dx = 1. \tag{2.4.29}$$

Then, for every test function $\phi \in \mathscr{D}(\mathbb{R})$, there exists a test function $\phi_1 \in \mathscr{D}(\mathbb{R})$ such that $\phi = K\phi_0 + \phi_1$, where

$$K = \int_{-\infty}^{\infty} \phi(x)\, dx \quad \text{and} \quad \int_{-\infty}^{\infty} \phi_1(x)\, dx = 0.$$

Suppose $F \in \mathscr{D}'(\mathbb{R})$. We define a functional G on $\mathscr{D}(\mathbb{R})$ by

$$(G, \phi) = (G, K\phi_0 + \phi_1) = CK - (F, \psi),$$

where C is a constant and ψ is a test defined by

$$\psi(x) = \int_{-\infty}^{x} \phi_1(t)\, dt.$$

Then ,G is a distribution and $G' = F$.

We close this section by adding an example of application to partial differential equations.

Consider a partial differential operator L of order m in N variables

$$L = \sum_{|\alpha| \le m} A_\alpha D^\alpha , \tag{2.4.30}$$

where $\alpha = (\alpha_1, \alpha_2, \cdots, \alpha_N)$ is a multi-index, the α_n's are nonnegative integers, $|\alpha| = \alpha_1 + \alpha_2 + \cdots + \alpha_N$, $A_\alpha = A_{\alpha_1, \alpha_2, \cdots, \alpha_N}(x_1, x_2, \cdots, x_N)$ are functions on $\mathbb{R}^N$ (possibly constant), and

$$D^\alpha = \left(\frac{\partial}{\partial x_1}\right)^{\alpha_1} \cdots \left(\frac{\partial}{\partial x_N}\right)^{\alpha_N} = \frac{\partial^{|\alpha|}}{\partial x_1^{\alpha_1} \cdots \partial x_N^{\alpha_N}}. \tag{2.4.31}$$

Equations of the form

$$LG = \delta \tag{2.4.32}$$

are of particular interest. Suppose G is a distribution which satisfies (2.3.32). Then, for any distribution f with compact support, the convolution $(f * G)$ is a solution of the partial differential equation

$$Lu = f. \tag{2.4.33}$$

We have

$$\begin{aligned} L(f * G) &= \sum_{|\alpha| \le m} A_\alpha D^\alpha (f * G) \\ &= \sum_{|\alpha| \le m} A_\alpha (f * D^\alpha G) \\ &= f * \left(\sum_{|\alpha| \le m} A_\alpha D^\alpha G \right) = f * LG \\ &= f * \delta = f. \end{aligned}$$

This explains the importance of the equation $Lu = \delta$, at least in the context of the existence of solutions of partial differential equations.

2.5 Definition and Examples of an Inner Product Space

Definition 2.5.1 (Inner Product Space). A (real or complex) inner product space is a (real or complex) vector space X with an *inner product* defined in X as a mapping

$$(\cdot,\cdot) : X \times X \to \mathbb{C}$$

such that, for any $x, y, z \in X$ and $\alpha, \beta \in \mathbb{C}$ (a set of complex numbers), the following conditions are satisfied:

(a) $(x, y) = \overline{(y, x)}$ (the bar denotes the complex conjugate),

(b) $(\alpha x + \beta y, z) = \alpha(x, z) + \beta(y, z)$,

(c) $(x, x) \geq 0$, and $(x, x) = 0$ implies $x = 0$.

Clearly, an inner product space is a vector space with an inner product specified. Often, an inner product space is called a *pre-Hilbert space* or a *unitary space*.

According to the above definition, the inner product of two vectors is a complex number. The reader should be aware that other symbols are sometimes used to denote the inner product: $\langle x, y \rangle$ or $\langle x / y \rangle$. Instead of $\bar{z}$, the symbol z^* is also used for the complex conjugate. In this book, we will use (x, y) and $\bar{z}$.

By (a), $(x, x) = \overline{(x, x)}$ which means that (x, x) is a real number for every $x \in X$. It follows from (b) that

$$(x, \alpha y + \beta z) = \overline{(\alpha y + \beta z, x)} = \overline{\alpha(y, x) + \beta(z, x)} = \bar{\alpha}(x, y) + \bar{\beta}(x, z).$$

In particular,

$$(\alpha x, y) = \alpha(x, y) \quad \text{and} \quad (x, \alpha y) = \bar{\alpha}(x, y).$$

Hence, if $\alpha = 0$,

$$(0, y) = (x, 0) = 0.$$

The algebraic properties (a)-(b) are generally the same as those governing the scalar product in ordinary vector algebra with which the reader should be familiar. The only property that is not obvious is that in a complex space the inner product is not linear but *conjugate linear* with respect to the second factor; that is, $(x, \alpha y) = \bar{\alpha}(x, y)$.

Example 2.5.1 The simplest but important example of an inner product space is the space of complex numbers $\mathbb{C}$. The inner product in $\mathbb{C}$ is defined by $(x, y) = x\bar{y}$.

Example 2.5.2 The space $\mathbb{C}^N$ of ordered N-tuples $x = (x_1, \ldots, x_N)$ of complex numbers, with the inner product defined by

$$(x, y) = \sum_{k=1}^{N} x_k \bar{y}_k, \quad x = (x_1, \ldots, x_N), \quad y = (y_1, \ldots, y_N),$$

is an inner product space.

Example 2.5.3 The space l^2 of all infinite sequences of complex numbers $\{x_n\}$ such that $\sum_{k=1}^{\infty} |x_k|^2 < \infty$ with the inner product defined by

$$(x, y) = \sum_{k=1}^{\infty} x_k \bar{y}_k, \quad \text{where } x = (x_1, x_2, x_3, \ldots), \quad y = (y_1, y_2, y_3, \ldots),$$

is an infinite dimensional inner product space. As we will see later, this space is one of the most important examples of an inner product space.

Example 2.5.4 Consider the space of infinite sequences $\{x_n\}$ of complex numbers such that only a finite number of terms are nonzero. This is an inner product space with the inner product defined as in Example 2.5.3.

Example 2.5.5 The space $\mathscr{C}([a, b])$ of all continuous complex-valued functions on the interval $[a, b]$ with the inner product

$$(f, g) = \int_a^b f(x) \overline{g(x)} \, dx \tag{2.5.1}$$

is an inner product space.

Example 2.5.6 (The Space of Square Integrable Functions). The function space $L^2([a, b])$ of all complex-valued Lebesgue square integrable functions on the interval $[a, b]$ with the inner product defined by (2.5.1) is an inner product space.

Similarly, the function space $L^2(\mathbb{R})$ is also an inner product space with the inner product defined by

$$(f,g) = \int_{-\infty}^{\infty} f(x)\,\overline{g(x)}\,dx, \tag{2.5.2}$$

where $f,g \in L^2(\mathbb{R})$.

Since

$$fg = \frac{1}{4}\left[(f+g)^2 - (f-g)^2\right],$$

and

$$|fg| \le \frac{1}{2}\left(|f|^2 + |g|^2\right),$$

it follows that $f,g \in L^1(\mathbb{R})$.

Furthermore,

$$|f+g|^2 \le |f|^2 + 2|fg| + |g|^2.$$

Integrating this inequality over $\mathbb{R}$ shows that $(f+g) \in L^2(\mathbb{R})$.

It can be shown that $L^2(\mathbb{R})$ is a complete normed space with the norm induced by (2.5.2), that is,

$$\|f\|_2 = \left\{\int_{\mathbb{R}} |f|^2\,dx\right\}^{\frac{1}{2}}. \tag{2.5.3}$$

This is exactly the L^2-norm defined by (2.3.1). Both spaces $L^2([a,b])$ and $L^2(\mathbb{R})$ are of special importance in theory and applications.

Example 2.5.7 Suppose D is a compact set in $\mathbb{R}^3$ and $X = C^2(D)$ is the space of complex-valued functions that have continuous second partial derivatives in D. If $u \in D$, we assume

$$\nabla u = \left(\frac{\partial u}{\partial x_1}, \frac{\partial u}{\partial x_2}, \frac{\partial u}{\partial x_3}\right). \tag{2.5.4}$$

We define the inner product by the integral

$$(u,v) = \int_D \left[u\overline{v} + \frac{\partial u}{\partial x_1}\cdot\frac{\partial \overline{v}}{\partial x_1} + \frac{\partial u}{\partial x_2}\cdot\frac{\partial \overline{v}}{\partial x_2} + \frac{\partial u}{\partial x_3}\cdot\frac{\partial \overline{v}}{\partial x_3}\right] dx, \tag{2.5.5}$$

where $x = (x_1, x_2, x_3)$.

Clearly, this is linear in u and also $(u,v)=\overline{(v,u)}$ and $(u,u)\geq 0$. Furthermore, if $(u,u)=0$, then $\int_D |u|^2\,dx=0$. Since u is continuous, this means that $u=0$. Hence, (2.5.5) defines an inner product in the space X. Obviously, the norm is given by

$$\|u\|=\left(\int_D \left(|u|^2+|\nabla u|^2\right)dx\right)^{\frac{1}{2}}, \tag{2.5.6}$$

where

$$|\nabla u|^2=\left|\frac{\partial u}{\partial x_1}\right|^2+\left|\frac{\partial u}{\partial x_2}\right|^2+\left|\frac{\partial u}{\partial x_3}\right|^2. \tag{2.5.7}$$

2.6 Norm in an Inner Product Space

An inner product space is a vector space with an inner product. It turns out that every inner product space is also a normed space with the norm defined by

$$\|x\|=\sqrt{(x,x)}.$$

First notice that the norm is well defined because (x,x) is always a nonnegative (real) number. Condition (c) of Definition 2.5.1 implies that $\|x\|=0$ if and only if $x=0$. Moreover,

$$\|\lambda x\|=\sqrt{(\lambda x,\lambda x)}=\sqrt{\lambda\bar{\lambda}\,(x,x)}=|\lambda|\|x\|.$$

It thus remains to prove the triangle inequality. This is not as simple as the first two conditions. We first prove the so-called *Schwarz's inequality*, which will be used in the proof of the triangle inequality.

Theorem 2.6.1 (Schwarz's Inequality). For any two elements x and y of an inner product space, we have

$$|(x,y)|\leq\|x\|\|y\|. \tag{2.6.1}$$

The equality $|(x,y)|=\|x\|\|y\|$ holds if and only if x and y are linearly dependent.

Proof. If $y = 0$, then (2.6.1) is satisfied because both sides are equal to zero. Assume then $y \neq 0$. By (c) in Definition 2.5.1, we have

$$0 \leq (x + \alpha y,\ x + \alpha y) = (x, x) + \overline{\alpha}(x, y) + \alpha(y, x) + |\alpha|^2 (y, y). \quad (2.6.2)$$

Now, put $\alpha = -(x, y)/(y, y)$ in (2.6.2) and then multiply by (y, y) to obtain

$$0 \leq (x, x)(y, y) - |(x, y)|^2.$$

This gives Schwarz's inequality.

If x and y are linearly dependent, then $y = \alpha x$ for some $\alpha \in \mathbb{C}$. Hence,

$$|(x, y)| = |(x, \alpha x)| = |\overline{\alpha}|(x, x) = |\alpha| \|x\| \|x\| = \|x\| \|\alpha x\| = \|x\| \|y\|.$$

Now, let x and y be vectors such that $|(x, y)| = \|x\| \|y\|$. Or, equivalently,

$$(x, y)(y, x) = (x, x)(y, y). \quad (2.6.3)$$

We next show that $(y, y)x - (x, y)y = 0$, which shows that x and y are linearly dependent. Indeed, by (2.6.3), we have

$$((y, y)x - (x, y)y,\ (y, y)x - (x, y)y)$$
$$= (y, y)^2 (x, x) - (y, y)(y, x)(x, y) - (x, y)(y, y)(y, x) + (x, y)(y, x)(y, y) = 0.$$

Thus, the proof is complete.

Corollary 2.6.1 (Triangle Inequality). For any two elements x and y of an inner product space X, we have

$$\|x + y\| \leq \|x\| + \|y\|. \quad (2.6.4)$$

Proof. When $\alpha = 1$, equality (2.6.2) can be written as

$$\begin{aligned} \|x + y\|^2 &= (x + y, x + y) = (x, x) + 2\,\mathrm{Re}(x, y) + (y, y) \\ &\leq (x, x) + 2|(x, y)| + (y, y) \\ &\leq \|x\|^2 + 2\|x\|\|y\| + \|y\|^2 \qquad \text{(by Schwarz's inequality)} \\ &= (\|x\| + \|y\|)^2, \end{aligned} \quad (2.6.5)$$

where $\mathrm{Re}\, z$ denotes the real part of $z \in \mathbb{C}$. This proves the triangle inequality.

Definition 2.6.1 (Norm in an Inner Product Space). By the *norm* in an inner product space X, we mean the functional defined by

$$\|x\| = \sqrt{(x, x)}. \tag{2.6.6}$$

We have proved that every inner product space is a normed space. It is only natural to ask whether every normed space is an inner product space. More precisely, is it possible to define in a normed space $(X, \|\cdot\|)$ with an inner product $(\cdot,\cdot)$ such that $\|x\| = \sqrt{(x, x)}$ for every $x \in X$? In general, the answer is negative. In the following theorem, we prove a property of the norm in an inner product space that is a necessary and sufficient condition for a normed space to be an inner product space.

The next theorem is usually called the parallelogram law because of its remarkable geometric interpretation, which reveals that the sum of the squares of the diagonals of a parallelogram is the sum of the squares of the sides. This characterizes the norm in a Hilbert space.

Theorem 2.6.2 (Parallelogram Law). For any two elements x and y of an inner product space X, we have

$$\|x + y\|^2 + \|x - y\|^2 = 2\left(\|x\|^2 + \|y\|^2\right). \tag{2.6.7}$$

Proof. We have

$$\|x + y\|^2 = (x + y, x + y) = (x, x) + (x, y) + (y, x) + (y, y)$$

and hence,

$$\|x + y\|^2 = \|x\|^2 + (x, y) + (y, x) + \|y\|^2. \tag{2.6.8}$$

Now, replace y by $-y$ to obtain

$$\|x - y\|^2 = \|x\|^2 - (x, y) - (y, x) + \|y\|^2. \tag{2.6.9}$$

By adding (2.6.8) and (2.6.9), we obtain the parallelogram law (2.6.7).

One of the most important consequences of having the inner product is the possibility of defining orthogonality of vectors. This makes the theory of Hilbert spaces so much different from the general theory of Banach spaces.

Definition 2.6.2 (Orthogonal Vectors). Two vectors x and y in an inner product space are called *orthogonal* (denoted by $x \perp y$) if $(x, y) = 0$.

Theorem 2.6.3 (Pythagorean Formula). For any pair of orthogonal vectors x and y, we have

$$\|x + y\|^2 = \|x\|^2 + \|y\|^2. \tag{2.6.10}$$

Proof. If $x \perp y$, then $(x, y) = 0$ and thus the equality (2.6.10) follows immediately from (2.6.8).

In the definition of the inner product space, we assume that X is a complex vector space. However, it is possible to define a real inner product space. Then condition (b) in Definition 2.5.1 becomes $(x, y) = (y, x)$. All of the above theorems hold in the real inner product space. If in Examples 2.5.1-2.5.6, the word *complex* is replaced by *real* and $\mathbb{C}$ by $\mathbb{R}$, we obtain a number of examples of real inner product spaces. A finite-dimensional real inner product space is called a *Euclidean space.*

If $x = (x_1, x_2, \ldots, x_N)$ and $y = (y_1, y_2, \ldots, y_N)$ are vectors in $\mathbb{R}^N$, then the inner product $(x, y) = \sum_{k=1}^{N} x_k y_k$ can be defined equivalently by

$$(x, y) = \|x\| \|y\| \cos\theta,$$

where θ is the angle between vectors x and y. In this case, Schwarz's inequality becomes

$$|\cos\theta| = \frac{|(x, y)|}{\|x\| \|y\|} \le 1.$$

2.7 Definition and Examples of Hilbert Spaces

Definition 2.7.1 (Hilbert Space). A complete inner product space is called a *Hilbert space.*

By the completeness of an inner product space X, we mean the completeness of X as a normed space. Now, we discuss completeness of the inner product spaces and also give some new examples of inner product spaces and Hilbert spaces.

Example 2.7.1 Since the space $\mathbb{C}$ is complete, it is a Hilbert space.

Example 2.7.2 Clearly, both $\mathbb{R}^N$ and $\mathbb{C}^N$ are Hilbert spaces.

In $\mathbb{R}^N$, the inner product is defined by $(x, y) = \sum_{k=1}^{N} x_k y_k$.

In $\mathbb{C}^N$, the inner product is given by $(x, y) = \sum_{k=1}^{N} x_k \bar{y}_k$.

In both cases, the norm is defined by

$$\|x\| = \sqrt{(x, x)} = \left(\sum_{k=1}^{n} |x_k|^2 \right)^{\frac{1}{2}}.$$

Since these spaces are complete, they are Hilbert spaces.

Example 2.7.3 The sequence space l^2 defined in Example 2.5.3 is a Hilbert space.

Example 2.7.4 The space X described in Example 2.5.4 is an inner product space which is not a Hilbert space because it is not complete. The sequence

$$x_n = \left(1, \frac{1}{2}, \frac{1}{3}, \ldots, \frac{1}{n}, 0, 0, \ldots\right)$$

is a Cauchy sequence because

$$\lim_{n,m\to\infty} \|x_n - x_m\| = \lim_{n,m\to\infty} \left[\sum_{k=m+1}^{n} \frac{1}{k^2} \right]^{1/2} = 0 \qquad \text{for } m < n.$$

However, the sequence does not converge in X because its limit $\left(1, \frac{1}{2}, \frac{1}{3}, \ldots\right)$ is not in X. However, this sequence $\{x_n\}$ converges in l^2.

Example 2.7.5 The space defined in Example 2.5.5 is another example of an incomplete inner product space. In fact, we consider the following sequence of functions in $\mathscr{C}([0,1])$ (see Figure 2.3):

$$f_n(x) = \begin{cases} 1 & \text{if } 0 \le x \le \dfrac{1}{2}, \\ 1 - 2n\left(x - \dfrac{1}{2}\right) & \text{if } \dfrac{1}{2} \le x \le \left(\dfrac{1}{2n} + \dfrac{1}{2}\right), \\ 0 & \text{if } \left(\dfrac{1}{2n} + \dfrac{1}{2}\right) \le x \le 1. \end{cases}$$

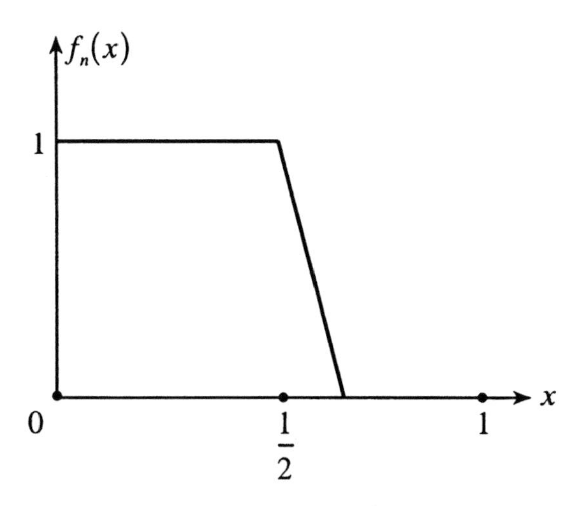

Figure 2.3. Sequence of functions $f_n(x)$.

Evidently, the $f_n(x)$ are continuous. Moreover,

$$\|f_n - f_m\| \le \left(\frac{1}{n} + \frac{1}{m}\right)^{1/2} \to 0 \qquad \text{as } m, n \to \infty.$$

Thus, $\{f_n\}$ is a Cauchy sequence. It is easy to check that this sequence converges to the limit function

$$f(x) = \begin{cases} 1 & \text{if } 0 \le x \le \dfrac{1}{2}, \\ 0 & \text{if } \dfrac{1}{2} < x \le 1. \end{cases}$$

The limit function is not continuous and hence is not an element of $\mathscr{C}([0,1])$. Consequently, $\mathscr{C}([0,1])$ is not a Hilbert space.

Example 2.7.6 The function space $L^2([a,b])$ is a Hilbert space. Since $L^2([a,b])$ is a normed space, it suffices to prove it is complete. Let $\{f_n\}$ be a Cauchy sequence in $L^2([a,b])$, that is,

$$\int_a^b |f_m - f_n|^2\, dx \to 0 \qquad \text{as } m,n \to \infty.$$

Schwarz's inequality imples that as $m,n \to \infty$

$$\int_a^b |f_m - f_n|\, dx \le \sqrt{\int_a^b dx}\ \sqrt{\int_a^b |f_m - f_n|^2\, dx} = \sqrt{b-a}\ \sqrt{\int_a^b |f_m - f_n|^2\, dx} \to 0\ .$$

Thus, $\{f_n\}$ is a Cauchy sequence in $L^1([a,b])$ and hence converges to a function f in $L^1([a,b])$, that is,

$$\int_a^b |f - f_n|\, dx \to 0 \qquad \text{as } n \to \infty.$$

By Riesz's theorem, there exists a subsequence $\{f_{p_n}\}$ convergent to f almost everywhere. Clearly, given an $\varepsilon > 0$, we have

$$\int_a^b |f_{p_m} - f_{p_n}|^2\, dx < \varepsilon$$

for sufficiently large m and n. Hence, by letting $n \to \infty$, we obtain

$$\int_a^b |f_{p_m} - f|^2\, dx \le \varepsilon$$

by Fatou's lemma (see Theorem 2.8.5 in Debnath and Mikusinski (1999), p. 60). This proves that $f \in L^2([a,b])$. Moreover,

$$\int_a^b |f - f_n|^2 \, dx \le \int_a^b \left|f - f_{p_n}\right|^2 \, dx + \int_a^b \left|f_{p_n} - f_n\right|^2 \, dx < 2\varepsilon$$

for sufficiently large n. This shows that the sequence $\{f_n\}$ converges to f in $L^2([a,b])$. Thus, the completeness is proved.

Example 2.7.7 Consider the space $\mathscr{C}_0(\mathbb{R})$ of all complex-valued continuous functions that vanish outside some finite interval. This is an inner product space with the inner product

$$(f,g) = \int_{-\infty}^{\infty} f(x) \, \overline{g(x)} \, dx .$$

Note that there is no problem with the existence of the integral because the product $f(x)\,\overline{g(x)}$ vanishes outside a bounded interval.

We now show that $\mathscr{C}_0(\mathbb{R})$ is not complete. We define

$$f_n(x) = \begin{cases} (\sin \pi x)/(1+|x|) & \text{if } |x| \le n, \\ 0 & \text{if } |x| > n. \end{cases}$$

Clearly, $f_n \in \mathscr{C}_0(\mathbb{R})$ for every $n \in \mathbb{N}$. For $n > m$, we have

$$\|f_n - f_m\|^2 = \int_{-\infty}^{\infty} |f_n(x) - f_m(x)|^2 \, dx \le 2 \int_m^n \frac{dx}{\left(1+|x|^2\right)} \to 0 \qquad \text{as } m \to \infty .$$

This shows that $\{f_n\}$ is a Cauchy sequence. On the other hand, it follows directly from the definition of f_n that

$$\lim_{n\to\infty} f_n(x) = \frac{\sin \pi x}{(1+|x|)},$$

which does not belong to $\mathscr{C}_0(\mathbb{R})$.

Example 2.7.8 We denote by $L^{2,\rho}([a,b])$ the space of all complex-valued square integrable functions on $[a,b]$ with a weight function ρ which is positive almost everywhere, that is, $f \in L^{2,\rho}([a,b])$ if

$$\int_a^b |f(x)|^2 \, \rho(x) \, dx < \infty .$$

Note that there is no problem with the existence of the integral because the product $f(x)\,\overline{g(x)}$ vanishes outside a bounded interval.

We now show that $\mathscr{C}_o(\mathbb{R})$ is not complete. We define

$$f_x(x) = \begin{cases} (\sin \pi x)/(1+|x|) & \text{if } |x| \le n, \\ 0 & \text{if } |x| > n. \end{cases}$$

Clearly, $f_n \in \mathscr{C}_0(\mathbb{R})$ for every $n \in \mathbb{N}$. For $n > m$, we have

$$\|f_n - f_m\|^2 = \int_{-\infty}^{\infty} |f_n(x) - f_m(x)|^2\, dx \le 2 \int_m^n \frac{dx}{\left(1+|x|^2\right)} \to 0 \quad \text{as } m \to \infty.$$

This shows that $\{f_n\}$ is a Cauchy sequence. On the other hand, it follows directly from the definition of f_n that

$$\lim_{n\to\infty} f_n(x) = \frac{\sin \pi x}{(1+|x|)},$$

which does not belong to $\mathscr{C}_0(\mathbb{R})$.

Example 2.7.8 We denote by $L^{2,\rho}([a,b])$ the space of all complex-valued square integrable functions on $[a,b]$ with a weight function ρ which is positive almost everywhere, that is, $f \in L^{2,\rho}([a,b])$ if

$$\int_a^b |f(x)|^2\ \rho(x)\, dx < \infty.$$

This is a Hilbert space with the inner product

$$(f,g) = \int_a^b f(x)\,\overline{g(x)}\,\rho(x)\, dx. \tag{2.7.1}$$

Example 2.7.9 (Sobolev Space). Let Ω be an open set in $\mathbb{R}^N$. Denote by $\tilde{H}^m(\Omega)$, $m = 1,2,\ldots$, the space of all complex-valued functions $f \in \mathscr{C}^m(\Omega)$ such that $D^\alpha f \in L^2(\Omega)$ for all $|\alpha| \le m$, where

$$D^\alpha f = \frac{\partial^{|\alpha|} f}{\partial x_1^{\alpha_1} \partial x_2^{\alpha_2} \cdots \partial x_N^{\alpha_N}}, \qquad |\alpha| = \alpha_1 + \cdots + \alpha_N, \quad \text{and} \quad \alpha_1, \ldots, \alpha_N \ge 0.$$

For example, if $N = 2$, $\alpha = (2,1)$, we have

$$D^{\alpha} f = \frac{\partial^3 f}{\partial x_1^2 \partial x_2}.$$

For $f \in \mathscr{C}^m(\Omega)$, we thus have

$$\int_{\Omega} \left| \frac{\partial^{|\alpha|} f}{\partial x_1^{\alpha_1} \partial x_2^{\alpha_2} \cdots \partial x_N^{\alpha_N}} \right|^2 < \infty$$

for every multi-index $\alpha = (\alpha_1, \alpha_2, \ldots, \alpha_N)$ such that $|\alpha| \le m$. The inner product in $\tilde{H}^m(\Omega)$ is defined by

$$(f, g) = \int_{\Omega} \sum_{|\alpha| \le m} D^{\alpha} f \, \overline{D^{\alpha} g}. \tag{2.7.2}$$

In particular, if $\Omega \subset \mathbb{R}^2$, then the inner product in $\tilde{H}^2(\Omega)$ is given by

$$(f, g) = \int_{\Omega} \left(f \overline{g} + f_x \overline{g}_x + f_y \overline{g}_y + f_{xx} \overline{g}_{xx} + f_{yy} \overline{g}_{yy} + f_{xy} \overline{g}_{xy} \right). \tag{2.7.3}$$

Or, if $\Omega = (a, b) \subset \mathbb{R}$, the inner product in $\tilde{H}^m(a, b)$ is

$$(f, g) = \int_a^b \sum_{n=1}^{m} \frac{d^n f}{dx^n} \cdot \overline{\frac{d^n g}{dx^n}}. \tag{2.7.4}$$

The function space $\tilde{H}^m(\Omega)$ is an inner product space, but it is not a Hilbert space because it is not complete. The completion of $\tilde{H}^m(\Omega)$, denoted by $H^m(\Omega)$, is a Hilbert space. The function space $H^m(\Omega)$ can be defined directly if D^{α} in the above is understood as the distributional derivative. This approach is often used in more advanced textbooks and treatises.

The space $H^m(\Omega)$ is a particular case of a general class of spaces denoted by $W_p^m(\Omega)$ and introduced by S.L. Sobolev. We have $H^m(\Omega) = W_2^m(\Omega)$. Because of the applications to partial differential equations, space $H^m(\Omega)$ is one of the most important examples of Hilbert spaces.

2.8 Strong and Weak Convergences

Since every inner product space is a normed space, it is equipped with a convergence, and the convergence is defined by the norm. This convergence is called the *strong convergence*. Moreover, the norm induces a topology in the space. Thus, a normed space is, in a natural way, a metric space and hence a topological space.

Definition 2.8.1 (Strong Convergence). A sequence $\{x_n\}$ of vectors in an inner product space X is called *strongly convergent* to a vector x in X if

$$\|x_n - x\| \to 0 \qquad \text{as } n \to \infty.$$

The word "strong" is added in order to distinguish "strong convergence" from "weak convergence".

Definition 2.8.2 (Weak Convergence). A sequence $\{x_n\}$ of vectors in an inner product space X is called *weakly convergent* to a vector x in X if

$$(x_n, y) \to (x, y) \qquad \text{as } n \to \infty, \qquad \text{for every } y \in X.$$

The condition in the above definition can also be stated as $(x_n - x, y) \to 0$ as $n \to \infty$, for every $y \in X$.

It is convenient to reserve the notation "$x_n \to x$" for the strong convergence and use "$x_n \xrightarrow{w} x$" to denote weak convergence.

Theorem 2.8.1 A strongly convergent sequence is weakly convergent (to the same limit), that is, $x_n \to x$ implies $x_n \xrightarrow{w} x$.

Proof. Suppose that the sequence $\{x_n\}$ converges strongly to x. This means

$$\|x_n - x\| \to 0 \qquad \text{as } n \to \infty.$$

By Schwarz's inequality, we have

$$|(x_n - x, y)| \le \|x_n - x\| \|y\| \to 0 \qquad \text{as } n \to \infty,$$

and thus,

$$(x_n - x, y) \to 0 \qquad \text{as } n \to \infty, \qquad \text{for every } y \in X.$$

This proves the theorem.

For any fixed y in an inner product space X, the mapping $(\cdot\,, y) : X \to \mathbb{C}$ is a linear functional on X. Theorem 2.8.1 states that such a functional is continuous for every $y \in X$. Obviously, the mapping $(x, \cdot) : X \to \mathbb{C}$ is also continuous.

In general, the converse of Theorem 2.8.1 is not true. A suitable example will be given in Section 2.9. On the other hand, we have the following theorem.

Theorem 2.8.2 If $x_n \xrightarrow{w} x$ and $\|x_n\| \to \|x\|$, then $x_n \to x$.

Proof. By the definition of weak convergence, we have

$$(x_n, y) \to (x, y) \qquad \text{as } n \to \infty, \qquad \text{for all } y.$$

Hence,

$$(x_n, x) \to (x, x) = \|x\|^2.$$

Now,

$$\begin{aligned}\|x_n - x\|^2 &= (x_n - x, x_n - x)\\ &= (x_n, x_n) - (x_n, x) - (x, x_n) + (x, x)\\ &= \|x_n\|^2 - 2\,\mathrm{Re}(x_n, x) + \|x\|^2 \to \|x\|^2 - 2\|x\|^2 + \|x\|^2 = 0 \qquad \text{as } n \to \infty.\end{aligned}$$

The sequence $\{x_n\}$ is thus strongly convergent to x.

Theorem 2.8.3 Suppose that the sequence $\{x_n\}$ converges weakly to x in a Hilbert space H. If, in addition,

$$\|x\| = \lim_{n\to\infty} \|x_n\|, \tag{2.8.1}$$

then $\{x_n\}$ converges strongly to x in H.

Proof. We assume that $(x_n, y) \to (x, y)$ and hence, $(y, x_n) \to (y, x)$. We have the result

$$\|x - x_n\|^2 = (x - x_n, x - x_n) = \|x\|^2 + \|x_n\|^2 - (x, x_n) - (x_n, x). \tag{2.8.2}$$

In view of the assumption (2.8.1), result (2.8.2) gives

$$\lim_{n\to\infty} \|x - x_n\|^2 = \|x\|^2 + \|x\|^2 - \|x\|^2 - \|x\|^2 = 0.$$

This proves the theorem.

We next state an important theorem (without proof) that describes an important property of weakly convergent sequences.

Theorem 2.8.4 Weakly convergent sequences are bounded, that is, if $\{x_n\}$ is a weakly convergent sequence, then there exists a number M such that $\|x_n\| \le M$ for all $n \in \mathbb{N}$.

2.9 Orthogonal and Orthonormal Systems

By a basis of a vector space X, we mean a linearly independent family $\mathcal{B}$ of vectors from X such that any vector $x \in X$ can be written as $x = \sum_{n=1}^{m} \lambda_n x_n$, where $x_n \in \mathcal{B}$ and the λ_n's are scalars. In inner product spaces, orthonormal bases are of much greater importance. Instead of finite combinations $\sum_{n=1}^{m} \lambda_n x_n$, infinite sums are allowed, and the condition of linear independence is replaced by orthogonality. One of the immediate advantages of these changes is that in all important examples it is possible to describe orthonormal bases. For example, $L^2([a,b])$ has countable orthonormal bases consisting of simple functions (see Example 2.9.2), whereas every basis of $L^2([a,b])$ is uncountable and we can only prove that such a basis exists without being able to describe its elements. In this section and the next, we give all necessary definitions and discuss basic properties of orthonormal bases.

Definition 2.9.1 (Orthogonal and Orthonormal Systems). Let X be an inner product space. A family S of nonzero vectors in X is called an *orthogonal system* if $x \perp y$ for any two distinct elements of S. If, in addition, $\|x\| = 1$ for all $x \in S$, S is called an *orthonormal system*.

Every orthogonal set of nonzero vectors can be normalized. If S is an orthogonal system, then the family

$$S_1 = \left\{ \frac{x}{\|x\|} : x \in S \right\}$$

is an orthonormal system. Both systems are equivalent in the sense that they span the same subspace of X.

Note that if x is orthogonal to each of $y_1, \ldots, y_n$, then x is orthogonal to every linear combination of vectors $y_1, \ldots, y_n$. In fact, we have

$$(x, y) = \left(x, \sum_{k=1}^{n} \lambda_k y_k \right) = \sum_{k=1}^{n} \bar{\lambda}_k (x, y_k) = 0.$$

Theorem 2.9.1 Orthogonal systems are linearly independent.

Proof. Let S be an orthogonal system. Suppose that $\sum_{k=1}^{n} \alpha_k x_k = 0$, for some $x_1, \ldots, x_n \in S$ and $\alpha_1, \ldots, \alpha_n \in \mathbb{C}$. Then,

$$0 = \left(\sum_{k=1}^{n} \alpha_k x_k, \sum_{k=1}^{n} \alpha_k x_k \right) = \sum_{k=1}^{n} |\alpha_k|^2 \|x_k\|^2 .$$

This means that $\alpha_k = 0$ for each $k \in N$. Thus, $x_1, \ldots, x_n$ are linearly independent.

Definition 2.9.2 (Orthonormal Sequence). A finite or infinite sequence of vectors which forms an orthonormal system is called an *orthonormal sequence*.

The condition of orthogonality of a sequence $\{x_n\}$ can be expressed in terms of the Kronecker delta symbol:

$$(x_m, x_n) = \delta_{mn} = \begin{cases} 0 & \text{if } m \neq n, \\ 1 & \text{if } m = n. \end{cases} \tag{2.9.1}$$

Example 2.9.1 For $e_n = (0, \ldots, 0, 1, 0, \ldots)$ with 1 in the nth position, the set $S = \{e_1, e_2, \ldots\}$ is an orthonormal system in the sequence space l^2.

Example 2.9.2 (Trigonometric Functions). The sequence $\phi_n(x) = e^{inx}/\sqrt{2\pi}$, $n = 0, \pm 1, \pm 2, \ldots$ is an orthonormal system in $L^2([-\pi,\pi])$. Indeed, for $m \neq n$, we have

$$(\phi_m, \phi_n) = \frac{1}{2\pi} \int_{-\pi}^{\pi} e^{i(m-n)x} dx = \frac{e^{\pi i(m-n)} - e^{-\pi i(m-n)}}{2\pi i(m-n)} = 0.$$

On the other hand,

$$(\phi_n, \phi_n) = \frac{1}{2\pi} \int_{-\pi}^{\pi} e^{i(n-n)x} dx = 1.$$

Thus, $(\phi_m, \phi_n) = \delta_{mn}$ for every pair of integers m and n.

For the real Hilbert space $L^2([-\pi,\pi])$, we can use the real and imaginary parts of the sequence $\{\phi_n\}$ and find that functions

$$\frac{1}{\sqrt{2\pi}} \cos nx, \quad \frac{1}{\sqrt{2\pi}} \sin nx, \quad (n = 0, 1, 2, \ldots)$$

form an orthonormal sequence.

Example 2.9.3 The *Legendre polynomials* defined by

$$P_0(x) = 1, \tag{2.9.2a}$$

$$P_n(x) = \frac{1}{2^n n!} \frac{d^n}{dx^n} \left(x^2 - 1\right)^n, \quad n = 1, 2, 3, \ldots, \tag{2.9.2b}$$

form an orthogonal system in the space $L^2([-1,1])$. It is convenient to write $\left(x^2 - 1\right)^n = p_n(x)$ so that

$$\int_{-1}^{1} P_n(x)\, x^m dx = \frac{1}{2^n n!} \int_{-1}^{1} p_n^{(n)}(x)\, x^m dx. \tag{2.9.3}$$

We evaluate this integral for $m < n$ by recursion. First, we note that

$$p_n^{(k)}(x) = 0$$

for $x = \pm 1$ and $k = 0, 1, 2, \ldots, (n-1)$. Hence, by integrating (2.9.3) by parts, we obtain

$$\int_{-1}^{1} p_n^{(n)}(x)\, x^m dx = -m \int_{-1}^{1} p_n^{(n-1)}(x)\, x^{m-1} dx.$$

Repeated application of this operation ultimately leads to

$$m!(-1)^m \int_{-1}^{1} p_n^{(n-m)}(x)\, dx = m!(-1)^m \left[p_n^{(n-m-1)}(x)\right]_{-1}^{1} = 0 \qquad (m < n).$$

Consequently,

$$\int_{-1}^{1} P_n(x)\, x^m dx = 0 \qquad \text{for } m < n. \tag{2.9.4}$$

Since P_m is a polynomial of degree m, it follows that

$$(P_n, P_m) = \int_{-1}^{1} P_n(x)\, P_m(x)\, dx = 0 \qquad \text{for } n \neq m. \tag{2.9.5}$$

This proves the orthogonality of the Legendre polynomials. To obtain an orthonormal system from the Legendre polynomials, we have to evaluate the norm of P_n in $L^2([-1,1])$:

$$\|P_n\| = \sqrt{\int_{-1}^{1} (P_n(x))^2\, dx}\,.$$

By repeated integration by parts, we first obtain

$$\begin{aligned}
\int_{-1}^{1} \left(1 - x^2\right)^n dx &= \int_{-1}^{1} (1 - x)^n (1 + x)^n dx \\
&= \frac{n}{n+1} \int_{-1}^{1} (1 - x)^{n-1} (1 + x)^{n+1} dx = \cdots \\
&= \frac{n(n-1)\cdots\cdots 2\cdot 1}{(n+1)(n+2)\cdots\cdots 2n} \int_{-1}^{1} (1 + x)^{2n} dx \\
&= \frac{(n!)^2\, 2^{2n+1}}{(2n)!(2n+1)}.
\end{aligned} \tag{2.9.6}$$

A similar procedure gives

$$\int_{-1}^{1} \left\{ p_n^{(n)}(x) \right\}^2 dx = 0 - \int_{-1}^{1} p_n^{(n-1)}(x)\, p_n^{(n+1)}(x)\, dx = \cdots$$

$$= (-1)^n \int_{-1}^{1} p_n(x)\, p_n^{(2n)}(x)\, dx$$

$$= (2n)! \int_{-1}^{1} (1-x)^n (1+x)^n\, dx, \tag{2.9.7}$$

where we have used the fact that the $2n$th derivative of $p_n(x) = \left(x^2 - 1\right)^n$ is the same as the derivative of the term of exponent $2n$. The $2n$th derivatives of all the other terms of the sum are zero. From (2.9.2), (2.9.6), and (2.9.7), we obtain

$$\int_{-1}^{1} \left\{ P_n(x) \right\}^2 dx = \frac{1}{\left(2^n n!\right)^2} (2n)! \frac{(n!)^2\, 2^{2n+1}}{(2n)!(2n+1)} = \frac{2}{2n+1}. \tag{2.9.8}$$

Thus, the polynomials $\sqrt{n + \frac{1}{2}}\; P_n(x)$ form an orthonormal system in the space $L^2([-1,1])$.

Example 2.9.4 We denote by H_n the Hermite polynomials of degree n, that is,

$$H_n(x) = (-1)^n e^{x^2} \frac{d^n}{dx^n} e^{-x^2}. \tag{2.9.9}$$

The functions $\phi_n(x) = e^{-x^2/2} H_n(x)$ form an orthogonal system in $L^2(\mathbb{R})$. The inner product

$$(\phi_n, \phi_m) = (-1)^{n+m} \int_{-\infty}^{\infty} e^{x^2} \frac{d^n}{dx^n} e^{-x^2} \frac{d^m}{dx^m} e^{-x^2}\, dx$$

can be evaluated by integrating by parts, which gives

$$(-1)^{n+m} (\phi_n, \phi_m) = \left[e^{x^2} \frac{d^n}{dx^n} e^{-x^2} \frac{d^{m-1}}{dx^{m-1}} e^{-x^2} \right]_{-\infty}^{\infty}$$

$$- \int_{-\infty}^{\infty} \frac{d}{dx} \left[e^{x^2} \frac{d^n}{dx^n} e^{-x^2} \right] \frac{d^{m-1}}{dx^{m-1}} e^{-x^2}\, dx, \tag{2.9.10}$$

and hence, all terms under the differential sign contain the factor e^{-x^2}. Since, for any $k \in \mathbb{N}$, we have

$$x^k e^{-x^2} \to 0 \qquad \text{as } x \to \infty,$$

the first term in (2.9.10) vanishes. Therefore, repeated integration by parts gives the result

$$(\phi_n, \phi_m) = 0 \qquad \text{as } n \neq m. \tag{2.9.11}$$

To obtain an orthonormal system, we evaluate the norm:

$$\|\phi_n\|^2 = \int_{-\infty}^{\infty} e^{-x^2} (H_n(x))^2 \, dx = \int_{-\infty}^{\infty} e^{-x^2} \left[e^{x^2} \frac{d^n}{dx^n} e^{-x^2} \right]^2 dx.$$

Integrating by parts n times yields

$$\|\phi_n\|^2 = (-1)^n \int_{-\infty}^{\infty} e^{-x^2} \left[e^{x^2} \frac{d^n}{dx^n} e^{-x^2} \right] dx.$$

Since $H_n(x)$ is a polynomial of degree n, direct differentiation gives

$$e^{x^2} \frac{d^n}{dx^n} e^{-x^2} = (-2x)^n + \cdots$$

and

$$\frac{d^n}{dx^n} \left[e^{x^2} \frac{d^n}{dx^n} e^{-x^2} \right] = \frac{d^n}{dx^n} \left\{ (-2x)^n + \cdots \right\} = (-1)^n \, 2^n n!.$$

Consequently,

$$\|\phi_n\|^2 = 2^n \, n! \int_{-\infty}^{\infty} e^{-x^2} \, dx = 2^n \, n! \sqrt{\pi}. \tag{2.9.12}$$

Thus, the functions

$$\psi_n(x) = \frac{1}{\sqrt{2^n \, n! \sqrt{\pi}}} \, e^{-\frac{x^2}{2}} H_n(x)$$

form an orthonormal system in the Hilbert space $L^2(\mathbb{R})$.

In the preceding examples, the original sequence of functions is orthogonal but not orthonormal. Although the calculations involved might be complicated, it is always possible to normalize the functions and obtain an orthonormal sequence. It turns out that the same is possible if the original sequence of functions (or, in general, a sequence of vectors in an inner product space) is linearly independent, not necessarily orthogonal. The method of transforming such a sequence into an orthonormal sequence is called the *Gram-Schmidt orthonormalization process*. The process can be described as follows.

Given a sequence $\{y_n\}$ of linearly independent vectors in an inner product space, define sequences $\{w_n\}$ and $\{x_n\}$ inductively by

$$w_1 = y_1, \qquad x_1 = \frac{w_1}{\|w_1\|},$$

$$w_k = y_k - \sum_{n=1}^{k-1} (y_k, x_n) x_n, \qquad x_k = \frac{w_k}{\|w_k\|}, \qquad \text{for } k = 1, 2, \ldots$$

The sequence $\{w_n\}$ is orthogonal. Indeed,

$$(w_2, w_1) = (y_2 - (y_2, x_1)x_1, y_1) = (y_2, y_1) - (y_2, x_1)(x_1, y_1)$$

$$= (y_2, y_1) - \frac{(y_2, y_1)(y_1, y_1)}{\|y_1\|^2} = 0.$$

Assume now that $w_1, \ldots, w_{k-1}$ are orthogonal. Then, for any $m < k$,

$$(w_k, w_m) = (y_k, w_m) - \frac{\sum_{n=1}^{k-1} (y_k, w_n)(w_n, w_m)}{\|w_m\|^2}$$

$$= (y_k, w_m) - \frac{(y_k, w_m)(w_m, w_m)}{\|w_m\|^2} = 0.$$

Therefore, vectors $w_1, \ldots, w_k$ are orthogonal. It follows, by induction, that the sequence $\{w_n\}$ is orthogonal and thus, $\{x_n\}$ is orthonormal. It is easy to check that any linear combination of vectors $x_1, \ldots, x_n$ is also a linear combination of $y_1, \ldots, y_n$ and vice versa. In other words, span $\{x_1, \ldots, x_n\} =$ span $\{y_1, \ldots, y_n\}$ for every $n \in \mathbb{N}$.

2.10 Properties of Orthonormal Systems

In Section 2.6, we proved that the Pythagorean formula holds for any pair of orthogonal vectors in an inner product space X. It turns out that it can be generalized to any finite number of orthogonal vectors.

Theorem 2.10.1 (Pythagorean Formula). If $x_1, \dots, x_n$ are orthogonal vectors in an inner product space X, then

$$\left\| \sum_{k=1}^{n} x_k \right\|^2 = \sum_{k=1}^{n} \|x_k\|^2 . \tag{2.10.1}$$

Proof. If $x_1 \perp x_2$, then $\|x_1 + x_2\|^2 = \|x_1\|^2 + \|x_2\|^2$ by (2.6.10). Thus, the theorem is true for $n = 2$. Assume now that the (2.10.1) holds for $n - 1$, that is,

$$\left\| \sum_{k=1}^{n-1} x_k \right\|^2 = \sum_{k=1}^{n-1} \|x_k\|^2 .$$

Set $x = \sum_{k=1}^{n-1} x_k$ and $y = x_n$. Since $x \perp y$, we have

$$\left\| \sum_{k=1}^{n} x_k \right\|^2 = \|x + y\|^2 = \|x\|^2 + \|y\|^2 = \sum_{k=1}^{n-1} \|x_k\|^2 + \|x_n\|^2 = \sum_{k=1}^{n} \|x_k\|^2 .$$

This proves the theorem.

Theorem 2.10.2 (Bessel's Equality and Inequality). Let $x_1, \dots, x_n$ be an orthonormal set of vectors in an inner product space X. Then, for every $x \in X$, we have

$$\left\| x - \sum_{k=1}^{n} (x, x_k) x_k \right\|^2 = \|x\|^2 - \sum_{k=1}^{n} |(x, x_k)|^2 \tag{2.10.2}$$

and

$$\sum_{k=1}^{n} |(x, x_k)|^2 \le \|x\|^2 . \tag{2.10.3}$$

Proof. In view of the Pythagorean formula (2.10.1), we have

$$\left\| \sum_{k=1}^{n} \alpha_k x_k \right\|^2 = \sum_{k=1}^{n} \|\alpha_k x_k\|^2 = \sum_{k=1}^{n} |\alpha_k|^2$$

for any arbitrary complex numbers $\alpha_1, \dots, \alpha_n$. Hence,

$$\left\| x - \sum_{k=1}^{n} \alpha_k x_k \right\|^2 = \left(x - \sum_{k=1}^{n} \alpha_k x_k,\ x - \sum_{k=1}^{n} \alpha_k x_k \right)$$
$$= \|x\|^2 - \left(x, \sum_{k=1}^{n} \alpha_k x_k \right) - \left(\sum_{k=1}^{n} \alpha_k x_k, x \right) + \sum_{k=1}^{n} |\alpha_k|^2 \|x_k\|^2$$
$$= \|x\|^2 - \sum_{k=1}^{n} \overline{\alpha_k} (x, x_k) - \sum_{k=1}^{n} \alpha_k \overline{(x, x_k)} + \sum_{k=1}^{n} \alpha_k \overline{\alpha_k}$$
$$= \|x\|^2 - \sum_{k=1}^{n} |(x, x_k)|^2 + \sum_{k=1}^{n} |(x, x_k) - \alpha_k|^2. \tag{2.10.4}$$

In particular, if $\alpha_k = (x, x_k)$, this result yields (2.10.2). From (2.10.2), it follows that

$$0 \le \|x\|^2 - \sum_{k=1}^{n} |(x, x_k)|^2,$$

which gives (2.10.3). Thus, the proof is complete.

Remarks.

1. Note that expression (2.10.4) is minimized by taking $\alpha_k = (x, x_k)$. This choice of α_k's minimizes $\left\| x - \sum_{k=1}^{n} \alpha_k x_k \right\|$ and thus provides the best approximation of x by a linear combination of vectors $x_1, \ldots, x_n$.

2. If $\{x_n\}$ is an orthonormal sequence of vectors in an inner product space X, then, from (2.10.2), by letting $n \to \infty$, we obtain

$$\sum_{k=1}^{\infty} |(x, x_k)|^2 \le \|x\|^2. \tag{2.10.5}$$

This shows that the series $\sum_{k=1}^{\infty} |(x, x_k)|^2$ converges for every $x \in X$. In other words, the sequence $\{(x, x_n)\}$ is an element of l^2. We can say that an orthonormal sequence in X induces a mapping from X into l^2. The expansion

$$x \sim \sum_{n=1}^{\infty} (x, x_n)\, x_n \tag{2.10.6}$$

is called a *generalized Fourier series* of x. The scalars $\alpha_n = (x, x_n)$ are called the *generalized Fourier coefficients* of x with respect to the orthonormal sequence $\{x_n\}$. It may be observed that this set of coefficients gives the best approximation. In general, we do not know whether the series in (2.10.6) is convergent. However, as the next theorem shows, the completeness of the space ensures the convergence.

Theorem 2.10.3 Let $\{x_n\}$ be an orthonormal sequence in a Hilbert space H and let $\{\alpha_n\}$ be a sequence of complex numbers. Then, the series $\sum_{n=1}^{\infty} \alpha_n x_n$ converges if and only if $\sum_{n=1}^{\infty} |\alpha_n|^2 < \infty$ and in that case

$$\left\| \sum_{n=1}^{\infty} \alpha_n x_n \right\|^2 = \sum_{n=1}^{\infty} |\alpha_n|^2. \tag{2.10.7}$$

Proof. For every $m > k > 0$, we have

$$\left\| \sum_{n=k}^{m} \alpha_n x_n \right\|^2 = \sum_{n=k}^{m} |\alpha_n|^2 \quad \text{by (2.10.1).} \tag{2.10.8}$$

If $\sum_{n=1}^{\infty} |\alpha_n|^2 < \infty$, then the sequence $s_m = \sum_{n=1}^{\infty} \alpha_n x_n$ is a Cauchy sequence by (2.10.8). This implies convergence of the series $\sum_{n=1}^{\infty} \alpha_n x_n$ because of the completeness of H.

Conversely, if the series $\sum_{n=1}^{\infty} \alpha_n x_n$ converges, then the same formula (2.10.8) implies the convergence of $\sum_{n=1}^{\infty} |\alpha_n|^2$ because the sequence of numbers $\sigma_m = \sum_{n=1}^{m} |\alpha_n|^2$ is a Cauchy sequence in $\mathbb{R}$.

To obtain (2.10.7), it is enough to take $k = 1$ and let $m \to \infty$ in (2.10.8).

The above theorem and (2.10.5) imply that in a Hilbert space H the series $\sum_{n=1}^{\infty} (x, x_n)\, x_n$ converges for every $x \in H$. However, it may happen that it converges to an element different from x.

Example 2.10.1 Let $H = L^2([-\pi,\pi])$, and let $x_n(t) = \frac{1}{\sqrt{\pi}} \sin nt$ for $n = 1,2,\ldots$. The sequence $\{x_n\}$ is an orthonormal set in H. On the other hand, for $x(t) = \cos t$, we have

$$\sum_{n=1}^{\infty} (x, x_n) x_n(t) = \sum_{n=1}^{\infty} \left[\frac{1}{\sqrt{\pi}} \int_{-\pi}^{\pi} \cos t \, \sin nt \, dt \right] \frac{\sin nt}{\sqrt{\pi}}$$

$$= \sum_{n=1}^{\infty} 0 \cdot \sin nt = 0 \neq \cos t.$$

If $\{x_n\}$ is an orthonormal sequence in an inner product space X, then, for every $x \in X$, we have

$$\sum_{n=1}^{\infty} |(x, x_n)|^2 < \infty,$$

and consequently,

$$\lim_{n\to\infty} (x, x_n) = 0.$$

Therefore, orthonormal sequences are weakly convergent to zero. On the other hand, since $\|x_n\| = 1$ for all $n \in \mathbb{N}$, orthonormal sequences are not strongly convergent.

Definition 2.10.1 (Complete Orthonormal Sequence). An orthonormal sequence $\{x_n\}$ in an inner product space X is said to be *complete* if, for every $x \in X$, we have

$$x = \sum_{n=1}^{\infty} (x, x_n) \, x_n. \tag{2.10.9}$$

It is important to remember that since the right-hand side of (2.10.9) is an infinite series, the equality means

$$\lim_{n\to\infty} \left\| x - \sum_{k=1}^{n} (x, x_k) x_k \right\| = 0,$$

where $\|\cdot\|$ is the norm in X. For example, if $X = L^2([-\pi,\pi])$ and $\{f_n\}$ is an orthonormal sequence in X, then by

$$f = \sum_{n=1}^{\infty} (f, f_n) f_n$$

we mean

$$\lim_{n\to\infty} \int_{-\pi}^{\pi} \left| f(t) - \sum_{k=1}^{n} \alpha_k f_k(t) \right|^2 dt = 0, \qquad \text{where } \alpha_k = \int_{-\pi}^{\pi} f(t) \overline{f_k(t)}\, dt.$$

This, in general, does not imply pointwise convergence: $f(x) = \sum_{n=1}^{\infty} \alpha_n f_n(x)$.

Definition 2.10.2 (Orthonormal Basis). An orthonormal system S in an inner product space X is called an *orthonormal basis* if every $x \in X$ has a unique representation

$$x = \sum_{n=1}^{\infty} \alpha_n x_n ,$$

where $\alpha_n \in \mathbb{C}$ and x_n's are distinct elements of S.

Remarks.

1. Note that a complete orthonormal sequence $\{x_n\}$ in an inner product space X is an orthonormal basis in X. It suffices to show the uniqueness. Indeed, if

$$x = \sum_{n=1}^{\infty} \alpha_n x_n \quad \text{and} \quad x = \sum_{n=1}^{\infty} \beta_n x_n ,$$

then

$$0 = \|x - x\|^2 = \left\| \sum_{n=1}^{\infty} \alpha_n x_n - \sum_{n=1}^{\infty} \beta_n x_n \right\|^2 = \left\| \sum_{n=1}^{\infty} (\alpha_n - \beta_n) x_n \right\|^2 = \sum_{n=1}^{\infty} |\alpha_n - \beta_n|^2$$

by Theorem 2.10.3. This means that $\alpha_n = \beta_n$ for all $n \in \mathbb{N}$. This proves the uniqueness.

2. If $\{x_n\}$ is a complete orthonormal sequence in an inner product space X, then the set

$$\operatorname{span}\{x_1, x_2, \ldots\} = \left\{ \sum_{k=1}^{n} \alpha_k x_k \,:\, n \in \mathbb{N},\ \alpha_1, \ldots, \alpha_k \in \mathbb{C} \right\}$$

is dense in X.

The following two theorems give important characterizations of complete orthonormal sequences in Hilbert spaces.

Theorem 2.10.4 An orthonormal sequence $\{x_n\}$ in a Hilbert space H is complete if and only if $(x, x_n) = 0$ for all $n \in \mathbb{N}$ implies $x = 0$.

Proof. Suppose $\{x_n\}$ is a complete orthonormal sequence in H. Then, every $x \in H$ has the representation

$$x = \sum_{n=1}^{\infty} (x, x_n) x_n .$$

Thus, if $(x, x_n) = 0$ for every $n \in \mathbb{N}$, then $x = 0$.

Conversely, suppose $(x, x_n) = 0$ for all $n \in \mathbb{N}$ implies $x = 0$. Let x be an element of H. We define

$$y = \sum_{n=1}^{\infty} (x, x_n) x_n .$$

The sum y exists in H by (2.10.5) and Theorem 2.10.3. Since, for every $n \in \mathbb{N}$,

$$\begin{aligned}(x - y,\, x_n) &= (x, x_n) - \left(\sum_{k=1}^{\infty} (x, x_k) x_k, x_n \right) \\ &= (x, x_n) - \sum_{k=1}^{\infty} (x, x_k)(x_k, x_n) \\ &= (x, x_n) - (x, x_n) = 0,\end{aligned}$$

we have $x - y = 0$ and hence,

$$x = \sum_{n=1}^{\infty} (x, x_n) x_n .$$

Theorem 2.10.5 (Parseval's Formula). An orthonormal sequence $\{x_n\}$ in a Hilbert space H is complete if and only if

$$\|x\|^2 = \sum_{n=1}^{\infty} |(x, x_n)|^2 \tag{2.10.10}$$

for every $x \in H$.

Proof. Let $x \in H$. By (2.10.2), for every $n \in \mathbb{N}$, we have

$$\left\| x - \sum_{k=1}^{n} (x, x_k) x_k \right\|^2 = \|x\|^2 - \sum_{k=1}^{n} |(x, x_k)|^2. \tag{2.10.11}$$

If $\{x_n\}$ is a complete sequence, then the expression on the left-hand side in (2.10.11) converges to zero as $n \to \infty$. Hence,

$$\lim_{n\to\infty} \left[\|x\|^2 - \sum_{k=1}^{n} |(x, x_k)|^2 \right] = 0.$$

Therefore, (2.10.10) holds.

Conversely, if (2.10.10) holds, then the expression on the right-hand side of (2.10.11) converges to zero as $n \to \infty$ and thus,

$$\lim_{n\to\infty} \left\| x - \sum_{k=1}^{n} (x, x_k) x_k \right\|^2 = 0.$$

This proves that $\{x_n\}$ is a complete sequence.

Example 2.10.2 The orthonormal system

$$\phi_n(x) = \frac{e^{inx}}{\sqrt{2\pi}}, \qquad n = 0, \pm 1, \pm 2, \ldots,$$

given in Example 2.9.2, is complete in the space $L^2([-\pi, \pi])$. The proof of completeness is not simple. It will be discussed in Section 2.11.

A simple change of scale allows us to represent a function $f \in L^2([0, a])$ in the form

$$f(x) = \sum_{n=-\infty}^{\infty} \beta_n e^{2n\pi i x/a},$$

where

$$\beta_n = \frac{1}{a} \int_0^a f(t) e^{-2n\pi i t/a} dt.$$

Example 2.10.3 The sequence of functions

$$\frac{1}{\sqrt{2\pi}}, \frac{\cos x}{\sqrt{\pi}}, \frac{\sin x}{\sqrt{\pi}}, \frac{\cos 2x}{\sqrt{\pi}}, \frac{\sin 2x}{\sqrt{\pi}}, \dots$$

is a complete orthonormal system in $L^2([-\pi,\pi])$. The orthogonality follows from the following identities by simple integration:

$$2\cos nx \cos mx = \cos(n+m)x + \cos(n-m)x,$$

$$2\sin nx \sin mx = \cos(n-m)x - \cos(n+m)x,$$

$$2\cos nx \sin mx = \sin(n+m)x - \sin(n-m)x.$$

Since

$$\int_{-\pi}^{\pi} \cos^2 nx\, dx = \int_{-\pi}^{\pi} \sin^2 mx\, dx = \pi,$$

the sequence is also orthonormal. The completeness follows from the completeness of the sequence in Example 2.10.2 in view of the following identities:

$$e^0 = 1 \quad \text{and} \quad e^{inx} = (\cos nx + i \sin nx).$$

Example 2.10.4 Each of the following two sequences of functions is a complete orthonormal system in the space $L^2([-\pi,\pi])$:

$$\frac{1}{\sqrt{\pi}}, \sqrt{\frac{2}{\pi}}\cos x, \sqrt{\frac{2}{\pi}}\cos 2x, \sqrt{\frac{2}{\pi}}\cos 3x, \dots,$$

$$\sqrt{\frac{2}{\pi}}\sin x, \sqrt{\frac{2}{\pi}}\sin 2x, \sqrt{\frac{2}{\pi}}\sin 3x, \dots.$$

Example 2.10.5 (Rademacher Functions and Walsh Functions). Rademacher functions $R(m,x)$ can be introduced in many different ways. We will use the definition based on the sine function,

$$R(m,x) = \operatorname{sgn}\left(\sin\left(2^m \pi x\right)\right), \quad m = 0,1,2,\dots, \quad x \in [0,1],$$

where sgn denotes the signum function defined by

$$\operatorname{sgn}(x)=\begin{cases}1 & \text{if } x>0,\\ 0 & \text{if } x=0,\\ -1 & \text{if } x<0.\end{cases}$$

Rademacher functions form an orthonormal system in $L^2([0,1])$. Obviously,

$$\int_0^1 |R(m,x)|^2\,dx = 1 \qquad \text{for all } m.$$

To show that for $m \neq n$, we have

$$\int_0^1 R(m,x)\,\overline{R(n,x)}\,dx = 0.$$

First, notice that $\int_a^b R(m,x)\,dx = 0$ whenever $2^m(b-a)$ is an even number. Thus, for $m > n \geq 0$, we have

$$\begin{aligned}\int_0^1 R(m,x)\,\overline{R(n,x)}\,dx &= \int_0^1 R(m,x)\,R(n,x)\,dx\\ &= \sum_{k=1}^{2^n}\int_{\frac{k-1}{2^n}}^{\frac{k}{2^n}} R(m,x)\,R(n,x)\,dx\\ &= \sum_{k=1}^{2^n}\operatorname{sgn}\left(R\left(n,\frac{2k-1}{2}\right)\right)\int_{\frac{k-1}{2^n}}^{\frac{k}{2^n}} R(m,x)\,dx = 0\end{aligned}$$

because all of the integrals vanish.

The sequence of Rademacher functions is not complete. Indeed, consider the function

$$f(x)=\begin{cases}0 & \text{if } 0\leq x<\frac{1}{4},\\ 1 & \text{if } \frac{1}{4}\leq x\leq\frac{3}{4},\\ 0 & \text{if } \frac{3}{4}<x\leq 1.\end{cases}$$

Then

$$\int_0^1 R(0,x)\,f(x)\,dx = \frac{1}{2} \quad \text{and} \quad \int_0^1 R(m,x)\,f(x)\,dx = 0 \quad \text{for } m\geq 1,$$

but $f(x) \neq \frac{1}{2} R(0,x)$.

Rademacher functions can be used to construct Walsh functions, which form a complete orthonormal system. Walsh (1923) functions are denoted by $W(m,x)$, $m = 0,1,2,\ldots$. For $m = 0$, we set $W(0,x) = 1$. For other values of m, we first represent m as a binary number, that is,

$$m = \sum_{k=1}^{n} 2^{k-1} a_k = a_1 + 2^1 a_2 + 2^2 a_3 + \cdots + 2^{n-1} a_n,$$

where $a_1, a_2, \ldots, a_n = 0$ or 1. Then, we define

$$W(m,x) = \prod_{k=1}^{n} (R(k,x))^{a_k} = (R(1,x))^{a_1} (R(2,x))^{a_2} \cdots (R(n,x))^{a_n},$$

where $(R(m,x))^0 \equiv 1$. For instance, since 53 is written as 110101 in binary form, we have

$$W(53,x) = R(1,x)\ R(3,x)\ R(5,x)\ R(6,x).$$

Clearly, we have

$$R(n,x) = W\left(2^{n-1}, x\right), \qquad n \in \mathbb{N}.$$

Several Walsh functions are shown in Figure 2.4.

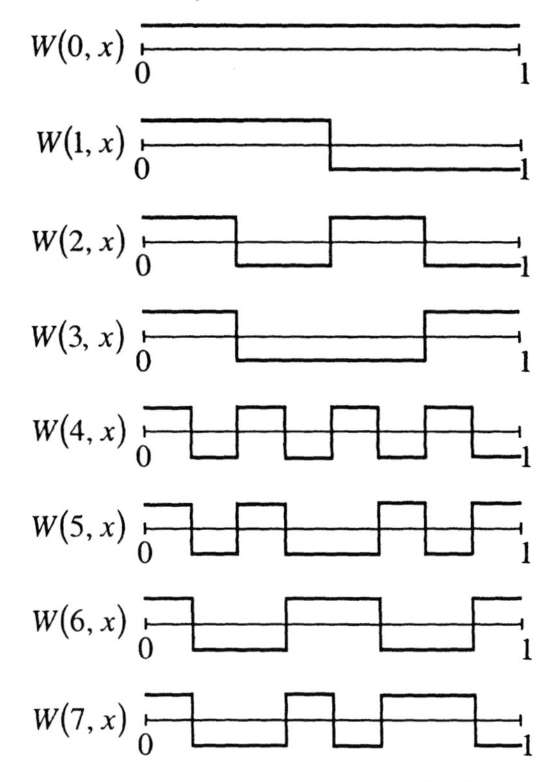

Figure 2.4. Walsh functions $W(n, x)$.

2.11 Trigonometric Fourier Series

In this section, we prove that the sequence

$$\phi_n(x) = \frac{e^{inx}}{\sqrt{2\pi}}, \quad n = 0, \pm 1, \pm 2, \ldots,$$

is a complete orthonormal sequence in $L^2([-\pi, \pi])$. The orthogonality has been established in Example 2.9.2. The proof of completeness is much more complicated. For the purpose of this proof, it is convenient to identify elements of the space $L^1([-\pi, \pi])$ with 2π-periodic locally integrable functions on $\mathbb{R}$ due to the fact that

$$\int_{-\pi}^{\pi} f(t)\,dt = \int_{-\pi-x}^{\pi-x} f(t)\,dt = \int_{-\pi}^{\pi} f(t-x)\,dt$$

for any $f \in L^1([-\pi,\pi])$ and any $x \in \mathbb{R}$.

Let $f \in L^1([-\pi,\pi])$ and

$$f_n = \sum_{k=-n}^{n} (f,\phi_k)\,\phi_k, \quad n = 0,1,2,\ldots.$$

Then

$$f_n(x) = \sum_{k=-n}^{n} \frac{1}{2\pi} \int_{-\pi}^{\pi} f(t) e^{-ikt} dt\, e^{ikx} = \sum_{k=-n}^{n} \frac{1}{2\pi} \int_{-\pi}^{\pi} f(t) e^{ik(x-t)} dt.$$

We next show that, for every $f \in L^1([-\pi,\pi])$, we have

$$\lim_{n\to\infty} \frac{f_0 + f_1 + \cdots + f_n}{n+1} = f$$

in the $L^1([-\pi,\pi])$ norm. We first observe that

$$\begin{aligned}
\frac{f_0(x) + f_1(x) + \cdots + f_n(x)}{n+1} &= \sum_{k=-n}^{n} \left(1 - \frac{|k|}{n+1}\right) (f,\phi_k)\phi_k(x) \\
&= \sum_{k=-n}^{n} \frac{1}{2\pi} \left(1 - \frac{|k|}{n+1}\right) \int_{-\pi}^{\pi} f(t) e^{-ikt} dt\, e^{ikx} \\
&= \frac{1}{2\pi} \int_{-\pi}^{\pi} f(t) \left(\sum_{k=-n}^{n} \left(1 - \frac{|k|}{n+1}\right) e^{ik(x-t)} \right) dt. \quad (2.11.1)
\end{aligned}$$

Lemma 2.11.1 For every $n \in \mathbb{N}$ and $x \in \mathbb{R}$, we have

$$\sum_{k=-n}^{n} \left(1 - \frac{|k|}{n+1}\right) e^{ikx} = \frac{1}{n+1} \frac{\sin^2 \frac{(n+1)x}{2}}{\sin^2 \frac{x}{2}}.$$

Proof. We have

$$\sin^2 \frac{x}{2} = \frac{1}{2}(1 - \cos x) = -\frac{1}{4} e^{-ix} + \frac{1}{2} - \frac{1}{4} e^{ix}.$$

Then, direct calculation gives

$$\left(-\frac{1}{4}e^{-ix}+\frac{1}{2}-\frac{1}{4}e^{ix}\right)\sum_{k=-n}^{n}\left(1-\frac{|k|}{n+1}\right)e^{ikx}$$
$$=\frac{1}{n+1}\left(-\frac{1}{4}e^{-i(n+1)x}+\frac{1}{2}-\frac{1}{4}e^{i(n+1)x}\right).$$

This proves the lemma.

Lemma 2.11.2 The sequence of functions

$$K_n(t)=\sum_{k=-n}^{n}\left(1-\frac{|k|}{n+1}\right)e^{ikt}$$

is a Fejér summability kernel.

Proof. Since $\int_{-\pi}^{\pi} e^{ikt}dt = 2\pi$ if $k=0$ and $\int_{-\pi}^{\pi} e^{ikt}dt = 0$ for any other integer k, we obtain

$$\int_{-\pi}^{\pi} K_n(t)\,dt=\sum_{k=-n}^{n}\left(1-\frac{|k|}{n+1}\right)\int_{-\pi}^{\pi} e^{ikt}dt = 2\pi .$$

From Lemma 2.11.1, it follows that $K_n \geq 0$ and hence

$$\int_{-\pi}^{\pi} |K_n(t)|\,dt = \int_{-\pi}^{\pi} K_n(t)\,dt = 2\pi .$$

Finally, let $\delta \in (0,\pi)$. For $t \in (\delta, 2\pi-\delta)$, we have $\sin\frac{t}{2} \geq \sin\frac{\delta}{2}$ and therefore

$$K_n(t)=\frac{1}{n+1}\frac{\sin^2\frac{(n+1)x}{2}}{\sin^2\frac{x}{2}} \leq \frac{1}{(n+1)\sin^2\frac{\delta}{2}}.$$

Thus,

$$\int_{\delta}^{2\pi-\delta} K_n(t)dt \leq \frac{2\pi}{(n+1)\sin^2\frac{\delta}{2}}.$$

For a fixed δ, the right-hand side tends to 0 as $n \to \infty$. This proves the lemma.

Theorem 2.11.1 If $f \in L^2([-\pi,\pi])$ and $|(f,\phi_n)| = 0$ for all integers n, then $f = 0$ a.e.

Proof. If

$$\int_{-\pi}^{\pi} f(t) e^{-int} dt = 0$$

for all integers n, then

$$f_n(x) = \sum_{k=-n}^{n} \frac{1}{2\pi} \int_{-\pi}^{\pi} f(t) e^{ik(x-t)} dt = 0.$$

Consequently,

$$\frac{f_0(x) + f_1(x) + \cdots + f_n(x)}{n+1} = \frac{1}{2\pi} \int_{-\pi}^{\pi} f(t) \left(\sum_{k=-n}^{n} \left(1 - \frac{|k|}{n+1}\right) e^{ik(x-t)} \right) dt = 0.$$

On the other hand, since f and all the functions e^{ikx} are 2π-periodic, we have

$$\frac{1}{2\pi} \int_{-\pi}^{\pi} f(t) \left(\sum_{k=-n}^{n} \left(1 - \frac{|k|}{n+1}\right) e^{ik(x-t)} \right) dt$$

$$= \frac{1}{2\pi} \int_{-\pi}^{\pi} f(x-t) \left(\sum_{k=-n}^{n} \left(1 - \frac{|k|}{n+1}\right) e^{ikt} \right) dt$$

and hence, by Theorem 3.8.1 (see Debnath and Mikusinski, 1999) and Lemma 2.11.2,

$$\lim_{n\to\infty} \frac{f_0 + f_1 + \cdots + f_n}{n+1} = f$$

in the $L^1([-\pi,\pi])$ norm. Therefore, $f = 0$ a.e.

Theorem 2.11.2 The sequence of functions

$$\phi_n(x) = \frac{e^{inx}}{\sqrt{2\pi}}, \qquad n = 0, \pm 1, \pm 2, \ldots,$$

is complete.

Proof. If $f \in L^2([-\pi,\pi])$, then $f \in L^1([-\pi,\pi])$. Thus, by Theorem 2.11.1 if $(f,\phi_n) = 0$ for all integers n, then $f = 0$ a.e., that is, $f = 0$ in $f \in L^2([-\pi,\pi])$. This proves completeness of the sequence by Theorem 2.10.4.

Theorem 2.11.2 implies that, for every $f \in L^2([-\pi,\pi])$, we have

$$f = \sum_{k=-\infty}^{\infty} \alpha_n \phi_n, \tag{2.11.2}$$

where

$$\phi_n(x) = \frac{e^{inx}}{\sqrt{2\pi}} \quad \text{and} \quad \alpha_n = \frac{1}{\sqrt{2\pi}} \int_{-\pi}^{\pi} f(t) e^{-ikt} dt.$$

In this case, Parseval's formula yields

$$\|f\|^2 = \int_{-\pi}^{\pi} |f(x)|^2 dx = \sum_{n=-\infty}^{\infty} |\alpha_n|^2.$$

The series (2.11.2) is called the *Fourier series* of f, and the numbers α_n are called the *Fourier coefficients* of f. It is important to point out that, in general, (2.11.2) does not imply pointwise convergence. The problem of pointwise convergence of Fourier series is much more difficult. In 1966, L. Carleson proved that Fourier series of functions in $L^2([-\pi,\pi])$ converge almost everywhere.

2.12 Orthogonal Complements and the Projection Theorem

By a subspace of a Hilbert space H, we mean a vector subspace of H. A subspace of a Hilbert space is an inner product space. If we additionally assume that S is a closed subspace of H, then S is a Hilbert space itself because a closed subspace of a complete normed space is complete.

Definition 2.12.1 (Orthogonal Complement). Let S be a nonempty subset of a Hilbert space H. An element $x \in H$ is said to be orthogonal to S, denoted by $x \perp S$, if $(x, y) = 0$ for every $y \in S$. The set of all elements of H orthogonal to S, denoted by $S^\perp$, is called the *orthogonal complement* of S. In symbols,

$$S^\perp = \{x \in H,\ x \perp S\}.$$

The orthogonal complement of $S^\perp$ is denoted by $S^{\perp\perp} = (S^\perp)^\perp$.

Remarks. If $x \perp y$ for every $y \in H$, then $x = 0$. Thus $H^{\perp} = \{0\}$. Similarly, $\{0\}^{\perp} = H$. Two subsets A and B of a Hilbert space are said to be *orthogonal* if $x \perp y$ for every $x \in A$ and $y \in B$. This is denoted by $A \perp B$. Note that if $A \perp B$, then $A \cap B = \{0\}$ or $\emptyset$.

Theorem 2.12.1 (Orthogonal Complement). For any subset of S of a Hilbert space H, the set $S^{\perp}$ is a closed subspace of H.

Proof. If $\alpha, \beta \in \mathbb{C}$ and $x, y \in S^{\perp}$, then

$$(\alpha x + \beta y, z) = \alpha(x, z) + \beta(y, z) = 0$$

for every $z \in S$. Thus, $S^{\perp}$ is a vector subspace of H. We next prove that $S^{\perp}$ is closed.

Let $\{x_n\} \in S^{\perp}$ and $x_n \to x$ for some $x \in H$. From the continuity of the inner product, we find

$$(x, y) = \left(\lim_{n\to\infty} x_n, y\right) = \lim_{n\to\infty} (x_n, y) = 0,$$

for every $y \in S$. This shows that $x \in S^{\perp}$, and thus, $S^{\perp}$ is closed.

The above theorem implies that $S^{\perp}$ is a Hilbert space for any subset S of H. Note that S does not have to be a vector space. Since $S \perp S^{\perp}$, we have $S \cap S^{\perp} = \{0\}$ or $S \cap S^{\perp} = \emptyset$.

Definition 2.12.2 (Convex Sets). A set S in a vector space is called *convex* if, for any $x, y \in S$ and $\alpha \in (0,1)$, we have $\alpha x + (1-\alpha) y \in S$.

Note that a vector subspace is a convex set.

The following theorem concerning the minimization of the norm is of fundamental importance in approximation theory.

Theorem 2.12.2 (The Closest Point Property). Let S be a closed convex subset of a Hilbert space H. For every point $x \in H$, there exists a unique point $y \in S$ such that

$$\|x - y\| = \inf_{z \in S} \|x - z\|. \tag{2.12.1}$$

Proof. Let $\{y_n\}$ be a sequence in S such that

$$\lim_{n\to\infty} \|x - y_n\| = \inf_{z \in S} \|x - z\|.$$

Denote $d = \inf_{z \in S} \|x - z\|$. Since $\frac{1}{2}(y_m + y_n) \in S$, we have

$$\left\|x - \frac{1}{2}(y_m + y_n)\right\| \geq d, \qquad \text{for all } m, n \in \mathbb{N}.$$

Moreover, by the parallelogram law (2.6.7),

$$\begin{aligned}\|y_m - y_n\|^2 &= 4\left\|x - \frac{1}{2}(y_m + y_n)\right\|^2 + \|y_m - y_n\|^2 - 4\left\|x - \frac{1}{2}(y_m + y_n)\right\|^2 \\ &= \|(x - y_m) + (x - y_n)\|^2 + \|(x - y_m) - (x - y_n)\|^2 - 4\left\|x - \frac{1}{2}(y_m + y_n)\right\|^2 \\ &= 2\left(\|x - y_m\|^2 + \|x - y_n\|^2\right) - 4\left\|x - \frac{1}{2}(y_m + y_n)\right\|^2.\end{aligned}$$

Since

$$2\left(\|x - y_m\|^2 + \|x - y_n\|^2\right) \to 4d^2, \qquad \text{as } m, n \to \infty,$$

and

$$\left\|x - \frac{1}{2}(y_m + y_n)\right\|^2 \geq d^2,$$

it follows that $\|y_m - y_n\|^2 \to 0$ as $m, n \to \infty$. Thus, $\{y_n\}$ is a Cauchy sequence. Since H is complete and S is closed, $\lim_{n\to\infty} y_n = y$ exists and $y \in S$. It follows from the continuity of the norm that

$$\|x - y\| = \left\|x - \lim_{n\to\infty} y_n\right\| = \lim_{n\to\infty} \|x - y_n\| = d.$$

We have proved that there exists a point in S satisfying (2.12.1). It remains to prove the uniqueness. Suppose there is another point y_1 in S satisfying (2.12.1). Then, since $\frac{1}{2}(y + y_1) \in S$, we have

$$\|y - y_1\|^2 = 4d^2 - 4\left\|x - \frac{y + y_1}{2}\right\|^2 \leq 0.$$

This can only be true if $y = y_1$.

Remark. Theorem 2.12.2 gives an existence and uniqueness result which is crucial for optimization problems. However, it does not tell us how to find that optimal point. The characterization of the optimal point in the case of a real Hilbert space stated in the following theorem is often useful in such problems.

Theorem 2.12.3 Let S be a closed convex subset of a real Hilbert space H, $y \in S$, and let $x \in H$. Then, the following conditions are equivalent:

(a) $$\|x - y\| = \inf_{z \in S} \|x - z\|,$$

(b) $$(x - y,\, z - y) \le 0 \qquad \text{for all } z \in S.$$

Proof. Let $z \in S$. Since S is convex, $\lambda z + (1 - \lambda) y \in S$ for every $\lambda \in (0,1)$. Then, by (a), we have

$$\|x - y\| \le \|x - \lambda z - (1 - \lambda) y\| = \|(x - y) - \lambda (z - y)\|.$$

Since H is a real Hilbert space, we get

$$\|x - y\|^2 \le \|x - y\|^2 - 2\lambda (x - y,\, z - y) + \lambda^2 \|z - y\|^2.$$

Consequently,

$$|(x - y,\, z - y)| \le \frac{\lambda}{2} \|z - y\|^2.$$

Thus, (b) follows by letting $\lambda \to 0$.

Conversely, if $x \in H$ and $y \in S$ satisfy (b), then, for every $z \in S$, we have

$$\|x - y\|^2 - \|x - z\|^2 = 2(x - y, z - y) - \|z - y\|^2 \le 0.$$

Thus, x and y satisfy (a).

If $H = \mathbb{R}^2$ and S is a closed convex subset of $\mathbb{R}^2$, then condition (b) has an important geometric meaning: the angle between the line through x and y and the line through z and y is always *obtuse*, as shown in Figure 2.5.

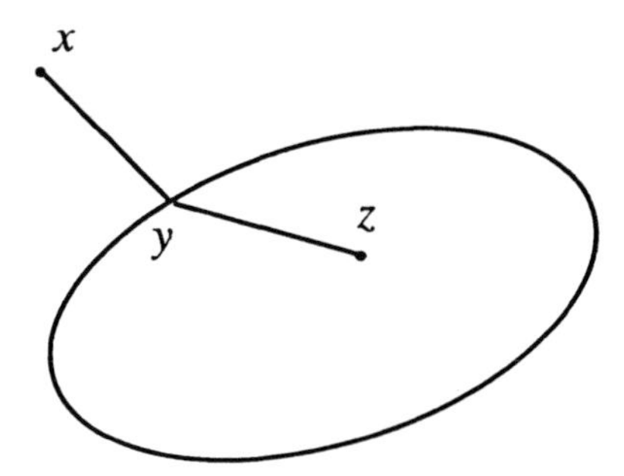

Figure 2.5. Angle between two lines.

Theorem 2.12.4 (Orthogonal Projection). If S is a closed subspace of a Hilbert space H, then every element $x \in H$ has a unique decomposition in the form $x = y + z$, where $y \in S$ and $z \in S^{\perp}$.

Proof. If $x \in S$, then the obvious decomposition is $x = x + 0$. Suppose now that $x \notin S$. Let y be the unique point of S satisfying $\|x - y\| = \inf_{w \in S} \|x - w\|$, as in Theorem 2.12.2. We show that $x = y + (x - y)$ is the desired decomposition.

If $w \in S$ and $\lambda \in \mathbb{C}$, then $y + \lambda w \in S$ and

$$\|x - y\|^2 \le \|x - y - \lambda w\|^2 = \|x - y\|^2 - 2\mathcal{R}\lambda \,|(w,\, x - y)| + |\lambda|^2 \|w\|^2 .$$

Hence,

$$-2\mathcal{R}\lambda\,(w,\, x - y) + |\lambda|^2 \|w\|^2 \ge 0 .$$

If $\lambda > 0$, then dividing by λ and letting $\lambda \to 0$ gives

$$\mathcal{R}(w, x - y) \le 0 . \tag{2.12.2}$$

Similarly, replacing λ by $-i\lambda\,(\lambda > 0)$, dividing by λ, and letting $\lambda \to 0$ yields

$$\mathcal{T}(w, x - y) \le 0 . \tag{2.12.3}$$

Since $y \in S$ implies $-y \in S$, inequalities (2.12.2) and (2.12.3) hold also with $-w$ instead of w. Therefore, $(w, x - y) = 0$ for every $w \in S$, which means $x - y \in S^{\perp}$.

To prove the uniqueness, note that if $x = y_1 + z_1$, $y_1 \in S$, and $z_1 \in S^\perp$, then $y - y_1 \in S$ and $z - z_1 \in S^\perp$. Since $y - y_1 = z_1 - z$, we must have $y - y_1 = z_1 - z = 0$.

Remarks.

1. According to Theorem 2.12.4, every element of H can be uniquely represented as the sum of an element of S and an element of $S^\perp$. This can be stated symbolically as

$$H = S \oplus S^\perp. \tag{2.12.4}$$

We say that H is the direct sum of S and $S^\perp$. Equality (2.12.4) is called an *orthogonal decomposition* of H. Note that the union of a basis of S and a basis of $S^\perp$ is a basis of H.

2. Theorem 2.12.2 allows us to define a mapping $P_S(x) = y$, where y is as in (2.12.1). The mapping P_S is called the *orthogonal projection* onto S.

Example 2.12.1 Let $H = \mathbb{R}^2$. Figure 2.6 exhibits the geometric meaning of the orthogonal decomposition in $\mathbb{R}^2$. Here, $x \in \mathbb{R}^2$, $x = y + z$, $y \in S$, and $z \in S^\perp$. Note that if s_0 is a unit vector in S, then $y = (x, s_0)\, s_0$.

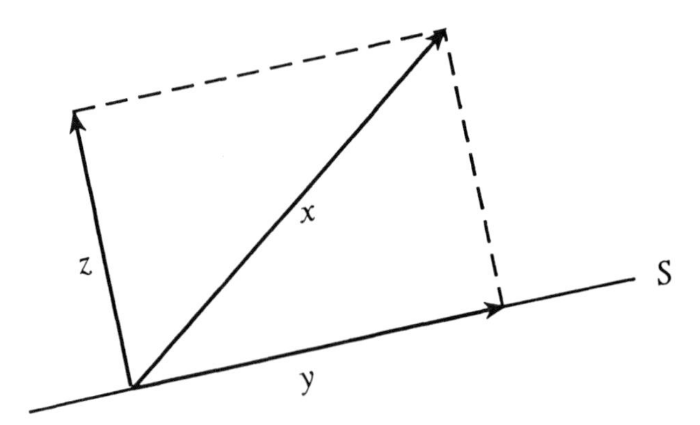

Figure 2.6. Orthogonal decomposition in $\mathbb{R}^2$.

Example 2.12.2 If $H = \mathbb{R}^3$, given a plane P, any vector x can be projected onto the plane P. Figure 2.7 illustrates this example.

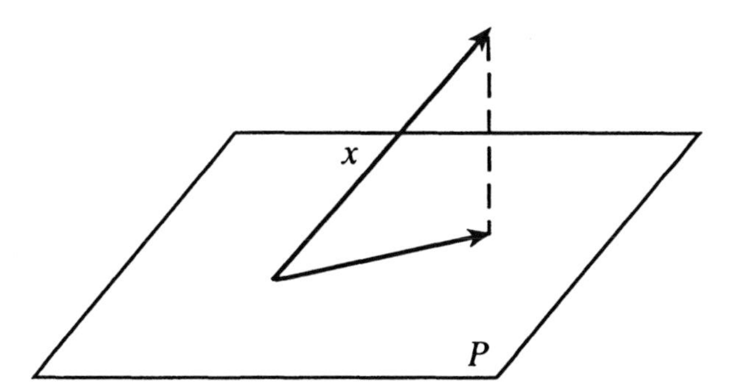

Figure 2.7. Orthogonal projection onto a plane.

Theorem 2.12.5 If S is a closed subspace of a Hilbert space H, then $S^{\perp\perp} = S$.

Proof. If $x \in S$, then for every $z \in S^{\perp}$ we have $(x,z) = 0$, which means $x \in S^{\perp\perp}$. Thus, $S \subset S^{\perp\perp}$. To prove that $S^{\perp\perp} \subset S$ consider an $x \in S^{\perp\perp}$. Since S is closed, $x = y + z$ for some $y \in S$ and $z \in S^{\perp}$. In view of the inclusion $S \subset S^{\perp\perp}$, we have $y \in S^{\perp\perp}$ and thus, $z = x - y \in S^{\perp\perp}$ because $S^{\perp\perp}$ is a vector subspace. But $z \in S^{\perp}$, so we must have $z = 0$, which means $x = y \in S$. This shows that $S^{\perp\perp} \subset S$. This completes the proof.

2.13 Linear Functionals and the Riesz Representation Theorem

In Section 2.7, we have remarked that for any fixed vector x_0 in an inner product space X, the formula $f(x) = (x, x_0)$ defines a bounded linear functional on X. It turns out that if X is a Hilbert space, then every bounded linear functional is of this form. Before proving this result, known as the Riesz representation theorem, we discuss some examples and prove a lemma.

Example 2.13.1 Let $H = L^2((a,b))$, $-\infty < a < b < \infty$. Define a linear functional f on H by the formula

$$f(x) = \int_a^b x(t)\,dt.$$

If x_0 denotes the constant function 1 on (a,b), then clearly $f(x) = (x, x_0)$ and thus, f is a bounded functional.

Example 2.13.2 Let $H = L^2(a,b)$ and let t_0 be a fixed point in (a,b). Let f be a functional on H defined by $f(x) = x(t_0)$. This functional is linear, but it is not bounded.

Example 2.13.3 Let $H = \mathbb{C}^n$ and let $n_0 \in \{1, 2, \ldots, n\}$. Define f by the formula

$$f((x_1, \ldots, x_n)) = x_{n_0}.$$

We have

$$f((x_1, \ldots, x_n)) = ((x_1, \ldots, x_n), e_{n_0}),$$

where e_{n_0} is the vector which has 1 on the n_0-th place and zeros in the remaining places. Thus, f is a bounded linear functional.

Lemma 2.13.1 Let f be a bounded linear functional on an inner product space X. Then, $\dim \mathcal{N}(f)^\perp \le 1$.

Proof. If $f = 0$, then $\mathcal{N}(f) = X$ and $\dim \mathcal{N}(f)^\perp = 0 \le 1$. It remains to show that $\dim \mathcal{N}(f)^\perp = 1$ when f is not zero. Continuity of f implies that $\mathcal{N}(f)$ is a closed subspace of X and thus $\mathcal{N}(f)^\perp$ is not empty. Let $x_1, x_2 \in \mathcal{N}(f)^\perp$ be nonzero vectors. Since $f(x_1) \ne 0$ and $f(x_2) \ne 0$, there exists a scalar $a \ne 0$ such that $f(x_1) + a\, f(x_2) = 0$ or $f(x_1 + ax_2) = 0$. Thus, $x_1 + ax_2 \in \mathcal{N}(f)$. On the other hand, since $\mathcal{N}(f)^\perp$ is a vector space and $x_1, x_2 \in \mathcal{N}(f)^\perp$, we must have $x_1 + ax_2 \in \mathcal{N}(f)^\perp$. This is only possible if $x_1 + ax_2 = 0$ which shows that x_1 and x_2 are linearly dependent because $a \ne 0$.

Theorem 2.13.1 (The Riesz Representation Theorem). Let f be a bounded linear functional on a Hilbert space H. There exists exactly one $x_0 \in H$ such that $f(x) = (x, x_0)$ for all $x \in H$. Moreover, we have $\|f\| = \|x_0\|$.

Proof. If $f(x) = 0$ for all $x \in H$, then $x_0 = 0$ has the desired properties. Assume now that f is a nonzero functional. Then, $\dim \mathcal{N}(f)^{\perp} = 1$, by Lemma 2.13.1. Let z_0 be a unit vector in $\mathcal{N}(f)^{\perp}$. Then, for every $x \in H$, we have

$$x = x - (x, z_0) z_0 + (x, z_0) z_0 .$$

Since $(x, z_0) z_0 \in \mathcal{N}(f)^{\perp}$, we must have $x - (x, z_0) z_0 \in \mathcal{N}(f)$, which means that

$$f\left(x - (x, z_0) z_0\right) = 0 .$$

Consequently,

$$f(x) = f\left((x, z_0) z_0\right) = (x, z_0)\, f(z_0) = \left(x, \overline{f(z_0)}\, z_0\right)$$

Therefore, if we put

$$x_0 = \overline{f(z_0)}\, z_0 ,$$

then $f(x) = (x, x_0)$ for all $x \in H$.

Suppose now that there is another point x_1 such that $f(x) = (x, x_1)$ for all $x \in H$. Then $(x, x_0 - x_1) = 0$ for all $x \in H$ and thus $(x_0 - x_1, x_0 - x_1) = 0$. This is only possible if $x_0 = x_1$.

Finally, we have

$$\|f\| = \sup_{\|x\|=1} |f(x)| = \sup_{\|x\|=1} |(x, x_0)| \le \sup_{\|x\|=1} \left(\|x\| \|x_0\|\right) = \|x_0\|$$

and

$$\|x_0\|^2 = (x_0, x_0) = |f(x_0)| \le \|f\| \|x_0\| .$$

Therefore,

$$\|f\| = \|x_0\| .$$

The collection H' of all bounded linear functionals on a Hilbert space H is a Banach space. The Riesz representation theorem states that $H' = H$ or, more precisely, that H' and H are isomorphic. The element x_0 corresponding to a functional f is sometimes called the *representer* of f.

Note that the functional f defined by $f(x) = (x_0, x)$, where $x_0 \neq 0$ is a fixed element of a complex Hilbert space H, is not linear. Indeed, we have $f(\alpha x + \beta y) = \overline{\alpha} f(x) + \overline{\beta} f(y)$. Such functionals are often called *anti-linear* or *conjugate-linear*.

2.14 Separable Hilbert Spaces

Definition 2.14.1 (Separable Hilbert Space). A Hilbert space is called *separable* if it contains a complete orthonormal sequence. Finite-dimensional Hilbert spaces are considered separable.

Example 2.14.1 The Hilbert space $L^2([-\pi, \pi])$ is separable. Example 2.10.2 shows a complete orthonormal sequence in $L^2([-\pi, \pi])$.

Example 2.14.2 The sequence space l^2 is separable.

Example 2.14.3 (Nonseparable Hilbert Space). Let H be the space of all complex-valued functions defined on $\mathbb{R}$ which vanish everywhere except a countable number of points in $\mathbb{R}$ and such that

$$\sum_{f(x) \neq 0} |f(x)|^2 < \infty.$$

The inner product in H can be defined as

$$(f, g) = \sum_{f(x)g(x) \neq 0} f(x)\overline{g(x)}.$$

This space is not separable because, for any sequence of functions $f_n \in H$, there are nonzero functions f such that $(f, f_n) = 0$ for all $n \in \mathbb{N}$.

Recall that a set S in a Banach space X is called dense in X if every element of X can be approximated by a sequence of elements of S. More precisely, for every $x \in X$, there exist $x_n \in S$ such that $\|x - x_n\| \to 0$ as $n \to \infty$.

Theorem 2.14.1 Every separable Hilbert space contains a countable dense subset.

Proof. Let $\{x_n\}$ be a complete orthonormal sequence in a Hilbert space H. The set

$$S = \{(\alpha_1 + i\beta_1)x_1 + \cdots + (\alpha_n + i\beta_n)x_n : \alpha_1, \ldots, \alpha_n; \beta_1, \ldots, \beta_n \in \mathbb{Q}, n \in \mathbb{N}\}$$

is obviously countable. Since, for every $x \in H$,

$$\left\| \sum_{k=1}^{n} (x, x_k) x_k - x \right\| \to 0 \qquad \text{as } n \to \infty,$$

the set S is dense in H.

The statement in the preceding theorem is often used as a definition of separability.

Theorem 2.14.2 Every orthogonal set in a separable Hilbert space is countable.

Proof. Let S be an orthogonal set in a separable Hilbert space H, and let S_1 be the set of normalized vectors from S, that is, $S_1 = \{x/\|x\| : x \in S\}$. For any distinct $x, y \in S_1$, we have

$$\begin{aligned} \|x - y\|^2 &= (x - y, x - y) \\ &= (x, x) - (x, y) - (y, x) + (y, y) \\ &= 1 - 0 - 0 + 1 \qquad \text{(by the orthogonality)} \\ &= 2. \end{aligned}$$

This means that the distance between any two distinct elements of S_1 is $\sqrt{2}$.

Now, consider the collection of $(1/\sqrt{2})$–neighborhoods about every element of S_1. Clearly, no two of these neighborhoods can have a common point. Since every dense subset of H must have at least one point in every neighborhood and H has a countable dense subset, S_1 must be countable. Thus, S is countable.

Definition 2.14.2 (Hilbert Space Isomorphism). A Hilbert space H_1 is said to be *isomorphic* to a Hilbert space H_2 if there exists a one-to-one linear mapping T from H_1 onto H_2 such that

$$(T(x),T(y))=(x,y) \tag{2.14.1}$$

for every $x,y\in H_1$. Such a mapping T is called a *Hilbert space isomorphism* of H_1 onto H_2.

Note that (2.14.1) implies $\|T\|=1$ because $\|T(x)\|=\|x\|$ for every $x\in H_1$.

Theorem 2.14.3 Let H be a separable Hilbert space.

(a) If H is infinite-dimensional, then it is isomorphic to the space l^2;

(b) If H has dimension N, then it is isomorphic to the space $\mathbb{C}^N$.

Proof. Let $\{x_n\}$ be a complete orthonormal sequence in H. If H is infinite dimensional, then $\{x_n\}$ is an infinite sequence. Let x be an element of H. Define $T(x)=\{\alpha_n\}$, where $\alpha_n=(x,x_n)$, $n=1,2,\ldots$. By Theorem 2.10.3, T is a one-to-one mapping from H onto l^2. It is clearly a linear mapping. Moreover, for $\alpha_n=(x,x_n)$ and $\beta_n=(y,x_n)$, $x,y\in H$, $n\in\mathbb{N}$, we have

$$\begin{aligned}(T(x),T(y)) &= ((\alpha_1,\alpha_2,\cdots),(\beta_1,\beta_2,\cdots))\\ &= \sum_{n=1}^{\infty}\alpha_n\overline{\beta_n}=\sum_{n=1}^{\infty}(x,x_n)\overline{(y,x_n)}\\ &= \sum_{n=1}^{\infty}\left(x,(y,x_n)\,x_n\right)=\left|\left(x,\sum_{n=1}^{\infty}(y,x_n)x_n\right)\right|=(x,y).\end{aligned}$$

Thus, T is an isomorphism from H onto l^2.

The proof of (b) is left as an exercise.

It is easy to check that isomorphism of Hilbert spaces is an equivalence relation.

Since any infinite dimensional separable Hilbert space is isomorphic to the space l^2, it follows that any two such spaces are isomorphic. The same is true for real Hilbert spaces; any real infinite dimensional separable Hilbert space is isomorphic to the real space l^2. In some sense, there is only one real and one complex infinite dimensional separable Hilbert space.

2.15 Linear Operators on Hilbert Spaces

The concept of an operator (or transformation) on a Hilbert space is a natural generalization of the idea of a function of a real variable. Indeed, it is fundamental in mathematics, science, and engineering. Linear operators on a Hilbert space are widely used to represent physical quantities, and hence, they are more important and useful. The most important operators include differential, integral, and matrix operators. In signal processing and wavelet analysis, almost all algorithms are mainly based on linear operators.

Definition 2.15.1 (Linear Operator). An operator T of a vector space X into another vector space Y, where X and Y have the same scalar field, is called a *linear operator* if

$$T(ax_1 + bx_2) = a\,Tx_1 + b\,Tx_2 \tag{2.15.1}$$

for all scalars a, b and for all $x_1, x_2 \in X$.

Otherwise, it is called a *nonlinear operator.*

Example 2.15.1 (Integral Operator). One of the most important operators is the *integral operator* T defined by

$$Tx(s) = \int_a^b K(s,t)\; x(t)\; dt, \tag{2.15.2}$$

where a and b are finite or infinite. The function K is called the *kernel* of the operator.

Example 2.15.2 (Differential Operator). Another important operator is called the *differential* operator

$$(Df)(x) = \frac{d\,f(x)}{dx} = f'(x) \tag{2.15.3}$$

defined on the space of all differentiable functions on some interval $[a,b] \subset \mathbb{R}$, which is a linear subspace of $L^2([a,b])$.

Example 2.15.3 (Matrix Operator). Consider an operator T on $\mathbb{C}^n$, and let $\{e_1, e_2, \ldots, e_n\}$ be the standard base in $\mathbb{C}^n$, that is, $e_1 = (1,0,0,\cdots,0)$, $e_2 = (0,1,0,\cdots,0)$, $\cdots$, $e_n = (0,0,\cdots,1)$.

We define

$$a_{ij} = (Te_j, e_i) \text{ for all } i, j \in \{1,2,\ldots,n\}.$$

Then, for $x = \sum_{j=1}^{n} a_j e_j \in \mathbb{C}^n$, we have

$$Tx = \sum_{j=1}^{n} a_j Te_j \tag{2.15.4}$$

and hence

$$(Tx, e_i) = \sum_{j=1}^{n} a_j (Te_j, e_i) = \sum_{j=0}^{n} a_{ij} a_j. \tag{2.15.5}$$

Thus, every operator T on the space $\mathbb{C}^n$ is defined by an $n \times n$ matrix.

Conversely, for every $n \times n$ matrix (a_{ij}), formula (2.15.5) defines an operator on $\mathbb{C}^n$. We thus have a one-to-one correspondence between operators on an n-dimensional vector space and $n \times n$ matrices.

Definition 2.15.2 (Bounded Operator). An operator $T : X \to X$ is called *bounded* if there exists a number K such that

$$\|Tx\| \le K \|x\| \qquad \text{for every } x \in X.$$

The norm of an operator T is defined as the least of all such number K or, equivalently, by

$$\|T\| = \sup_{\|x\|=1} \|T\|.$$

It follows from this definition that

$$\|Tx\| \le \|T\| \|x\|.$$

If operator T is defined by the matrix (a_{ij}) in Example 2.15.3, then

$$\|T\| \le \sqrt{\sum_{i=1}^{n}\sum_{j=1}^{n}|a_{ij}|^2}. \tag{2.15.6}$$

This means that every operator on $\mathbb{C}^n$, and thus also every operator on any finite dimensional Hilbert space is bounded.

The differential operator defined in Example 2.15.2 is *unbounded*. Consider the sequence of functions $f_n(x) = \sin nx$, $n = 1,2,3,\ldots$, defined on $[-\pi, \pi]$. Then,

$$\|f_n\| = \left\{\int_{-\pi}^{\pi} \sin^2 nx\, dx\right\}^{\frac{1}{2}} = \sqrt{\pi}$$

and

$$\|D f_n\| = \left\{\int_{-\pi}^{\pi} (n \cos nx)^2\, dx\right\}^{\frac{1}{2}} = n\sqrt{\pi}.$$

Thus,

$$\|D f_n\| = n\|f_n\| \to \infty \quad \text{as} \quad n \to \infty.$$

Definition 2.15.3 (Continuous Operator). A linear operator $T : X \to Y$, where X and Y are normed spaces, is *continuous* at a point $x_0 \in X$, if, for any sequence $\{x_n\}$ of elements in X convergent to x_0, the sequence $\{T(x_n)\}$ converges to $T(x_0)$. In other words, T is continuous at x_0 if $\|x_n - x_0\| \to 0$ implies $\|T(x_n) - T(x_0)\| \to 0$. If T is continuous at every point $x \in X$, we simply say that T is continuous in X.

Theorem 2.15.1 A linear operator is continuous if and only if it is bounded.

The proof is fairly simple (see Debnath and Mikusinski, 1999, p. 22) and omitted here.

Two operators T and S on a vector space X are said to be *equal*, $T = S$, if $Tx = Sx$ for every $x \in X$. The set of all operators forms a vector space with the addition and multiplication by a scalar defined by

$$(T+S)x = Tx + Sx,$$

$$(\alpha T)x = \alpha Tx.$$

The product TS of operators T and S is defined by

$$(TS)(x) = T(Sx).$$

In general, $TS \neq ST$. Operators T and S for which $TS = ST$ are called *commuting operators.*

Example 2.15.4 Consider the space of differentiable functions on $\mathbb{R}$ and the operators

$$T f(x) = x f(x) \quad \text{and} \quad D = \frac{d}{dx}.$$

It is easy to check that $TD \neq DT$.

The square of an operator T is defined as $T^2 x = T(Tx)$. Using the principle of induction, we can define any power of T by

$$T^n x = T\left(T^{n-1}x\right).$$

Theorem 2.15.2 The product TS of bounded operators T and S is bounded and

$$\|TS\| \leq \|T\| \|S\|.$$

Proof. Suppose T and S are two bounded operators on a normed space X; $\|T\| = k_1$ and $\|S\| = k_2$. Then,

$$\|TS\,x\| \leq k_1 \|Sx\| \leq k_1 k_2 \|x\| \quad \text{for every } x \in X.$$

This proves the theorem.

Theorem 2.15.3 A bounded operator on a separable infinite dimensional Hilbert space can be represented by an infinite matrix.

Proof. Suppose T is a bounded operator on a Hilbert space H and $\{e_n\}$ is a complete orthonormal sequence in H. For $i, j \in \mathbb{N}$, define

$$a_{ij} = \left(Te_j, e_i\right).$$

For any $x \in H$, we have

$$Tx = T\left(\lim_{n\to\infty} \sum_{j=1}^{n} \left(x, e_j\right) e_j\right)$$

$$= \lim_{n\to\infty} T\left(\sum_{j=1}^{n} \left(x, e_j\right) e_j\right), \qquad \text{by continuity of } T$$

$$= \lim_{n\to\infty} \left(\sum_{j=1}^{n} \left(x, e_j\right) Te_j\right), \qquad \text{by linearity of } T$$

$$= \sum_{j=1}^{n} \left(x, e_j\right) Te_j.$$

Now,

$$\left(T\,x,\, e_i\right) = \left(\sum_{j=1}^{n} \left(x, e_j\right) Te_j, e_i\right) = \sum_{j=1}^{\infty} \left(Te_j, e_i\right)\left(x, e_j\right) = \sum_{j=1}^{n} a_{ij}\left(x, e_j\right).$$

This shows that T is represented by the matrix $\left(a_{ij}\right)$.

Suppose T is a bounded operator on a Hilbert space H. For every fixed $x_0 \in H$, the functional f defined on H by

$$f(x) = \left(Tx, x_0\right)$$

is a bounded linear functional on H. Thus, by the Riesz representation theorem, there exists a unique $y_0 \in H$ such that $f(x) = \left(x, y_0\right)$ for all $x \in H$. Or, equivalently, $\left(Tx, x_0\right) = \left(x, y_0\right)$ for all $x \in H$. If we denote by T^* the operator which to every $x_0 \in H$ assigns that unique y_0, then we have

$$\left(Tx,\, y\right) = \left(x,\, T^* y\right) \qquad \text{for all } x, y, \in H.$$

Definition 2.15.4 (Adjoint Operator). If T is a bounded linear operator on a Hilbert space H, the operator $T^* : H \to H$ defined by

$$\left(Tx,\, y\right) = \left(x, T^* y\right) \qquad \text{for all } x, y, \in H$$

is called the *adjoint operator* of T.

The following are immediate consequences of the preceding definition.

$$(T+S)^* = T^* + S^*, \qquad (\alpha T)^* = \overline{\alpha}\, T^*,$$

$$\left(T^*\right)^* = T, \qquad I^* = I, \qquad (TS)^* = S^* T^*,$$

for arbitrary operators T and S, I is the identity operator and for any scalar α.

Theorem 2.15.4 The adjoint operator T^* of a bounded operator T is bounded. Moreover,

$$\|T\| = \|T^*\| \text{ and } \|T^*T\| = \|T\|^2.$$

Proof. The reader is referred to Debnath and Mikusinski (1999, p. 151).

In general, $T \neq T^*$. For example, suppose $H = \mathbb{C}^2$, and suppose T is defined by

$$T(z_1, z_2) = (0, z_1).$$

Then,

$$\left(T(x_1, x_2), (y_1, y_2)\right) = x_1 \overline{y}_2 \text{ and } \left((x_1, x_2), T(y_1, y_2)\right) = x_2 \overline{y}_1.$$

However, operators for which $T = T^*$ are of special interest.

Definition 2.15.5 (Self-Adjoint Operator). If $T = T^*$, that is, $(Tx, y) = (x, Ty)$ for all $x, y \in H$, then T is called *self-adjoint* (or *Hermitian*).

Example 2.15.5 Suppose $H = \mathbb{C}^n$ and that $\{e_1, e_2, \ldots, e_n\}$ is a standard orthonormal base in H. Suppose T is an operator represented by the matrix (a_{ij}), where $a_{ij} = (Te_j, e_i)$ (see Example 2.15.3). Then, the adjoint operator T^* is represented by the matrix $b_{kj} = (T^* e_j, e_k)$. Consequently,

$$b_{kj} = (e_j, Te_k) = \overline{(Te_k, e_j)} = \overline{a_{jk}}.$$

Therefore, the operator T is self-adjoint if and only if $a_{ij} = \overline{a_{ji}}$. A matrix that satisfies this condition is often called *Hermitian*.

Example 2.15.6 Suppose H is a separable, infinite-dimensional Hilbert space, and suppose $\{e_n\}$ is a complete orthonormal sequence in H. If T is a bounded operator on H represented by an infinite matrix $\left(a_{ij}\right)$, the operator T is self-adjoint if and only if $a_{ij} = \overline{a_{ji}}$ for all $i, j \in \mathbb{N}$.

Example 2.15.7 Suppose T is a Fredholm operator on $L^2([a,b])$ defined by (2.15.2), where the kernel K is defined on $[a,b] \times [a,b]$ such that

$$\int_a^b \int_a^b |K(s,t)|^2 \, ds \, dt < \infty.$$

This condition is satisfied if K is continuous. We have

$$\begin{aligned}
(Tx, y) &= \int_a^b \int_a^b K(s,t) \, x(t) \, \overline{y(s)} \, ds \, dt \\
&= \overline{\int_a^b \int_a^b \overline{K(s,t) \, x(t)} \, y(s) \, ds \, dt} \\
&= \int_a^b x(t) \, \overline{\int_a^b \overline{K(s,t)} \, y(s) \, ds \, dt} \\
&= \left(x, \int_a^b \overline{K(s,t)} \, y(t) \, dt \right).
\end{aligned}$$

This shows that

$$(T^* x)(s) = \int_a^b \overline{K(t,s)} \, x(t) \, dt.$$

Thus, the Fredholm operator is self-adjoint if its kernel satisfies the equality $K(s,t) = \overline{K(t,s)}$.

Example 2.15.8 The operator T on $L^2([a,b])$ defined by $(Tx)(t) = t\,x(t)$ is self-adjoint.

We have

$$(Tx, y) = \int_a^b t\,x(t) \, \overline{y(t)} \, dt = \int_a^b x(t) \, \overline{t\,y(t)} \, dt = (x, Ty).$$

Example 2.15.9 The operator T defined on $L^2(\mathbb{R})$ defined by $(Tx)(t) = e^{-|t|}\,x(t)$ is bounded and self-adjoint.

The fact that T is self-adjoint follows from

$$(Tx, y) = \int_{-\infty}^{\infty} e^{-|t|}\,x(t)\,\overline{y(t)}\,dt = \int_{-\infty}^{\infty} x(t)\,\overline{e^{-|t|}\,y(t)}\,dt = (x, Ty).$$

The proof of boundedness is left as an exercise.

Theorem 2.15.5 If T is a bounded operator on a Hilbert space H, the operators $A = T + T^*$ and $B = T^*T$ are self-adjoint.

Proof. For all $x, y \in H$, we have

$$(Ax, y) = \left(\left(T + T^*\right) x, y\right) = \left(x, \left(T + T^*\right)^* y\right) = \left(x, \left(T + T^*\right) y\right) = (x, Ax)$$

and

$$(Bx, y) = \left(T^*Tx, y\right) = (Tx, Ty) = \left(x, T^*Ty\right) = (x, By).$$

Theorem 2.15.6 The product of two self-adjoint operators is self-adjoint if and only if they commute.

Proof. Suppose T and S are two self-adjoint operators. Then,

$$(TSx, y) = (Sx, Tx) = (x, STy).$$

Thus, if $TS = ST$, then TS is self-adjoint.

Conversely, if TS is self-adjoint, then the above implies $TS = (TS)^* = ST$.

Example 2.15.10 Consider the differential operator D in the space of all differentiable functions on $\mathbb{R}$ vanishing at infinity. Then,

$$(Dx, y) = \int_{-\infty}^{\infty} \frac{d}{dt} x(t) \cdot \overline{y(t)}\, dt = -\int_{-\infty}^{\infty} x(t) \cdot \frac{d}{dt} \overline{y(t)}\, dt$$

$$= \int_{-\infty}^{\infty} x(t) \cdot \overline{\left(-\frac{d}{dt} y(t)\right)}\, dt = (x, -Dy).$$

Thus, $-D$ is the adjoint of the operator D.

Example 2.15.11 Consider the operator $T = i\,\frac{d}{dt}$ in the space of all differentiable functions on $\mathbb{R}$ vanishing at infinity.

We have

$$(Tx, y) = \int_{-\infty}^{\infty} i\, \frac{d}{dt} x(t) \cdot \overline{y(t)}\, dt = -i \int_{-\infty}^{\infty} x(t) \cdot \frac{d}{dt} \overline{y(t)}\, dt$$

$$= \int_{-\infty}^{\infty} x(t) \overline{\left(i\, \frac{d}{dt} y(t)\right)}\, dt = (x, Ty).$$

Therefore, T is a self-adjoint operator.

Theorem 2.15.7 For every bounded operator T on a Hilbert space H, there exist unique self-adjoint operators A and B such that $T = A + iB$ and $T^* = A - iB$.

Proof. Suppose T is a bounded operator on H. Define

$$A = \frac{1}{2}\left(T + T^*\right) \qquad \text{and} \qquad B = \frac{1}{2i}\left(T - T^*\right).$$

Evidently, A and B are self-adjoint and $T = A + iB$. Moreover, for any $x, y \in H$, we have

$$(Tx, y) = ((A + iB)\, x, y) = (Ax, y) + i\,(Bx, y)$$

$$= (x, Ay) + i\,(x, By) = (x, (A - iB)\, y).$$

Hence, $T^* = A - iB$.

The proof of uniqueness is left as an exception.

In particular, if T is self-adjoint, then $T = A$ and $B = 0$. This implies that self-adjoint operators are like real numbers in $\mathbb{C}$.

We next discuss projection operators and their properties.

According to the projection theorem 2.12.4, if S is a closed subspace of a Hilbert space H, then for every $x \in H$, there exists a unique element $y \in S$ such that $x = y + z$ and $z \in S^{\perp}$. Thus, every closed subspace induces an operator on H which assigns to x that unique y.

Definition 2.15.6 (Anti-Hermitian Operator). An operator A is called *anti-Hermitian* if $A = -A^*$.

The operator in Example 2.15.10 is anti-Hermitian.

Definition 2.15.7 (Inverse Operator). Let T be an operator defined on a vector subspace of X. An operator S defined on $R(T)$ is called the *inverse* of T if $TSx = x$ for all $x \in R(T)$ and $STx = x$ for all $x \in D(T)$. An operator which has an inverse is called *invertible*. The inverse of T will be denoted by T^{-1}.

If an operator has an inverse, then it is unique. Indeed, suppose S_1 and S_2 are inverses of T. Then

$$S_1 = S_1 I = S_1 T S_2 = I S_2 = S_2 .$$

Note also that

$$D(T^{-1}) = R(T) \qquad \text{and} \qquad R(T^{-1}) = D(T).$$

First, we recall some simple algebraic properties of invertible operators.

Theorem 2.15.8

(a) The inverse of a linear operator is a linear operator.

(b) An operator T is invertible if and only if $Tx = 0$ implies $x = 0$.

(c) If an operator T is invertible and vectors $x_1, \dots, x_n$ are linearly independent, then $Tx_1, \dots, Tx_n$ are linearly independent.

(d) If operators T and S are invertible, then the operator TS is invertible and we have $(TS)^{-1} = S^{-1} T^{-1}$.

Proof. (a) For any $x, y \in R(T)$ and $\alpha, \beta \in \mathbb{C}$, we have

$$T^{-1}(\alpha x + \beta y) = T^{-1}\left(\alpha TT^{-1}x + \beta TT^{-1}y\right)$$

$$= T^{-1}T\left(\alpha T^{-1}x + \beta T^{-1}y\right) = \alpha T^{-1}x + \beta T^{-1}y.$$

(b) If T is invertible and $Tx = 0$, then $x = T^{-1}Tx = T^{-1}0 = 0$. Assume now that $Tx = 0$ implies $x = 0$. If $Tx_1 = Tx_2$, then $T(x_1 - x_2) = 0$ and thus $x_1 - x_2 = 0$. Consequently, $x_1 - x_2 = 0$, which proves that T is invertible.

(c) Suppose $\alpha_1 Tx_1 + \cdots + \alpha_n Tx_n = 0$. Then, $T(\alpha_1 x_1 + \cdots + \alpha_n x_n) = 0$, and since T is invertible, $\alpha_1 x_1 + \cdots + \alpha_n x_n = 0$. Linear independence of $x_1, \ldots, x_n$ implies $\alpha_1 = \cdots = \alpha_n = 0$. Thus, vectors $Tx_1, \ldots, Tx_n$ are linearly independent.

(d) In view of (b), if $T(Sx) = 0$, then $Sx = 0$ since T is invertible. If $Sx = 0$, then $x = 0$, since S is invertible. Thus, TS is invertible by (b). Moreover,

$$\left(S^{-1}T^{-1}\right)(TS) = S^{-1}\left(T^{-1}T\right)S = S^{-1}S = I.$$

Similarly, $(TS)\left(S^{-1}T^{-1}\right) = I$. This proves that $(TS)^{-1} = S^{-1}T^{-1}$.

It follows from part (c) in the preceding theorem that if X is a finite dimensional vector space and T is a linear invertible operator on X, then $R(T) = X$. As the following example shows, in infinite dimensional vector spaces this is not necessarily true.

Example 2.15.12 Let $X = l^2$. Define an operator T on X by

$$T(x_1, x_2, \ldots) = (0, x_1, x_2, \ldots).$$

Clearly, this is a linear invertible operator on l^2 whose range is a proper subspace of l^2.

The next example shows that the inverse of a bounded operator is not necessarily bounded.

Example 2.15.13 Let $X = l^2$. Define an operator T on X by

$$T(x_1, x_2, \ldots) = \left(x_1, \frac{x_2}{2}, \frac{x_3}{3}, \ldots, \frac{x_n}{n}, \ldots\right).$$

Since

$$\|T(x_1,x_2,\ldots)\| = \sqrt{\sum_{n=1}^{\infty} \frac{|x_n|^2}{n^2}} \leq \sqrt{\sum_{n=1}^{\infty} |x_n|^2} = \|(x_1,x_2,\ldots)\|,$$

T is a bounded operator. T is also invertible:

$$T^{-1}(x_1,x_2,\ldots) = (x_1,2x_2,3x_3,\ldots,nx_n,\ldots).$$

However, T^{-1} is not bounded. In fact, consider the sequence $\{e_n\}$ of elements of l^2, where $\{e_n\}$ is the sequence whose nth term is 1 and all the remaining terms are 0. Then, $\|e_n\| = 1$ and $\|T^{-1} e_n\| = n$. Therefore, T^{-1} is unbounded.

If X is finite dimensional, then the inverse of any invertible operator on X is bounded because every operator on a finite dimensional space is bounded.

Theorem 2.15.9 Let T be a bounded operator on a Hilbert space H such that $R(T) = H$. If T has a bounded inverse, then the adjoint T^* is invertible and $(T^*)^{-1} = (T^{-1})^*$.

Proof. It suffices to show that

$$(T^{-1})^* T^* x = T^* (T^{-1})^* x = x \tag{2.15.7}$$

for every $x \in H$. Indeed, for any $y \in H$, we have

$$\left(y, (T^{-1})^* T^* x\right) = \left(T^{-1}y, T^* x\right) = \left(TT^{-1}y, x\right) = (y,x)$$

and

$$\left(y, T^* (T^{-1})^* x\right) = \left(Ty, (T^{-1})^* x\right) = \left(T^{-1}Ty, x\right) = (y,x).$$

Thus,

$$\left(y, (T^{-1})^* T^* x\right) = \left(y, T^* (T^{-1})^* x\right) = (y,x) \qquad \text{for all } y \in H. \tag{2.15.8}$$

This implies (2.15.7).

Corollary 2.15.1 If a bounded self-adjoint operator T has bounded inverse T^{-1}, then T^{-1} is self-adjoint.

Proof. $(T^{-1})^* = (T^*)^{-1} = T^{-1}$.

Definition 2.15.8 (Isometric Operator). A bounded operator T on a Hilbert space H is called an *isometric operator* if $\|Tx\| = \|x\|$ for all $x \in H$.

Example 2.15.14 Let $\{e_n\}$, $n \in \mathbb{N}$, be a complete orthonormal sequence in a Hilbert space H. There exists a unique operator T such that $Te_n = e_{n+1}$ for all $n \in \mathbb{N}$. In fact, if $x = \sum_{n=1}^{\infty} \alpha_n e_n$, then $Tx = \sum_{n=1}^{\infty} \alpha_n e_{n+1}$. Clearly, T is linear and $\|Tx\|^2 = \sum_{n=1}^{\infty} |\alpha_n|^2 = \|x\|^2$. Therefore, T is an isometric operator. The operator T is called a *one-sided shift operator*.

Theorem 2.15.9 A bounded operator T defined on a Hilbert space H is isometric if and only if $T^* T = I$ on H.

Proof. If T is isometric, then for every $x \in H$ we have $\|Tx\|^2 = \|x\|^2$ and hence,

$$(T^* Tx, x) = (Tx, Tx) = (x, x) \qquad \text{for all } x \in H.$$

This implies that $T^* T = I$. Similarly, if $T^* T = I$, then

$$\|Tx\| = \sqrt{(Tx, Tx)} = \sqrt{(T^* Tx, x)} = \sqrt{(x, x)} = \|x\|.$$

Note that isometric operators "preserve inner product": $(Tx, Ty) = (x, y)$. In particular, $x \perp y$ if and only if $Tx \perp Ty$. The operator in Example 2.15.12 is an isometric operator.

Definition 2.15.9 (Unitary Operator). A bounded operator T on a Hilbert space H is called a *unitary operator* if $T^* T = TT^* = I$ on H.

In the above definition it is essential that the domain and the range of T be the entire space H.

Theorem 2.15.10 An operator T is unitary if and only if it is invertible and $T^{-1} = T^*$.

Proof. Assume that T is an invertible operator on a Hilbert space H such that $T^{-1} = T^*$. Then, $T^* T = T^{-1}T = I$ and $TT^* = TT^{-1} = I$. Therefore, T is a unitary operator. The proof of the converse is similar.

Theorem 2.15.11 Suppose T is a unitary operator. Then

(a) T is isometric,

(b) T^{-1} and T^* are unitary.

Proof. (a) follows from Theorem 2.15.9. To prove (b), note that

$$\left(T^{-1}\right)^* T^{-1} = T^{**}T^{-1} = TT^{-1} = I.$$

Similarly, $T^{-1}\left(T^{-1}\right)^* = I$, and thus, T^{-1} is unitary. Since $T^* = T^{-1}$, by Theorem 2.15.10, T^* is also unitary.

Example 2.15.15 Let H be the Hilbert space of all sequences of complex numbers $x = \{\dots, x_{-1}, x_0, x_1, \dots\}$ such that $\|x\| = \sum_{n=-\infty}^{\infty} |x_n|^2 < \infty$. The inner product is defined by

$$(x, y) = \sum_{n=-\infty}^{\infty} x_n \overline{y_n}.$$

Define an operator T by $T(x_n) = (x_{n-1})$. T is a unitary operator and hence, T is invertible and

$$(Tx, y) = \sum_{n=-\infty}^{\infty} x_{n-1} \overline{y_n} = \sum_{n=-\infty}^{\infty} x_n \overline{y_{n+1}} = \left(x, T^{-1}y\right).$$

This implies that $T^* = T^{-1}$.

Example 2.15.16 Let $H = L^2([0,1])$. Define an operator T on H by $(Tx)(t) = x(1-t)$. This operator is a one-to-one mapping of H onto H. Moreover, we have $T = T^* = T^{-1}$. Thus, T is a unitary operator.

Definition 2.15.10 (Positive Operator). An operator T is called *positive* if it is self-adjoint and $(Tx, x) \geq 0$ for all $x \in H$.

Example 2.15.17 Let ϕ be a nonnegative continuous function on $[a,b]$. The *multiplication operator* T on $L^2([a,b])$ defined by $Tx = \phi x$ is positive. In fact, for any $x \in L^2([a,b])$, we have

$$(Tx, x) = \int_a^b \phi(t)\, x(t)\, \overline{x(t)}\, dt = \int_a^b \phi(t)\, |x(t)|^2\, dt \geq 0.$$

Example 2.15.18 Let K be a positive continuous function defined on $[a,b]\times[a,b]$. The integral oeprator T on $L^2([a,b])$ defined by

$$(Tx)(s) = \int_a^b K(s,t)\, x(t)\, dt$$

is positive. Indeed, we have

$$(Tx,x) = \int_a^b\int_a^b K(s,t)\, x(t)\, \overline{x(t)}\, dt\, ds = \int_a^b\int_a^b K(s,t)\, |x(t)|^2\, dt\, ds \geq 0$$

for all $x \in L^2([a,b])$.

Theorem 2.15.12 For any bounded operator A on a Hilbert space H, the operators A^*A and AA^* are positive.

Proof. For any $x \in H$, we have

$$(A^* Ax, x) = (Ax, Ax) = \|Ax\|^2 \geq 0$$

and

$$(AA^* x, x) = (A^* x, A^* x) = \|A^* x\|^2 \geq 0.$$

Theorem 2.15.13 If A is an invertible positive operator on a Hilbert space H, then its inverse A^{-1} is positive.

Proof. If $y \in D(A^{-1})$, then $y = Ax$ for some $x \in H$, and then

$$(A^{-1}y, y) = (A^{-1}Ax, Ax) = (x, Ax) \geq 0.$$

To indicate that A is a positive operator, we write $A \geq 0$. If the difference $A - B$ of two self-adjoint operators is a positive operator, that is, $A - B \geq 0$, then we write $A \geq B$. Consequently,

$$A \geq B \quad \text{if and only if} \quad (Ax, x) \geq (Bx, x) \quad \text{for all } x \in H.$$

This relation has the following natural properties:

If $A \geq B$ and $C \geq D$, then $A + C \geq B + D$;

If $A \geq 0$ and $\alpha \geq 0 (\alpha \in \mathbb{R})$, then $\alpha A \geq 0$;

If $A \geq B$ and $B \geq C$, then $A \geq C$.

Proofs are left as exercises.

Theorem 2.15.14 If T is a self-adjoint operator on H and $\|T\| \leq 1$, then $T \leq I$.

Proof. If $\|T\| \leq 1$, then

$$(Tx, x) \leq \|T\| \|x\|^2 \leq (x, x) = (Ix, x)$$

for all $x \in H$.

Definition 2.15.11 (Orthogonal Projection Operator). If S is a closed subspace of a Hilbert space H, the operator P on H defined by

$$Px = y \quad \text{if} \quad x = y + z, \quad y \in S \text{ and } z \in S^{\perp}, \tag{2.15.9}$$

is called the *orthogonal projection operator* onto S, or simply, the *projection operator* onto S. The vector y is called the *projection* of x onto S.

Since the decomposition $x = y + z$ is unique, it follows that projection operators are linear. The Pythagorean formula implies that

$$\|Px\|^2 = \|y\|^2 = \|x\|^2 - \|z\|^2 \le \|x\|^2 .$$

This shows that projection operators are bounded and $\|P\| \le 1$. The zero operator is a projection operator onto the zero subspace. If P is a nonzero projection operator, then $\|P\| = 1$ because, for every $x \in S$, we have $Px = x$. The identity operation I is the projection operator onto the whole space H.

Moreover, it follows from (2.15.9) that

$$(Px,\ x - Px) = 0 \qquad \text{for every} \qquad x \in H .$$

Example 2.15.19 If S is a closed subspace of a Hilbert space H and $\{e_n\}$ is a complete orthonormal system in S, then the projection operator P onto S can be defined by

$$Px = \sum_{n=1}^{\infty} (x, e_n)\, e_n .$$

In particular, if the dimension of S is unity and $u \in S$, $\|u\| = 1$, then $Px = (x, u)\, u$.

Example 2.15.20 Suppose that $H = L^2([-\pi, \pi])$. Every $x \in H$ can be represented as $x = y + z$, where y is an even function and z is an odd function. The operator defined by $Px = y$ is the projection operator onto the subspace of all even functions. This operator can also be defined as in Example 2.15.19:

$$Px = \sum_{n=0}^{\infty} (x, \phi_n)\, \phi_n ,$$

where $\phi_0(t) = \dfrac{1}{\sqrt{2\pi}}$ and $\phi_n(t) = \dfrac{1}{\sqrt{\pi}} \cos nt$, $n = 1, 2, 3, \ldots$.

Example 2.15.21 Let $H = L^2([-\pi, \pi])$ and P be an operator defined by

$$(Px)(t) = \begin{Bmatrix} 0, & t \le 0, \\ x(t), & t > 0 \end{Bmatrix}.$$

Then, P is the projection operator onto the space of all functions that vanish for $t \le 0$.

Definition 2.15.12 (Idempotent Operator). An operator T is called *idempotent* if $T^2 = T$.

Every projection operator is idempotent. In fact, if P is the projection operator onto a subspace S, then P is the identity operator on S. Since $Px \in S$ for every $x \in H$, it follows that $P^2 x = P(Px)$ for all $x \in H$.

Example 2.15.22 Consider the operator T on $\mathbb{C}^2$ defined by $T(x,y) = (x - y,\ 0)$. Obviously, T is idempotent. On the other hand, since

$$\left(T(x,y),\ (x,y) - T(x,y)\right) = x\,\bar{y} - |y|^2,$$

$T(x,y)$ need not be orthogonal to $(x,y) - T(x,y)$ and thus T is not a projection.

Definition 2.15.13 (Compact Operator). An operator T on a Hilbert space H is called a *compact operator* (or *completely continuous operator*) if, for every bounded sequence $\{x_n\}$ in H, the sequence $\{Tx_n\}$ contains a convergent subsequence.

Compact operators constitute an important class of bounded operators. The concept originated from the theory of integral equations of the second kind. Compact operators also provide a natural generalization of operators with finite-dimensional range.

Example 2.15.23 Every operator on a finite dimensional Hilbert space is compact. Indeed, if T is an operator on $\mathbb{C}^N$, then it is bounded. Therefore, if $\{x_n\}$ is a bounded sequence, then $\{Tx_n\}$ is a bounded sequence in $\mathbb{C}^N$. By the Bolzano-Weierstrass theorem, $\{Tx_n\}$ contains a convergent subsequence.

Theorem 2.15.15 Compact operators are bounded.

Proof. If an operator T is not bounded, then there exists a sequence $\{x_n\}$ such that $\|x_n\| = 1$, for all $n \in \mathbb{N}$, and $\|Tx_n\| \to \infty$. Then, $\{Tx_n\}$ does not contain a convergent subsequence, which means that A is not compact.

Not every bounded operator is compact.

Example 2.15.24 The identity operator I on an infinite dimensional Hilbert space H is not compact, although it is bounded. In fact, consider an orthonormal sequence $\{e_n\}$ in H. Then, the sequence $I e_n = e_n$ does not contain a convergent subsequence.

Example 2.15.25 Let y and z be fixed elements of a Hilbert space H. Define

$$Tx = (x, y)z.$$

Let $\{x_n\}$ be a bounded sequence, that is, $\|x_n\| \le M$ for some $M > 0$ and all $n \in \mathbb{N}$. Since

$$|(x_n, y)| \le \|x_n\|\|y\| \le M\|y\|,$$

the sequence $\{(x_n, y)\}$ contains a convergent subsequence $\{(x_{p_n}, y)\}$. Denote the limit of that subsequence by α. Then,

$$Tx_{p_n} = (x_{p_n}, y)\, z \to \alpha\, z \qquad \text{as } n \to \infty.$$

Therefore, T is a compact operator.

Example 2.15.26 Important examples of compact operators are integral operators T on $L^2([a,b])$ defined by

$$(Tx)(s) = \int_a^b K(s,t)\, x(t)\, dt,$$

where a and b are finite and K is continuous.

Example 2.15.27 Let S be a finite-dimensional subspace of a Hilbert space H. The projection operator P_S is a compact operator.

Theorem 2.15.16 Let A be a compact operator on a Hilbert space H, and let B be a bounded operator on H. Then, AB and BA are compact.

Proof. Let $\{x_n\}$ be a bounded sequence in H. Since B is bounded, the sequence $\{Bx_n\}$ is bounded. Next, since A is compact, the sequence $\{ABx_n\}$ contains a convergent subsequence, which means that the operator AB is compact. Similarly, since A is compact, the sequence $\{Ax_n\}$ contains a convergent subsequence $\{Ax_{p_n}\}$. Now, since B is bounded (and thus continuous), the sequence $\{BAx_{p_n}\}$ converges. Therefore, the operator BA is compact.

The operator defined in Example 2.15.27 is a special case of a finite-dimensional operator.

Definition 2.15.14 (Finite-Dimensional Operator). An operator is called *finite-dimensional* if its range is of finite dimension.

Theorem 2.15.17 Finite-dimensional bounded operators are compact.

Proof. Let A be a finite-dimensional bounded operator and let $\{z_1,\ldots,z_k\}$ be an orthonormal basis of the range of A. Define

$$T_n x = (Ax, z_n)\, z_n$$

for $n = 1,\ldots,k$. Since

$$T_n x = (Ax,\ z_n)\ z_n = (x,\ A^* z_n)\ z_n,$$

the operators T_n are compact, as proved in Example 2.15.25. Since

$$A = \sum_{n=1}^{k} T_n,$$

A is compact because the collection of all compact operators on a Hilbert space H is a vector space.

Theorem 2.15.18 If $T_1,\ T_2,\ \ldots$ are compact operators on a Hilbert space H and $\|T_n - T\| \to 0$ as $n \to \infty$ for some operator T on H, then T is compact.

Proof. Let $\{x_n\}$ be a bounded sequence in H. Since T_1 is compact, there exists a subsequence $\{x_{1,n}\}$ of $\{x_n\}$ such that $\{T_1 x_{1,n}\}$ is convergent. Similarly, the sequence $\{T_2 x_{1,n}\}$ contains a convergent subsequence $\{T_2 x_{2,n}\}$. In general, for $k \geq 2$, let $\{x_{k,n}\}$ be a subsequence of $\{x_{k-1,n}\}$ such that $\{T_k x_{k,n}\}$ is convergent. Consider the sequence $\{x_{n,n}\}$. Since it is a subsequence of $\{x_n\}$, we can put $x_{p_n} = x_{n,n}$, where $\{p_n\}$ is an increasing sequence of positive integers. Obviously, the sequence $\{T_k x_{p_n}\}$ converges for every $k \in \mathbb{N}$. We will show that the sequence $\{Tx_{p_n}\}$ also converges.

Let $\varepsilon > 0$. Since $\|T_n - T\| \to 0$, there exists $k \in \mathbb{N}$ such that $\|T_k - T\| < \dfrac{\varepsilon}{3M}$, where M is a constant such that $\|x_n\| \leq M$ for all $n \in \mathbb{N}$. Next, let $k_1 \in \mathbb{N}$ be such that

$$\left\|T_k x_{p_n} - T_k x_{p_m}\right\| < \frac{\varepsilon}{3}$$

for all $n, m > k_1$. Then,

$$\begin{aligned}\left\|Tx_{p_n} - Tx_{p_m}\right\| &\leq \left\|Tx_{p_n} - T_k x_{p_n}\right\| + \left\|T_k x_{p_n} - T_k x_{p_m}\right\| + \left\|T_k x_{p_m} - Tx_{p_m}\right\| \\ &< \frac{\varepsilon}{3} + \frac{\varepsilon}{3} + \frac{\varepsilon}{3} = \varepsilon\end{aligned}$$

for sufficiently large n and m. Thus, $\{Tx_{p_n}\}$ is a Cauchy sequence in H. Completeness of H implies that $\{Tx_{p_n}\}$ is convergent.

Corollary 2.15.3 The limit of a convergent sequence of finite-dimensional operators is a compact operator.

Proof. Finite-dimensional operators are compact.

Theorem 2.15.19 The adjoint of a compact operator is compact.

Proof. Let T be a compact operator on a Hilbert space H, and let $\{x_n\}$ be a bounded sequence in H, that is, $\|x_n\| \leq M$ for some M for all $n \in \mathbb{N}$. Define

$y_n = T^* x_n$, $n = 1, 2, \ldots$. Since T^* is bounded, the sequence $\{y_n\}$ is bounded. It thus contains a subsequence $\{y_{k_n}\}$ such that the sequence $\{T y_{k_n}\}$ converges in H. Now, for any $m, n \in \mathbb{N}$, we have

$$\begin{aligned}
\left\| y_{k_m} - y_{k_n} \right\|^2 &= \left\| T^* x_{k_m} - T^* x_{k_n} \right\|^2 \\
&= \left(T^* \left(x_{k_m} - x_{k_n} \right), T^* \left(x_{k_m} - x_{k_n} \right) \right) \\
&= \left(TT^* \left(x_{k_m} - x_{k_n} \right), \left(x_{k_m} - x_{k_n} \right) \right) \\
&\le \left\| TT^* \left(x_{k_m} - x_{k_n} \right) \right\| \left\| x_{k_m} - x_{k_n} \right\| \\
&\le 2M \left\| T y_{k_m} - T y_{k_n} \right\| \to 0, \qquad \text{as } m, n \to \infty.
\end{aligned}$$

Therefore, $\{y_{k_n}\}$ is a Cauchy sequence in H, which implies that $\{y_{k_n}\}$ converges. This proves that T^* is a compact operator.

In the next theorem, we characterize compactness of operators in terms of weakly convergent sequences. Recall that we write "$x_n \to x$" to denote strong convergence and "$x_n \xrightarrow{w} x$" to denote weak convergence.

Theorem 2.15.20 An operator T on a Hilbert space H is compact if and only if $x_n \xrightarrow{w} x$ implies $Tx_n \to Tx$.

For a proof of this theorem, the reader is referred to Debnath and Mikusinski (1999).

Corollary 2.15.4 If T is a compact operator on a Hilbert space H and $\{x_n\}$ is an orthonormal sequence in H, then $\lim_{n\to\infty} Tx_n = 0$.

Proof. Orthonormal sequences are weakly convergent to 0.

It follows from the above theorem that the inverse of a compact operator on an infinite-dimensional Hilbert space, if it exists, is unbounded.

It has already been noted that compactness of operators is a stronger condition than boundedness. For operators, boundedness is equivalent to continuity. Bounded operators are exactly those operators that map strongly convergent sequences into strongly convergent sequences. Theorem 2.15.20 states that compact operators on a Hilbert space can be characterized as those operators which map weakly convergent sequences into strongly convergent sequences. From this point of view, compactness of operators is a stronger type of continuity. For this reason, compact operators are sometimes called *completely continuous operators.* The above condition has been used by F. Riesz as the definition of compact operators. Hilbert used still another (equivalent) definition of compact oeprators: an operator T defined on a Hilbert space H is compact if $x_n \to x$ weakly and $y_n \to y$ weakly implies $(Tx_n, y_n) \to (Tx, y)$ strongly.

2.16 Eigenvalues and Eigenvectors of an Operator

This section deals with concepts of eigenvalues and eigenvectors which play a central role in the theory of operators.

Definition 2.16.1 (Eigenvalue). Let T be an operator on a complex vector space X. A complex number λ is called an *eigenvalue* of T if there is a nonzero vector $u \in X$ such that

$$Tu = \lambda u. \tag{2.16.1}$$

Every vector u satisfying (2.16.1) is called an *eigenvector* of T corresponding to the eigenvalue λ. If X is a function space, eigenvectors are often called *eigenfunctions.*

Example 2.16.1 Let S be a linear subspace of an inner product space X, and T be the projection on S. The only eigenvalues of T are 0 and 1. Indeed, if, for some $\lambda \in \mathbb{C}$ and $0 \neq u \in X$, we have $Tu = \lambda u$, then

$$\lambda u = \lambda^2 u,$$

because $T^2 = T$. Therefore, $\lambda = 0$ or $\lambda = 1$. The eigenvectors corresponding to 0 are the vectors of X which are orthogonal to S. The eigenvectors corresponding to 1 are all elements of S.

It is important to note that every eigenvector corresponds to exactly one eigenvalue, but there are always infinitely many eigenvectors corresponding to an eigenvalue. Indeed, every multiple of an eigenvector is an eigenvector. Moreover, several linearly independent vectors may correspond to the same eigenvalue. We have the following simple theorem.

Theorem 2.16.1 The collection of all eigenvectors corresponding to one particular eigenvalue of an operator is a vector space.

The easy proof is left as an exercise.

Definition 2.16.2 (Eigenvalue Space). The set of all eigenvectors corresponding to one particular eigenvalue λ is called the ***eigenvalue space*** of λ. The dimension of that space is called the *multiplicity* of λ. An eigenvalue of multiplicity one is called *simple* or *nondegenerate*. In such a case, the number of linearly independent eigenvectors is also called the *degree of degeneracy*.

Example 2.16.2 Consider the integral operator $T: L^2([0,2\pi]) \to L^2([0,2\pi])$ defined by

$$(Tu)(t) = \int_0^{2\pi} \cos(t-y)\, u(y)\, dy. \tag{2.16.2}$$

We will show that T has exactly one nonzero eigenvalue $\lambda = \pi$, and its eigenfunctions are

$$u(t) = a\cos t + b\sin t$$

with arbitrary a and b.

The eigenvalue equation is

$$(Tu)(t) = \int_0^{2\pi} \cos(t-y)\, u(y)\, dy = \lambda u(t).$$

Or,

$$\cos t \int_0^{2\pi} u(y) \cos y\, dy + \sin t \int_0^{2\pi} u(y) \sin y\, dy = \lambda u(t). \tag{2.16.3}$$

This means that, for $\lambda \neq 0$, u is a linear combination of cosine and sine functions, that is,

$$u(t) = a\cos t + b\sin t, \tag{2.16.4}$$

where $a, b \in \mathbb{C}$. Substituting this into (2.16.3), we obtain

$$\pi a = \lambda a \qquad \text{and} \qquad \pi b = \lambda b. \tag{2.16.5}$$

Hence, $\lambda = \pi$, which means that T has exactly one nonzero eigenvalue and its eigenfunctions are given by (2.16.4). This is a two-dimensional eigenspace, so the multiplicity of the eigenvalue is 2.

Equation (2.16.3) reveals that $\lambda = 0$ is also an eigenvalue of T. The corresponding eigenfunctions are all the functions orthogonal to $\cos t$ and $\sin t$. Therefore, $\lambda = 0$ is an eigenvalue of infinite multiplicity.

Note that if λ is not an eigenvalue of T, then the operator $T - \lambda I$ is invertible, and conversely. If space X is finite dimensional and λ is not an eigenvalue of T, then the operator $(T - \lambda I)^{-1}$ is bounded because all operators on a finite-dimensional space are bounded. The situation for infinite dimensional spaces is more complicated.

Definitions 2.16.3 (Resolvent, Spectrum). Let T be an operator on a normed space X. The operator

$$T_\lambda = (T - \lambda I)^{-1}$$

is called the *resolvent* of T. The values λ for which T_λ is defined on the whole space X and is bounded are called *regular points* of T. The set of all λ's which are not regular is called the *spectrum* of T.

Every eigenvalue belongs to the spectrum. The following example shows that the spectrum may contain points that are not eigenvalues. In fact, a nonempty spectrum may contain no eigenvalues at all.

Example 2.16.3 Let X be the space $C([a,b])$ of continuous functions on the interval $[a,b]$. For a fixed $u \in C([a,b])$, consider the operator T defined by

$$(Tx)(t) = u(t)\, x(t).$$

Since

$$(T - \lambda I)^{-1} x(t) = \frac{x(t)}{u(t) - \lambda},$$

the spectrum of T consists of all λ's such that $\lambda - u(t) = 0$ for some $t \in [a,b]$. This means that the spectrum of T is exactly the range of u. If $u(t) = c$ is a constant function, then $\lambda = c$ is an eigenvalue of T. On the other hand, if u is a strictly increasing function, then T has no eigenvalues. The spectrum of T in such a case is the interval $[u(a), u(b)]$.

The problem of finding eigenvalues and eigenvectors is called the eigenvalue problem. One of the main sources of eigenvalue problems in mechanics is the theory of oscillating systems. The state of a given system at a given time t may be represented by an element $u(t) \in H$, where H is an appropriate Hilbert space of functions. The equation of motion in classical mechanics is

$$\frac{d^2u}{dt^2} = Tu, \tag{2.16.6}$$

where T is an operator in H. If the system oscillates, the time dependence of u is sinusoidal, so that $u(t) = v \sin \omega t$, where v is a fixed element of H. If T is linear, then (2.16.6) becomes

$$Tv = \left(-\omega^2\right) v. \tag{2.16.7}$$

This means that $-\omega^2$ is an eigenvalue of T. Physically, the eigenvalues of T correspond to possible frequencies of oscillations. In atomic systems, the frequencies of oscillations are visible as bright lines in the spectrum of light they emit. Thus, the name spectrum arises from physical considerations.

The following theorems describe properties of eigenvalues and eigenvectors for some special classes of operators. Our main interest is in self-adjoint, unitary, and compact operators.

Theorem 2.16.2 Let T be an invertible operator on a vector space X, and let A be an operator on X. The operators A and TAT^{-1} have the same eigenvalues.

Proof. Let λ be an eigenvalue of A. This means that there exists a nonzero vector u such that $Au = \lambda u$. Since T is invertible, $Tu \neq 0$ and

$$TAT^{-1}(Tu) = TAu = T(\lambda u) = \lambda Tu.$$

Thus, λ is an eigenvalue of TAT^{-1}.

Assume now that λ is an eigenvalue of TAT^{-1}, that is, $TAT^{-1}u = \lambda u$ for some nonzero vector $u = Tv$. Since $AT^{-1}u = \lambda T^{-1}u$ and $T^{-1}u \neq 0$, hence, λ is an eigenvalue of A.

Theorem 2.16.3 All eigenvalues of a self-adjoint operator on a Hilbert space are real.

Proof. Let λ be an eigenvalue of a self-adjoint operator T, and let u be a nonzero eigenvector of λ. Then,

$$\lambda(u,u) = (\lambda u, u) = (Tu, u) = (u, Tu) = (u, \lambda u) = \bar{\lambda}(u,u).$$

Since $(u,u) > 0$, we conclude $\lambda = \bar{\lambda}$.

Theorem 2.16.4 All eigenvalues of a positive operator are nonnegative. All eigenvalues of a strictly positive operator are positive.

Proof. Let T be a positive operator, and let $Tx = \lambda x$ for some $x \neq 0$. Since T is self-adjoint, we have

$$0 \leq (Tx, x) = \lambda(x,x) = \lambda \|x\|^2. \tag{2.16.8}$$

Thus, $\lambda \geq 0$. The proof of the second part of the theorem is obtained by replacing $\leq$ by $<$ in (2.16.8).

Theorem 2.16.5 All eigenvalues of a unitary operator on a Hilbert space are complex numbers of modulus 1.

Proof. Let λ be an eigenvalue of a unitary operator T, and let u be an eigenvector of λ, $u \neq 0$. Then,

$$(Tu, Tu) = (\lambda u, \lambda u) = |\lambda|^2 \|u\|^2.$$

On the other hand,

$$(Tu, Tu) = (u, T^* Tu) = (u,u) = \|u\|^2.$$

Thus, $|\lambda| = 1$.

Theorem 2.16.6 Eigenvectors corresponding to distinct eigenvalues of a self-adjoint or unitary operator on a Hilbert space are orthogonal.

Proof. Let T be a self-adjoint operator, and let u_1 and u_2 be eigenvectors corresponding to distinct eigenvalues λ_1 and λ_2, that is, $Tu_1 = \lambda_1 u_1$ and $Tu_2 = \lambda_2 u_2$, $\lambda_1 \neq \lambda_2$. By Theorem 2.16.3, λ_1 and λ_2 are real. Then

$$\lambda_1 (u_1, u_2) = (Tu_1, u_2) = (u_1, Tu_2) = (u_1, \lambda_2 u_2) = \overline{\lambda}_2 (u_1, u_2) = \lambda_2 (u_1, u_2),$$

and hence,

$$(\lambda_1 - \lambda_2)(u_1, u_2) = 0.$$

Since $\lambda_1 \neq \lambda_2$, we have $(u_1, u_2) = 0$, that is, u_1 and u_2 are orthogonal.

Suppose now that T is a unitary operator on a Hilbert space H. Then, $TT^* = T^*T = I$ and $\|Tu\| = \|u\|$ for all $u \in H$. First, note that $\lambda_1 \neq \lambda_2$ implies $\lambda_1 \overline{\lambda}_2 \neq 1$. Indeed, if $\lambda_1 \overline{\lambda}_2 = 1$, then

$$\lambda_2 = \lambda_1 \overline{\lambda}_2 \lambda_2 = \lambda_1 |\lambda_2|^2 = \lambda_1,$$

because $|\lambda_2| = 1$ by Theorem 2.16.5. Now,

$$\lambda_1 \overline{\lambda}_2 (u_1, u_2) = (\lambda_1 u_1, \lambda_2 u_2) = (Tu_1, Tu_2) = (u_1, T^* Tu_2) = (u_1, u_2).$$

Since $\lambda_1 \overline{\lambda}_2 \neq 1$, we get $(u_1, u_2) = 0$. This proves that the eigenvectors u_1 and u_2 are orthogonal.

Theorem 2.16.7 For every eigenvalue λ of a bounded operator T, we have $|\lambda| \leq \|T\|$.

Proof. Let u be a nonzero eigenvector corresponding to λ. Since $Tu = \lambda u$, we have

$$\|\lambda u\| = \|Tu\|,$$

and thus,

$$|\lambda|\|u\| = \|Tu\| \le \|T\|\|u\|.$$

This implies that $|\lambda| \le \|T\|$.

If the eigenvalues are considered as points in the complex plane, the preceding result implies that all the eigenvalues of a bounded operator T lie inside the circle of radius $\|T\|$.

Corollary 2.16.1 All eigenvalues of a bounded, self-adjoint operator T satisfy the inequality

$$|\lambda| \le \sup_{\|x\| \le 1} |(Tx, x)|. \tag{2.16.9}$$

The proof follows immediately from Theorem 2.16.5, proved by Debnath and Mikusinski (1999).

Theorem 2.16.8 If T is a nonzero, compact, self-adjoint operator on a Hilbert space H, then it has an eigenvalue λ equal to either $\|T\|$ or $-\|T\|$.

Proof. Let $\{u_n\}$ be a sequence of elements of H such that $\|u_n\| = 1$, for all $n \in \mathbb{N}$, and

$$\|Tu_n\| \to \|T\| \qquad \text{as } n \to \infty. \tag{2.16.10}$$

Then

$$\begin{aligned}
\left\|T^2 u_n - \|Tu_n\|^2 u_n\right\|^2 &= \left(T^2 u_n - \|Tu_n\|^2 u_n,\, T^2 u_n - \|Tu_n\|^2 u_n\right) \\
&= \left\|T^2 u_n\right\|^2 - 2\|Tu_n\|^2 \left(T^2 u_n,\, u_n\right) + \|Tu_n\|^4 \|u_n\|^2 \\
&= \left\|T^2 u_n\right\|^2 - \|Tu_n\|^4 \\
&\le \|T\|^2 \|Tu_n\|^2 - \|Tu_n\|^4 \\
&= \|Tu_n\|^2 \left(\|T\|^2 - \|Tu_n\|^2\right).
\end{aligned}$$

Since $\|Tu_n\|$ converges to $\|T\|$, we obtain

$$\left\|T^2 u_n - \|Tu_n\|^2 u_n\right\| \to 0 \qquad \text{as } n \to \infty. \tag{2.16.11}$$

The operator T^2, being the product of two compact operators, is also compact. Hence, there exists a subsequence $\{u_{p_n}\}$ of $\{u_n\}$ such that $\{T^2 u_{p_n}\}$ converges. Since $\|T\| \neq 0$, the limit can be written in the form $\|T\|^2 v$, $v \neq 0$. Then, for every $n \in \mathbb{N}$, we have

$$\left\| \|T\|^2 v - \|T\|^2 u_{p_n} \right\| \leq \left\| \|T\|^2 v - T^2 u_{p_n} \right\| + \left\| T^2 u_{p_n} - \|T u_{p_n}\|^2 u_{p_n} \right\| + \left\| \|T u_{p_n}\|^2 u_{p_n} - \|T\|^2 u_{p_n} \right\|.$$

Thus, by (2.16.10) and (2.16.11), we have

$$\left\| \|T^2\| v - \|T\|^2 u_{p_n} \right\| \to 0 \qquad \text{as } n \to \infty.$$

Or,

$$\left\| \|T\|^2 \left(v - u_{p_n} \right) \right\| \to 0 \qquad \text{as } n \to \infty.$$

This means that the sequence $\{u_{p_n}\}$ converges to v and therefore

$$T^2 v = \|T\|^2 v.$$

The above equation can be written as

$$(T - \|T\| I)(T + \|T\| I) v = 0.$$

If $w = (T + \|T\| I) v \neq 0$, then $(T - \|T\| I) w = 0$, and thus $\|T\|$ is an eigenvalue of T. On the other hand, if $w = 0$, then $-\|T\|$ is an eigenvalue of T.

Corollary 2.16.2 If T is a nonzero compact, self-adjoint operator on a Hilbert space H, then there is a vector w such that $\|w\| = 1$ and

$$|(Tw, w)| = \sup_{\|x\| \leq 1} |(Tx, x)|.$$

Proof. Let w, $\|w\| = 1$, be an eigenvector corresponding to an eigenvalue λ such that $|\lambda| = \|T\|$. Then

$$|(Tw, w)| = |(\lambda w, w)| = |\lambda| \|w\|^2 = |\lambda| = \|T\| = \sup_{\|x\| \leq 1} |(Tx, x)|$$

by Theorem 4.4.5, proved by Debnath and Mikusinski (1999).

Theorem 2.16.8 guarantees the existence of at least one nonzero eigenvalue but no more in general. The corollary gives a useful method for finding that eigenvalue by maximizing certain quadratic expressions.

Theorem 2.16.9 The set of distinct non-zero eigenvalues $\{\lambda_n\}$ of a self-adjoint compact operator is either finite or $\lim_{n\to\infty} \lambda_n = 0$.

Proof. Suppose T is a self-adjoint, compact operator that has infinitely many distinct eigenvalues λ_n, $n \in \mathbb{N}$. Let u_n be an eigenvector corresponding to λ_n such that $\|u_n\| = 1$. By Theorem 2.16.6, $\{u_n\}$ is an orthonormal sequence. Since orthonormal sequences are weakly convergent to 0, Theorem 2.15.13 implies

$$\begin{aligned} 0 &= \lim_{n\to\infty} \|Tu_n\|^2 = \lim_{n\to\infty} (Tu_n, Tu_n) \\ &= \lim_{n\to\infty} (\lambda_n u_n, \lambda_n u_n) = \lim_{n\to\infty} \lambda_n^2 \|u_n\|^2 = \lim_{n\to\infty} \lambda_n^2 . \end{aligned}$$

Example 2.16.4 We determine the eigenvalues and eigenfunctions of the operator T on $L^2([0,2\pi])$ defined by

$$(Tu)(x) = \int_0^{2\pi} k(x-t)u(t)\, dt,$$

where k is a periodic function with period 2π and square integrable on $[0,2\pi]$.

As a trial solution, we take

$$u_n(x) = e^{inx}$$

and note that

$$(Tu_n)(x) = \int_0^{2\pi} k(x-t)e^{int}\, dt = e^{inx} \int_{x-2\pi}^{x} k(s)\, e^{ins}\, ds.$$

Thus,

$$Tu_n = \lambda_n u_n, \qquad n \in \mathbb{Z},$$

where

$$\lambda_n = \int_0^{2\pi} k(s) e^{ins} \, ds .$$

The set of functions $\{u_n\}$, $n \in \mathbb{Z}$ is a complete orthogonal system in $L^2([0, 2\pi])$. Note that T is self-adjoint if $k(x) = k(-x)$ for all x, but the sequence of eigenfunctions is complete even if T is not self-adjoint.

Theorem 2.16.10 Let $\{P_n\}$ be a sequence of pairwise orthogonal projection operators on a Hilbert space H, and let $\{\lambda_n\}$ be a sequence of numbers such that $\lambda_n \to 0$ as $n \to \infty$. Then,

(a) $\sum_{n=1}^{\infty} \lambda_n P_n$ converges in $B(H, H)$ and thus, defines a bounded operator;

(b) For each $n \in \mathbb{N}$, λ_n is an eigenvalue of the operator $T = \sum_{n=1}^{\infty} \lambda_n P_n$, and the only other possible eigenvalue of T is 0.

(c) If all λ_n's are real, then T is self-adjoint.

(d) If all projections P_n are finite-dimensional, then T is compact.

For a proof of this theorem, the reader is referred to Debnath and Mikusinski (1999).

Definition 2.16.4 (Approximate Eigenvalue). Let T be an operator on a Hilbert space H. A scalar λ is called an *approximate eigenvalue* of T if there exists a sequence of vectors $\{x_n\}$ such that $\|x_n\| = 1$ for all $n \in \mathbb{N}$ and $\|Tx_n - \lambda x_n\| \to 0$ as $n \to \infty$.

Obviously, every eigenvalue is an approximate eigenvalue.

Example 2.16.5 Let $\{e_n\}$ be a complete orthonormal sequence in a Hilbert space H. Let λ_n be a strictly decreasing sequence of scalars convergent to some λ. Define an operator T on H by

$$Tx = \sum_{n=1}^{\infty} \lambda_n (x, e_n) \, e_n .$$

It is easy to see that every λ_n is an eigenvalue of T, but λ is not. On the other hand,

$$\|Te_n - \lambda e_n\| = \|\lambda_n e_n - \lambda e_n\| = \|(\lambda_n - \lambda)\, e_n\| = |\lambda_n - \lambda| \to 0 \qquad \text{as } n \to \infty .$$

Thus, λ is an approximate eigenvalue of T. Note that the same is true if we just assume that $\lambda_n \to \lambda$ and $\lambda_n \neq \lambda$ for all $n \in \mathbb{N}$.

For further properties of approximate eigenvalues, see the exercises at the end of this chapter.

The rest of this section is concerned with several theorems involving spectral decomposition.

Let H be a finite-dimensional Hilbert space, say $H = \mathbb{C}^N$. It is known from linear algebra that eigenvectors of a self-adjoint operator on H form an orthogonal basis of H. The following theorems generalize this result to infinite-dimensional spaces.

Theorem 2.16.11 (Hilbert-Schmidt Theorem). For every self-adjoint, compact operator T on an infinite-dimensional Hilbert space H, there exists an orthonormal system of eigenvectors $\{u_n\}$ corresponding to nonzero eigenvalues $\{\lambda_n\}$ such that every element $x \in H$ has a unique representation in the form

$$x = \sum_{n=1}^{\infty} \alpha_n u_n + v, \tag{2.16.12}$$

where $\alpha_n \in \mathbb{C}$ and v satisfies the equation $Tv = 0$. If T has infinitely many distinct eigenvalues $\lambda_1, \lambda_2, \ldots$, then $\lambda_n \to 0$ as $n \to \infty$.

For a proof of this theorem, the reader is referred to Debnath and Mikusinski (1999).

Theorem 2.16.12 (Spectral Theorem for Self-Adjoint, Compact Operators). Let T be a self-adjoint, compact operator on an infinite-dimensional Hilbert space H. Then, there exists in H a complete orthonormal system (an orthonormal basis) $\{v_1, v_2, \ldots\}$ consisting of eigenvectors of T. Moreover, for every $x \in H$,

$$Tx = \sum_{n=1}^{\infty} \lambda_n (x, v_n)\, v_n , \tag{2.16.13}$$

where λ_n is the eigenvalue corresponding to v_n.

Proof. Most of this theorem is already contained in Theorem 2.16.11. To obtain a complete orthonormal system $\{v_1, v_2, \ldots\}$, we must add an arbitrary orthonormal basis of $S^\perp$ to the system $\{u_1, u_2, \ldots\}$ (defined in the proof of Theorem 2.16.11). All of the eigenvalues corresponding to those vectors from $S^\perp$ are all equal to zero. Equality (2.16.13) follows from the continuity of T.

Theorem 2.16.13 For any two commuting, self-adjoint, compact operators A and B on a Hilbert space H, there exists a complete orthonormal system of common eigenvectors.

Proof. Let λ be an eigenvalue of A, and let X be the corresponding eigenspace. For any $x \in X$, we have

$$A\,Bx = B\,Ax = B(\lambda x) = \lambda\,Bx.$$

This means that Bx is an eigenvector of A corresponding to λ, provided $Bx \neq 0$. In any case, $Bx \in X$ and hence B maps X into itself. Since B is a self-adjoint, compact operator, by Theorem 2.16.12, X has an orthonormal basis consisting of eigenvalues of B, but these vectors are also eigenvectors of A because they belong to X. If we repeat the same procedure with every eigenspace of A, then the union of all of these eigenvectors will be an orthonormal basis of H.

Theorem 2.16.14 Let T be a self-adjoint, compact operator on a Hilbert space H with a complete orthonormal system of eigenvectors $\{v_1, v_2, \ldots\}$ corresponding to eigenvalues $\{\lambda_1, \lambda_2, \ldots\}$. Let P_n be the projection operator onto the one-dimensional space spanned by v_n. Then, for all $x \in H$,

$$x = \sum_{n=1}^{\infty} P_n x \tag{2.16.14}$$

and

$$T = \sum_{n=1}^{\infty} \lambda_n P_n. \tag{2.16.15}$$

Proof. From the spectral theorem 2.16.12, we have

$$x = \sum_{n=1}^{\infty} (x, v_n)\, v_n . \tag{2.16.16}$$

For every $k \in \mathbb{N}$, the projection operator P_k onto the one-dimensional subspace S_k spanned by v_k is given by

$$P_k x = (x, v_k)\, v_k .$$

Now, (2.16.16) can be written as

$$x = \sum_{n=1}^{\infty} P_n x ,$$

and thus, by Theorem 2.16.2,

$$Tx = \sum_{n=1}^{\infty} \lambda_n (x, v_n)\, v_n = \sum_{n=1}^{\infty} \lambda_n P_n x .$$

Hence, for all $x \in H$,

$$Tx = \left(\sum_{n=1}^{\infty} \lambda_n P_n \right) x .$$

This proves (2.16.15) since convergence of $\sum_{n=1}^{\infty} \lambda_n P_n$ is guaranteed by Theorem 2.16.10.

Theorem 2.16.15 is another version of the spectral theorem. This version is important in the sense that it can be extended to noncompact operators. It is also useful because it leads to an elegant expression for powers and more general functions of an operator.

Theorem 2.16.15 If eigenvectors $u_1, u_2, \ldots$ of a self-adjoint operator T on a Hilbert space H form a complete orthonormal system in H and all eigenvalues are positive (or nonnegative), then T is strictly positive (or positive).

Proof. Suppose $u_1, u_2, \ldots$ is a complete orthonormal system of eigenvalues of T corresponding to real eigenvalues $\lambda_1, \lambda_2, \ldots$. Then, any nonzero vector $u \in H$ can be represented as $u = \sum_{n=1}^{\infty} \alpha_n u_n$, and we have

$$(Tu, u) = \left(Tu, \sum_{n=1}^{\infty} \alpha_n u_n \right) = \sum_{n=1}^{\infty} \overline{\alpha_n} \, (Tu, u_n) = \sum_{n=1}^{\infty} \overline{\alpha_n} \, (u, Tu_n)$$

$$= \sum_{n=1}^{\infty} \overline{\alpha_n} \, (u, \lambda_n u_n) = \sum_{n=1}^{\infty} \lambda_n \overline{\alpha_n} \, (u, u_n) = \sum_{n=1}^{\infty} \lambda_n \overline{\alpha_n} \, \alpha_n$$

$$= \sum_{n=1}^{\infty} \lambda_n |\alpha_n|^2 \geq 0,$$

if all eigenvalues are nonnegative. If all λ_n's are positive, then the last inequality becomes strict.

2.17 Exercises

1. Show that on any inner product space X

   (a) $(x, \alpha y + \beta z) = \overline{\alpha}(x, y) + \overline{\beta}(x, z)$ for all $\alpha, \beta \in \mathbb{C}$,

   (b) $2[(x, y) + (y, x)] = \|x + y\|^2 - \|x - y\|^2$.

2. Prove that the space $C_0(\mathbb{R})$ of all complex-valued continuous functions that vanish outside some finite interval is an inner product space with the inner product

$$(f, g) = \int_{-\infty}^{\infty} f(x) \, \overline{g(x)} \, dx.$$

3. (a) Show that the space $C^1([a, b])$ of all continuously differentiable complex-valued functions on $[a, b]$ is not an inner product space with the inner product

$$(f, g) = \int_a^b f'(x) \, \overline{g'(x)} \, dx.$$

(b) If $f \in C^1([a,b])$ with $f(a) = 0$, show that $C^1([a,b])$ is an inner product space with the inner product defined in (a).

4. (a) Show that the space $C([a,b])$ of real or complex-valued functions is a normed space with the norm $\|f\| = \max_{a \le x \le b} |f(x)|$.

(b) Show that the space $C([a,b])$ is a complete metric space with the metric induced by the norm in (a), that is,

$$d(f,g) = \|f - g\| = \max_{a \le x \le b} |f(x) - g(x)|.$$

5. Prove that the space $C_0^1(\mathbb{R})$ of all continuously differentiable complex-valued continuous functions that vanish outside some finite interval is an inner product space with the inner product

$$(f,g) = \int_{-\infty}^{\infty} f'(x)\,\overline{g'(x)}\,dx.$$

6. Prove that the norm in an inner product space is strictly convex, that is, if

$x \ne y$ and $\|x\| = \|y\| = 1$, then $\|x + y\| \le 2$.

7. (a) Show that the space $C([-\pi,\pi])$ of continuous functions with the norm defined by (2.2.4) is an incomplete normed space.

(b) In the Banach space $L^2([-\pi,\pi])$,

$$f(x) = \sum_{n=1}^{\infty} \frac{1}{n} \sin nx,$$

where $f(x) = -\frac{\pi}{4}$ in $(-\pi,0)$ and $f(x) = \frac{\pi}{4}$ in $(0,\pi)$. Show that f is not continuous in $C([-\pi,\pi])$, but the series converges in $L^2([-\pi,\pi])$.

8. Show that, in any inner product space X,

$$\|x - y\| + \|y - z\| = \|x - z\|$$

if and only if $y = \alpha x + (1-\alpha) z$ for some α in $0 \le \alpha \le 1$.

9. (a) Prove that the *polarization identity*

$$(x,y) = \frac{1}{4}\left(\|x + y\|^2 - \|x - y\|^2 + i\|x + iy\|^2 - i\|x - iy\|^2\right)$$

holds in any complex inner product space.

(b) In any real inner product space, show that

$$(x, y) = \frac{1}{4}\left(\|x + y\|^2 - \|x - y\|^2\right).$$

10. Prove that, for any x in a Hilbert space, $\|x\| = \sup_{\|y\|=1} |(x, y)|$.

11. Show that $L^2([a,b])$ is the only inner product space among the spaces $L^p([a,b])$.

12. Show that the *Apollonius identity* in an inner product space is

$$\|z - x\|^2 + \|z - y\|^2 = \frac{1}{2}\|x - y\|^2 + 2\left\|z - \frac{x+y}{2}\right\|^2.$$

13. Prove that any finite-dimensional inner product space is a Hilbert space.

14. Let $X = \{f \in C^1([a,b]) : f(a) = 0\}$ and

$$(f, g) = \int_a^b f'(x)\overline{g'(x)}\, dx.$$

Is X a Hilbert space?

15. Is the space $C_0^1(\mathbb{R})$ with the inner product

$$(f, g) = \int_{-\infty}^{\infty} f'(x)\overline{g'(x)}\, dx$$

a Hilbert space?

16. Let X be an incomplete inner product space. Let H be the completion of X. Is it possible to extend the inner product from X onto H such that H would become a Hilbert space?

17. Suppose $x_n \to x$ and $y_n \to y$ as $n \to \infty$ in a Hilbert space, and $\alpha_n \to \alpha$ in $\mathbb{C}$. Prove that

(a) $x_n + y_n \to x + y$,

(b) $\alpha_n x_n \to \alpha x$,

(c) $(x_n, y_n) \to (x, y)$,

(d) $\|x_n\| \to \|x\|$.

18. Suppose $x_n \xrightarrow{w} x$ and $y_n \xrightarrow{w} y$ as $n \to \infty$ in a Hilbert space, and $\alpha_n \to \alpha$ in $\mathbb{C}$. Prove or give a counterexample:

(a) $x_n + y_n \xrightarrow{w} x + y$,

(b) $\alpha_n x_n \xrightarrow{w} \alpha x$,

(c) $(x_n, y_n) \to (x, y)$,

(d) $\|x_n\| \to \|x\|$,

(e) If $x_n = y_n$ for all $n \in \mathbb{N}$, then $x = y$.

19. Show that, in a finite-dimensional Hilbert space, weak convergence implies strong convergence.

20. Is it always possible to find a norm on an inner product space X which would define the weak convergence in X?

21. If $\sum_{n=1}^{\infty} u_n = u$, show that

$$\sum_{n=1}^{\infty} (u_n, x) = (u, x)$$

for any x in an inner product space X.

22. Let $\{x_1, \ldots, x_n\}$ be a finite orthonormal set in a Hilbert space H. Prove that for any $x \in H$ the vector

$$x - \sum_{k=1}^{n} (x, x_k) x_k$$

is orthogonal to x_k for every $k = 1, \ldots, n$.

23. In the pre-Hilbert space $\mathscr{C}([-\pi, \pi])$, show that the following sequences of functions are orthogonal

(a) $x_k(t) = \sin kt, \quad k = 1, 2, 3, \ldots,$

(b) $y_n(t) = \cos nt, \quad n = 0, 1, 2, \ldots.$

24. Show that the application of the Gram-Schmidt process to the sequence of functions

$$f_0(t) = 1,\ f_1(t) = t,\ f_2(t) = t^2, \ldots, f_n(t) = t^n, \ldots$$

(as elements of $L^2([-1,1])$) yields the Legendre polynomials.

25. Show that the application of the Gram-Schmidt process to the sequence of functions

$$f_0(t) = e^{-t^2/2}, f_1(t) = te^{-t^2/2}, f_2(t) = t^2e^{-t^2/2}, \dots, f_n(t) = t^n e^{-t^2/2}, \dots$$

(as elements of $L^2(\mathbb{R})$) yields the orthonormal system discussed in Example 2.9.4.

26. Apply the Gram-Schmidt process to the sequence of functions

$$f_0(t) = 1, f_1(t) = t, f_2(t) = t^2, \dots, f_n(t) = t^n, \dots$$

defined on $\mathbb{R}$ with the inner product

$$(f, g) = \int_{-\infty}^{\infty} f(t)\overline{g(t)} \exp(-t^2)\, dt.$$

Compare the result with Example 2.9.4.

27. Apply the Gram-Schmidt process to the sequence of functions

$$f_0(t) = 1, f_1(t) = t, f_2(t) = t^2, \dots, f_n(t) = t^n, \dots$$

defined on $[0, \infty)$ with the inner product

$$(f, g) = \int_0^{\infty} f(t)\overline{g(t)}\, e^{-t}\, dt.$$

The resulting polynomials are called the *Laguerre polynomials*.

28. Let T_n be the *Chebyshev polynomial* of degree n, that is,

$$T_0(x) = 1, \quad T_n(x) = 2^{1-n}\cos(n \arccos x).$$

Show that the functions

$$\phi_n(x) = \frac{2^n}{\sqrt{2\pi}}\, T_n(x), \qquad n = 0, 1, 2, \dots,$$

form an orthonormal system in $L^2[(-1,1)]$ with respect to the inner product

$$(f, g) = \int_{-1}^{1} \frac{1}{\sqrt{1-x^2}}\, f(x)\overline{g(x)}\, dx.$$

29. Prove that for any polynomial

$$p_n(x) = x^n + a_{n-1}x^{n-1} + \dots + a_0,$$

we have

$$\max_{[-1,1]} |p_n(x)| \geq \max_{[-1,1]} |T_n(x)|,$$

where T_n denotes the Chebyshev polynomial of degree n.

30. Show that the complex functions

$$\phi_n(z) = \sqrt{\frac{n}{\pi}}\, z^{n-1}, \qquad n = 1,2,3,\ldots,$$

form an orthonormal system in the space of continuous complex functions defined in the unit disk $D = \{z \in \mathbb{C} : |z| \leq 1\}$ with respect to the inner product

$$(f,g) = \int_D f(z)\overline{g(z)}\, dz.$$

31. Prove that the complex functions

$$\psi_n(z) = \frac{1}{\sqrt{2\pi}}\, z^{n-1}, \qquad n = 1,2,3,\ldots$$

form an orthonormal system in the space of continuous complex functions defined on the unit circle $C = \{z \in \mathbb{C} : |z| = 1\}$ with respect to the inner product

$$(f,g) = \int_C f(z)\overline{g(z)}\, dz.$$

32. With respect to the inner product

$$(f,g) = \int_{-1}^{1} f(x)\overline{g(x)}\, \omega(x)\, dx,$$

where $\omega(x) = (1-x)^{\alpha}(1+x)^{\beta}$ and α, $\beta > -1$, show that the *Jacobi polynomials*

$$P_n^{(\alpha\beta)}(x) = \frac{(-1)^n}{n!2^n}(1-x)^{-\alpha}(1+x)^{-\beta}\frac{d^n}{dx^n}\left[(1-x)^{\alpha}(1+x)^{\beta}\left(1-x^2\right)^n\right]$$

form an orthogonal system.

33. Show that the *Gegenbauer polynomials*

$$C_n^{\gamma}(x) = \frac{(-1)^n}{n!2^n}\left(1-x^2\right)^{\frac{1}{2}-\gamma}\frac{d^n}{dx^n}\left(1-x^2\right)^{n+\gamma-\frac{1}{2}},$$

where $\gamma > \frac{1}{2}$ form an orthonormal system with respect to the inner product

$$(f,g)=\int_{-1}^{1} f(x)\overline{g(x)}\left(1-x^2\right)^{\frac{1}{2}-\gamma}dx.$$

Note that Gegenbauer polynomials are a special case of Jacobi polynomials if $\alpha=\beta=\gamma-\frac{1}{2}$.

34. If x and x_k $(k=1,\ldots,n)$ belong to a real Hilbert space, show that

$$\left\|x-\sum_{k=1}^{n} a_k x_k\right\|^2=\|x\|^2-\sum_{k=1}^{n} a_k\left(x,x_k\right)+\sum_{k=1}^{n}\sum_{l=1}^{n} a_k a_l\left(x_k,x_l\right).$$

Also show that this expression is minimum when $Aa=b$ where $a=\left(a_1,\ldots,a_n\right)$, $b=\left(\left(x,x_1\right),\ldots,\left(x,x_n\right)\right)$, and the matrix $A=\left(a_{kl}\right)$ is defined by $a_{kl}=\left(x_k,x_l\right)$.

35. If $\{a_n\}$ is an orthonormal sequence in a Hilbert space H and $\{\alpha_n\}$ is a sequence in the space l^2, show that there exists $x\in H$ such that

$$\left(x,a_n\right)=\alpha_n \text{ and } \left\|\{\alpha_n\}\right\|=\|x\|,$$

where $\left\|\{\alpha_n\}\right\|$ denotes the norm in the sequence space l^2.

36. If α_n and β_n $(n=1,2,3,\ldots)$ are generalized Fourier coefficients of vectors x and y with respect to a complete orthonormal sequence in a Hilbert space, show that

$$(x,y)=\sum_{k=1}^{\infty}\alpha_k\overline{\beta_k}.$$

37. If $\{x_n\}$ is an orthonormal sequence in a Hilbert space H such that the only element orthogonal to all the x_n's is the null element, show that the sequence $\{x_n\}$ is complete.

38. Let $\{x_n\}$ be an orthonormal sequence in a Hilbert space H. Show that $\{x_n\}$ is complete if and only if $\operatorname{cl}\left(\operatorname{span}\{x_1,x_2,\ldots\}\right)=H$. In other words, $\{x_n\}$ is complete if and only if every element of H can be approximated by a sequence of finite combinations of x_n's.

39. Show that the sequence of functions

$$\phi_n(x) = \frac{e^{-x/2}}{n!} L_n(x), \quad n = 0, 1, 2, \ldots,$$

where L_n is the Laguerre polynomial of degree n, that is,

$$L_n(x) = e^x \frac{d^n}{dx^n}\left(x^n e^{-x}\right),$$

form a complete orthonormal system in $L^2(0, \infty)$.

40. Let

$$\phi_n(x) = \frac{e^{inx}}{\sqrt{2\pi}}, \quad n = 0, \pm 1, \pm 2, \ldots,$$

and let $f \in L^1([-\pi, \pi])$. Define

$$f_n(x) = \sum_{k=-n}^{n} (f, \phi_k)\phi_k, \quad \text{for } n = 0, 1, 2, \ldots.$$

Show that

$$\frac{f_0(x) + f_1(x) + \cdots + f_n(x)}{n+1} = \sum_{k=-n}^{n}\left(1 - \frac{|k|}{n+1}\right)(f, \phi_k)\phi_k(x).$$

41. Show that the sequence of functions

$$\frac{1}{\sqrt{2\pi}}, \frac{\cos x}{\sqrt{\pi}}, \frac{\sin x}{\sqrt{\pi}}, \frac{\cos 2x}{\sqrt{\pi}}, \frac{\sin 2x}{\sqrt{\pi}}, \ldots$$

is a complete orthonormal sequence in $L^2([-\pi, \pi])$.

42. Show that the following sequence of functions is a complete orthonormal system in $L^2([0, \pi])$:

$$\frac{1}{\sqrt{\pi}}, \sqrt{\frac{2}{\pi}} \cos x, \sqrt{\frac{2}{\pi}} \cos 2x, \sqrt{\frac{2}{\pi}} \cos 3x, \ldots.$$

43. Show that the following sequence of functions is a complete orthonormal system in $L^2([0, \pi])$:

$$\sqrt{\frac{2}{\pi}} \sin x, \sqrt{\frac{2}{\pi}} \sin 2x, \sqrt{\frac{2}{\pi}} \sin 3x, \ldots.$$

44. Show that the sequence of functions defined by

$$f_n(x) = \frac{1}{\sqrt{2a}} \exp\left(\frac{in\pi x}{a}\right), \quad n = 0, \pm 1, \pm 2, \ldots$$

is a complete orthonormal system in $L^2([-a, a])$.

45. Show that the sequence of functions

$$\frac{1}{\sqrt{2a}}, \frac{1}{\sqrt{a}}\cos\left(\frac{n\pi x}{a}\right), \frac{1}{\sqrt{a}}\sin\left(\frac{n\pi x}{a}\right), \dots$$

is a complete orthonormal system in $L^2([-a,a])$.

46. Show that each of the following sequences of functions is a complete orthonormal system in $L^2([0,a])$:

$$\frac{1}{\sqrt{a}}, \sqrt{\frac{2}{a}}\cos\left(\frac{\pi x}{a}\right), \sqrt{\frac{2}{a}}\cos\left(\frac{2\pi x}{a}\right), \dots, \sqrt{\frac{2}{a}}\cos\left(\frac{n\pi x}{a}\right), \dots$$

$$\sqrt{\frac{2}{a}}\sin\left(\frac{\pi x}{a}\right), \sqrt{\frac{2}{a}}\sin\left(\frac{2\pi x}{a}\right), \dots, \sqrt{\frac{2}{a}}\sin\left(\frac{n\pi x}{a}\right), \dots.$$

47. Let X be the Banach space $\mathbb{R}^2$ with the norm $\|(x,y)\| = \max\{|x|,|y|\}$. Show that X does not have the closest-point property.

48. Let S be a closed subspace of a Hilbert space H and let $\{e_n\}$ be a complete orthonormal sequence in S. For an arbitrary $x \in H$, there exists $y \in S$ such that $\|x-y\| = \inf_{z \in S} \|x-z\|$. Define y in terms of $\{e_n\}$.

49. If S is a closed subspace of a Hilbert space H, then $H = S \oplus S^{\perp}$. Is this true in every inner product space?

50. Show that the functional in Example 2.13.2 is unbounded.

51. The Riesz representation theorem states that for every bounded linear functional $f \in H'$ on a Hilbert space H, there exists a representer $x_f \in H$ such that $f(x) = (x, x_f)$ for all $x \in H$. Let $T : H' \to H$ be the mapping that assigns x_f to f. Prove the following properties of T:

   (a) T is onto,

   (b) $T(f+g) = T(f) + T(g)$,

   (c) $T(\alpha f) = \overline{\alpha}\, T(f)$,

   (d) $\|T(f)\| = \|f\|$,

   where $f, g \in H'$ and $\alpha \in \mathbb{C}$.

52. Let f be a bounded linear functional on a closed subspace X of a Hilbert space H. Show that there exists a bounded linear functional g on H such that $\|f\| = \|g\|$ and $f(x) = g(x)$ whenever $x \in X$.

53. Show that the space l^2 is separable.

54. (a) Show that the sequence of Gaussian functions on $\mathbb{R}$ defined by

$$f_n(x) = \frac{n}{\sqrt{\pi}} \exp\left(-n^2 x^2\right), \quad n = 1, 2, 3, \ldots$$

converges to the Dirac delta distribution $\delta(x)$.

(b) Show that the sequence of functions on $\mathbb{R}$ defined by

$$f_n(x) = \frac{\sin nx}{\pi x}, \quad n = 1, 2, \ldots$$

converges to the Dirac delta distribution.

55. Show that the sequence of functions on $\mathbb{R}$ defined by

$$f_n(x) = \begin{Bmatrix} 0, & \text{for } x < -\frac{1}{2n}, \\ n, & \text{for } -\frac{1}{2n} \le x \le \frac{1}{2n}, \\ 0, & \text{for } x > \frac{1}{2n} \end{Bmatrix}$$

converges to the Dirac delta distribution.

56. If f is a locally integrable function on $\mathbb{R}^N$, show that the functional F on $\mathcal{D}$ defined by

$$(F, \phi) = \int_{\mathbb{R}^N} f \phi$$

is a distribution.

57. If $f_n(x) = \sin nx$, show that $f_n \to 0$ in the distributional sense.

58. Find the nth distributional derivative of $f(x) = |x|$.

59. Verify which functions belong to $L^1(\mathbb{R})$ and which do not belong to $L^1(\mathbb{R})$. Find their $L^1(\mathbb{R})$ norms when they exist.

(a) $f(x) = \left(a^2 + x^2\right)^{-1}$, (b) $f(x) = x\left(a^2 + x^2\right)^{-1}$,

(c) $f(x) = \begin{Bmatrix} 1, & |x| \le 1, \\ |x|^{-r}, & |x| > 1 \end{Bmatrix}$, (d) $f(x) = x^{-1}$.

60. Let $\{e_n\}$ be a complete orthonormal sequence in a Hilbert space H, and let $\{\lambda_n\}$ be a sequence of scalars.

   (a) Show that there exists a unique operator T on H such that $Te_n = \lambda_n e_n$.

   (b) Show that T is bounded if and only if the sequence $\{\lambda_n\}$ is bounded.

   (c) For a bounded sequence $\{\lambda_n\}$, find the norm of T.

61. Let $T: \mathbb{R}^2 \to \mathbb{R}^2$ be defined by $T(x,y) = (x+2y,\ 3x+2y)$. Find the eigenvalues and eigenvectors of T.

62. Let $T: \mathbb{R}^2 \to \mathbb{R}^2$ be defined by $T(x,y) = (x+3y,\ 2x+y)$. Show that $T^* \neq T$.

63. Let $T: \mathbb{R}^3 \to \mathbb{R}^3$ be given by $T(x,y,z) = (3x-z,\ 2y,\ -x+3z)$. Show that T is self-adjoint.

64. Compute the adjoint of each of the following operators:

   (a) $A: \mathbb{R}^3 \to \mathbb{R}^3$, $\quad A(x,y,z) = (-y+z,\ -x+2z,\ x+2y)$,

   (b) $B: \mathbb{R}^3 \to \mathbb{R}^3$,
   $B(x,y,z) = (x+y-z,\ -x+2y+2z,\ x+2y+3z)$,

   (c) $C: P_2(\mathbb{R}) \to P_2(\mathbb{R}),\ C\{p(x)\} = x\frac{d}{dx}p(x) - \frac{d}{dx}(xp(x))$,

   where $P_2(\mathbb{R})$ is the space of all polynomials on $\mathbb{R}$ of degree less than or equal to 2.

65. If A is a self-adjoint operator and B is a bounded operator, show that B^*AB is self-adjoint.

66. Prove that the representation $T = A + iB$ in Theorem 2.15.7 is unique.

67. If $A^*A + B^*B = 0$, show that $A = B = 0$.

68. If T is self-adjoint and $T \neq 0$, show that $T^n \neq 0$ for all $n \in \mathbb{N}$.

69. Let T be a self-adjoint operator. Show that

   (a) $\|Tx + ix\|^2 = \|Tx\|^2 + \|x\|^2$,

   (b) the operator $U = (T - iI)(T + iI)^{-1}$ is unitary. (U is called the *Cayley transform* of T).

70. Show that the limit of a convergent sequence of self-adjoint operators is a self-adjoint operator.

71. If T is a bounded operator on H with one-dimensional range, show that there exists vectors $y, z \in H$ such that $Tx = (x, z)\, y$ for all $x \in H$. Hence, show that

    (a) $T^* x = (x, y)\, z$ for all $x \in H$,

    (b) $T^2 = \lambda T$, where λ is a scalar,

    (c) $\|T\| = \|y\| \|z\|$,

    (d) $T^* = T$ if and only if $y = \alpha z$ for some real scalar α.

72. Let T be a bounded self-adjoint operator on a Hilbert space H such that $\|T\| \le 1$. Prove that $(x, Tx) \ge (1 - \|T\|)\, \|x\|^2$ for all $x \in H$.

73. If A is a positive operator and B is a bounded operator, show that $B^* AB$ is positive.

74. If A and B are positive operators and $A + B = 0$, show that $A = B = 0$.

75. Show that, for any self-adjoint operator A, there exists positive operators S and T such that $A = S - T$ and $ST = 0$.

76. If P is self-adjoint and P^2 is a projection operator, is P a projection operator?

77. Let T be a multiplication operator on $L^2([a, b])$. Find necessary and sufficient conditions for T to be a projection.

78. Show that P is a projection if and only if $P = P^* P$.

79. If P, Q, and $P + Q$ are projections, show that $PQ = 0$.

80. Show that every projection P is positive and $0 \le P \le I$.

81. Show that, for projections P and Q, the operator $P + Q - PQ$ is a projection if and only if $PQ = QP$.

82. Show that the projection onto a closed subspace X of a Hilbert space H is a compact operator if and only if X is finite dimensional.

83. Show that the operator $T : l^2 \to l^2$ defined by $T(x_n) = (2^{-n} x_n)$ is compact.

84. Prove that the collection of all eigenvectors corresponding to one particular eigenvalue of an operator is a vector space.

85. Show that the space of all eigenvectors corresponding to one particular eigenvalue of a compact operator is finite dimensional.

86. Show that a self-adjoint operator T is compact if and only if there exists a sequence of finite-dimensional operators strongly convergent to T.

87. Show that eigenvalues of a symmetric operator are real and eigenvectors corresponding to different eigenvalues are orthogonal.

88. Give an example of a self-adjoint operator that has no eigenvalues.

89. Show that a non-zero vector x is an eigenvector of an operator T if and only if $|(Tx,x)| = \|Tx\|\|x\|$.

90. Show that if the eigenvectors of a self-adjoint operator T form a complete orthogonal system and all eigenvalues are nonnegative (or positive), then T is positive (or strictly positive).

91. If λ is an approximate eigenvalue of an operator T, show that $|\lambda| \leq \|T\|$.

92. Show that if T has an approximate eigenvalue λ such that $|\lambda| = \|T\|$, then $\sup_{\|x\| \leq 1} |(Tx,x)| = \|T\|$.

93. If λ is an approximate eigenvalue of T, show that $\lambda + \mu$ is an approximate eigenvalue of $T + \mu I$ and $\lambda \mu$ is an approximate eigenvalue of μT.

94. For every approximate eigenvalue λ of an isometric operator, show that we have $|\lambda| = 1$.

95. Show that every approximate eigenvalue of a self-adjoint operator is real.

Chapter 3

Fourier Transforms and Their Applications

"The profound study of nature is the most fertile source of mathematical discoveries."

Joseph Fourier

"Fourier was motivated by the study of heat diffusion, which is governed by a linear differential equation. However, the Fourier transform diagonalizes all linear time-invariant operators, which are building blocks of signal processing. It is therefore not only the starting point of our exploration but the basis of all further developments."

Stéphane Mallat

3.1 Introduction

This chapter deals with Fourier transforms in $L^1(\mathbb{R})$ and in $L^2(\mathbb{R})$ and their basic properties. Special attention is given to the convolution theorem and summability kernels including Cesáro, Fejér, and Gaussian kernels. Several important results including the approximate identity theorem, general Parseval's relation, and Plancherel theorem are proved. This is followed by the Poisson summation formula, Gibbs' phenomenon, the Shannon sampling theorem, and Heisenberg's uncertainty principle. Many examples of applications of the Fourier transforms to mathematical statistics, signal processing, ordinary differential equations, partial differential equations, and integral equations are discussed. Included are some examples of applications of multiple Fourier transforms to important partial differential equations and Green's functions.

Before we discuss Fourier transforms, we define the *translation*, *modulation*, and *dilation* operators respectively, by

$$T_a f(x) = f(x-a) \qquad \text{(Translation)},$$

$$M_b f(x) = e^{ibx} f(x) \qquad \text{(Modulation)},$$

$$D_c f(x) = \frac{1}{\sqrt{|c|}} f\left(\frac{x}{c}\right) \qquad \text{(Dilation)},$$

where $a, b, c \in \mathbb{R}$ and $c \neq 0$.

In particular, D_{-1} is called the *parity operator*, P so that $P f(x) = D_{-1} f(x) = f(-x)$.

The operators T_a, M_a, D_a preserve the L^2*-norm* defined by (2.3.5), that is,

$$\|f\|_2 = \|T_a f\| = \|M_a f\| = \|D_a f\|.$$

Each of these operators is a unitary operator from $L^2(\mathbb{R})$ onto itself. The following results can easily be verified:

$$T_a M_b f(x) = \exp\{ib(x-a)\} f(x-a),$$

$$M_b T_a f(x) = \exp(ibx) f(x-a),$$

$$D_c T_a f(x) = \frac{1}{\sqrt{|c|}} f\left(\frac{x-a}{c}\right),$$

$$T_a D_c f(x) = \frac{1}{\sqrt{|c|}} f\left(\frac{x-a}{c}\right),$$

$$M_b D_c f(x) = \frac{1}{\sqrt{|c|}} \exp\left(i\frac{b}{c}x\right) f\left(\frac{x}{c}\right),$$

$$D_c M_b f(x) = \frac{1}{\sqrt{|c|}} \exp\left(i\frac{b}{c}x\right) f\left(\frac{x}{c}\right).$$

Using the inner product (2.5.2) on $L^2(\mathbb{R})$, the following results can also be verified:

$$(f,\ T_a\, g) = (T_{-a} f,\ g),$$

$$(f,\ M_b\, g) = (M_{-b} f,\ g),$$

$$(f,\ D_c\, g) = \left(D_{\frac{1}{c}} f,\ g\right).$$

3.2 Fourier Transforms in $L^1(\mathbb{R})$

Suppose f is a Lebesgue integrable function on $\mathbb{R}$. Since $\exp(-i\omega t)$ is continuous and bounded, the product $\exp(-i\omega t)\ f(t)$ is locally integrable for any $\omega \in \mathbb{R}$. Also, $|\exp(-i\omega t)| \le 1$ for all ω and t on $\mathbb{R}$. Consider the integral

$$\left(f,\ e^{i\omega t}\right) = \int_{-\infty}^{\infty} f(t)\ e^{-i\omega t}\, dt, \qquad \omega \in \mathbb{R}. \tag{3.2.1}$$

Clearly,

$$\left|\int_{-\infty}^{\infty} e^{-i\omega t}\ f(t)\ dt\right| \le \int_{-\infty}^{\infty} |f(t)|\, dt = \|f\|_1 < \infty. \tag{3.2.2}$$

This means that the integral in (3.2.1) exists for all $\omega \in \mathbb{R}$. Thus, we give the following definition.

***Definition 3.2.1** (The Fourier Transform in $L^1(\mathbb{R})$).* Let $f \in L^1(\mathbb{R})$. The Fourier transform of $f(t)$ is denoted by $\hat{f}(\omega)$ and defined by

$$\hat{f}(\omega) = \mathcal{F}\{f(t)\} = \int_{-\infty}^{\infty} e^{-i\omega t} f(t)\, dt. \tag{3.2.3}$$

Physically, the Fourier integral (3.2.3) measures oscillations of f at the frequency ω, and $\hat{f}(\omega)$ is called the *frequency spectrum* of a signal or waveform $f(t)$. It seems equally justified to refer to $f(t)$ as the waveform in the time domain and $\hat{f}(\omega)$ as the waveform in the frequency domain. Such terminology describes the duality and the equivalence of waveform representations.

In some books, the Fourier transform is defined with the factor $\frac{1}{\sqrt{2\pi}}$ in integral (3.2.3). Another modification is the definition without the minus sign in the kernel $\exp(-i\omega t)$. In electrical engineering, usually t and ω represent the time and the frequency respectively. In quantum physics and fluid mechanics, it is convenient to use the space variable x, the wavenumber k instead of t and ω respectively. All of these changes do not alter the theory of Fourier transforms at all. We shall use freely both symbols $\hat{f}(\omega)$ and $\mathscr{F}\{f(t)\}$ in this book.

Example 3.2.1 (a) $\hat{f}(\omega) = \mathscr{F}\{\exp(-a^2t^2)\} = \frac{\sqrt{\pi}}{a}\exp\left(-\frac{\omega^2}{4a^2}\right)$ for $a > 0$.

We have, by definition,

$$\hat{f}(\omega) = \int_{-\infty}^{\infty} \exp\left[-\left(i\omega t + a^2t^2\right)\right] dt = \int_{-\infty}^{\infty} \exp\left[-a^2\left(t + \frac{i\omega}{2a^2}\right)^2 - \frac{\omega^2}{4a^2}\right] dt$$

$$= \exp\left(-\frac{\omega^2}{4a^2}\right)\int_{-\infty}^{\infty} \exp\left(-a^2y^2\right) dy = \frac{\sqrt{\pi}}{a}\exp\left(-\frac{\omega^2}{4a^2}\right), \tag{3.2.4}$$

in which the change of variable $y = \left(t + \frac{i\omega}{2a^2}\right)$ is used. Even though $\left(\frac{i\omega}{2a^2}\right)$ is a complex number, the above result is correct. The change of variable can be justified by the method of complex analysis. The graphs of $f(t)$ and $\hat{f}(\omega)$ are drawn in Figure 3.1.

In particular, when $a^2 = \frac{1}{2}$ and $a = 1$, we obtain the following results

(b) $$\mathscr{F}\left\{\exp\left(-\frac{t^2}{2}\right)\right\} = \sqrt{2\pi}\exp\left(-\frac{\omega^2}{2}\right), \tag{3.2.5}$$

(c) $$\mathscr{F}\left\{\exp\left(-t^2\right)\right\} = \sqrt{\pi}\exp\left(-\frac{\omega^2}{4}\right). \tag{3.2.6}$$

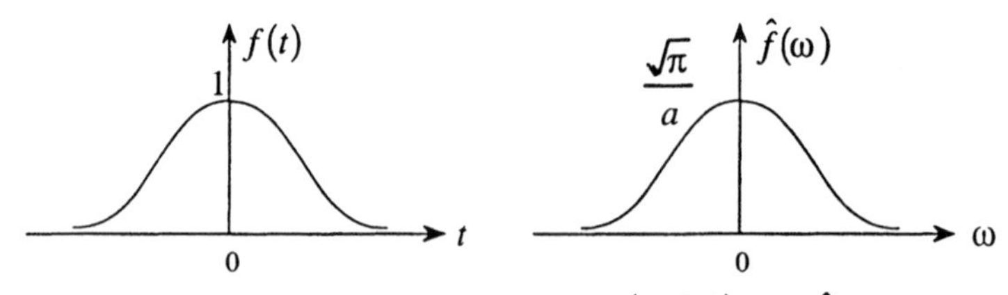

Figure 3.1. Graphs of $f(t) = \exp(-a^2t^2)$ and $\hat{f}(\omega)$.

Example 3.2.2 $\mathscr{F}\{\exp(-a|t|)\} = \dfrac{2a}{(a^2+\omega^2)}, \quad a>0.$

We have

$$\mathscr{F}\{\exp(-a|t|)\} = \int_{-\infty}^{\infty} \exp(-a|t| - i\omega t)\, dt$$
$$= \int_{-\infty}^{0} e^{(a-i\omega)t} dt + \int_{0}^{\infty} e^{-(a+i\omega)t} dt$$
$$= \frac{1}{a-i\omega} + \frac{1}{a+i\omega} = \frac{2a}{(a^2+\omega^2)}. \tag{3.2.7}$$

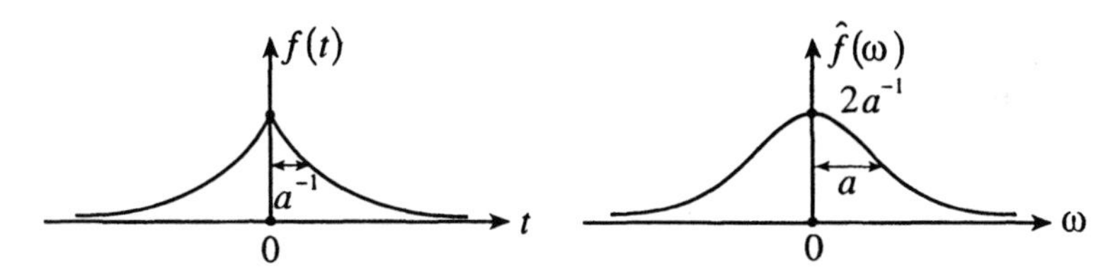

Figure 3.2. Graphs of $f(t) = \exp(-a|t|)$ and $\hat{f}(\omega)$.

Example 3.2.3 (Characteristic Function). This function is defined by

$$\chi_\tau(t) = \begin{Bmatrix} 1, & -\tau < t < \tau \\ 0, & \text{otherwise} \end{Bmatrix}. \tag{3.2.8}$$

In science and engineering, this function is often called a *rectangular pulse* or *gate function*. Its Fourier transform is

$$\hat{\chi}_\tau(\omega) = \mathscr{F}\{\chi_\tau(t)\} = \left(\frac{2}{\omega}\right) \sin(\omega\tau). \tag{3.2.9}$$

We have

$$\hat{\chi}_\tau(\omega) = \int_{-\infty}^{\infty} \chi_\tau(t) \exp(-i\omega t)\, dt = \int_{-\tau}^{\tau} \exp(-i\omega t)\, dt = \left(\frac{2}{\omega}\right) \sin(\omega\tau).$$

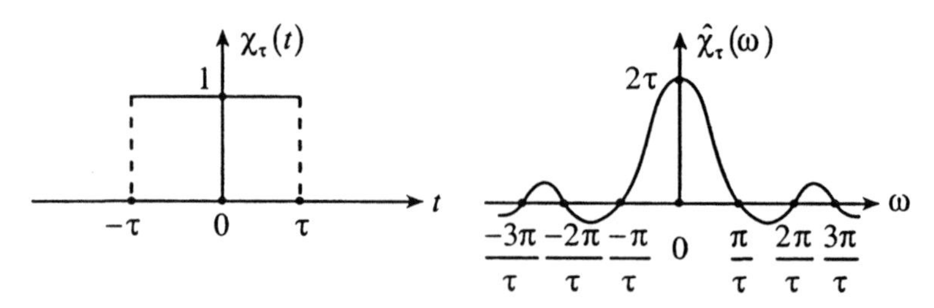

Figure 3.3. Graphs of $\chi_\tau(t)$ and $\hat{\chi}_\tau(\omega)$.

Note that $\chi_\tau(t) \in L^1(\mathbb{R})$, but its Fourier transform $\hat{\chi}_\tau(\omega) \notin L^1(\mathbb{R})$.

Example 3.2.4 $\mathscr{F}\{f(t)\} = \mathscr{F}\left\{\left(a^2 + t^2\right)^{-1}\right\} = \dfrac{\pi}{a}\exp\left(-a|\omega|\right), \quad a > 0.$

This can easily be verified and hence is left to the reader.

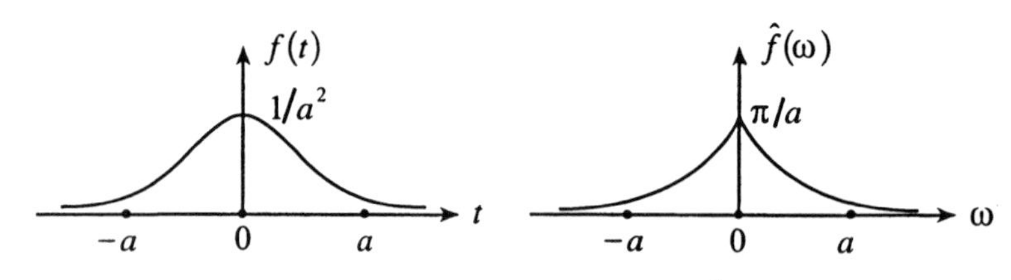

Figure 3.4. Graphs of $f(t)$ and $\hat{f}(\omega)$.

Example 3.2.5 (Triangular Pulse). This function is defined by

$$\Delta_\tau(t) = \left\{\begin{array}{cc} 1 + \dfrac{t}{\tau}, & -\tau \le t < 0, \\ 1 - \dfrac{t}{\tau}, & 0 \le t < \tau, \\ 0, & |t| > \tau \end{array}\right\}. \tag{3.2.10}$$

It can easily be verified that

$$\hat{\Delta}_\tau(\omega) = \tau \cdot \frac{\sin^2\left(\dfrac{\omega\tau}{2}\right)}{\left(\dfrac{\omega\tau}{2}\right)^2}. \tag{3.2.11}$$

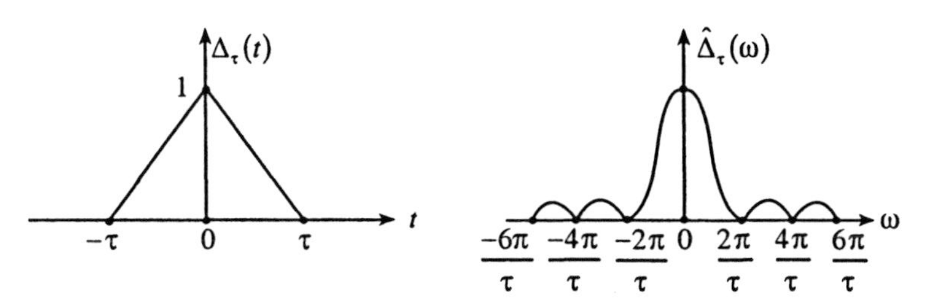

Figure 3.5. Graphs of $\Delta_\tau(t)$ and $\hat{\Delta}_\tau(\omega)$.

Note that $\Delta'_\tau(t)$ exists except at $t=-\tau,0,\tau$, and it represents a piecewise constant function

$$\Delta'_\tau(t) = \begin{Bmatrix} -\frac{1}{\tau}\operatorname{sgn} t, & -\tau < t < \tau, \\ 0, & |t| > \tau \end{Bmatrix}. \tag{3.2.12}$$

Or, equivalently,

$$\Delta'_\tau(t) = \frac{1}{\tau}\left\{\Delta_{\frac{\tau}{2}}\left(t+\frac{\tau}{2}\right) - \Delta_{\frac{\tau}{2}}\left(t-\frac{\tau}{2}\right)\right\} \text{ except for } t=-\tau,0,\tau.$$

Thus, $\Delta'_\tau(t) \in L^1(\mathbb{R})$, and its Fourier transform is given by

$$\mathscr{F}\{\Delta'_\tau(t)\} = \left(\frac{4i}{\omega}\right)\sin^2\left(\frac{\omega\tau}{2}\right). \tag{3.2.13}$$

Remarks.

1. It is important to point out that several elementary functions, such as the constant function c, $\sin\omega_0 t$, and $\cos\omega_0 t$, do not belong to $L^1(\mathbb{R})$ and hence they do not have Fourier transforms. However, when these functions are multiplied by the characteristic function $\chi_\tau(t)$, the resulting functions belong to $L^1(\mathbb{R})$ and have Fourier transforms (see Section 3.14 Exercises).

2. In general, the Fourier transform $\hat{f}(\omega)$ is a complex function of a real variable ω. From a physical point of view, the polar representation of the Fourier transform is often convenient. The Fourier transform $\hat{f}(\omega)$ can be expressed in the polar form

$$\hat{f}(\omega) = R(\omega) + i\,X(\omega) = A(\omega)\exp\{i\theta(\omega)\}, \tag{3.2.14}$$

where $A(\omega) = \left|\hat{f}(\omega)\right|$ is called the *amplitude spectrum* of the signal $f(t)$, and $\theta(\omega) = \arg\left\{\hat{f}(\omega)\right\}$ is called the *phase spectrum* of $f(t)$.

The nature of $A(\omega)$ can be explained by using Example 3.2.3 which shows $A(\omega) = |\hat{\chi}_\tau(\omega)| \le \frac{2}{\omega}$. This means that the amplitude spectrum is very low at high frequencies when τ is very large, $A(\omega) = \frac{2}{\omega}\sin(\omega\tau)$ is very high at low frequencies.

3.3 Basic Properties of Fourier Transforms

Theorem 3.3.1 (***Linearity***). If $f(t), g(t) \in L^1(\mathbb{R})$ and α, β are any two complex constants, then

$$\mathscr{F}\{\alpha f(t) + \beta g(t)\} = \alpha\, \mathscr{F}\{f(t)\} + \beta\, \mathscr{F}\{g(t)\}. \tag{3.3.1}$$

The proof follows readily from Definition 2.3.1 and is left as an exercise.

Theorem 3.3.2. If $f(t) \in L^1(\mathbb{R})$, then the following results hold:

(a) (**Shifting**) $$\mathscr{F}\{T_a f(t)\} = M_{-a}\hat{f}(\omega), \tag{3.3.2}$$

(b) (**Scaling**) $$\mathscr{F}\left\{D_{\frac{1}{a}} f(t)\right\} = D_a \hat{f}(\omega), \tag{3.3.3}$$

(c) (**Conjugation**) $$\mathscr{F}\left\{\overline{D_{-1} f(t)}\right\} = \overline{\hat{f}(\omega)}, \tag{3.3.4}$$

(d) (**Modulation**) $$\mathscr{F}\{M_a f(t)\} = T_a \hat{f}(\omega), \tag{3.3.5}$$

Proof *(a)***.** It follows from definition 3.2.1 that

$$\mathscr{F}\{T_a f(t)\} = \mathscr{F}\{f(t-a)\} = \int_{-\infty}^{\infty} e^{-i\omega t} f(t-a)\,dt = \int_{-\infty}^{\infty} e^{-i\omega(x+a)} f(x)\,dx = M_{-a}\hat{f}(\omega),$$

in which a change of variable $t - a = x$ was used.

The proofs of (b)-(d) follow easily from definition (3.2.3) and are left as exercises.

Example 3.3.1 **(*Modulated Gaussian Function*).**

If $f(t) = \exp\left(i\omega_0 t - \frac{1}{2} t^2\right)$, then

$$\hat{f}(\omega) = \exp\left\{-\frac{1}{2}(\omega - \omega_0)^2\right\}. \tag{3.3.6}$$

This easily follows from Example 3.2.1 combined with the shifting property (3.3.2). The graphs of $\text{Re}\{f(t)\}$ and $\hat{f}(\omega)$ are shown in Figure 3.6.

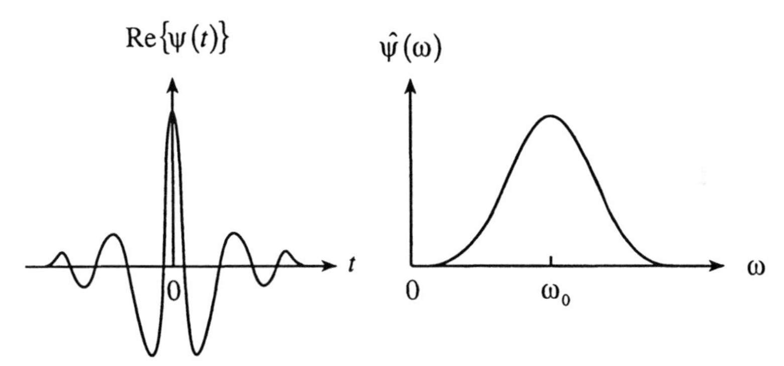

Figure 3.6. Graphs of $f(t)$ and $\hat{f}(\omega)$.

Theorem 3.3.3 **(*Continuity*).** If $f(t) \in L^1(\mathbb{R})$, then $\hat{f}(\omega)$ is continuous on $\mathbb{R}$.

Proof. For any $\omega, h \in \mathbb{R}$, we have

$$\left|\hat{f}(\omega + h) - \hat{f}(\omega)\right| = \left|\int_{-\infty}^{\infty} e^{-i\omega t}\left(e^{-iht} - 1\right) f(t)\, dt\right| \leq \int_{-\infty}^{\infty}\left|e^{-iht} - 1\right|\left|f(t)\right| dt. \tag{3.3.7}$$

Since

$$\left|e^{-iht} - 1\right|\left|f(t)\right| \leq 2\left|f(t)\right| \text{ and } \lim_{h\to 0}\left|e^{-iht} - 1\right| = 0$$

for all $t \in R$, we conclude that as $h \to 0$

$$\left|\hat{f}(\omega + h) - \hat{f}(\omega)\right| \to 0,$$

which is independent of ω by the Lebesgue dominated convergence theorem (see Debnath and Mikusinski, 1999). This proves that $\hat{f}(\omega)$ is continuous in $\mathbb{R}$. Since (3.3.7) is independent of ω, $\hat{f}(\omega)$ is, in fact, uniformly continuous on $\mathbb{R}$.

Theorem 3.3.4 (Derivatives of Fourier Transforms). If both $f(t)$ and $t\,f(t)$ belong to $L^1(\mathbb{R})$, then $\frac{d}{d\omega}\hat{f}(\omega)$ exists and is given by

$$\frac{d}{d\omega}\hat{f}(\omega) = (-i)\,\mathscr{F}\{t\,f(t)\}. \tag{3.3.8}$$

Proof. We have

$$\frac{d\hat{f}}{d\omega} = \lim_{h\to 0}\frac{1}{h}\left[\hat{f}(\omega+h)-\hat{f}(\omega)\right] = \lim_{h\to 0}\left[\int_{-\infty}^{\infty} e^{-i\omega t} f(t)\left(\frac{e^{-iht}-1}{h}\right)dt\right]. \tag{3.3.9}$$

Note that

$$\left|\frac{1}{h}\left(e^{-iht}-1\right)\right| = \frac{1}{|h|}\left|e^{-\frac{iht}{2}}\left(e^{\frac{iht}{2}}-e^{-\frac{iht}{2}}\right)\right| = 2\left|\frac{\sin\left(\frac{ht}{2}\right)}{h}\right| \le |t|.$$

Also,

$$\lim_{h\to 0}\left(\frac{e^{-iht}-1}{h}\right) = -it.$$

Thus, result (3.3.9) becomes

$$\frac{d\hat{f}}{d\omega} = \int_{-\infty}^{\infty} e^{-i\omega t} f(t)\lim_{h\to 0}\left(\frac{e^{-iht}-1}{h}\right)dt$$

$$= (-i)\int_{-\infty}^{\infty} t\,f(t)\,e^{-i\omega t}dt = (-i)\,\mathscr{F}\{t\,f(t)\}.$$

This proves the theorem.

Corollary 3.3.1 (The nth Derivative of $\hat{f}(\omega)$). If $f \in L^1(\mathbb{R})$ such that $t^n f(t)$ is integrable for finite $n \in \mathbb{N}$, then the nth derivative of $\hat{f}(\omega)$ exists and is given by

$$\frac{d^n \hat{f}}{d\omega^n} = (-i)^n \mathscr{F}\left\{t^n f(t)\right\}. \tag{3.3.10}$$

Proof. This corollary follows from Theorem 3.3.4 combined with the mathematical induction principle. In particular, putting $\omega = 0$ in (3.3.10) gives

$$\left[\frac{d^n \hat{f}(\omega)}{d\omega^n}\right]_{\omega=0} = (-i)^n \int_{-\infty}^{\infty} t^n f(t)\, dt = (-i)^n m_n, \tag{3.3.11}$$

where m_n represents the *nth moment* of $f(t)$. Thus, the moments $m_1, m_2, m_3, \cdots, m_n$ can be calculated from (3.3.11).

Theorem 3.3.5 (The Riemann-Lebesgue Lemma). If $f(t) \in L^1(\mathbb{R})$, then

$$\lim_{|\omega| \mapsto \infty} \left|\hat{f}(\omega)\right| = 0. \tag{3.3.12}$$

Proof. Since $e^{-i\omega t} = -\exp\left\{-i\omega\left(t + \frac{\pi}{\omega}\right)\right\}$, we have

$$\hat{f}(\omega) = -\int_{-\infty}^{\infty} \exp\left\{-i\omega\left(t + \frac{\pi}{\omega}\right)\right\} f(t)\, dt = -\int_{-\infty}^{\infty} e^{-i\omega x} f\left(x - \frac{\pi}{\omega}\right) dx.$$

Thus,

$$\begin{aligned}\hat{f}(\omega) &= \frac{1}{2}\left[\int_{-\infty}^{\infty} e^{-i\omega t} f(t)\, dt - \int_{-\infty}^{\infty} e^{-i\omega t} f\left(t - \frac{\pi}{\omega}\right) dt\right] \\ &= \frac{1}{2}\int_{-\infty}^{\infty} e^{-i\omega t}\left[f(t) - f\left(t - \frac{\pi}{\omega}\right)\right] dt.\end{aligned}$$

Clearly,

$$\lim_{|\omega| \to \infty} \left|\hat{f}(\omega)\right| \le \frac{1}{2} \lim_{|\omega| \to \infty} \int_{-\infty}^{\infty} \left| f(t) - f\left(t - \frac{\pi}{\omega}\right)\right| dt = 0.$$

This completes the proof.

Observe that the space $C_0(\mathbb{R})$ of all continuous functions on $\mathbb{R}$ which decay at infinity, that is, $f(t) \to 0$ as $|t| \to \infty$, is a normed space with respect to the norm defined by

$$\| f \| = \sup_{t \in \mathbb{R}} | f(t) |. \tag{3.3.13}$$

It follows from above theorems that the Fourier transform is a continuous linear operator from $L^1(\mathbb{R})$ into $C_0(\mathbb{R})$.

Theorem 3.3.6 (a) If $f(t)$ is a continuously differentiable function, $\lim_{|t| \to \infty} f(t) = 0$ and both $f,\ f' \in L^2(\mathbb{R})$, then

(a)
$$\mathscr{F}\{f'(t)\} = i\omega\, \mathscr{F}\{f(t)\} = (i\omega)\, \hat{f}(\omega). \tag{3.3.14}$$

(b) If $f(t)$ is continuously n-times differentiable, $f, f', \cdots, f^{(n)} \in L^1(\mathbb{R})$ and

$$\lim_{|t| \to \infty} f^{(r)}(t) = 0 \quad \text{for} \quad r = 0, 1, \ldots, n-1,$$

then

$$\mathscr{F}\left\{f^{(n)}(t)\right\} = (i\omega)^n\, \mathscr{F}\{f(t)\} = (i\omega)^n\, \hat{f}(\omega). \tag{3.3.15}$$

Proof. We have, by definition,

$$\mathscr{F}\{f'(t)\} = \int_{-\infty}^{\infty} e^{-i\omega t} f'(t)\, dt,$$

which is, integrating by parts,

$$= \left[e^{-i\omega t} f(t)\right]_{-\infty}^{\infty} + (i\omega) \int_{-\infty}^{\infty} e^{-i\omega t} f(t)\, dt$$

$$= (i\omega)\, \hat{f}(\omega).$$

This proves part (a) of the theorem.

A repeated application of (3.3.10) to higher-order derivatives gives result (3.3.15).

We next calculate the Fourier transform of partial derivatives. If $u(x,t)$ is continuously n times differentiable and $\dfrac{\partial^r u}{\partial x^r} \to 0$ as $|x| \to \infty$ for $r = 1, 2, 3, \ldots, (n-1)$, then, the Fourier transform of $\dfrac{\partial^n u}{\partial x^n}$ with respect to x is

$$\mathscr{F}\left\{\frac{\partial^n u}{\partial x^n}\right\} = (ik)^n\, \mathscr{F}\{u(x,t)\} = (ik)^n\, \hat{u}(k,t). \tag{3.3.16}$$

It also follows from the definition (3.2.3) that

$$\mathscr{F}\left\{\frac{\partial u}{\partial t}\right\}=\frac{d\hat{u}}{dt},\quad \mathscr{F}\left\{\frac{\partial^2 u}{\partial t^2}\right\}=\frac{d^2\hat{u}}{dt^2},\ \cdots,\ \mathscr{F}\left\{\frac{\partial^n u}{\partial t^n}\right\}=\frac{d^n\hat{u}}{dt^n}. \tag{3.3.17}$$

Definition 3.3.1 (Inverse Fourier Transform). If $f \in L^1(\mathbb{R})$ and its Fourier transform $\hat{f} \in L^1(\mathbb{R})$, then the inverse Fourier transform of $\hat{f}(\omega)$ is defined by

$$f(t)=\mathscr{F}^{-1}\left\{\hat{f}(\omega)\right\}=\frac{1}{2\pi}\int_{-\infty}^{\infty} e^{i\omega t}\hat{f}(\omega)\,d\omega \tag{3.3.18a}$$

for almost every $t \in \mathbb{R}$. If f is continuous, then (3.3.18) holds for every t. In general, f can be reconstructed from $\hat{f}$ at each point $t \in \mathbb{R}$, where f is continuous.

Using the polar form (3.2.14) of the Fourier spectrum $\hat{f}(\omega)$, the function (or signal) $f(t)$ can be expressed as

$$f(t)=\frac{1}{2\pi}\int_{-\infty}^{\infty} A(\omega)\exp\left[i\left\{\omega t+\theta(\omega)\right\}\right]d\omega, \tag{3.3.18b}$$

where $A(\omega)$ is the amplitude spectrum and $\theta(\omega)$ is the phase spectrum of the signal $f(t)$. This integral shows that the signal $f(t)$ is represented as a superposition of the infinite number of sinusoidal oscillations of infinitesimal amplitude $A(\omega)\,d\omega$ and of phase $\theta(\omega)$.

Physically, (3.3.18) implies that any signal $f(t)$ can be regarded as a superposition of an infinite number of sinusoidal oscillations with different frequencies $\omega = 2\pi\nu$ so that

$$f(t)=\int_{-\infty}^{\infty}\hat{f}(\nu)\,e^{2\pi i\nu t}d\nu. \tag{3.3.19a}$$

Equation (3.3.18a) or (3.3.19b) is called the *spectral resolution* of the signal f, and $\hat{f}$ is called the *spectral density* represented by

$$\hat{f}(\nu)=\int_{-\infty}^{\infty} e^{-2\pi i\nu t}f(t)\,dt. \tag{3.3.19b}$$

Thus, the symmetrical form (3.3.19ab) is often used as the alternative definition of the Fourier transform pair. This symmetry does not have a simple physical explanation in signal analysis. There seems to be no a priori reason for the symmetrical form of the waveform in the time domain and in the frequency domain. Mathematically, the symmetry seems to be associated with the fact that $\mathbb{R}$ is self-dual as a locally compact Abelian group. Physically, (3.3.19a) can be considered as the synthesis of a signal (or waveform) f from its individual components, whereas (3.3.19b) represents the resolution of the signal (or the waveform) into frequency components.

The *convolution* of two functions $f, g \in L^1(\mathbb{R})$ is defined by

$$(f * g)(t) = \int_{-\infty}^{\infty} f(t-\tau)\, g(\tau)\, d\tau. \tag{3.3.20}$$

We next prove the convolution theorem of the Fourier transform.

Theorem 3.3.7 (Convolution Theorem). If $f, g \in L^1(\mathbb{R})$, then

$$\mathscr{F}\{(f * g)(t)\} = \mathscr{F}\{f(t)\}\, \mathscr{F}\{g(t)\} = \hat{f}(\omega)\, \hat{g}(\omega). \tag{3.3.21}$$

Or, equivalently,

$$(f * g)(t) = \mathscr{F}^{-1}\left\{\hat{f}(\omega)\, \hat{g}(\omega)\right\}. \tag{3.3.22}$$

Or

$$\int_{-\infty}^{\infty} f(t-\tau)\, g(\tau)\, d\tau = \frac{1}{2\pi} \int_{-\infty}^{\infty} e^{i\omega t} \hat{f}(\omega)\, \hat{g}(\omega)\, d\omega. \tag{3.3.23}$$

Proof. Since $f * g \in L^1(\mathbb{R})$, we apply the definition of the Fourier transform to obtain

$$\begin{aligned}
\mathscr{F}\{(f * g)(t)\} &= \int_{-\infty}^{\infty} e^{-i\omega t} dt \int_{-\infty}^{\infty} f(t-\tau)\, g(\tau)\, d\tau \\
&= \int_{-\infty}^{\infty} g(\tau) \int_{-\infty}^{\infty} e^{-i\omega t} f(t-\tau)\, dt\, d\tau \\
&= \int_{-\infty}^{\infty} e^{-i\omega\tau} g(\tau)\, d\tau \int_{-\infty}^{\infty} e^{-i\omega u} f(u)\, du, \quad (t-\tau = u) \\
&= \hat{f}(\omega)\, \hat{g}(\omega),
\end{aligned}$$

in which Fubini's theorem was utilized.

Corollary 3.3.2 If $f, g, h \in L^1(\mathbb{R})$ such that

$$h(x) = \int_{-\infty}^{\infty} g(\omega)\, e^{i\omega x} d\omega, \tag{3.3.24}$$

then

$$(f * h)(x) = \int_{-\infty}^{\infty} g(\omega)\, \hat{f}(\omega)\, e^{i\omega x} d\omega.$$

Proof. We have

$$\begin{aligned}(f * h)(x) &= \int_{-\infty}^{\infty} h(x-t)\, f(t)\, dt \\ &= \int_{-\infty}^{\infty}\left[\int_{-\infty}^{\infty} g(\omega) e^{i(x-t)\omega} d\omega\right] f(t)\, dt \\ &= \int_{-\infty}^{\infty} g(\omega)\, e^{i\omega x} \hat{f}(\omega)\, d\omega.\end{aligned}$$

Example 3.2.3 shows that if $f \in L^1(\mathbb{R})$, it does not necessarily imply that its Fourier transform $\hat{f}$ also belongs to $L^1(\mathbb{R})$, so that the Fourier integral

$$\int_{-\infty}^{\infty} \hat{f}(\omega)\, e^{i\omega t} d\omega \tag{3.3.25}$$

may not exist as a Lebesgue integral. However, we can introduce a function $K_\lambda(\omega)$ in the integrand and formulate general conditions on $K_\lambda(\omega)$ and its Fourier transform so that the following result holds:

$$\lim_{\lambda\to\infty} \int_{-\lambda}^{\lambda} \hat{f}(\omega)\, K_\lambda(\omega)\, e^{i\omega t} d\omega = f(t) \tag{3.3.26}$$

for almost every t. This kernel $K_\lambda(\omega)$ is called a *convergent factor* or a *summability kernel* on $\mathbb{R}$ which can formally be defined as follows.

Definition 3.3.2 (Summability Kernel). A summability kernel on $\mathbb{R}$ is a family $\{K_\lambda,\ \lambda > 0\}$ of continuous functions with the following properties:

(i) $\int_{\mathbb{R}} K_\lambda(x)\,dx = 1$ for all $\lambda > 0$,

(ii) $\int_{\mathbb{R}} |K_\lambda(x)\,dx| \le M$ for all $\lambda > 0$ and for a constant M,

(iii) $\lim_{\lambda \to \infty} \int_{|x|>\delta} |K_\lambda(x)|\,dx = 0$ for all $\delta > 0$.

A simple construction of a summability on $\mathbb{R}$ is as follows. Suppose F is a continuous Lebesque integrable function so that

$$\int_{\mathbb{R}} F(x)\,dx = 1.$$

Then, we set

$$K_\lambda(x) = \lambda\, F(\lambda\, x), \qquad \text{for } \lambda > 0 \text{ and } x \in \mathbb{R}. \tag{3.3.27}$$

Evidently, it follows that

$$\int_{\mathbb{R}} K_\lambda(x)\,dx = \int_{\mathbb{R}} \lambda\, F(\lambda\, x)\,dx = \int_{\mathbb{R}} F(x)\,dx = 1,$$

$$\int_{\mathbb{R}} |K_\lambda(x)\,dx| = \int_{\mathbb{R}} |F(x)|\,dx = \|F\|_1,$$

and for $\delta > 0$

$$\int_{|x|>\delta} |K_\lambda(x)|\,dx = \int_{|x|>\delta} |F(x)|\,dx \to 0 \text{ as } \lambda \to \infty.$$

Obviously, the family $\{K_\lambda(x),\ \lambda > 0\}$ defined by (3.3.27) is a summability kernel on $\mathbb{R}$.

Example 3.3.2 **(*The Fejér Kernel*).** We may take (see Example 3.2.5) $\Delta_\lambda(x) = 1 - \left|\frac{x}{\lambda}\right|$ for $\left|\frac{x}{\lambda}\right| < 1$ and $\Delta_\lambda(x) = 0$ for $\left|\frac{x}{\lambda}\right| > 1$. This function is called the *Cesáro kernel*. Its Fourier transform represents a family

$$\hat{\Delta}_\lambda(\omega) = F_\lambda(\omega) = \lambda\; F(\lambda\,\omega), \tag{3.3.28}$$

where

$$F(x) = \frac{\sin^2\left(\dfrac{x}{2}\right)}{\left(\dfrac{x}{2}\right)^2} \tag{3.3.29}$$

is called the *Fejér kernel* on $\mathbb{R}$.

Example 3.3.3 (The Gaussian Kernel). The family of functions

$$G_\lambda(x) = \lambda\, G(\lambda x), \quad \lambda > 0, \tag{3.3.30}$$

where

$$G(x) = \frac{1}{\sqrt{\pi}} \exp\left(-x^2\right), \tag{3.3.31}$$

is called the *Gaussian kernel* on $\mathbb{R}$.

Example 3.2.1 shows that $\hat{G}_\lambda(\omega) = \exp\left(-\dfrac{\omega^2}{4\lambda^2}\right)$.

Lemma 3.3.1 For the Fejér kernel defined by (3.3.29), we have

$$F_\lambda(t) = \lambda\, F(\lambda t) = \lambda \frac{\sin^2\left(\dfrac{\lambda t}{2}\right)}{\left(\dfrac{\lambda t}{2}\right)^2} = \int_{-\lambda}^{\lambda} \left(1 - \frac{|\omega|}{\lambda}\right) e^{i\omega t} d\omega. \tag{3.3.32}$$

Proof. It follows that

$$\begin{aligned}
\int_{-\lambda}^{\lambda} \left(1 - \frac{|\omega|}{\lambda}\right) e^{i\omega t} d\omega &= \int_{-\lambda}^{0} \left(1 - \frac{|\omega|}{\lambda}\right) e^{i\omega t} d\omega + \int_{0}^{\lambda} \left(1 - \frac{\omega}{\lambda}\right) e^{i\omega t} d\omega \\
&= \int_{0}^{\lambda} \left(1 - \frac{\omega}{\lambda}\right) e^{-i\omega t} d\omega + \int_{0}^{\lambda} \left(1 - \frac{\omega}{\lambda}\right) e^{i\omega t} d\omega \\
&= 2\int_{0}^{\lambda} \cos(\omega t)\, d\omega - \frac{2}{\lambda} \int_{0}^{\lambda} \omega \cos(\omega t)\, d\omega \\
&= \frac{2}{\lambda t^2}\left(1 - \cos \lambda t\right) = \lambda \frac{\sin^2\left(\dfrac{\lambda t}{2}\right)}{\left(\dfrac{\lambda^2 t^2}{4}\right)} = F_\lambda(t).
\end{aligned}$$

This completes the proof of the lemma.

The idea of a summability kernel helps establish the so-called approximate identity theorem.

Theorem 3.3.8 **(*Approximate Identity Theorem*).** If $f \in L^1(\mathbb{R})$ and $\{K_\lambda, \lambda > 0\} \in L^1(\mathbb{R})$ is a summability kernel, then

$$\lim_{\lambda\to\infty} \left\| (f * K_\lambda) - f \right\| = 0. \tag{3.3.33a}$$

Or, equivalently,

$$\lim_{\lambda\to\infty} \left[(f * K_\lambda)(t) \right] = f(t). \tag{3.3.33b}$$

Proof. We have, by definition of the convolution (3.3.20),

$$(f * K_\lambda)(t) = \int_{-\infty}^{\infty} f(t-u)\, K_\lambda(u)\, du,$$

so that

$$\begin{aligned}
\left| (f * K_\lambda)(t) - f(t) \right| &= \left| \int_{-\infty}^{\infty} \{ f(t-u)\, K_\lambda(u)\, du - f(t) \} \right| \\
&= \left| \int_{-\infty}^{\infty} \{ f(t-u) - f(t) \}\, K_\lambda(u)\, du \right|, \text{ by definition 3.3.4(i),} \\
&\le \int_{-\infty}^{\infty} \left| K_\lambda(u) \right| \left| f(t-u) - f(t) \right| du.
\end{aligned}$$

Given $\varepsilon > 0$, we can choose $\delta > 0$ such that if $0 \le |u| < \delta$, then $|f(t-u) - f(t)| < \frac{\varepsilon}{M}$, where $\| K_\lambda \|_1 \le M$. Consequently,

$$\begin{aligned}
\left\| (f * K_\lambda)(t) - f(t) \right\| &= \int_{\mathbb{R}} \left| f * K_\lambda(t) - f(t) \right| dt \\
&\le \int_{\mathbb{R}} dt \int_{-\infty}^{\infty} \left| K_\lambda(u) \right| \left| f(t-u) - f(t) \right| du \\
&= \int_{-\infty}^{\infty} \left| K_\lambda(u) \right| \sigma_f(u)\, du,
\end{aligned}$$

where

$$\sigma_f(u) = \int_{\mathbb{R}} \left| f(t-u) - f(t) \right| dt \le C.$$

Thus,

$$\left\| \left(f * K_{\lambda}\right)(t) - f(t) \right\| \le \int_{|u|<\delta} \left| K_{\lambda}(u) \right| \sigma_f(u)\, du + \int_{|u|>\delta} \left| K_{\lambda}(u) \right| \sigma_f(u)\, du$$

$$\le \varepsilon + C \int_{|u|>\delta} \left| K_{\lambda}(u) \right| du = \varepsilon,$$

since the integral on the right-hand side tends to zero for $\delta > 0$ by definition 3.3.3 (iii). This completes the proof.

Corollary 3.3.3 If $f \in L^1(\mathbb{R})$ and $F_{\lambda}(t)$ is the Fejér kernel for $\lambda > 0$, then

$$\lim_{\lambda \to \infty} \left[\left(F_{\lambda} * f\right)(t) \right] = \lim_{\lambda \to \infty} \int_{-\lambda}^{\lambda} \left(1 - \frac{|\omega|}{\lambda}\right) \hat{f}(\omega)\, e^{i\omega t} d\omega = f(t). \qquad (3.3.34)$$

Proof. By Lemma 3.3.1, we have

$$F_{\lambda}(t) = \int_{-\lambda}^{\lambda} \left(1 - \frac{|\omega|}{\lambda}\right) e^{i\omega t} d\omega.$$

Then, by Corollary 3.3.2,

$$\left(F_{\lambda} * f\right)(t) = \int_{-\lambda}^{\lambda} \left(1 - \frac{|\omega|}{\lambda}\right) \hat{f}(\omega)\, e^{i\omega t} d\omega.$$

Taking the limit as $\lambda \to \infty$ and using result (3.3.33a) completes the proof.

Corollary 3.3.4 (Uniqueness). If $f \in L^1(\mathbb{R})$ such that $\hat{f}(\omega) = 0$, for all $\omega \in \mathbb{R}$, then $f = 0$ almost everywhere.

Proof. It follows from (3.3.34) that

$$f(t) = \lim_{\lambda \to \infty} \int_{-\lambda}^{\lambda} \left(1 - \frac{|\omega|}{\lambda}\right) \hat{f}(\omega)\, e^{i\omega t} d\omega = 0$$

almost everywhere. This completes the proof.

We now ask a question. Does there exist a function $g \in L^1(\mathbb{R})$ such that

$$f * g = f, \qquad (3.3.35)$$

where $f \in L^1(\mathbb{R})$?

If (3.3.35) is valid, then the convolution theorem 3.3.7 will give

$$\hat{f}(\omega)\,\hat{g}(\omega)=\hat{f}(\omega),$$

so that

$$\hat{g}(\omega)=1. \tag{3.3.36}$$

This contradicts the Riemann-Lebesgue lemma and $g \notin L^1(\mathbb{R})$. Therefore, the answer to the above question is negative. However, an approximation of the convolution identity (3.3.33) seems to be very useful in the theory of Fourier transforms. For example, we introduce a sequence $\{g_n(t)\} \in L^1(\mathbb{R})$ of ordinary functions such that

$$\hat{g}_n(\omega)\to 1 \text{ as } n\to\infty \text{ for all } \omega\in\mathbb{R}. \tag{3.3.37}$$

We normalize $\hat{g}_n$ by setting $\hat{g}_n(0)=1$. Or, equivalently,

$$\int_{-\infty}^{\infty} g_n(t)\,dt = 1. \tag{3.3.38}$$

A good example of such a sequence $g_n(t)$ is a sequence of Gaussian functions given by

$$g_n(t)=\sqrt{\frac{n}{\pi}}\,\exp\left(-nt^2\right), \qquad n=1,2,3,\ldots. \tag{3.3.39}$$

This sequence of functions was drawn in Figure 2.1 (see Chapter 2, p 36) for $n=1,2,3,\ldots$.

Its Fourier transform is given by

$$\hat{g}_n(\omega)=\exp\left(-\frac{\omega^2}{4n}\right). \tag{3.3.40}$$

Clearly, $\hat{g}_n(\omega)$ satisfies the conditions (3.3.37).

Even though the Dirac delta function $\delta \notin L^1(\mathbb{R})$, formally (2.4.1) represents $f*\delta=f$ which means that δ plays the role of an identity element in $L^1(\mathbb{R})$ space under the convolution operation. Also,

$$\hat{f}(\omega)\,\hat{\delta}(\omega)=\hat{f}(\omega),$$

so that

$$\hat{\delta}(\omega)=\mathscr{F}\{\delta(t)\}=1, \tag{3.3.41}$$

and, by Definition 3.3.1 of the inverse Fourier transform,

$$\delta(t) = \frac{1}{2\pi} \int_{-\infty}^{\infty} e^{i\omega t} d\omega . \tag{3.3.42}$$

This is an integral representation of the Dirac delta function extensively utilized in quantum mechanics.

Finally, results (3.3.41) and (3.3.42) can be used to carry out the following formal calculation to derive the inversion formula for the Fourier transform. Hence,

$$\begin{aligned} f(t) &= \int_{-\infty}^{\infty} f(x)\, \delta(t-x)\, dx \\ &= \int_{-\infty}^{\infty} f(x) \left[\frac{1}{2\pi} \int_{-\infty}^{\infty} \exp\left[i\omega (t-x) \right] d\omega \right] dx \\ &= \frac{1}{2\pi} \int_{-\infty}^{\infty} e^{i\omega t} \left[\int_{-\infty}^{\infty} f(x)\, e^{-i\omega x} dx \right] d\omega \\ &= \frac{1}{2\pi} \int_{-\infty}^{\infty} e^{i\omega t} \hat{f}(\omega)\, d\omega . \end{aligned}$$

Theorem 3.3.9 (General Modulation). If $\mathscr{F}\{f(t)\} = \hat{f}(\omega)$ and $\mathscr{F}\{g(t)\} = \hat{g}(\omega)$, where $\hat{f}$ and $\hat{g}$ belong to $L^1(\mathbb{R})$, then

$$\mathscr{F}\{f(t)\, g(t)\} = \frac{1}{2\pi} \left(\hat{f} * \hat{g} \right)(\omega) . \tag{3.3.43}$$

Or, equivalently,

$$\int_{-\infty}^{\infty} e^{-i\omega t} f(t)\, g(t)\, dt = \frac{1}{2\pi} \int_{-\infty}^{\infty} \hat{f}(x)\, \hat{g}(\omega - x)\, dx . \tag{3.3.44}$$

This can be regarded as the convolution theorem with respect to the frequency variable.

Proof. Using the inverse Fourier transform, we can rewrite the left-hand side of (3.3.43) as

$$\mathscr{F}\{f(t)\,g(t)\} = \int_{-\infty}^{\infty} e^{-i\omega t} f(t)\,g(t)\,dt$$

$$= \frac{1}{2\pi}\int_{-\infty}^{\infty} e^{-i\omega t} g(t)\,dt \int_{-\infty}^{\infty} e^{ixt}\hat{f}(x)\,dx$$

$$= \frac{1}{2\pi}\int_{-\infty}^{\infty} \hat{f}(x)\,dx \int_{-\infty}^{\infty} e^{-it(\omega-x)} g(t)\,dt$$

$$= \frac{1}{2\pi}\int_{-\infty}^{\infty} \hat{f}(x)\,\hat{g}(\omega - x)\,dx$$

$$= \frac{1}{2\pi}\left(\hat{f} * \hat{g}\right)(\omega).$$

This completes the proof.

In particular, if, result (3.3.43) reduces to the modulation property (3.3.5).

The definition of the Fourier transform shows that a sufficient condition for $f(t)$ to have a Fourier transform is that $f(t)$ is absolutely integrable in $-\infty < t < \infty$. This existence condition is too strong for many practical applications. Many simple functions, such as a constant function, $\sin \omega t$ and $t^n\,H(t)$, do not have Fourier transforms, even though they occur frequently in applications. The above definition of the Fourier transform has been extended for a more general class of functions to include the above and other functions. We simply state the fact that there is a sense, useful in practical applications, in which the above stated functions and many others do have Fourier transforms. The following are examples of such functions and their Fourier transforms.

$$\mathscr{F}\{H(a - |t|)\} = \left(\frac{\sin a\omega}{\omega}\right), \tag{3.3.45}$$

where $H(t)$ is the Heaviside unit step function.

$$\mathscr{F}\{\delta(t - a)\} = \exp(-ia\omega). \tag{3.3.46}$$

We have

$$\mathscr{F}\{H(t - a)\} = \pi\left[\frac{1}{i\pi\omega} + \delta(\omega)\right]. \tag{3.3.47}$$

Example 3.3.4 Use the definition to show that

$$\mathscr{F}\{e^{-at}H(t)\} = \frac{(a - i\omega)}{(a^2 + \omega^2)}, \quad a > 0. \tag{3.3.48}$$

We have, by definition,

$$\mathscr{F}\{e^{-at}H(t)\} = \int_0^\infty \exp\{-t(a + i\omega)\}\, dt = \frac{1}{a + i\omega}.$$

Example 3.3.5 Apply the definition to prove that

$$\mathscr{F}\{f_a(t)\} = \frac{-2i\omega}{(a^2 + \omega^2)}, \tag{3.3.49}$$

where $f_a(t) = e^{-at}\, H(t) - e^{at}\, H(-t)$. Hence, find the Fourier transform of $\operatorname{sgn}(t)$.

We have, by definition,

$$\mathscr{F}\{f_a(t)\} = \int_{-\infty}^{0} \exp\{(a - i\omega)t\}\, dt + \int_0^\infty \exp\{-(a + i\omega)t\}\, dt$$

$$= \frac{1}{(a + i\omega)} - \frac{1}{(a - i\omega)} = \frac{-2i\omega}{(a^2 + \omega^2)}.$$

In the limit as $a \to 0$, $f_a(t) \to \operatorname{sgn}(t)$, and hence,

$$\mathscr{F}\{\operatorname{sgn}(t)\} = \left(\frac{2}{i\omega}\right). \tag{3.3.50}$$

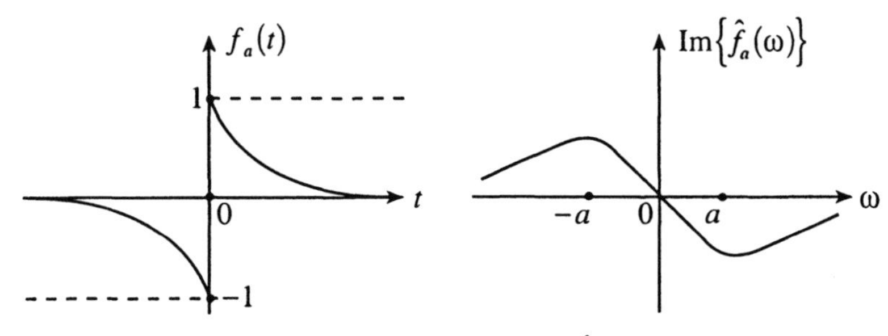

Figure 3.7. Graphs of $f_a(t)$ and $\hat{f}_a(\omega)$.

3.4 Fourier Transforms in $L^2(\mathbb{R})$

In this section, we discuss the extension of the Fourier transform onto $L^2(\mathbb{R})$. It turns out that if $f \in L^2(\mathbb{R})$, then the Fourier transform $\hat{f}$ of f is also in $L^2(\mathbb{R})$ and $\|\hat{f}\|_2 = \sqrt{2\pi}\,\|f\|_2$, where

$$\|f\|_2 = \left\{\int_{-\infty}^{\infty} |f(t)|^2\,dt\right\}^{\frac{1}{2}}. \tag{3.4.1}$$

The factor $\sqrt{2\pi}$ involved in the above result can be avoided by defining the Fourier transform as

$$\hat{f}(\omega) = \frac{1}{\sqrt{2\pi}} \int_{-\infty}^{\infty} e^{-i\omega t} f(t)\,dt. \tag{3.4.2}$$

We denote the norm in $L^2(\mathbb{R})$ by $\|\cdot\|_2$ and this norm is defined by

$$\|f\|_2 = \left\{\int_{-\infty}^{\infty} |f(t)|^2\,dt\right\}^{\frac{1}{2}}, \tag{3.4.3}$$

where $f \in L^2(\mathbb{R})$.

Theorem 3.4.1 Suppose f is a continuous function on $\mathbb{R}$ vanishing outside a bounded interval. Then , $\hat{f} \in L^2(\mathbb{R})$ and

$$\|f\|_2 = \|\hat{f}\|_2. \tag{3.4.4}$$

Proof. We assume that f vanishes outside the interval $[-\pi, \pi]$. We use the Parseval formula for the orthonormal sequence of functions on $[-\pi, \pi]$,

$$\phi_n(t) = \frac{1}{\sqrt{2\pi}} \exp(int), \quad n = 0, \pm 1, \pm, 2, \ldots,$$

to obtain

$$\|f\|_2^2 = \sum_{n=-\infty}^{\infty} \left| \frac{1}{\sqrt{2\pi}} \int_{-\infty}^{\infty} e^{-int} f(t)\,dt \right|^2 = \sum_{n=-\infty}^{\infty} |\hat{f}(n)|^2.$$

Since this result also holds for $g(t)=e^{-ixt}f(t)$ instead of $f(t)$, and $\|f\|_2^2=\|g\|_2^2$, then, $\|f\|_2^2=\sum_{n=-\infty}^{\infty}\left|\hat{f}(n+x)\right|^2$.

Integrating this result with respect to x from 0 to 1 gives

$$\|f\|_2^2=\sum_{n=-\infty}^{\infty}\int_0^1\left|\hat{f}(n+x)\right|^2dx=\sum_{n=-\infty}^{\infty}\int_n^{n+1}\left|\hat{f}(y)\right|^2dy,\ (y=n+x)$$

$$=\int_{-\infty}^{\infty}\left|\hat{f}(y)\right|^2dy=\left\|\hat{f}\right\|_2^2.$$

If f does not vanish outside $[-\pi,\pi]$, then we take a positive number a for which the function $g(t)=f(at)$ vanishes outside $[-\pi,\pi]$. Then,

$$\hat{g}(\omega)=\frac{1}{a}\hat{f}\left(\frac{\omega}{a}\right).$$

Thus, it turns out that

$$\|f\|_2^2=a\|g\|_2^2=a\int_{-\infty}^{\infty}\left|\frac{1}{a}\hat{f}\left(\frac{\omega}{a}\right)\right|^2d\omega=\int_{-\infty}^{\infty}\left|\hat{f}(\omega)\right|^2d\omega=\left\|\hat{f}\right\|_2^2.$$

This completes the proof.

The space of all continuous functions on $\mathbb{R}$ with compact support is dense in $L^2(\mathbb{R})$. Theorem 3.4.1 shows that the Fourier transform is a continuous mapping from that space into $L^2(\mathbb{R})$. Since the mapping is linear, it has a unique extension to a linear mapping from $L^2(\mathbb{R})$ into itself. This extension will be called the Fourier transform on $L^2(\mathbb{R})$.

Definition 3.4.1 (Fourier Transform in $L^2(\mathbb{R})$). If $f\in L^2(\mathbb{R})$ and $\{\phi_n\}$ is a sequence of continuous functions with compact support convergent to f in $L^2(\mathbb{R})$, that is, $\|f-\phi_n\|\to 0$ as $n\to\infty$, then the Fourier transform of f is defined by

$$\hat{f}=\lim_{n\to\infty}\phi_n, \tag{3.4.5}$$

where the limit is taken with respect to the norm in $L^2(\mathbb{R})$.

Theorem 3.4.1 ensures that the limit exists and is independent of a particular sequence approximating f. It is important to note that the convergence in $L^2(\mathbb{R})$

does not imply pointwise convergence, and therefore the Fourier transform of a square integrable function is not defined at a point, unlike the Fourier transform of an integrable function. We can assert that the Fourier transform $\hat{f}$ of $f \in L^2(\mathbb{R})$ is defined almost everywhere on $\mathbb{R}$ and $\hat{f} \in L^2(\mathbb{R})$. For this reason, we cannot say that, if $f \in L^1(\mathbb{R}) \cap L^2(\mathbb{R})$, the Fourier transform defined by (3.2.3) and the one defined by (3.4.5) are equal. To be more precise, we should state that the transform defined by (3.2.3) belongs to the equivalence class of square integrable functions defined by (3.4.5). In spite of this difference, we shall use the same symbol to denote both transforms.

An immediate consequence of Definition 3.3.2 and Theorem 3.4.1 leads to the following theorem.

Theorem 3.4.2 (Parseval's Relation). If $f \in L^2(\mathbb{R})$, then

$$\| f \|_2 = \| \hat{f} \|_2. \tag{3.4.6}$$

In physical problems, the quantity $\| f \|_2$ is a measure of energy, and $\| \hat{f} \|_2$ represents the power spectrum of a signal f. More precisely, the total energy of a signal (or waveform) is defined by

$$E = \|f\|_2^2 = \int_{-\infty}^{\infty} |f(t)|^2 \, dt. \tag{3.4.7}$$

Theorem 3.4.3 If $f \in L^2(\mathbb{R})$, then

$$\hat{f}(\omega) = \lim_{n \to \infty} \frac{1}{\sqrt{2\pi}} \int_{-n}^{n} e^{-i\omega t} f(t) \, dt, \tag{3.4.8}$$

where the convergence is with respect to the L^2-norm.

Proof. For $n = 1, 2, 3, \ldots$, we define

$$f_n(t) = \begin{Bmatrix} f(t), & \text{for } |t| < n \\ 0, & \text{for } |t| \geq n \end{Bmatrix}. \tag{3.4.9}$$

Clearly, $\| f - f_n \|_2 \to 0$ and hence, $\| \hat{f} - \hat{f}_n \|_2 \to 0$ as $n \to \infty$.

Theorem 3.4.4 If $f, g \in L^2(\mathbb{R})$, then

$$\left(f, \overline{\hat{g}}\right) = \int_{-\infty}^{\infty} f(t)\, \hat{g}(t)\, dt = \int_{-\infty}^{\infty} \hat{f}(t)\, g(t)\, dt = \left(\hat{f}, \overline{g}\right). \tag{3.4.10}$$

Proof. We define both $f_n(t)$ and $g_n(t)$ by (3.4.9) for $n = 1, 2, 3, \ldots$.

Since

$$\hat{f}_m(t) = \frac{1}{\sqrt{2\pi}} \int_{-\infty}^{\infty} e^{-ixt} f_m(x)\, dx,$$

we obtain

$$\int_{-\infty}^{\infty} \hat{f}_m(t)\, g_n(t)\, dt = \frac{1}{\sqrt{2\pi}} \int_{-\infty}^{\infty} g_n(t) \int_{-\infty}^{\infty} e^{-ixt} f_m(x)\, dx\, dt.$$

The function $\exp(-i\,x\,t)\, g_n(t)\, f_m(x)$ is integrable over $\mathbb{R}^2$ and hence, the Fubini Theorem can be used to rewrite the above integral in the form

$$\int_{-\infty}^{\infty} \hat{f}_m(t)\, g_n(t)\, dt = \frac{1}{\sqrt{2\pi}} \int_{-\infty}^{\infty} f_m(x) \int_{-\infty}^{\infty} e^{-ixt} g_n(t)\, dt\, dx$$

$$= \int_{-\infty}^{\infty} f_m(x)\, \hat{g}_n(x)\, dx.$$

Since $\| g - g_n \|_2 \to 0$ and $\| \hat{g} - \hat{g}_n \|_2 \to 0$, letting $n \to \infty$ combined with the continuity of the inner product yields

$$\int_{-\infty}^{\infty} \hat{f}_m(t)\, g(t)\, dt = \int_{-\infty}^{\infty} f_m(t)\, \hat{g}(t)\, dt.$$

Similarly, letting $m \to 0$ gives the desired result (3.4.10).

Definition 3.4.2 (Autocorrelation Function). The autocorrelation function of a signal $f \in L^2(\mathbb{R})$ is defined by

$$F(t) = \int_{-\infty}^{\infty} f(t + \tau)\, \overline{f(\tau)}\, d\tau. \tag{3.4.11}$$

In view of the Schwarz inequality, the integrand in (3.4.11) belongs to $L^1(\mathbb{R})$, so $F(t)$ is finite for each $t \in \mathbb{R}$. The *normalized autocorrelation* function is defined by

$$\gamma(t) = \frac{\int_{-\infty}^{\infty} f(t+\tau)\,\overline{f(\tau)}\,d\tau}{\int_{-\infty}^{\infty} f(\tau)\,\overline{f(\tau)}\,d\tau} = \frac{F(t)}{F(0)}. \tag{3.4.12}$$

The Fourier transform of $F(t)$ is

$$\hat{F}(\omega) = \left|\hat{f}(\omega)\right|^2. \tag{3.4.13}$$

This can be verified as follows:

$$\begin{aligned}
F(t) &= \int_{-\infty}^{\infty} f(t+\tau)\,\overline{f(\tau)}\,d\tau \\
&= \int_{-\infty}^{\infty} f(t-u)\,\overline{f(-u)}\,du, \quad (\tau = -u) \\
&= \int_{-\infty}^{\infty} f(t-u)\,g(u)\,du, \quad \left(g(u) = \overline{f(-u)}\right) \\
&= f(t) * g(t) \\
&= \mathscr{F}^{-1}\left\{\hat{f}(\omega)\,\hat{g}(\omega)\right\} \\
&= \frac{1}{\sqrt{2\pi}} \int_{-\infty}^{\infty} \hat{f}(\omega)\,\hat{g}(\omega)\,e^{i\omega t} d\omega \\
&= \frac{1}{\sqrt{2\pi}} \int_{-\infty}^{\infty} \hat{f}(\omega)\,\overline{\hat{f}(\omega)}\,e^{i\omega t} d\omega, \quad \hat{g}(\omega) = \overline{\hat{f}(\omega)} \qquad \text{by } (3.3.4) \\
&= \mathscr{F}^{-1}\left\{\left|\hat{f}(\omega)\right|^2\right\}.
\end{aligned}$$

This leads to result (3.4.13).

Lemma 3.4.1 If $f \in L^2(\mathbb{R})$ and $g = \overline{\hat{f}}$, then $f = \overline{\hat{g}}$.

Proof. In view of Theorems 3.4.2 and 3.4.4 and the assumption $g = \hat{\bar{f}}$, we find

$$\left(f, \bar{\hat{g}}\right) = \left(\hat{f}, \bar{g}\right) = \left(\hat{f}, \hat{f}\right) = \left\| \hat{f} \right\|_2^2 = \| f \|_2^2. \tag{3.4.14}$$

Also, we have

$$\overline{\left(f, \bar{\hat{g}}\right)} = \overline{\left(\hat{f}, \hat{f}\right)} = \| f \|_2^2. \tag{3.4.15}$$

Finally, by Parseval's relation,

$$\| \hat{g} \|_2^2 = \| g \|_2^2 = \left\| \hat{f} \right\|_2^2 = \| f \|_2^2. \tag{3.4.16}$$

Using (3.4.14)-(3.4.16) gives the following

$$\left\| f - \bar{\hat{g}} \right\|_2^2 = \left(f - \bar{\hat{g}}, f - \bar{\hat{g}}\right) = \| f \|_2^2 - \left(f, \bar{\hat{g}}\right) - \overline{\left(f, \bar{\hat{g}}\right)} + \| \hat{g} \|_2^2 = 0.$$

This proves the result $f = \bar{\hat{g}}$.

Example 3.4.1 (The Haar Function). The Haar function is defined by

$$f(t) = \left\{ \begin{array}{cc} 1, & \text{for } 0 \le t < \frac{1}{2}, \\ -1, & \text{for } \frac{1}{2} \le t < 1, \\ 0, & \text{otherwise} \end{array} \right\}. \tag{3.4.17}$$

Evidently,

$$\hat{f}(\omega) = \int_{-\infty}^{\infty} e^{-i\omega t} f(t)\, dt = \left[\int_0^{\frac{1}{2}} e^{-i\omega t} dt - \int_{\frac{1}{2}}^{1} e^{-i\omega t} dt \right]$$

$$= \frac{1}{i\omega} \left(1 - 2e^{-\frac{i\omega}{2}} + e^{-i\omega} \right)$$

$$= \frac{\exp\left(-\frac{i\omega}{2}\right)}{(i\omega)} \left(e^{\frac{i\omega}{2}} - 2 + e^{-\frac{i\omega}{2}}\right)$$

$$= i \exp\left(-\frac{i\omega}{2}\right) \frac{\sin^2\left(\frac{\omega}{4}\right)}{\frac{\omega}{4}}. \tag{3.4.18}$$

The graphs of $f(t)$ and $\hat{f}(\omega)$ are shown in Figure 3.8.

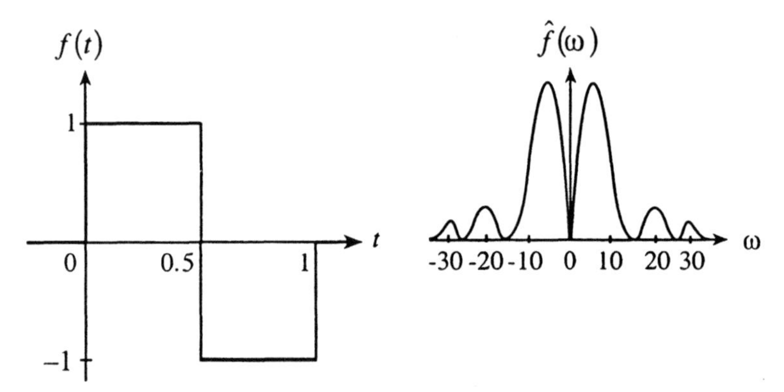

Figure 3.8. The graphs of $f(t)$ and $\hat{f}(\omega)$.

Example 3.4.2 (The Second Derivative of the Gaussian Function). If

$$f(t) = \left(1 - t^2\right) \exp\left(-\frac{1}{2} t^2\right), \tag{3.4.19}$$

then

$$\hat{f}(\omega) = \omega^2 \exp\left(-\frac{1}{2} \omega^2\right). \tag{3.4.20}$$

We have

$$\hat{f}(\omega) = \mathscr{F}\left\{\left(1 - t^2\right) \exp\left(-\frac{1}{2} t^2\right)\right\} = -\mathscr{F}\left\{\frac{d^2}{dt^2} \exp\left(-\frac{1}{2} t^2\right)\right\}$$

$$= -(i\omega)^2 \mathscr{F}\left\{\exp\left(-\frac{1}{2} t^2\right)\right\} = \omega^2 \exp\left(-\frac{1}{2} \omega^2\right).$$

Both $f(t)$ and $\hat{f}(\omega)$ are plotted in Figure 3.9.

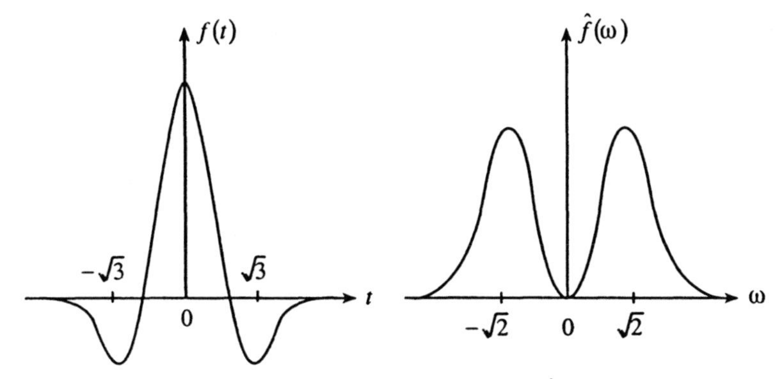

Figure 3.9. Graphs of $f(t)$ and $\hat{f}(\omega)$.

Example 3.4.3 (The Shannon Function). The Shannon function $f(t)$ is defined by

$$f(t)=\frac{1}{\pi t}(\sin 2\pi t-\sin \pi t)=\left(\frac{2}{\pi t}\right)\sin\left(\frac{\pi t}{2}\right)\cos\left(\frac{3\pi t}{2}\right). \qquad (3.4.21)$$

Its Fourier transform $\hat{f}(\omega)$ is given by

$$\hat{f}(\omega)=\begin{Bmatrix}1, & \text{for } \pi<|\omega|<2\pi, \\ 0, & \text{otherwise}\end{Bmatrix}. \qquad (3.4.22)$$

Both the Shannon function $f(t)$ and its Fourier transform $\hat{f}(\omega)$ are shown in Figure 3.10.

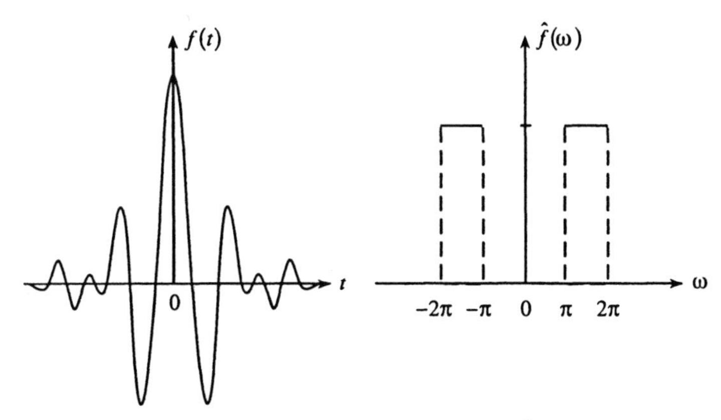

Figure 3.10. Graphs of $f(t)$ and $\hat{f}(\omega)$.

Example 3.4.4 (Fourier Transform of Hermite Functions). The Hermite functions are given by

$$h_n(x) = \frac{1}{n!} \exp\left(-\frac{x^2}{2}\right) H_n(x), \tag{3.4.23}$$

where $H_n(x)$ is the Hermite polynomial of degree n defined by

$$H_n(x) = (-1)^n \exp\left(x^2\right)\left(\frac{d}{dx}\right)^n \exp\left(-x^2\right). \tag{3.4.24}$$

Then,

$$\hat{h}_n(\omega) = (-i)^n \, h_n(\omega), \tag{3.4.25}$$

where the Fourier transform is defined by (3.4.2).

We have

$$h_0(x) = \exp\left(-\frac{x^2}{2}\right),$$

then, by (3.2.5),

$$\hat{h}_0(\omega) = \exp\left(-\frac{\omega^2}{2}\right) = h_0(\omega).$$

Differentiating (3.4.23) with respect to x gives

$$\begin{aligned}
h_n'(x) &= -x\, h_n(x) + \frac{1}{n!} \exp\left(-\frac{x^2}{2}\right) H_n'(x) \\
&= -x\, h_n(x) + \frac{1}{n!} \exp\left(-\frac{x^2}{2}\right)(-1)^n \left[2x \exp\left(x^2\right)\left(\frac{d}{dx}\right)^n \exp\left(-\frac{x^2}{2}\right)\right. \\
&\qquad\qquad \left. + \exp\left(x^2\right)\left(\frac{d}{dx}\right)^{n+1} \exp\left(-x^2\right)\right] \\
&= -x\, h_n(x) + 2x\, h_n(x) - \frac{(n+1)(-1)^{n+1}}{(n+1)\,n!} \cdot \exp\left(x^2\right)\left(\frac{d}{dx}\right)^{n+1} \left\{\exp\left(-x^2\right)\right\} \\
&= x\, h_n(x) - (n+1)\, h_{n+1}(x).
\end{aligned} \tag{3.4.26}$$

Using (3.3.9) and the fact that

$$\frac{d}{d\omega} \hat{f}(\omega) = (-i)\, \hat{g}(\omega), \quad \text{where } g(t) = t\, f(t),$$

the Fourier transform of result (3.4.26) is

$$(i\omega)\,\hat{h}_n(\omega) = i\frac{d}{d\omega}\,\hat{h}_n(\omega) - (n+1)\,\hat{h}_{n+1}(\omega),$$

so that

$$\hat{h}_{n+1}(\omega) = -\frac{1}{(n+1)}\left[i\omega\,\hat{h}_n(\omega) - i\frac{d}{d\omega}\,\hat{h}_n(\omega)\right].$$

If result (3.4.21) is true for n, then, for $n+1$, we obtain

$$\begin{aligned}
\hat{h}_{n+1}(\omega) &= -\frac{1}{(n+1)}\left[-(-i)^{n+1}\sqrt{2\pi}\;\omega\,h_n(\omega) + (-i)^{n+1}\sqrt{2\pi}\;h_n'(\omega)\right] \\
&= (-i)^{n+1}\left[-\frac{1}{(n+1)}\left\{-\omega\,h_n(\omega) + h_n'(\omega)\right\}\right] \\
&= (-i)^{n+1}\;h_{n+1}(\omega).
\end{aligned}$$

Thus, result (3.4.25) is true for all n.

Theorem 3.4.5 If $f \in L^2(\mathbb{R})$, then

(i) $$(F_\lambda * f)(t) = \int_{-\infty}^{\infty} \phi(x)\,\hat{f}(x)\,dx, \tag{3.4.27}$$

(ii) $$\lim_{\lambda\to\infty}(F_\lambda * f)(t) = f(t), \tag{3.4.28}$$

where

$$(F_\lambda * f)(t) = \int_{-\lambda}^{\lambda}\left(1 - \frac{|x|}{\lambda}\right)\hat{f}(x)\,e^{ixt}dt, \tag{3.4.29}$$

and

$$\phi(x) = \Delta_\lambda(x)\,e^{ixt}. \tag{3.4.30}$$

Proof. We have, by (3.3.30),

$$\hat{\phi}(\omega) = \int_{-\infty}^{\infty}\Delta_\lambda(x)\,\exp\{ix(\omega - t)\}\,dx = F_\lambda(\omega - t).$$

Then,

$$
\begin{aligned}
(F_\lambda * f)(t) &= \int_{-\infty}^{\infty} F_\lambda(t-x)\, f(x)\, dx \\
&= \int_{-\infty}^{\infty} F_\lambda(x-t)\, f(x)\, dx \\
&= \int_{-\infty}^{\infty} \hat{\phi}(x)\, f(x)\, dx \\
&= \int_{-\infty}^{\infty} \phi(x)\, \hat{f}(x)\, dx, \qquad \text{by (3.4.10).}
\end{aligned}
$$

An argument similar to that of the Approximate Identity Theorem 3.3.8 gives (ii).

Theorem 3.4.6 (Inversion of Fourier Transforms in $L^2(\mathbb{R})$). If $f \in L^2(\mathbb{R})$, then

$$
f(t) = \lim_{n\to\infty} \frac{1}{2\pi} \int_{-n}^{n} e^{i\omega t} \hat{f}(\omega)\, d\omega, \tag{3.4.31}
$$

where the convergence is with respect to the $L^2(\mathbb{R})$-norm.

Proof. If $f \in L^2(\mathbb{R})$, and $g = \bar{\hat{f}}$, then, by Lemma 3.4.1,

$$
\begin{aligned}
f(t) = \overline{\hat{g}(t)} &= \overline{\lim_{n\to\infty} \frac{1}{2\pi} \int_{-n}^{n} e^{-i\omega t} g(\omega)\, d\omega} \\
&= \lim_{n\to\infty} \frac{1}{2\pi} \int_{-n}^{n} \overline{e^{-i\omega t} g(\omega)}\, d\omega \\
&= \lim_{n\to\infty} \frac{1}{2\pi} \int_{-n}^{n} e^{i\omega t}\, \overline{g(\omega)}\, d\omega \\
&= \lim_{n\to\infty} \frac{1}{2\pi} \int_{-n}^{n} e^{i\omega t} \hat{f}(\omega)\, d\omega.
\end{aligned}
$$

Corollary 3.4.1 If $f \in L^1(\mathbb{R}) \cap L^2(\mathbb{R})$, then the following formula

$$
f(t) = \frac{1}{2\pi} \int_{-n}^{n} e^{i\omega t} \hat{f}(\omega)\, d\omega \tag{3.4.32}
$$

holds almost everywhere in $\mathbb{R}$.

The formula (3.4.31) is called the *inverse Fourier transform.* If we use the factor $\left(1/\sqrt{2\pi}\right)$ in the definition of the Fourier transform, then the Fourier transform and its inverse are symmetrical in form, that is,

$$\hat{f}(\omega)=\frac{1}{\sqrt{2\pi}}\int_{-\infty}^{\infty} e^{-i\omega t} f(t)\, dt, \quad f(t)=\frac{1}{\sqrt{2\pi}}\int_{-\infty}^{\infty} e^{i\omega t}\hat{f}(\omega)\, d\omega. \qquad (3.4.33\text{a,b})$$

Theorem 3.4.7 (General Parseval's Relation). If $f, g \in L^2(\mathbb{R})$, then

$$(f,g)=\int_{-\infty}^{\infty} f(t)\,\overline{g(t)}\, dt = \int_{-\infty}^{\infty} \hat{f}(\omega)\,\overline{\hat{g}(\omega)}\, d\omega = \left(\hat{f},\hat{g}\right), \qquad (3.4.34)$$

where the symmetrical form (3.4.33a,b) of the Fourier transform and its inverse is used.

Proof. It follows from (3.4.4) that

$$\|f+g\|_2^2 = \left\|\hat{f}+\hat{g}\right\|_2^2.$$

Or, equivalently,

$$\int_{-\infty}^{\infty} |f+g|^2\, dt = \int \left|\hat{f}+\hat{g}\right|^2 d\omega,$$

$$\int_{-\infty}^{\infty} (f+g)\left(\bar{f}+\bar{g}\right) dt = \int \left(\hat{f}+\hat{g}\right)\left(\bar{\hat{f}}+\bar{\hat{g}}\right) d\omega.$$

Simplifying both sides gives

$$\int_{-\infty}^{\infty} |f|^2\, dt + \int_{-\infty}^{\infty}\left(f\bar{g}+g\bar{f}\right) dt + \int_{-\infty}^{\infty} |g|^2\, dt$$

$$= \int_{-\infty}^{\infty}\left|\hat{f}\right|^2 d\omega + \int_{-\infty}^{\infty}\left(\hat{f}\bar{\hat{g}}+\hat{g}\bar{\hat{f}}\right) d\omega + \int_{-\infty}^{\infty} |\hat{g}|^2\, d\omega.$$

Applying (3.4.4) to the above identity leads to

$$\int_{-\infty}^{\infty}\left(f\bar{g}+g\bar{f}\right) dt = \int_{-\infty}^{\infty}\left(\hat{f}\bar{\hat{g}}+\hat{g}\bar{\hat{f}}\right) d\omega. \qquad (3.4.35)$$

Since g is an arbitrary element of $L^2(\mathbb{R})$, we can replace $g, \hat{g}$ by $ig, i\hat{g}$ respectively, in (3.4.35) to obtain

$$\int_{-\infty}^{\infty}\left[f\left(\overline{ig}\right)+(ig)\,\bar{f}\right]dt=\int_{-\infty}^{\infty}\left[\hat{f}\left(\overline{i\hat{g}}\right)+(i\hat{g})\,\bar{\hat{f}}\right]d\omega.$$

Or

$$-i\int_{-\infty}^{\infty} f\,\bar{g}\,dt+i\int_{-\infty}^{\infty} g\,\bar{f}\,dt=-i\int_{-\infty}^{\infty}\hat{f}\,\bar{\hat{g}}\,d\omega+i\int_{-\infty}^{\infty}\hat{g}\,\bar{\hat{f}}\,d\omega,$$

which is, multiplying by i,

$$\int_{-\infty}^{\infty} f\,\bar{g}\,dt-\int_{-\infty}^{\infty} g\,\bar{f}\,dt=\int_{-\infty}^{\infty}\hat{f}\,\bar{\hat{g}}\,d\omega-\int_{-\infty}^{\infty}\hat{g}\,\bar{\hat{f}}\,d\omega. \tag{3.4.36}$$

Adding (3.4.35) and (3.4.36) gives

$$\int_{-\infty}^{\infty} f(t)\,\overline{g(t)}\,dt=\int_{-\infty}^{\infty}\hat{f}(\omega)\,\overline{\hat{g}(\omega)}\,d\omega.$$

This completes the proof.

Note 1. If $g=f$ in the above result, we retrieve result (3.4.4).

Note 2. A formal calculation easily establishes the result (3.4.34) as follows. The right-hand side of (3.4.34) is

$$\begin{aligned}\int_{-\infty}^{\infty}\hat{f}(\omega)\,\overline{\hat{g}(\omega)}\,d\omega &=\frac{1}{\sqrt{2\pi}}\int_{-\infty}^{\infty} f(t)\,e^{-i\omega t}\,\overline{\hat{g}(\omega)}\,d\omega\,dt\\ &=\int_{-\infty}^{\infty} f(t)\,\frac{1}{\sqrt{2\pi}}\,\overline{\int_{-\infty}^{\infty} e^{i\omega t}\,\hat{g}(\omega)\,d\omega}\,dt\\ &=\int_{-\infty}^{\infty} f(t)\,\overline{g(t)}\,dt.\end{aligned}$$

Note 3. If the Fourier transform pair is defined by (3.2.1) and (3.3.18), then the general Parseval formula (3.3.4) reads

$$(f,g)=\frac{1}{2\pi}\left(\hat{f},\hat{g}\right). \tag{3.4.37}$$

The following theorem summarizes the major results of this section. It is known as the Plancherel theorem.

Theorem 3.4.8 (Plancherel's Theorem). For every $f \in L^2(\mathbb{R})$, there exists $\hat{f} \in L^2(\mathbb{R})$ such that

(i) If $f \in L^1(\mathbb{R}) \cap L^2(\mathbb{R})$, then $\hat{f}(\omega) = \frac{1}{\sqrt{2\pi}} \int_{-\infty}^{\infty} e^{-i\omega t} f(t)\, dt$.

(ii) $$\left\| \hat{f}(\omega) - \frac{1}{\sqrt{2\pi}} \int_{-n}^{n} e^{-i\omega t} f(t)\, dt \right\|_2 \to 0 \text{ as } n \to \infty, \text{ and}$$

$$\left\| f(t) - \frac{1}{\sqrt{2\pi}} \int_{-n}^{n} e^{i\omega t} \hat{f}(\omega)\, d\omega \right\|_2 \to 0 \text{ as } n \to \infty.$$

(iii) $(f, g) = (\hat{f}, \hat{g})$,

(iv) $\|f\|_2 = \|\hat{f}\|_2$,

(v) The mapping $f \to \hat{f}$ is a Hilbert space isomorphism of $L^2(\mathbb{R})$ onto $L^2(\mathbb{R})$.

Proof. All parts of this theorem have been proved except the fact that the Fourier transform is "onto". If $f \in L^2(\mathbb{R})$, then we define

$$h = \bar{f} \quad \text{and} \quad g = \bar{\hat{h}}.$$

Then, by Lemma 3.4.1, we get

$$\bar{f} = h = \bar{\hat{g}}$$

and hence,

$$f = \hat{g}.$$

This ensures that every square integrable function is the Fourier transform of a square integrable function.

Theorem 3.4.9 The Fourier transform is a unitary operator on $L^2(\mathbb{R})$.

Proof. We first observe that

$$\hat{g}(\omega) = \frac{1}{\sqrt{2\pi}} \int_{-\infty}^{\infty} e^{-i\omega t} \overline{g(t)}\, dt = \frac{1}{\sqrt{2\pi}} \overline{\int_{-\infty}^{\infty} e^{i\omega t} g(t)\, dt} = \overline{\mathscr{F}^{-1}\{g(t)\}(\omega)}.$$

Using Theorem 3.4.4, we obtain

$$\left(\mathscr{F}\{f(t)\}, g\right) = \frac{1}{\sqrt{2\pi}} \int_{-\infty}^{\infty} \hat{f}(\omega)\, \overline{g(\omega)}\, d\omega = \frac{1}{\sqrt{2\pi}} \int_{-\infty}^{\infty} f(\omega)\, \hat{g}(\omega)\, d\omega$$

$$= \frac{1}{\sqrt{2\pi}} \int_{-\infty}^{\infty} f(\omega)\, \overline{\mathscr{F}^{-1}\{g(t)\}(\omega)}\, d\omega = \left(f, \mathscr{F}^{-1} g\right).$$

This proves that $\mathscr{F}^{-1} = \mathscr{F}^{*}$, and hence the Fourier transform operator is unitary.

Theorem 3.4.10 (Duality). If $\hat{f}(\omega) = \mathscr{F}\{f(t)\}$, then

$$\mathscr{F}\left\{\hat{f}(\omega)\right\} = f(-\omega). \tag{3.4.38}$$

Proof. We have, by definition (3.4.33a),

$$\mathscr{F}\left\{\hat{f}(\omega)\right\} = \frac{1}{\sqrt{2\pi}} \int_{-\infty}^{\infty} e^{-i\omega t}\, \hat{f}(\omega)\, d\omega$$

$$= \frac{1}{\sqrt{2\pi}} \int_{-\infty}^{\infty} e^{-i\omega t}\, \hat{f}(t)\, dt$$

$= f(-\omega)$, by the inversion formula (3.4.33b).

Or, equivalently,

$$\mathscr{F}\left\{\hat{f}(-\omega)\right\} = \hat{f}^{\wedge}(-\omega) = f(\omega).$$

Similarly,

$$\hat{f}^{\wedge\wedge}(\omega) = \mathscr{F}\left\{\hat{f}^{\wedge}(\omega)\right\} = \frac{1}{\sqrt{2\pi}} \int_{-\infty}^{\infty} e^{-i\omega t} f(-\omega)\, d\omega$$

$$= \frac{1}{\sqrt{2\pi}} \int_{-\infty}^{\infty} e^{i\sigma t} f(\sigma)\, d\sigma, \quad (\omega = -\sigma)$$

$$= \frac{1}{\sqrt{2\pi}} \int_{-\infty}^{\infty} e^{i\omega t} f(t)\, dt$$

$$= \hat{f}(-\omega).$$

Finally, it turns out that

$$\hat{f}^{\wedge\wedge\wedge}(\omega) = \mathscr{F}\left\{\hat{f}(-\omega)\right\} = f(\omega). \tag{3.4.39}$$

Corollary 3.4.2 (Eigenvalues and Eigenfunctions of the Fourier Transform). The eigenvalues of the operator $\mathscr{F}$ are $\lambda = 1, i, -1, -i$.

Proof. Consider the eigenvalue problem

$$\mathscr{F} f = \lambda f.$$

We have $\hat{f} = \lambda f$, $\hat{f}^{\wedge} = \lambda \hat{f} = \lambda^2 f$, $\hat{f}^{\wedge\wedge} = \lambda^2 \hat{f} = \lambda^3 f$. It follows from (3.4.39) that

$$f = \hat{f}^{\wedge\wedge\wedge} = \lambda^4 f.$$

Consequently, $\lambda^4 = 1$ giving the four eigenvalues $\pm 1, \pm i$ of $\mathscr{F}$.

It has already been shown in Example 3.4.2 that

$$\mathscr{F}\left\{h_n(x)\right\} = (-i)^n h_n(x).$$

Clearly, the Hermite functions $h_n(x)$ defined by (3.4.23) are the eigenvalues of the operator $\mathscr{F}$.

Example 3.4.5 Using the duality theorem 3.4.10, show that

$$\mathscr{F}\left\{\frac{1}{t}\right\} = -\pi i \operatorname{sgn}(\omega), \tag{3.4.40}$$

$$\mathscr{F}\left\{f(t)\right\} = \mathscr{F}\left\{\frac{t}{a^2+t^2}\right\} = \begin{Bmatrix} -\pi i e^{a\omega}, & \omega > 0, \\ \pi i e^{-a|\omega|}, & \omega < 0 \end{Bmatrix}. \tag{3.4.41}$$

The graphs of $f(t)$ and $\hat{f}(\omega)$ given in (3.4.41) are shown in Figure 3.11.

Finally, we find

$$\mathscr{F}\left\{t^n \operatorname{sgn}(t)\right\} = (-i)^n \frac{d^n}{d\omega^n} \mathscr{F}\left\{\operatorname{sgn}(t)\right\} = (-i)^n \frac{d^n}{d\omega^n}\left(-\frac{2}{i\omega}\right) = (-i)^{n+1} \frac{2n!}{\omega^{n+1}}, \tag{3.4.42}$$

$$\mathscr{F}\left\{|t|\right\} = \mathscr{F}\left\{t \operatorname{sgn}(t)\right\} = -\frac{2}{\omega^2}. \tag{3.4.43}$$

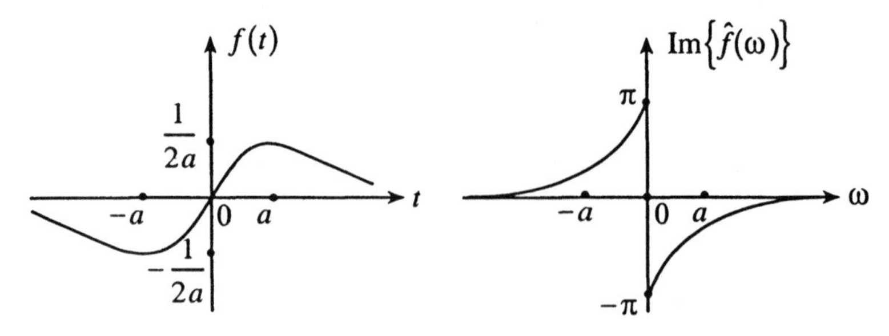

Figure 3.11. Graphs of $f(t)=t\left(a^2+t^2\right)^{-1}$ and $\hat{f}(\omega)$.

3.5 Poisson's Summation Formula

Although the theory of Fourier series is a very important subject, a detailed study is beyond the scope of this book. Without rigorous analysis, we establish a simple relation between Fourier transforms of functions in $L^1(\mathbb{R})$ and the Fourier series of related periodic functions in $L^1(0,2\pi)$ of period 2π.

If $f(t)\in L^1(0,2\pi)$ and $f(t)$ is a periodic function with period 2π, then the *Fourier series* of f is defined by

$$\sum_{n=-\infty}^{\infty} c_n e^{int}, \quad (0\le t\le 2\pi), \tag{3.5.1}$$

where the Fourier coefficients c_n is given by

$$c_n=\frac{1}{2\pi}\int_0^{2\pi} f(t)\, e^{-int}dt. \tag{3.5.2}$$

Theorem 3.5.1 If $f(t)\in L^1(\mathbb{R})$, then the series

$$\sum_{n=-\infty}^{\infty} f(t+2n\pi) \tag{3.5.3}$$

converges absolutely for almost all t in $(0,2\pi)$ and its sum $F(t)\in L^1(0,2\pi)$ with $F(t+2\pi)=F(t)$ for $t\in\mathbb{R}$.

If a_n denotes the Fourier coefficient of F, then

$$a_n = \frac{1}{2\pi}\int_0^{2\pi} F(t)\, e^{-int} dt = \frac{1}{2\pi}\int_{-\infty}^{\infty} f(t)\, e^{-int} dt = \frac{1}{2\pi}\hat{f}(n).$$

Proof. We have

$$\begin{aligned}\sum_{n=-\infty}^{\infty}\int_0^{2\pi} |f(t+2n\pi)|\, dt &= \lim_{N\to\infty}\sum_{n=-N}^{N}\int_0^{2\pi} |f(t+2\pi n)|\, dt \\ &= \lim_{N\to\infty}\sum_{n=-N}^{N}\int_{2n\pi}^{2(n+1)\pi} |f(x)|\, dx \\ &= \lim_{N\to\infty}\int_{-2\pi N}^{2\pi(N+1)} |f(x)|\, dx \\ &= \int_{-\infty}^{\infty} |f(x)|\, dx < \infty.\end{aligned}$$

It follows from Lebesgue's theorem on monotone convergence that

$$\int_0^{2\pi}\sum_{n=-\infty}^{\infty} |f(t+2n\pi)|\, dx = \sum_{n=-\infty}^{\infty}\int_0^{2\pi} |f(t+2n\pi)|\, dt < \infty.$$

Hence, the series $\sum_{n=-\infty}^{\infty} f(t+2n\pi)$ converges absolutely for almost all t. If $F_N(t) = \sum_{n=-N}^{N} f(t+2n\pi)$, $\lim_{N\to\infty} F_N(t) = F(t)$, where $F \in L^1(0,2\pi)$ and $F(t+2\pi) = F(t)$.

Furthermore,

$$\begin{aligned}\| F \|_1 = \int_0^{2\pi} |F(t)|\, dt &= \int_0^{2\pi}\left|\sum_{n=-\infty}^{\infty} f(t+2n\pi)\right| dt \\ &\le \int_0^{2\pi}\sum_{n=-\infty}^{\infty} |f(t+2n\pi)|\, dt \\ &= \sum_{n=-\infty}^{\infty}\int_0^{2\pi} |f(t+2n\pi)|\, dt = \int_{-\infty}^{\infty} |f(x)|\, dx = \| f \|_1.\end{aligned}$$

We consider the Fourier series of F given by

$$F(t) = \sum_{m=-\infty}^{\infty} a_m \, e^{imt},$$

where the Fourier coefficient a_m is

$$a_m = \frac{1}{2\pi} \int_0^{2\pi} F(t) \, e^{-imt} dt = \frac{1}{2\pi} \int_0^{2\pi} \left[\lim_{N \to \infty} F_N(t) \right] e^{-imt} dt$$

$$= \lim_{N \to \infty} \frac{1}{2\pi} \int_0^{2\pi} \sum_{n=-N}^{N} f(t + 2n\pi) \, e^{-imt} dt$$

$$= \lim_{N \to \infty} \frac{1}{2\pi} \sum_{n=-N}^{N} \int_0^{2\pi} f(t + 2n\pi) \, e^{-imt} dt$$

$$= \lim_{N \to \infty} \frac{1}{2\pi} \sum_{n=-N}^{N} \int_{2n\pi}^{2(n+1)\pi} f(x) \, e^{-imx} dx$$

$$= \lim_{N \to \infty} \frac{1}{2\pi} \int_{-2N\pi}^{2(N+1)\pi} f(x) \, e^{-imx} dx$$

$$= \frac{1}{2\pi} \int_{-\infty}^{\infty} f(x) \, e^{-imx} dx = \frac{1}{2\pi} \hat{f}(m).$$

Hence, if the Fourier series of $F(t)$ converges to $F(t)$, then, for $t \in \mathbb{R}$,

$$\sum_{n=-\infty}^{\infty} f(t + 2n\pi) = F(t) = \sum_{m=-\infty}^{\infty} \frac{1}{2\pi} \hat{f}(m) \, e^{imt}. \tag{3.5.4}$$

In particular, when $t = 0$, (3.5.4) becomes

$$\sum_{n=-\infty}^{\infty} f(2n\pi) = \frac{1}{2\pi} \sum_{n=-\infty}^{\infty} \hat{f}(n). \tag{3.5.5}$$

This is the so-called *Poisson summation formula.*

To obtain a more general result, we assume a is a given positive constant and write $g(t) = f(at)$ for all t. Then, $f\left(a \cdot \frac{2\pi u}{a}\right) = g\left(\frac{2\pi u}{a}\right)$ and

$$\hat{f}(n) = \int_{-\infty}^{\infty} e^{-int} f(t)\, dt = \int_{-\infty}^{\infty} e^{-int} f\left(a \cdot \frac{t}{a}\right) dt$$

$$= \int_{-\infty}^{\infty} e^{-int} g\left(\frac{t}{a}\right) dt$$

$$= a \int_{-\infty}^{\infty} g(x)\, e^{-i(an)x} dx = a\,\hat{g}(an).$$

Thus, (3.5.5) becomes

$$\sum_{n=-\infty}^{\infty} g\left(\frac{2\pi u}{a}\right) = \frac{a}{2\pi} \sum_{n=-\infty}^{\infty} \hat{g}(an). \tag{3.5.6}$$

Putting $b = \dfrac{2\pi}{a}$ in (3.5.6) gives

$$\sum_{n=-\infty}^{\infty} g(bn) = b^{-1} \sum_{n=-\infty}^{\infty} \hat{g}\left(2\pi b^{-1} n\right). \tag{3.5.7}$$

This reduces to (3.5.5) when $b = 1$.

We can apply the Poisson summation formula to derive the following important identities:

$$\sum_{n=-\infty}^{\infty} \exp\left(-\pi x n^2\right) = \frac{1}{\sqrt{x}} \sum_{n=-\infty}^{\infty} \exp\left(-\frac{\pi n^2}{x}\right), \tag{3.5.8}$$

$$\sum_{n=-\infty}^{\infty} \frac{1}{\left(n^2 + a^2\right)} = \left(\frac{\pi}{a}\right) \coth(\pi a), \tag{3.5.9}$$

$$\sum_{n=-\infty}^{\infty} \frac{1}{(t + n\pi)^2} = \operatorname{cosec}^2 t. \tag{3.5.10}$$

To prove (3.5.8), we choose the Gaussian kernel

$$G_\lambda(t) = \lambda\, G(\lambda t), \qquad G(t) = \exp\left(-t^2\right).$$

Hence,

$$\hat{G}_\lambda(\omega) = \sqrt{\pi}\, \exp\left(-\frac{\omega^2}{4\lambda^2}\right).$$

Replacing $f(t)$ by $G_\lambda(t)$ in (3.5.4) gives

$$\lambda \sum_{n=-\infty}^{\infty} \exp\left[-\lambda^2 (t + 2n\pi)^2\right] = \frac{1}{2\sqrt{\pi}} \sum_{n=-\infty}^{\infty} \exp\left(\text{int} - \frac{n^2}{4\lambda^2}\right).$$

If we put $t = 0$ and let $\lambda = \frac{1}{2}\sqrt{\frac{x}{\pi}}$, the above result reduces to

$$\sqrt{x}\sum_{n=-\infty}^{\infty}\exp\left(-\pi x n^2\right) = \sum_{n=-\infty}^{\infty}\exp\left(-\frac{n^2\pi}{x}\right).$$

This proves the identity (3.5.8) which is important in number theory and in the theory of elliptic functions. The function

$$\vartheta(x) = \sum_{n=-\infty}^{\infty}\exp\left(-\pi x n^2\right) \tag{3.5.11}$$

is called the *Jacobi theta function*. In terms of $\vartheta(x)$, the identity (3.5.8) becomes

$$\sqrt{x}\;\vartheta(x) = \vartheta\left(\frac{1}{x}\right). \tag{3.5.12}$$

To show (3.5.9), we let $f(t) = \left(t^2 + x^2\right)^{-1}$ so that $\hat{f}(\omega) = \left(\frac{\pi}{x}\right)\exp\left(-x|\omega|\right)$.

Consequently, formula (3.5.5) becomes

$$\sum_{n=-\infty}^{\infty}\frac{1}{\left(4n^2\pi^2 + x^2\right)} = \frac{1}{2x}\sum_{n=-\infty}^{\infty}\exp\left(-x|n|\right)$$

or

$$\sum_{n=-\infty}^{\infty}\frac{1}{\left(n^2 + a^2\right)} = \frac{\pi}{a}\sum_{n=-\infty}^{\infty}\exp\left(-2\pi a|n|\right), \quad (x = 2\pi a)$$

$$= \frac{\pi}{a}\left[\sum_{n=1}^{\infty}\exp\left(-2\pi a n\right) + \sum_{n=1}^{\infty}\exp\left(2\pi a n\right)\right]$$

which is, by setting $r = \exp(-2\pi a)$,

$$= \frac{\pi}{a}\left[\sum_{n=1}^{\infty} r^n + \sum_{n=1}^{\infty}\frac{1}{r^n}\right] = \frac{\pi}{a}\left(\frac{r}{1-r} + \frac{1}{1-r}\right)$$

$$= \frac{\pi}{a}\left(\frac{1+r}{1-r}\right) = \frac{\pi}{a}\coth(\pi a).$$

Thus, we have from (3.5.9),

$$\sum_{n=-\infty}^{\infty}\frac{1}{\left(n^2 + a^2\right)} = \left(\frac{\pi}{a}\right)\frac{\left(1 + e^{-2\pi a}\right)}{\left(1 - e^{-2\pi a}\right)}.$$

Or,

$$2\sum_{n=1}^{\infty}\frac{1}{\left(n^2+a^2\right)}+\frac{1}{a^2}=\left(\frac{\pi}{a}\right)\frac{\left(1+e^{-2\pi a}\right)}{\left(1-e^{-2\pi a}\right)}.$$

This gives

$$\sum_{n=1}^{\infty}\frac{1}{\left(n^2+a^2\right)}=\left(\frac{\pi}{2a}\right)\left[\frac{\left(1+e^{-2\pi a}\right)}{\left(1-e^{-2\pi a}\right)}-\frac{1}{\pi a}\right]$$

$$=\frac{\pi^2}{x}\left[\frac{\left(1+e^{-x}\right)}{\left(1-e^{-x}\right)}-\frac{2}{x}\right],\quad (2\pi a=x)$$

$$=\frac{\pi^2}{x^2}\left[\frac{x\left(1+e^{-x}\right)-2\left(1-e^{-x}\right)}{\left(1-e^{-x}\right)}\right]$$

$$=\left(\frac{\pi}{x}\right)^2\left[\frac{x^3\left(\frac{1}{2}-\frac{1}{3}\right)-\frac{x^4}{31}+\cdots}{\left(x-\frac{x^2}{21}+\frac{x^2}{31}-\cdots\right)}\right].$$

In the limit as $a\to 0$ (or $x\to 0$), we find the well-known result

$$\sum_{n=1}^{\infty}\frac{1}{n^2}=\frac{\pi^2}{6}. \tag{3.5.13}$$

3.6 The Shannon Sampling Theorem and Gibbs's Phenomenon

An analog signal $f(t)$ is a continuous function of time t defined in $-\infty<t<\infty$, with the exception of perhaps a countable number of jump discontinuities. Almost all analog signals $f(t)$ of interest in engineering have finite energy. By this we mean that $f\in L^2(-\infty,\infty)$. The norm of f defined by

$$\| f \|=\left[\int_{-\infty}^{\infty}|f(t)|^2\,dt\right]^{\frac{1}{2}} \tag{3.6.1}$$

represents the square root of the total energy content of the signal $f(t)$. The *spectrum* of a signal $f(t)$ is represented by its Fourier transform $\hat{f}(\omega)$, where

ω is called the *frequency*. The frequency is measured by $\nu = \frac{\omega}{2\pi}$ in terms of Hertz.

A signal $f(t)$ is called *band-limited* if its Fourier transform has a compact support, that is,

$$\hat{f}(\omega) = 0 \quad \text{for } |\omega| > \omega_0 \tag{3.6.2}$$

for some $\omega_0 > 0$. If ω_0 is the smallest value for which (3.6.2) holds, then it is called the *bandwidth* of the signal. Even if an analog signal $f(t)$ is not band-limited, we can reduce it to a band-limited signal by what is called an *ideal lowpass filtering*. To reduce $f(t)$ to a band-limited signal $f_{\omega_0}(t)$ with bandwidth less than or equal to ω_0, we consider

$$\hat{f}_{\omega_0}(\omega) = \begin{cases} \hat{f}(\omega) & \text{for } |\omega| \le \omega_0, \\ 0 & \text{for } |\omega| > \omega_0. \end{cases} \tag{3.6.3}$$

and we find the *low-pass filter function* $f_{\omega_0}(t)$ by the inverse Fourier transform

$$f_{\omega_0}(t) = \frac{1}{2\pi} \int_{-\infty}^{\infty} e^{i\omega t} \hat{f}_{\omega_0}(\omega)\, d\omega = \frac{1}{2\pi} \int_{-\omega_0}^{\omega_0} e^{i\omega t} \hat{f}_{\omega_0}(\omega)\, d\omega .$$

In particular, if

$$\hat{f}_{\omega_0}(\omega) = \begin{cases} 1 & \text{for } |\omega| \le \omega_0 \\ 0 & \text{for } |\omega| > \omega_0, \end{cases} \tag{3.6.4}$$

then $\hat{f}_{\omega_0}(\omega)$ is called the *gate function*, and its associated signal $f_{\omega_0}(t)$ is given by

$$f_{\omega_0}(t) = \frac{1}{2\pi} \int_{-\omega_0}^{\omega_0} e^{i\omega t}\, d\omega = \frac{\sin \omega_0 t}{\pi t}. \tag{3.6.5}$$

This function is called the *Shannon sampling function*. When $\omega_0 = \pi$, $f_\pi(t)$ is called the *Shannon scaling function*. Both $f_{\omega_o}(t)$ and $\hat{f}_{\omega_o}(\omega)$ are shown in Figure 3.12.

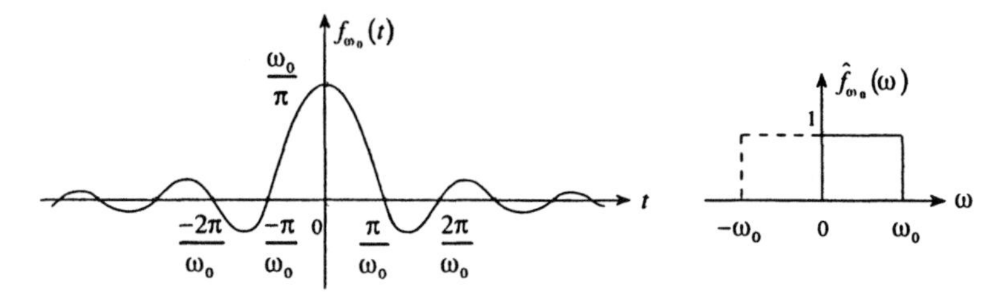

Figure 3.12. Shannon's functions.

In engineering, a linear analog filtering process is defined by the *time-domain convolution*. If $\phi(t)$ is the filter function, then the input-output relation of this filter is given by

$$g(t) = (\phi * f)(t) = \int_{-\infty}^{\infty} \phi(\tau) f(t-\tau)\, d\tau. \tag{3.6.6}$$

In the frequency domain, the filtering process is the Fourier transform of (3.6.6) and is represented by pointwise multiplication as

$$\hat{g}(\omega) = \hat{\phi}(\omega) \hat{f}(\omega), \tag{3.6.7}$$

where $\hat{\phi}(\omega)$ is the transfer function of the filter.

Consider the limit as $\omega_0 \to \infty$ of the Fourier integral

$$1 = \lim_{\omega_0 \to \infty} \hat{f}_{\omega_0}(\omega) = \lim_{\omega_0 \to \infty} \int_{-\infty}^{\infty} e^{-i\omega t} f_{\omega_0}(t)\, dt = \lim_{\omega_0 \to \infty} \int_{-\infty}^{\infty} e^{-i\omega t} \frac{\sin \omega_0 t}{\pi t}\, dt$$

$$= \int_{-\infty}^{\infty} e^{-i\omega t} \lim_{\omega_0 \to \infty} \left(\frac{\sin \omega_0 t}{\pi t} \right) dt = \int_{-\infty}^{\infty} e^{-i\omega t}\, \delta(t)\, dt.$$

Clearly, the delta function can be thought of as the limit of the sequence of signal functions $f_{\omega_0}(t)$. More explicitly,

$$\delta(t) = \lim_{\omega_0 \to \infty} \frac{\sin \omega_o t}{\pi t}.$$

The band-limited signal $f_{\omega_0}(t)$ has the representation

$$f_{\omega_0}(t) = \frac{1}{2\pi} \int_{-\omega_0}^{\omega} \hat{f}(\omega) e^{i\omega t} d\omega = \frac{1}{2\pi} \int_{-\infty}^{\infty} \hat{f}(\omega) \hat{f}_{\omega_0}(\omega) e^{i\omega t} d\omega, \tag{3.6.8}$$

which gives, by the convolution theorem,

$$f_{\omega_0}(t) = \int_{-\omega_0}^{\omega} f(\tau) f_{\omega_0}(t-\tau)\, d\tau = \int_{-\infty}^{\infty} \frac{\sin \omega_0 (t-\tau)}{\pi (t-\tau)} f(\tau)\, d\tau. \qquad (3.6.9)$$

This gives the *sampling integral representation* of a band-limited signal $f_{\omega_0}(t)$. Thus, $f_{\omega_0}(t)$ can be interpreted as the weighted average of $f(\tau)$ with the Fourier kernel $\dfrac{\sin \omega_0 (t-\tau)}{\pi (t-\tau)}$ as weight.

We now examine the so-called *Gibbs' jump phenomenon* which deals with the limiting behavior of $f_{\omega_0}(t)$ at a point of discontinuity of $f(t)$. This phenomenon reveals the intrinsic overshoot near a jump discontinuity of a function associated with the Fourier series. More precisely, the partial sums of the Fourier series overshoot the function near the discontinuity, and the overshoot continues no matter how many terms are taken in the partial sum. However, the Gibbs phenomenon does not occur if the partial sums are replaced by the Cesáro means, that is, the average of the partial sums.

In order to demonstrate the Gibbs phenomenon, we rewrite (3.6.8) in the form

$$f_{\omega_0}(t) = \int_{-\infty}^{\infty} f(\tau) \frac{\sin \omega_0 (t-\tau)}{\pi (t-\tau)}\, d\tau = \left(f * \delta_{\omega_0}\right)(t), \qquad (3.6.10)$$

where

$$\delta_{\omega_0}(t) = \frac{\sin \omega_0 t}{\pi t}. \qquad (3.6.11)$$

Clearly, at every point of continuity of $f(t)$, we have

$$\begin{aligned} \lim_{\omega_0 \to \infty} f_{\omega_0}(t) &= \lim_{\omega_0 \to \infty} \left(f * \delta_{\omega_0}\right)(t) = \lim_{\omega_0 \to \infty} \int_{-\infty}^{\infty} f(\tau) \frac{\sin \omega_0 (t-\tau)}{\pi (t-\tau)}\, d\tau \\ &= \int_{-\infty}^{\infty} f(\tau) \left[\lim_{\omega_0 \to \infty} \frac{\sin \omega_0 (t-\tau)}{\pi (t-\tau)} \right] d\tau \\ &= \int_{-\infty}^{\infty} f(\tau)\, \delta(t-\tau)\, d\tau = f(t). \end{aligned} \qquad (3.6.12)$$

We now consider the limiting behavior of $f_{\omega_0}(t)$ at the point of discontinuity $t = t_0$. To simplify the calculation, we set $t_0 = 0$ so that we can write $f(t)$ as a sum of a suitable step function in the form

$$f(t) = f_c(t) + [f(0+) - f(0-)]\, H(t). \tag{3.6.13}$$

Replacing $f(t)$ by the right hand side of (3.6.13) in equation (3.6.9) yields

$$\begin{aligned} f_{\omega_0}(t) &= \int_{-\infty}^{\infty} f_c(\tau) \frac{\sin \omega_0 (t-\tau)}{\pi (t-\tau)}\, d\tau + [f(0+) - f(0-)] \int_{-\infty}^{\infty} H(\tau) \frac{\sin \omega_0 (t-\tau)}{\pi (t-\tau)}\, d\tau \\ &= f_c(t) + [f(0+) - f(0-)]\, H_{\omega_0}(t), \end{aligned} \tag{3.6.14}$$

where

$$\begin{aligned} H_{\omega_0}(t) &= \int_{-\infty}^{\infty} H(\tau) \frac{\sin \omega_0 (t-\tau)}{\pi (t-\tau)}\, d\tau = \int_{0}^{\infty} \frac{\sin \omega_0 (t-\tau)}{\pi (t-\tau)}\, d\tau \\ &= \int_{-\infty}^{\omega_0 t} \frac{\sin x}{\pi x}\, dx \quad \left(\text{putting } \omega_0 (t-\tau) = x\right) \\ &= \left(\int_{-\infty}^{0} + \int_{0}^{\omega_0 t} \right) \left(\frac{\sin x}{\pi x} \right) dx = \left(\int_{0}^{\infty} + \int_{0}^{\omega_0 t} \right) \left(\frac{\sin x}{\pi x} \right) dx \\ &= \frac{1}{2} + \frac{1}{\pi}\, si(\omega_0 t), \end{aligned} \tag{3.6.15}$$

and the function $si(t)$ is defined by

$$si(t) = \int_0^t \frac{\sin x}{x}\, dx. \tag{3.6.16}$$

Note that

$$H_{\omega_0}\left(\frac{\pi}{\omega_0} \right) = \frac{1}{2} + \int_0^{\pi} \frac{\sin x}{\pi x}\, dx > 1,$$

$$H_{\omega_0}\left(-\frac{\pi}{\omega_0} \right) = \frac{1}{2} - \int_0^{\pi} \frac{\sin x}{\pi x}\, dx < 0.$$

Clearly, for a fixed ω_0, $\frac{1}{\pi}\, si(\omega_0 t)$ attains its maximum at $t = \frac{\pi}{\omega_0}$ in $(0, \infty)$ and minimum at $t = -\frac{\pi}{\omega_0}$, since for larger t the integrand oscillates with decreasing amplitudes. The function $H_{\omega_0}(t)$ is shown in Figure 3.13 since $H_{\omega_0}(0) = \frac{1}{2}$ and $f_c(0) = f(0-)$ and

$$f_{\omega_0}(0) = f_c(0) + \frac{1}{2}[f(0+) - f(0-)] = \frac{1}{2}[f(0+) + f(0-)].$$

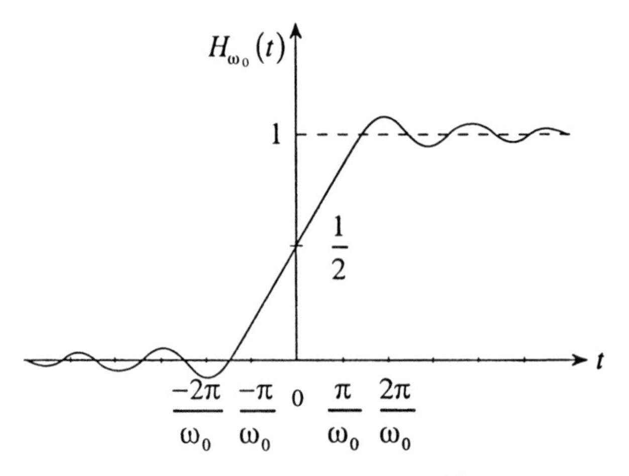

Figure 3.13. Graph of $H_{\omega_0}(t)$.

Thus, the graph of $H_{\omega_0}(t)$ shows that as ω_0 increases, the time scale changes, and the ripples remain the same. In the limit $\omega_0 \to \infty$, the convergence of $H_{\omega_0}(t) = (H * \delta_{\omega_0})(t)$ to $H(t)$ exhibits the intrinsic overshoot leading to the classical Gibbs phenomenon.

Next, we consider the Fourier series expansion of the Fourier transform $\hat{f}_{\omega_0}(\omega)$ of a band-limited signal $f_{\omega_0}(t)$ on the interval $-\omega_0 < \omega < \omega_0$ in terms of the orthogonal set of functions $\left[\exp\left(-\frac{in\tau\omega}{\omega_0}\right)\right]$ in the form

$$\hat{f}_{\omega_0}(\omega) = \sum_{n=-\infty}^{\infty} a_n \exp\left(-\frac{in\pi\omega}{\omega_0}\right), \tag{3.6.17}$$

where the Fourier coefficients a_n are given by

$$a_n = \frac{1}{2\pi}\int_{-\omega_0}^{\omega_0} \hat{f}_{\omega_0}(\omega) \exp\left(\frac{in\pi\omega}{\omega_0}\right) d\omega = \frac{1}{2\pi} f_{\omega_0}\left(\frac{n\pi}{\omega_0}\right).$$

Thus, the Fourier series expansion (3.6.17) reduces to the form

$$\hat{f}_{\omega_0}(\omega) = \frac{1}{2\pi}\sum_{n=-\infty}^{\infty} f_{\omega_0}\left(\frac{n\pi}{\omega_0}\right) \exp\left(-\frac{in\pi\omega}{\omega_0}\right). \tag{3.6.18}$$

Multiplying (3.6.18) by $e^{i\omega t}$ and integrating over $(-\omega_0, \omega_0)$ leads to the reconstruction of the signal $f_{\omega_0}(t)$ in the form

$$
\begin{aligned}
f_{\omega_0}(t) &= \int_{-\omega_0}^{\omega_0} \hat{f}_{\omega_0}(\omega)\, e^{i\omega t} d\omega \\
&= \frac{1}{2\pi} \int_{-\omega_0}^{\omega_0} \left[\sum_{n=-\infty}^{\infty} f_{\omega_0}\left(\frac{n\pi}{\omega_0}\right) \exp\left(-\frac{in\pi\omega}{\omega_0}\right)\right] d\omega \\
&= \frac{1}{2\pi} \sum_{n=-\infty}^{\infty} f_{\omega_0}\left(\frac{n\pi}{\omega_0}\right) \int_{-\omega_0}^{\omega_0} \exp\left[i\omega\left(t - \frac{n\pi}{\omega_0}\right)\right] d\omega \\
&= \sum_{n=-\infty}^{\infty} f_{\omega_0}\left(\frac{n\pi}{\omega_0}\right) \frac{\sin \omega_0\left(t - \dfrac{n\pi}{\omega_0}\right)}{\omega_0\left(t - \dfrac{n\pi}{\omega_0}\right)}. \qquad (3.6.19)
\end{aligned}
$$

This formula is referred to as the *Shannon sampling theorem*. It represents an expansion of a band-limited signal $f_{\omega_0}(t)$ in terms of its discrete values $f_{\omega_0}\left(\frac{n\pi}{\omega_0}\right)$. This is very important in practice because most systems receive discrete samples as an input.

Example 3.6.1 (Synthesis and Resolution of a Signal (or Waveform) and Physical Interpretation of Convolution). In science and engineering, a time-dependent electric, optical or electromagnetic pulse is usually called a *signal* (or *waveform*). Such a signal $f(t)$ can be regarded as a superposition of plane waves of all real frequencies, and so it can be represented by the inverse Fourier transform

$$f(t) = \mathscr{F}^{-1}\left\{\hat{f}(\omega)\right\} = \frac{1}{2\pi} \int_{-\infty}^{\infty} e^{i\omega t} \hat{f}(\omega)\, d\omega, \qquad (3.6.20)$$

where $\hat{f}(\omega)$ is the Fourier spectrum of the signal $f(t)$ given by

$$\hat{f}(\omega) = \int_{-\infty}^{\infty} e^{-i\omega t} f(t)\, dt. \qquad (3.6.21)$$

This represents the *resolution* of the signal (or waveform) into its angular frequency $(\omega = 2\pi\nu)$ components, and (3.6.20) gives a synthesis of the signal (or the waveform) from its individual components.

A continuous linear system is a device which transforms a signal $f(t) \in L^2(\mathbb{R})$ linearly. It has an associated *impulse response* $\phi(t)$ such that the *output* of the system $g(t)$ is defined by the convolution

$$g(t) = \phi(t) * f(t) = \int_{-\infty}^{\infty} \phi(t-\tau)\, f(\tau)\, d\tau. \tag{3.6.22}$$

Often, $f(t)$ and $g(t)$ are referred to as input and output signals, respectively. In science and engineering, filters, sensors, and amplifiers are common examples of linear systems.

Physically, the output of a system is represented by the integral superposition of an input modified by the system impulse function ϕ. Indeed, (3.6.22) is a fairly general mathematical representation of an output (effect) in terms of an input (cause) modified by the system impulse. According to the principle of causality, every effect has a cause. This principle is imposed by requiring

$$\phi(t-\tau) = 0 \qquad \text{for } \tau > t. \tag{3.6.23}$$

Consequently, (3.6.22) gives

$$g(t) = \int_{-\infty}^{t} f(\tau)\, \phi(t-\tau)\, d\tau. \tag{3.6.24}$$

In order to determine the significance of $\phi(t)$, we use the Dirac delta function as the input so that $f(t) = \delta(t)$ and

$$g(t) = \int_{-\infty}^{\infty} \delta(\tau)\, \phi(t-\tau)\, d\tau = \phi(t)\, H(t). \tag{3.6.25}$$

This recognized $\phi(t)$ as the output corresponding to a unit input at $t = 0$, and the Fourier transform of $\phi(t)$ is given by

$$\hat{\phi}(\omega) = \int_{0}^{\infty} \phi(t)\, e^{-i\omega t} dt, \tag{3.6.26}$$

where

$$\phi(t) = 0 \qquad \text{for } t < 0.$$

In general, the output can be best described by taking the Fourier transform of (3.6.22) so that

$$\hat{g}(\omega) = \hat{f}(\omega)\,\hat{\phi}(\omega), \tag{3.6.27}$$

where $\hat{\phi}(\omega)$ is called the *transfer function* of the system. Thus, the output can be calculated from (3.6.27) by the Fourier inversion formula

$$g(t) = \frac{1}{2\pi}\int_{-\infty}^{\infty} \hat{f}(\omega)\,\hat{\phi}(\omega)\,e^{i\omega t}d\omega. \tag{3.6.28}$$

Obviously, the transfer function $\hat{\phi}(\omega)$ is a characteristic of a linear system. A linear system is a filter if it passes signals of certain frequencies and attenuates others. If the transfer function

$$\hat{\phi}(\omega) = 0 \qquad \text{for} \qquad |\omega| \geq \omega_0, \tag{3.6.29}$$

then ϕ is called a *low-pass filter*.

On the other hand, if the transfer function

$$\hat{\phi}(\omega) = 0 \qquad \text{for} \qquad |\omega| \leq \omega_1, \tag{3.6.30}$$

then ϕ is a *high-pass filter*. A bandpass filter possesses a band $\omega_0 \leq |\omega| \leq \omega_1$.

It is often convenient to express the system transfer function $\hat{\phi}(\omega)$ in the complex form

$$\hat{\phi}(\omega) = A(\omega)\exp[-i\,\theta(\omega)], \tag{3.6.31}$$

where $A(\omega)$ is called the *amplitude* and $\theta(\omega)$ is called the *phase* of the transfer function.

Obviously, the system impulse response $\phi(t)$ is given by the inverse Fourier transform

$$\phi(t) = \frac{1}{2\pi}\int_{-\infty}^{\infty} A(\omega)\exp\left[i\left\{(\omega t - \theta(\omega))\right\}\right]d\omega. \tag{3.6.32}$$

For a unit step function as the input $f(t) = H(t)$, we have

$$\hat{f}(\omega) = \hat{H}(\omega) = \left(\pi\delta(\omega) + \frac{1}{i\omega}\right),$$

and the associated output $g(t)$ is then given by

$$g(t)=\frac{1}{2\pi}\int_{-\infty}^{\infty}\hat{\phi}(\omega)\,\hat{H}(\omega)\,e^{i\omega t}d\omega$$

$$=\frac{1}{2\pi}\int_{-\infty}^{\infty}\left(\pi\,\delta(\omega)+\frac{1}{i\,\omega}\right)A(\omega)\,\exp\left[i\left\{\omega\,t-\theta(\omega)\right\}\right]\,d\omega$$

$$=\frac{1}{2}\,A(0)+\frac{1}{2\pi}\int_{-\infty}^{\infty}\frac{A(\omega)}{\omega}\,\exp\left[i\left\{\omega\,t-\theta(\omega)-\frac{\pi}{2}\right\}\right]\,d\omega. \quad (3.6.33)$$

We next give another characterization of a filter in terms of the amplitude of the transfer function.

A filter is called *distortionless* if its output $g(t)$ to an arbitrary input $f(t)$ has the same form as the input, that is,

$$g(t)=A_0\,f(t-t_0). \quad (3.6.34)$$

Evidently,

$$\hat{g}(\omega)=A_0\,e^{-i\omega t_0}\hat{f}(\omega)=\hat{\phi}(\omega)\,\hat{f}(\omega),$$

where

$$\hat{\phi}(\omega)=A_0\,e^{-i\omega t_0}$$

represents the transfer function of the distortionless filter. It has a constant amplitude A_0 and a linear phase shift $\hat{\theta}(\omega)=\omega\,t_0$.

However, in general, the amplitude $A(\omega)$ of a transfer function is not constant, and the phase $\theta(\omega)$ is not a linear function.

A filter with a constant amplitude, $|\theta(\omega)|=A_0$ is called an *all-pass filter*. It follows from Parseval's formula given by Theorem 3.4.2 that the energy of the output of such a filter is proportional to the energy of its input.

A filter whose amplitude is constant for $|\omega|<\omega_0$ and zero for $|\omega|>\omega_0$ is called an *ideal low-pass filter*. More explicitly, the amplitude is given by

$$A(\omega)=A_0\,\hat{H}(\omega_0-|\omega|)=A_0\,\hat{\chi}_{\omega_0}(\omega), \quad (3.6.35)$$

where $\hat{\chi}_{\omega_0}(\omega)$ is a rectangular pulse. So, the transfer function of the low-pass filter is

$$\hat{\phi}(\omega)=A_0\,\hat{\chi}_{\omega_0}(\omega)\,\exp(-i\,\omega\,t_0). \quad (3.6.36)$$

Finally, the *ideal high-pass filter* is characterized by its amplitude given by

$$A(\omega)=A_0\,\hat{H}\left(|\omega|-\omega_0\right)=A_0\,\hat{\chi}_{\omega_0}(\omega), \tag{3.6.37}$$

where A_0 is a constant. Its transfer function is given by

$$\hat{\phi}(\omega)=A_0\left[1-\hat{\chi}_{\omega_0}(\omega)\right]\exp\left(-i\omega t_0\right). \tag{3.6.38}$$

Example 3.6.2 (Bandwidth and Bandwidth Equation). The Fourier spectrum of a signal (or waveform) gives an indication of the frequencies that exist during the total duration of the signal (or waveform). From the knowledge of the frequencies that are present, we can calculate the average frequency and the spread about that average. In particular, if the signal is represented by $f(t)$, we can define its Fourier spectrum by

$$\hat{f}(\nu)=\int_{-\infty}^{\infty}e^{-2\pi i\nu t}\,f(t)\,dt. \tag{3.6.39}$$

Using $\left|\hat{f}(\nu)\right|^2$ for the density in frequency, the average frequency is denoted by $\langle\nu\rangle$ and defined by

$$\langle\nu\rangle=\int_{-\infty}^{\infty}\nu\left|\hat{f}(\nu)\right|^2 d\nu. \tag{3.6.40}$$

The bandwidth is then the *root mean square* (RMS) deviation at about the average, that is,

$$B^2=\int_{-\infty}^{\infty}\left(\nu-\langle\nu\rangle\right)^2\left|\hat{f}(\nu)\right|^2 d\nu. \tag{3.6.41}$$

Expressing the signal in terms of its amplitude and phase

$$f(t)=a(t)\,\exp\left\{i\theta(t)\right\}, \tag{3.6.42}$$

the instantaneous frequency, $\nu_i(t)$ is the frequency at a particular time defined by

$$\nu_i(t)=\frac{1}{2\pi}\,\theta'(t). \tag{3.6.43}$$

Substituting (3.6.39) and (3.6.42) into (3.6.40) gives

$$\langle\nu\rangle=\frac{1}{2\pi}\int_{-\infty}^{\infty}\theta'(t)\,a^2(t)\,dt=\int_{-\infty}^{\infty}\nu_i(t)\,a^2(t)\,dt. \tag{3.6.44}$$

This formula states that the average frequency is the average value of the instantaneous frequency weighted by the absolute square of the amplitude of the signal.

We next derive the bandwidth equation in terms of the amplitude and phase of the signal in the form

$$B^2 = \frac{1}{(2\pi)^2}\int_{-\infty}^{\infty}\left[\frac{a'(t)}{a(t)}\right]^2 a^2(t)\,dt + \int_{-\infty}^{\infty}\left[\frac{1}{2\pi}\,\theta'(t) - \langle \nu \rangle\right]^2 a^2(t)\,dt. \tag{3.6.45}$$

A straightforward but lengthy way to derive it is to substitute (3.6.42) into (3.6.41) and simplify. However, we give an elegant derivation of (3.6.45) by representing the frequency by the operator

$$\nu \to \frac{1}{2\pi i}\,\frac{d}{dt}.$$

We calculate the average by sandwiching the operator in between the complex conjugate of the signal and the signal. Thus,

$$\begin{aligned}
\langle \nu \rangle &= \int_{-\infty}^{\infty} \nu \left|\hat{f}(\nu)\right|^2 d\nu = \int_{-\infty}^{\infty} \bar{f}(t)\left[\frac{1}{2\pi i}\,\frac{d}{dt}\right] f(t)\,dt \\
&= \frac{1}{2\pi}\int_{-\infty}^{\infty} a(t)\left\{-i\,a'(t) + a(t)\,\theta'(t)\right\} dt \\
&= \frac{1}{2\pi}\int_{-\infty}^{\infty} -\frac{1}{2}\,i\left[\frac{d}{dt}\,a^2(t)\right] dt + \frac{1}{2\pi}\int_{-\infty}^{\infty} a^2(t)\,\theta'(t)\,dt && (3.6.46) \\
&= \frac{1}{2\pi}\int_{-\infty}^{\infty} \theta'(t)\,a^2(t)\,dt, && (3.6.47)
\end{aligned}$$

provided the first integral in (3.6.46) vanishes if $a(t) \to 0$ as $|t| \to \infty$.

It follows from the definition (3.6.41) of the bandwidth that

$$B^2 = \int_{-\infty}^{\infty} (\nu - \langle \nu \rangle)^2 \left| \hat{f}(\nu) \right|^2 d\nu$$

$$= \int_{-\infty}^{\infty} \bar{f}(t) \left[\frac{1}{2\pi i} \frac{d}{dt} - \langle \nu \rangle \right]^2 f(t)\, dt$$

$$= \int_{-\infty}^{\infty} \left| \left[\frac{1}{2\pi i} \frac{d}{dt} - \langle \nu \rangle \right] f(t) \right|^2 dt$$

$$= \int_{-\infty}^{\infty} \left| \frac{1}{2\pi i} \frac{a'(t)}{a(t)} + \frac{1}{2\pi} \theta'(t) - \langle \nu \rangle \right|^2 a^2(t)\, dt$$

$$= \frac{1}{4\pi^2} \int_{-\infty}^{\infty} \left[\frac{a'(t)}{a(t)} \right]^2 a^2(t)\, dt + \int_{-\infty}^{\infty} \left[\frac{1}{2\pi} \theta'(t) - \langle \nu \rangle \right]^2 a^2(t)\, dt.$$

This completes the derivation.

Physically, the second term in equation (3.6.45) gives averages of all of the deviations of the instantaneous frequency from the average frequency. In electrical engineering literature, the spread of frequency about the *instantaneous frequency*, which is defined as an average of the frequencies that exist at a particular time, is called *instantaneous bandwidth*, given by

$$\sigma_{\nu/t}^2 = \frac{1}{(2\pi)^2} \left[\frac{a'(t)}{a(t)} \right]^2. \tag{3.6.48}$$

In the case of a chirp with a Gaussian envelope

$$f(t) = \left(\frac{\alpha}{\pi} \right)^{\frac{1}{4}} \exp\left[-\frac{1}{2} \alpha t^2 + \frac{1}{2} i \beta t^2 + 2\pi i \nu_0 t \right], \tag{3.6.49}$$

where its Fourier spectrum is given by

$$\hat{f}(\nu) = (\alpha\pi)^{\frac{1}{4}} \left(\frac{1}{\alpha - i\beta} \right)^{\frac{1}{2}} \exp\left[-2\pi^2 (\nu - \nu_0)^2 / (\alpha - i\beta) \right]. \tag{3.6.50}$$

The energy density spectrum of the signal is

$$\left| \hat{f}(\nu) \right|^2 = 2 \left(\frac{\alpha\pi}{\alpha^2 + \beta^2} \right)^{\frac{1}{2}} \exp\left[-\frac{4\alpha\pi^2 (\nu - \nu_0)^2}{\alpha^2 + \beta^2} \right]. \tag{3.6.51}$$

Finally, the average frequency $\langle \nu \rangle$ and the bandwidth square are respectively given by

$$\langle \nu \rangle = \nu_0 \quad \text{and} \quad B^2 = \frac{1}{8\pi^2}\left(\alpha + \frac{\beta^2}{\alpha}\right). \tag{3.6.52a,b}$$

A large bandwidth can be achieved in two very qualitatively different ways. The amplitude modulation can be made large by taking α large, and the frequency modulation can be made small by letting $\beta \to 0$. It is possible to make the frequency modulation large by making β large and α very small. These two extreme situations are physically very different even though they produce the same bandwidth.

3.7 Heisenberg's Uncertainty Principle

Heisenberg first formulated the uncertainty principle between the position and momentum in quantum mechanics. This principle has an important interpretation as an uncertainty of both the position and momentum of a particle described by a wave function $\psi \in L^2(\mathbb{R})$. In other words, it is impossible to determine the position and momentum of a particle exactly and simultaneously.

In signal processing, time and frequency concentrations of energy of a signal f are also governed by the Heisenberg uncertainty principle. The average or expectation values of time t and frequency ω, are respectively defined by

$$\langle t \rangle = \frac{1}{\|f\|_2^2}\int_{-\infty}^{\infty} t\,|f(t)|^2\,dt, \quad \langle \omega \rangle = \frac{1}{\|\hat{f}\|_2^2}\int_{-\infty}^{\infty} \omega\,|\hat{f}(\omega)|^2\,d\omega, \tag{3.7.1a,b}$$

where the energy of a signal f is well localized in time, and its Fourier transform $\hat{f}$ has an energy concentrated in a small frequency domain.

The variances around these average values are given respectively by

$$\sigma_t^2 = \frac{1}{\|f\|_2^2}\int_{-\infty}^{\infty} (t - \langle t \rangle)^2\,|f(t)|^2\,dt, \quad \sigma_\omega^2 = \frac{1}{2\pi\|\hat{f}\|_2^2}\int_{-\infty}^{\infty} (\omega - \langle \omega \rangle)^2\,|\hat{f}(\omega)|^2\,d\omega. \tag{3.7.2a,b}$$

Theorem 3.7.1 (Heisenberg's Inequality). If $f(t)$, $t\,f(t)$, and $\omega\,\hat{f}(\omega)$ belong to $L^2(\mathbb{R})$ and $\sqrt{t}\,|f(t)| \to 0$ as $|t| \to \infty$, then

$$\sigma_t^2\,\sigma_\omega^2 \geq \frac{1}{4}, \tag{3.7.3}$$

where σ_t is defined as a measure of time duration of a signal f, and σ_ω is a measure of frequency dispersion (or bandwidth) of its Fourier transform $\hat{f}$.

Equality in (3.7.3) holds only if $f(t)$ is a Gaussian signal given by $f(t) = C\,\exp\left(-bt^2\right),\ b > 0$.

Proof. If the average time and frequency localization of a signal f are $\langle t\rangle$ and $\langle \omega\rangle$, then the average time and frequency location of $\exp\left(-i\langle\omega\rangle t\right)\, f\left(t + \langle t\rangle\right)$ is zero. Hence, it is sufficient to prove the theorem around the zero mean values, that is, $\langle t\rangle = \langle\omega\rangle = 0$.

Since $\|f\|_2 = \|\hat{f}\|_2$, we have

$$\|f\|_2^4\,\sigma_t^2\,\sigma_\omega^2 = \frac{1}{2\pi}\int_{-\infty}^{\infty}\left|t\,f(t)\right|^2 dt \int_{-\infty}^{\infty}\left|\omega\,\hat{f}(\omega)\right|^2 d\omega.$$

Using $i\omega\,\hat{f}(\omega) = \mathscr{F}\{f'(t)\}$ and Parseval's formula

$$\|f'(t)\|_2 = \frac{1}{2\pi}\left\|i\,\omega\,\hat{f}(\omega)\right\|_2,$$

we obtain

$$\begin{aligned}
\|f\|_2^4\sigma_t^2\,\sigma_\omega^2 &= \int_{-\infty}^{\infty}\left|t\,f(t)\right|^2 dt \int_{-\infty}^{\infty}\left|f'(t)\right|^2\ dt \\
&\geq \left|\int_{-\infty}^{\infty}\left\{t\,f(t)\overline{f'(t)}\right\} dt\right|^2, \qquad \text{by Schwarz's inequality} \\
&\geq \left|\int_{-\infty}^{\infty} t\cdot\frac{1}{2}\left\{f'(t)\ \overline{f(t)} + \overline{f'(t)}\ f(t)\right\} dt\right|^2 \\
&= \frac{1}{4}\left[\int_{-\infty}^{\infty} t\left(\frac{d}{dt}|f|^2\right) dt\right]^2 = \frac{1}{4}\left\{\left[t\left|f(t)\right|^2\right]_{-\infty}^{\infty} - \int_{-\infty}^{\infty}|f|^2\ dt\right\}^2 \\
&= \frac{1}{4}\,\|f\|_2^4,
\end{aligned}$$

in which $\sqrt{t}\ f(t) \to 0$ as $|t| \to \infty$ was used to eliminate the integrated term.

This completes the proof of inequality (3.7.3).

If we assume $f'(t)$ is proportional to $t\,f(t)$, that is, $f'(t) = a\,t\,f(t)$, where a is a constant of proportionality, this leads to the Gaussian signal

$$f(t) = C\,\exp\left(-bt^2\right), \tag{3.7.4}$$

where C is a constant of integration and $b = -\dfrac{a}{2} > 0$.

Remarks.

1. In a time-frequency analysis of signals, the measure of the resolution of a signal f in the time or frequency domain is given by σ_t and σ_ω. Then, the joint resolution is given by the product $(\sigma_t)(\sigma_\omega)$ which is governed by the Heisenberg uncertainty principle. In other words, the product $(\sigma_t)(\sigma_\omega)$ cannot be arbitrarily small and is always greater than the minimum value $\dfrac{1}{2}$ which is attained only for the Gaussian signal.
2. In many applications in science and engineering, signals with a high concentration of energy in the time and frequency domains are of special interest. The uncertainty principle can also be interpreted as a measure of this concentration of the second moment of $f^2(t)$ and its energy spectrum $\hat{f}^2(\omega)$.

3.8 Applications of Fourier Transforms in Mathematical Statistics

In probability theory and mathematical statistics, the characteristic function of a random variable is defined by the Fourier transform or by the Fourier-Stieltjes transform of the distribution function of a random variable. Many important results in probability theory and mathematical statistics can be obtained, and their proofs can be simplified with rigor by using the methods of characteristic functions. Thus, the Fourier transforms play an important role in probability theory and statistics.

Definition 3.8.1 (Distribution Function). The *distribution* function $F(x)$ of a random variable X is defined as the probability, that is, $F(x) = P(X < x)$ for every real number x.

It is immediately evident from this definition that the distribution function satisfies the following properties:

(i) $F(x)$ is a nondecreasing function, that is, $F(x_1) \le F(x_2)$ if $x_1 < x_2$.

(ii) $F(x)$ is continuous only from the left at a point x, that is, $F(x-0) = F(x)$, but $F(x+0) \ne F(x)$.

(iii) $F(-\infty) = 0$ and $F(+\infty) = 1$.

If X is a continuous variable and if there exists a nonnegative function $f(x)$ such that for every real x relation

$$F(x) = \int_{-\infty}^{x} f(x)\, dx, \tag{3.8.1}$$

holds, where $F(x)$ is the distribution function of the random variable X, then the function $f(x)$ is called the *probability density* or simply the *density function* of the random variable X.

It is immediately obvious that every density function $f(x)$ satisfies the following properties:

(i) $$F(+\infty) = \int_{-\infty}^{\infty} f(x)\, dx = 1. \tag{3.8.2a}$$

(ii) For every real a and b, where $a < b$,

$$P(z \le X \le b) = F(b) - F(a) = \int_{a}^{b} f(x)\, dx. \tag{3.8.2b}$$

(iii) If $f(x)$ is continuous at some point x, then $F'(x) = f(x)$.

It is noted that every real function $f(x)$ which is nonnegative, integrable over the whole real line, and satisfies (3.8.2a,b) is the probability density function of a continuous random variable X. On the other hand, the function $F(x)$ defined by (3.8.1) satisfies all properties of a distribution function.

Definition 3.8.2 (***Characteristic Function***). If X is a continuous random variable with the density function $f(x)$, then the characteristic function $\phi(t)$ of the random variable X or the distribution function $F(x)$ is defined by the formula

$$\phi(t) = E\left[\exp(i t X)\right] = \int_{-\infty}^{\infty} f(x) \exp(i t x)\, dx, \tag{3.8.3}$$

where $E[g(X)]$ is called the *expected value* of the random variable $g(X)$.

In problems of mathematical statistics, it is convenient to define the Fourier transform of $f(x)$ and its inverse in a slightly different way by

$$\mathscr{F}\{f(x)\} = \phi(t) = \int_{-\infty}^{\infty} \exp(i t x)\, f(x)\, dx, \tag{3.8.4}$$

$$\mathscr{F}^{-1}\{\phi(t)\} = f(x) = \frac{1}{2\pi} \int_{-\infty}^{\infty} \exp(-i t x)\, \phi(t)\, dt. \tag{3.8.5}$$

Evidently, the characteristic function of $F(x)$ is the Fourier transform of the density function $f(x)$. The Fourier transform of the distribution function follows from the fact that

$$\mathscr{F}\{F'(x)\} = \mathscr{F}\{f(x)\} = \phi(t) \tag{3.8.6}$$

or, equivalently,

$$\mathscr{F}\{F(x)\} = i t^{-1} \phi(t).$$

The *composition of two distribution functions* $F_1(x)$ and $F_2(x)$ is defined by

$$F(x) = F_1(x) * F_2(x) = \int_{-\infty}^{\infty} F_1(x-y)\, F_2'(y)\, dy. \tag{3.8.7}$$

Thus, the Fourier transform of (3.8.7) gives

$$\begin{aligned} i t^{-1} \phi(t) &= \mathscr{F}\left\{\int_{-\infty}^{\infty} F_1(x-y)\, F_2'(y)\, dy\right\} \\ &= \mathscr{F}\{F_1(x)\}\, \mathscr{F}\{F_2'(x)\} = i t^{-1} \phi_1(t) \phi_2(t),\ F_2'(x) = f_2(x), \end{aligned}$$

whence an important result follows:

$$\phi(t) = \phi_1(t)\, \phi_2(t), \tag{3.8.8}$$

where $\phi_1(t)$ and $\phi_2(t)$ are the characteristic functions of the distribution functions $F_1(x)$ and $F_2(x)$, respectively.

The *nth moment* of a random variable X is defined by

$$m_n = E\left[X^n\right] = \int_{-\infty}^{\infty} x^n f(x)\,dx, \quad n = 1,2,3,\ldots, \tag{3.8.9}$$

provided this integral exists. The first moment m_1 (or simply m) is called the *expectation* of X and has the form

$$m = E(X) = \int_{-\infty}^{\infty} x\, f(x)\, dx. \tag{3.8.10}$$

Thus, the moment of any order n is calculated by evaluating the integral (3.8.9). However, the evaluation of the integral is, in general, a difficult task. This difficulty can be resolved with the help of the characteristic function defined by (3.8.4). Differentiating (3.8.4) n times and putting $t = 0$ gives a fairly simple formula,

$$m_n = \int_{-\infty}^{\infty} x^n\, f(x)\,dx = (-i)^n\, \phi^{(n)}(0), \tag{3.8.11}$$

where $n = 1,2,3,\ldots$.

When $n = 1$, the expectation of a random variable X becomes

$$m_1 = E(X) = \int_{-\infty}^{\infty} x\, f(x)\,dx = (-i)\phi'(0). \tag{3.8.12}$$

Thus, the simple formula (3.8.11) involving the derivatives of the characteristic function provides for the existence and the computation of the moment of any arbitrary order.

Similarly, the variance σ^2 of a random variable is given in terms of the characteristic function as

$$\begin{aligned}\sigma^2 &= \int_{-\infty}^{\infty} (x-m)^2\, f(x)\,dx = m_2 - m_1^2 \\ &= \{\phi'(0)\}^2 - \phi''(0).\end{aligned} \tag{3.8.13}$$

Example 3.8.1 Find the moments of the normal distribution defined by the density function

$$f(x)=\frac{1}{\sigma\sqrt{2\pi}}\exp\left\{-\frac{(x-m)^2}{2\sigma^2}\right\}. \tag{3.8.14}$$

The characteristic function of the normal distribution is the Fourier transform of $f(x)$ and is given by

$$\phi(t)=\frac{1}{\sigma\sqrt{2\pi}}\int_{-\infty}^{\infty}e^{itx}\exp\left\{-\frac{(x-m)^2}{2\sigma^2}\right\}dx.$$

We substitute $x-m=y$ and use Example 3.2.1 to obtain

$$\phi(t)=\frac{\exp(itm)}{\sigma\sqrt{2\pi}}\int_{-\infty}^{\infty}e^{ity}\exp\left(-\frac{y^2}{2\sigma^2}\right)dy=\exp\left(itm-\frac{1}{2}t^2\sigma^2\right). \tag{3.8.15}$$

Thus,

$$m_1=(-i)\phi'(0)=m,$$

$$m_2=-\phi''(0)=\left(m^2+\sigma^2\right),$$

$$m_3=m\left(m^2+3\sigma^2\right).$$

Finally, the variance of the normal distribution is

$$\sigma^2=m_2-m_1^2. \tag{3.8.16}$$

The preceding discussion reveals that characteristic functions are very useful for investigation of certain problems in mathematical statistics. We close this section by discussing some more properties of characteristic functions.

Theorem 3.8.1 (Addition Theorem). The characteristic function of the sum of a finite number of independent random variables is equal to the product of their characteristic functions.

Proof. Suppose $X_1, X_2, \ldots, X_n$ are n independent random variables and $Z=X_1+X_2+\cdots+X_n$. Further, suppose $\phi_1(t), \phi_2(t), \ldots, \phi_n(t)$, and $\phi(t)$ are the characteristic functions of $X_1, X_2, \ldots, X_n$, and Z, respectively.

Then we have

$$\phi(t)=E\left[\exp(itZ)\right]=E\left[\exp\left\{it\left(X_1+X_2+\cdots+X_n\right)\right\}\right],$$

which is, by the independence of the random variables,

$$= E\left(e^{itX_1}\right) E\left(e^{itX_2}\right) \cdots E\left(e^{itX_n}\right)$$

$$= \phi_1(t)\, \phi_2(t) \cdots \phi_n(t). \tag{3.8.17}$$

This proves the *addition theorem.*

Example 3.8.2 Find the expected value and the standard deviation of the sum of n independent normal random variables.

Suppose $X_1, X_2, \ldots, X_n$ are n independent random variables with the normal distributions $N(m_r, \sigma_r)$, where $r = 1, 2, \ldots, n$. The respective characteristic functions of these normal distributions are

$$\phi_r(t) = \exp\left[i t m_r - \frac{1}{2} t^2 \sigma_r^2\right], \qquad r = 1, 2, 3, \ldots, n. \tag{3.8.18}$$

Because of the independence of $X_1, X_2, \ldots, X_n$, the random variable $Z = X_1 + X_2 + \cdots + X_n$ has the characteristic function

$$\phi(t) = \phi_1(t)\phi_2(t)\cdots\phi_n(t)$$

$$= \exp\left[it(m_1 + m_2 + \cdots + m_n) - \frac{1}{2}\left(\sigma_1^2 + \sigma_2^2 + \cdots + \sigma_n^2\right) t^2\right]. \tag{3.8.19}$$

This represents the characteristic function of the normal distribution $N\left(m_1 + m_2 + \cdots + m_n, \sqrt{\sigma_1^2 + \sigma_2^2 + \cdots + \sigma_n^2}\right)$. Thus, the expected value of Z is $(m_1 + m_2 + \cdots + m_n)$ and its standard deviation is $\left(\sigma_1^2 + \sigma_2^2 + \cdots + \sigma_n^2\right)^{\frac{1}{2}}$.

Theorem 3.8.2 (The Central Limit Theorem). If f is a nonnegative function which belongs to $L^1(\mathbb{R})$,

$$\int_{-\infty}^{\infty} f(x)\,dx = 1, \quad \int_{-\infty}^{\infty} x\, f(x)\,dx = 0, \text{ and } \int_{-\infty}^{\infty} x^2 f(x)\,dx = 1, \tag{3.8.20}$$

and $f^n = (f * f * \cdots * f)$ is the n-times convolution of f with itself, then

$$\lim_{n\to\infty} \int_{a\sqrt{n}}^{b\sqrt{n}} f^n(x)\,dx = \int_a^b \frac{1}{\sqrt{2\pi}} \exp\left(-\frac{x^2}{2}\right) dx, \tag{3.8.21}$$

where $-\infty < a < b < \infty$.

Proof. Consider the characteristic function $\chi_{[a,b]}(x)$ defined by

$$\chi_{[a,b]}(x) = \begin{Bmatrix} 1, & \text{for } a \le x \le b \\ 0, & \text{otherwise} \end{Bmatrix}. \tag{3.8.22}$$

Consequently, we obtain

$$\int_{-\infty}^{\infty} \sqrt{n}\, f^n\left(x\sqrt{n}\right) \chi_{[a,b]}(x)\, dx = \int_{a}^{b} \sqrt{n}\, f^n\left(x\sqrt{n}\right) dx = \int_{a\sqrt{n}}^{b\sqrt{n}} f^n(y)\, dy$$

and

$$\frac{1}{\sqrt{2\pi}} \int_{-\infty}^{\infty} \exp\left(-\frac{1}{2}x^2\right) \chi_{[a,b]}(x)\, dx = \frac{1}{\sqrt{2\pi}} \int_{a}^{b} \exp\left(-\frac{x^2}{2}\right) dx.$$

It is sufficient to prove that

$$\lim_{n\to\infty} \int_{-\infty}^{\infty} \sqrt{n}\, f^n\left(x\sqrt{n}\right) \chi_{[a,b]}(x)\, dx = \int_{a}^{b} \frac{1}{\sqrt{2\pi}} \exp\left(-\frac{1}{2}x^2\right) dx.$$

Or, equivalently, it is sufficient to prove that

$$\lim_{n\to\infty} \int_{-\infty}^{\infty} \sqrt{n}\, f^n\left(x\sqrt{n}\right) h(x)\, dx = \frac{1}{\sqrt{2\pi}} \int_{-\infty}^{\infty} \exp\left(-\frac{1}{2}x^2\right) h(x)\, dx,$$

where $h(x)$ belongs to Schwartz's space of infinitely differentiable and rapidly decreasing functions.

We now use the Fourier inversion formula to express

$$\begin{aligned}
\int_{-\infty}^{\infty} \sqrt{n}\, f^n\left(x\sqrt{n}\right) h(x)\, dx &= \int_{-\infty}^{\infty} \sqrt{n}\, f^n\left(x\sqrt{n}\right) \left[\frac{1}{2\pi} \int_{-\infty}^{\infty} \hat{h}(\omega)\, e^{-i\omega x}\, d\omega\right] dx \\
&= \frac{1}{2\pi} \int_{-\infty}^{\infty} \hat{h}(\omega) \left[\int_{-\infty}^{\infty} \sqrt{n}\, f^n\left(x\sqrt{n}\right) e^{-i\omega x} dx\right] d\omega \\
&= \frac{1}{2\pi} \int_{-\infty}^{\infty} \hat{h}(\omega) \left[\int_{-\infty}^{\infty} f^n(t)\, e^{-i\left(\frac{\omega}{\sqrt{n}}\right)t} dt\right] d\omega \\
&= \frac{1}{2\pi} \int_{-\infty}^{\infty} \hat{h}(\omega) \left\{\hat{f}\left(\frac{\omega}{\sqrt{n}}\right)\right\}^n d\omega,
\end{aligned} \tag{3.8.23}$$

where $\hat{f}\left(\frac{\omega}{\sqrt{n}}\right)$ is the Fourier transform of $f(t)$ at the point $\left(\frac{\omega}{\sqrt{n}}\right)$.

Clearly,

$$\left|\hat{f}\left(\frac{\omega}{\sqrt{n}}\right)\right|^n \le \left(\| f \|_1\right)^n = 1 \quad \text{for } \omega \in \mathbb{R},\ \hat{f}(0) = 1, \tag{3.8.24}$$

and hence, for every fixed nonzero $\omega \in \mathbb{R}$, we have

$$\begin{aligned} \hat{f}\left(\frac{\omega}{\sqrt{n}}\right) &= \int_{-\infty}^{\infty} f(x) \exp\left(\frac{i\omega}{\sqrt{n}}\, x\right) dx \\ &= \int_{-\infty}^{\infty} \left\{1 + \frac{i\omega x}{\sqrt{n}} - \frac{\omega^2 x^2}{2n}\right\} \left(1 + \delta_n(x)\right) f(x)\, dx, \end{aligned} \tag{3.8.25}$$

where δ_n is bounded and tends to zero as $n \to \infty$.

In view of condition (3.8.20), result (3.8.25) leads to

$$\hat{f}\left(\frac{\omega}{\sqrt{n}}\right) = \left(1 - \frac{\omega^2}{2n}\right) \{1 + o(1)\} \qquad \text{as } n \to \infty. \tag{3.8.26}$$

Consequently,

$$\left\{\hat{f}\left(\frac{\omega}{\sqrt{n}}\right)\right\}^n = \left[1 - \frac{\omega^2}{2n} \{1 + o(1)\}\right]^n \to \exp\left(-\frac{\omega^2}{2}\right) \quad \text{as } n \to \infty, \tag{3.8.27}$$

in which the standard result

$$\lim_{n\to\infty} \left(1 - \frac{x}{n}\right)^n = e^{-x}$$

is used.

Finally, the use of (3.8.23) and (3.8.25) combined with the Lebesgue dominated convergence theorem leads us to obtain the final result

$$\begin{aligned} \lim_{n\to\infty} \int_{-\infty}^{\infty} \sqrt{n}\, f^n\left(x\sqrt{n}\right) h(x)\, dx &= \frac{1}{2\pi} \int_{-\infty}^{\infty} \hat{h}(\omega) \left[\lim_{n\to\infty} \left\{\hat{f}\left(\frac{\omega}{\sqrt{n}}\right)\right\}^n\right] d\omega \\ &= \frac{1}{2\pi} \int_{-\infty}^{\infty} \hat{h}(\omega) \exp\left(-\frac{\omega^2}{2}\right) d\omega, \qquad \text{by (3.8.6)} \\ &= \frac{1}{2\pi} \int_{-\infty}^{\infty} h(x) \sqrt{2\pi} \exp\left(-\frac{x^2}{2}\right) dx \quad \text{by (3.4.28).} \end{aligned}$$

This completes the proof.

Note. This theorem is perhaps the most significant result in mathematical statistics.

Finally, we state another version of the central limit theorem without proof.

Theorem 3.8.3 **(*The Lévy-Cramér Theorem*).** Suppose $\{X_n\}$ is a sequence of random variables, $F_n(x)$ and $\phi_n(t)$ are, respectively, the distribution and characteristic functions of X_n. Then, the sequence $\{F_n(x)\}$ is convergent to a distribution function $F(x)$ if and only if the sequence $\{\phi_n(t)\}$ is convergent at every point t on the real line to a function $\phi(t)$ continuous in some neighborhood of the origin. The limit function $\phi(t)$ is then the characteristic function of the limit distribution function $F(x)$, and the convergence $\phi_n(t) \to \phi(t)$ is uniform in every finite interval on the t-axis.

All of the ideas developed in this section can be generalized for the multi-dimensional distribution functions by the use of multiple Fourier transforms. We refer interested readers to Lukacs (1960).

3.9 Applications of Fourier Transforms to Ordinary Differential Equations

We consider the nth order linear nonhomogeneous ordinary differential equations with constant coefficients

$$Ly(x) = f(x), \tag{3.9.1}$$

where L is the nth order differential operator given by

$$L \equiv a_n D^n + a_{n-1} D^{n-1} + \cdots + a_1 D + a_0, \tag{3.9.2}$$

where $a_n, a_{n-1}, \ldots, a_1, a_0$ are constants, $D \equiv \frac{d}{dx}$, and $f(x)$ is a given function.

Application of the Fourier transform to both sides of (3.9.1) gives

$$\left[a_n (ik)^n + a_{n-1}(ik)^{n-1} + \cdots + a_1 (ik) + a_0\right] \hat{y}(k) = \hat{f}(k),$$

where $\mathscr{F}\{y(x)\} = \hat{y}(k)$ and $\mathscr{F}\{f(x)\} = \hat{f}(k)$.

Or, equivalently,

$$P(ik)\, \hat{y}(k) = \hat{f}(k),$$

where

$$P(z) = \sum_{r=0}^{n} a_r z^r .$$

Thus,

$$\hat{y}(k) = \frac{\hat{f}(k)}{P(ik)} = \hat{f}(k)\,\hat{q}(k), \tag{3.9.3}$$

where

$$\hat{q}(k) = \frac{1}{P(ik)}.$$

Applying the convolution theorem 3.3.7 to (3.9.3) gives the formal solution

$$y(x) = \int_{-\infty}^{\infty} f(\xi)\, q(x-\xi)\, d\xi, \tag{3.9.4}$$

provided $q(x) = \mathscr{F}^{-1}\{\hat{q}(k)\}$ is known explicitly.

In order to give a physical interpretation of the solution (3.9.4), we consider the differential equation with a suddenly applied impulse function $f(x) = \delta(x)$ in the form

$$L\{G(x)\} = \delta(x). \tag{3.9.5}$$

The solution of this equation can be written from the inversion of (3.9.3) in the form

$$G(x) = \mathscr{F}^{-1}\{\hat{q}(k)\} = q(x). \tag{3.9.6}$$

Thus, the solution (3.9.4) takes the form

$$y(x) = \int_{-\infty}^{\infty} f(\xi) G(x-\xi)\, d\xi. \tag{3.9.7}$$

Clearly, $G(x)$ behaves like a *Green's function*, that is, it is the response to a *unit impulse*. In any physical system, $f(x)$ usually represents the *input function*, while $y(x)$ is referred to as the *output* obtained by the superposition principle. The Fourier transform of $G(x)$ is $\hat{q}(k)$ which is called the *admittance*. In order to find the response to a given input, we determine the Fourier transform of the input function, multiply the result by the admittance, and then apply the inverse Fourier transform to the product so obtained. We illustrate these ideas by solving some simple problems.

Example 3.9.1 Find the solution of the ordinary differential equation

$$-\frac{d^2u}{dx^2}+a^2u=f(x), \quad -\infty<x<\infty \tag{3.9.8}$$

by the Fourier transform method.

Application of the Fourier transform to (3.9.8) gives

$$\hat{u}(k)=\frac{\hat{f}(k)}{k^2+a^2}.$$

This can readily be inverted by the Convolution Theorem 3.3.7 to obtain

$$u(x)=\int_{-\infty}^{\infty} f(\xi)\, g(x-\xi)\, d\xi, \tag{3.9.9}$$

where $g(x)=\mathscr{F}^{-1}\left\{\frac{1}{k^2+a^2}\right\}=\frac{1}{2a}\exp(-a|x|)$ by Example 3.3.2. Thus, the exact solution is

$$u(x)=\frac{1}{2a}\int_{-\infty}^{\infty} f(\xi)\, e^{-a|x-\xi|}\, d\xi. \tag{3.9.10}$$

Example 3.9.2 Solve the following ordinary differential equation

$$2u''(t)+t\,u'(t)+u(t)=0. \tag{3.9.11}$$

We apply the Fourier transform of $u(t)$ and result (3.3.9) to this equation to find

$$-2\omega^2\hat{u}+i\,\frac{d}{d\omega}\left[\mathscr{F}\{u'(t)\}\right]+\hat{u}=0.$$

Or

$$-2\omega^2\hat{u}+i\,\frac{d}{d\omega}\left[(-i)\,\omega\,\hat{u}(\omega)\right]+\hat{u}=0.$$

Thus,

$$\frac{d\hat{u}}{d\omega}=-2\omega\,\hat{u}.$$

The solution of this equation is

$$\hat{u}(\omega)=C\exp\left(-\omega^2\right),$$

where C is a constant of integration. The inverse Fourier transform gives the solution

$$u(t) = D\exp\left(-\frac{t^2}{4}\right), \tag{3.9.12}$$

where D is a constant.

Example 3.9.3 (***The Bernoulli-Euler Beam Equation***). We consider the vertical deflection $u(x)$ of an infinite beam on an elastic foundation under the action of a prescribed vertical load $W(x)$. The deflection $u(x)$ satisfies the ordinary differential equation

$$EI\frac{d^4u}{dx^4} + \kappa u = W(x), \qquad -\infty < x < \infty, \tag{3.9.13}$$

where EI is the flexural rigidity and κ is the foundation modulus of the beam. We find the solution assuming that $W(x)$ has a compact support and that u, u', u'', u''' all tend to zero as $|x| \to \infty$.

We first rewrite (3.9.13) as

$$\frac{d^4u}{dx^4} + a^4 u = w(x), \tag{3.9.14}$$

where $a^4 = \kappa/EI$ and $w(x) = W(x)/EI$. Using the Fourier transform to (3.9.14) gives

$$\hat{u}(k) = \frac{\hat{w}(k)}{k^4 + a^4}.$$

The inverse Fourier transform gives the solution

$$\begin{aligned} u(x) &= \frac{1}{2\pi}\int_{-\infty}^{\infty} \frac{\hat{w}(k)}{k^4 + a^4}\, e^{ikx}\, dk \\ &= \frac{1}{2\pi}\int_{-\infty}^{\infty} \frac{e^{ikx}}{k^4 + a^4}\, dk \int_{-\infty}^{\infty} w(\xi)\, e^{-ik\xi}\, d\xi \\ &= \int_{-\infty}^{\infty} w(\xi)\, G(x,\xi)\, d\xi, \end{aligned} \tag{3.9.15}$$

where

$$G(x,\xi)=\frac{1}{2\pi}\int_{-\infty}^{\infty}\frac{e^{ik(x-\xi)}}{k^4+a^4}\,dk=\frac{1}{\pi}\int_0^{\infty}\frac{\cos k(x-\xi)\,dk}{k^4+a^4}. \tag{3.9.16}$$

This integral can be evaluated by the theorem of residues or by using the table of Fourier integrals. We simply state the result

$$G(x,\xi)=\frac{1}{2a^3}\exp\left(-\frac{a}{\sqrt{2}}|x-\xi|\right)\sin\left[\frac{a(x-\xi)}{\sqrt{2}}+\frac{\pi}{4}\right]. \tag{3.9.17}$$

In particular, we find the explicit solution due to a concentrated load of unit strength acting at some point x_0; that is, $w(x)=\delta(x-x_0)$. Then, the solution for this case becomes

$$u(x)=\int_{-\infty}^{\infty}\delta(\xi-x_0)\,G(x,\xi)\,d\xi=G(x,x_0). \tag{3.9.18}$$

Thus, the kernel $G(x,\xi)$ involved in the solution (3.9.15) has the physical significance of being the deflection as a function of x due to a unit point load acting at ξ. Thus, the deflection due to a point load of strength $w(\xi)\,d\xi$ at ξ is $w(\xi)\,d\xi\cdot G(x,\xi)$ and hence (3.9.15) represents the superposition of all such incremental deflections.

The reader is referred to a more general dynamic problem of an infinite Bernoulli-Euler beam with damping and elastic foundation that has been solved by Stadler and Shreeves (1970) and also by Sheehan and Debnath (1972). These authors used the joint Fourier and Laplace transform method to determine the steady state and the transient solutions of the beam problem.

3.10 Solutions of Integral Equations

The method of Fourier transforms can be used to solve simple integral equations of the convolution type. We illustrate the method by examples.

We first solve the Fredholm integral equation with convolution kernel in the form

$$\int_{-\infty}^{\infty}f(t)\,g(x-t)\,dt+\lambda f(x)=u(x), \tag{3.10.1}$$

where $g(x)$ and $u(x)$ are given functions and λ is a known parameter.

Application of the Fourier transform defined by (3.2.3) to (3.10.1) gives

$$\hat{f}(k)\,\hat{g}(k) + \lambda\,\hat{f}(k) = \hat{u}(k).$$

Or

$$\hat{f}(k) = \frac{\hat{u}(k)}{\hat{g}(k) + \lambda}. \tag{3.10.2}$$

The inverse Fourier transform leads to a formal solution

$$f(x) = \frac{1}{2\pi}\int_{-\infty}^{\infty} \frac{\hat{u}(k)e^{ikx}dk}{\hat{g}(k) + \lambda}. \tag{3.10.3}$$

In particular, if $g(x) = \frac{1}{x}$ so that

$$\hat{g}(k) = -\pi i \,\mathrm{sgn}\, k,$$

then the solution becomes

$$f(x) = \frac{1}{2\pi}\int_{-\infty}^{\infty} \frac{\hat{u}(k)e^{ikx}dk}{(\lambda - \pi i\,\mathrm{sgn}\, k)}. \tag{3.10.4}$$

If $\lambda = 1$ and $g(x) = \frac{1}{2}\left(\frac{x}{|x|}\right)$ so that $\hat{g}(k) = \frac{1}{(ik)}$, solution (3.10.3) reduces to the form

$$\begin{aligned} f(x) &= \frac{1}{2\pi}\int_{-\infty}^{\infty} (ik)\,\frac{\hat{u}(k)e^{ikx}dk}{(1+ik)} \\ &= \frac{1}{2\pi}\int_{-\infty}^{\infty} \mathscr{F}\{u'(x)\}\,\mathscr{F}\{e^{-x}\}\,e^{ikx}\,dk \\ &= u'(x) * e^{-x} \\ &= \int_{-\infty}^{\infty} u'(\xi)\,\exp(\xi - x)\,d\xi. \end{aligned} \tag{3.10.5}$$

Example 3.10.1 Find the solution of the integral equation

$$\int_{-\infty}^{\infty} f(x-\xi)\,f(\xi)\,d\xi = \frac{1}{x^2 + a^2}. \tag{3.10.6}$$

Application of the Fourier transform gives

$$\hat{f}(k)\,\hat{f}(k) = \frac{e^{-a|k|}}{2a}.$$

Or

$$\hat{f}(k) = \frac{1}{\sqrt{2a}} \exp\left\{-\frac{1}{2}\,a|k|\right\}. \tag{3.10.7}$$

The inverse Fourier transform gives the solution

$$f(x) = \frac{1}{2\pi}\,\frac{1}{\sqrt{2a}} \int_{-\infty}^{\infty} \exp\left(i\,k\,x - \frac{1}{2}\,a|k|\right) dk$$

$$= \frac{1}{2\pi\sqrt{2a}} \left[\int_0^{\infty} \exp\left\{-k\left(\frac{a}{2} + i\,x\right)\right\} dk + \int_0^{\infty} \exp\left\{-k\left(\frac{a}{2} - i\,x\right)\right\} dk\right]$$

$$= \frac{1}{2\pi\sqrt{2a}} \left[\frac{4a}{\left(4x^2 + a^2\right)}\right] = \frac{1}{\pi} \cdot \frac{\sqrt{2a}}{\left(4x^2 + a^2\right)}.$$

Example 3.10.2 Solve the integral equation

$$\int_{-\infty}^{\infty} \frac{f(t)\,dt}{(x-t)^2 + a^2} = \frac{1}{\left(x^2 + b^2\right)}. \tag{3.10.8}$$

Applying the Fourier transform to both sides of (3.10.8), we obtain

$$\hat{f}(k)\,\mathscr{F}\left\{\frac{1}{x^2 + a^2}\right\} = \frac{e^{-b|k|}}{2b}.$$

Or,

$$\hat{f}(k) \cdot \frac{e^{-a|k|}}{2a} = \frac{e^{-b|k|}}{2b}.$$

Thus,

$$\hat{f}(k) = \left(\frac{a}{b}\right) \exp\{-|k|(b-a)\}. \tag{3.10.9}$$

The inverse Fourier transform leads to the final solution

$$f(x)=\frac{a}{2\pi b}\int_{-\infty}^{\infty}\exp\left[ikx-|k|(b-a)\right]\,dk$$

$$=\frac{a}{2\pi b}\left[\int_{0}^{\infty}\exp\left[-k\left\{(b-a)+ix\right\}\right]\,dk+\int_{0}^{\infty}\exp\left[-k\left\{(b-a)-ix\right\}\right]\right]dk$$

$$=\frac{a}{2\pi b}\left[\frac{1}{(b-a)+ix}+\frac{1}{(b-a)-ix}\right]$$

$$=\left(\frac{a}{\pi b}\right)\frac{(b-a)}{(b-a)^2+x^2}. \tag{3.10.10}$$

Example 3.10.3 Solve the integral equation

$$f(x)+4\int_{-\infty}^{\infty}e^{-a|x-t|}\,f(t)\,dt=g(x). \tag{3.10.11}$$

Application of the Fourier transform gives

$$\hat{f}(k)+4\hat{f}(k)\cdot\frac{2a}{\left(a^2+k^2\right)}=\hat{g}(k)$$

$$\hat{f}(k)=\frac{\left(a^2+k^2\right)}{a^2+k^2+8a}\,\hat{g}(k). \tag{3.10.12}$$

The inverse Fourier transform gives the solution

$$f(x)=\frac{1}{2\pi}\int_{-\infty}^{\infty}\frac{\left(a^2+k^2\right)\hat{g}(k)}{a^2+k^2+8a}\,e^{ikx}\,dk. \tag{3.10.13}$$

In particular, if $a=1$ and $g(x)=e^{-|x|}$ so that $\hat{g}(k)=\frac{2}{1+k^2}$, then the solution (3.10.13) becomes

$$f(x)=\frac{1}{\pi}\int_{-\infty}^{\infty}\frac{e^{ikx}}{k^2+3^2}\,dk. \tag{3.10.14}$$

For $x>0$, we use a semicircular closed contour in the lower half of the complex plane to evaluate (3.10.14). It turns out that

$$f(x)=\frac{1}{3}\,e^{-3x}. \tag{3.10.15}$$

Similarly, for $x<0$, a semicircular closed contour in the upper half of the complex plane is used to evaluate (3.10.14) so that

$$f(x) = \frac{1}{3}\, e^{3x}, \quad x < 0. \tag{3.10.16}$$

Thus, the final solution is

$$f(x) = \frac{1}{3}\, \exp(-3|x|). \tag{3.10.17}$$

3.11 Solutions of Partial Differential Equations

In this section, we present several examples of applications of Fourier transforms to partial differential equations.

Example 3.11.1 (Dirichlet's Problem in the Half Plane). We consider the solution of the Laplace equation in the half plane

$$u_{xx} + u_{yy} = 0, \quad -\infty < x < \infty, \quad y \ge 0, \tag{3.11.1}$$

with the boundary conditions

$$u(x,0) = f(x), \qquad -\infty < x < \infty, \tag{3.11.2}$$

$$u(x,y) \to 0 \qquad \text{as } |x| \to \infty, \quad y \to \infty. \tag{3.11.3}$$

We apply the Fourier transform with respect to x to the system (3.11.1)-(3.11.3) to obtain

$$\frac{d^2\hat{u}}{dy^2} - k^2\,\hat{u} = 0 \tag{3.11.4}$$

$$\hat{u}(k,0) = \hat{f}(k), \quad \text{and } \hat{u}(k,y) \to 0 \qquad \text{as } y \to \infty. \tag{3.11.5a,b}$$

Thus, the solution of this transformed system is

$$\hat{u}(k,y) = \hat{f}(k)\, \exp(-|k|\,y). \tag{3.11.6a,b}$$

Application of Theorem 3.3.7 gives the solution

$$u(x,y) = \int_{-\infty}^{\infty} f(\xi)\, g(x-\xi)\, d\xi, \tag{3.11.7}$$

where

$$g(x) = \mathscr{F}^{-1}\left\{e^{-|k|y}\right\} = \frac{1}{\pi}\, \frac{y}{(x^2 + y^2)}. \tag{3.11.8}$$

Consequently, solution (3.11.7) becomes

$$u(x,y) = \frac{y}{\pi}\int_{-\infty}^{\infty}\frac{f(\xi)\,d\xi}{(x-\xi)^2+y^2}, \quad y>0. \tag{3.11.9}$$

This is the well known *Poisson integral formula* in the half plane. It is noted that

$$\lim_{y\to 0^+} u(x,y) = \int_{-\infty}^{\infty} f(\xi)\left[\lim_{y\to 0^+}\frac{y}{\pi}\cdot\frac{1}{(x-\xi)^2+y^2}\right]d\xi = \int_{-\infty}^{\infty} f(\xi)\,\delta(x-\xi)\,d\xi, \tag{3.11.10}$$

in which Cauchy's definition of the delta function is used, that is,

$$\delta(x-\xi) = \lim_{y\to 0^+}\frac{y}{\pi}\cdot\frac{1}{(x-\xi)^2+y^2}. \tag{3.13.11}$$

This may be recognized as a solution of the Laplace equation for a dipole source at $(x,y)=(\xi,0)$.

In particular, when

$$f(x) = T_0\, H(a-|x|), \tag{3.11.12}$$

the solution (3.11.9) reduces to the integral form

$$\begin{aligned} u(x,y) &= \frac{yT_0}{\pi}\int_{-a}^{a}\frac{d\xi}{(\xi-x)^2+y^2} \\ &= \frac{T_0}{\pi}\left[\tan^{-1}\left(\frac{x+a}{y}\right)-\tan^{-1}\left(\frac{x-a}{y}\right)\right] \\ &= \frac{T_0}{\pi}\tan^{-1}\left(\frac{2ay}{x^2+y^2-a^2}\right). \end{aligned} \tag{3.11.13}$$

The curves in the upper half plane, for which the steady state temperature is constant, are known as *isothermal curves*. In this case, these curves represent a family of circular arcs

$$x^2+y^2-\alpha\, y = a^2 \tag{3.11.14}$$

with centers on the y-axis and fixed end points on the x-axis at $x=\pm a$, as shown in Figure 3.14.

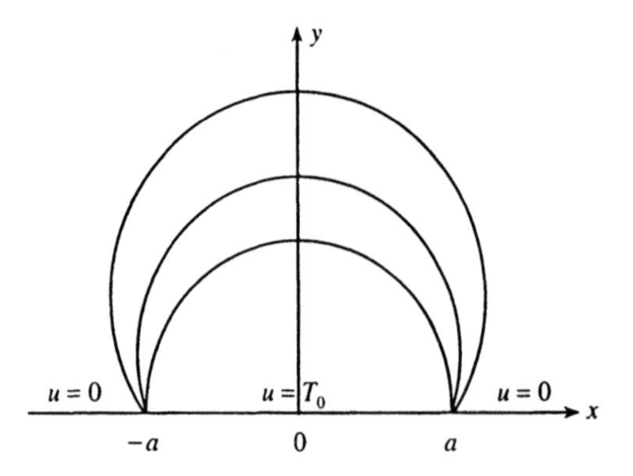

Figure 3.14. Isothermal curves representing a family of circular arcs.

Another special case deals with

$$f(x) = \delta(x). \tag{3.11.15}$$

The solution for this case follows from (3.11.9) and is given by

$$u(x,t) = \frac{y}{\pi} \int_{-\infty}^{\infty} \frac{\delta(\xi)\,d\xi}{(x-\xi)^2 + y^2} = \frac{y}{\pi} \frac{1}{(x^2+y^2)}. \tag{3.11.16}$$

Further, we can readily deduce the solution of the *Neumann problem* in the half plane from the solution of the Dirichlet problem.

Example 3.11.2 **(*Neumann's Problem in the Half Plane*).** Find a solution of the Laplace equation

$$u_{xx} + u_{yy} = 0, \qquad -\infty < x < \infty, \qquad y > 0, \tag{3.11.17}$$

with the Neumann boundary condition

$$u_y(x,0) = f(x), \quad -\infty < x < \infty. \tag{3.11.18}$$

This condition specifies the normal derivative on the boundary and physically describes the fluid flow or heat flux at the boundary.

We define a new function $\upsilon(x,y) = u_y(x,y)$ so that

$$u(x,y) = \int^{y} \upsilon(x,\eta)\,d\eta, \tag{3.11.19}$$

where an arbitrary constant can be added to the right-hand side. Clearly, the function υ satisfies the Laplace equation

$$\frac{\partial^2 \upsilon}{\partial x^2}+\frac{\partial^2 \upsilon}{\partial y^2}=\frac{\partial^2 u_y}{\partial x^2}+\frac{\partial^2 u_y}{\partial y^2}=\frac{\partial}{\partial y}\left(u_{xx}+u_{yy}\right)=0$$

with the boundary condition

$$\upsilon(x,0)=u_y(x,0)=f(x) \quad \text{for } -\infty<x<\infty .$$

Thus, $\upsilon(x,y)$ satisfies the Laplace equation with the Dirichlet condition on the boundary. Obviously, the solution is given by (3.11.9), that is,

$$\upsilon(x,y)=\frac{y}{\pi}\int_{-\infty}^{\infty}\frac{f(\xi)\,d\xi}{(x-\xi)^2+y^2}. \tag{3.11.20}$$

Then, the solution $u(x,y)$ can be obtained from (3.11.19) in the form

$$\begin{aligned}
u(x,y)&=\int^{y}\upsilon(x,\eta)\,d\eta=\frac{1}{\pi}\int^{y}\eta\,d\eta\int_{-\infty}^{\infty}\frac{f(\xi)\,d\xi}{(x-\xi)^2+\eta^2}\\
&=\frac{1}{\pi}\int_{-\infty}^{\infty}f(\xi)\,d\xi\int^{y}\frac{\eta\,d\eta}{(x-\xi)^2+\eta^2}, \quad y>0\\
&=\frac{1}{2\pi}\int_{-\infty}^{\infty}f(\xi)\,\log\left[(x-\xi)^2+y^2\right]d\xi,
\end{aligned} \tag{3.11.21}$$

where an arbitrary constant can be added to this solution. In other words, the solution of any Neumann problem is uniquely determined up to an arbitrary constant.

Example 3.11.3 **(*The Cauchy Problem for the Diffusion Equation*).** We consider the initial value problem for a one-dimensional diffusion equation with no sources or sinks involved

$$u_t=\kappa\,u_{xx}, \quad -\infty<x<\infty, \quad t>0, \tag{3.11.22}$$

where κ is a diffusivity constant with the initial condition

$$u(x,0)=f(x), \quad -\infty<x<\infty. \tag{3.11.23}$$

We solve this problem using the Fourier transform in the space variable x defined by (3.2.3). Application of this transform to (3.11.22)-(3.11.23) gives

$$\hat{u}_t=-\kappa\,k^2\hat{u}, \qquad t>0, \tag{3.11.24}$$

$$\hat{u}(k,0) = \hat{f}(k). \tag{3.11.25}$$

The solution of the transformed system is

$$\hat{u}(k,t) = \hat{f}(k)\,\exp\left(-\kappa k^2 t\right). \tag{3.11.26}$$

The inverse Fourier transform gives the solution

$$u(x,t) = \frac{1}{2\pi}\int_{-\infty}^{\infty}\hat{f}(k)\,\exp\left[\left(ikx - \kappa k^2 t\right)\right]dk$$

which is, by convolution theorem 3.3.7,

$$= \int_{-\infty}^{\infty} f(\xi)\, g(x-\xi)\, d\xi, \tag{3.11.27}$$

where

$$g(x) = \mathcal{F}^{-1}\left\{e^{-\kappa k^2 t}\right\} = \frac{1}{\sqrt{4\pi\kappa t}}\exp\left(-\frac{x^2}{4\kappa t}\right), \qquad \text{by (3.2.4).}$$

Thus, solution (3.11.27) becomes

$$u(x,t) = \frac{1}{\sqrt{4\pi\kappa t}}\int_{-\infty}^{\infty} f(\xi)\exp\left[-\frac{(x-\xi)^2}{4\kappa t}\right]d\xi. \tag{3.11.28}$$

The integrand involved in the solution consists of the initial value $f(x)$ and the *Green's function* (or *fundamental solution*) $G(x-\xi,t)$ of the diffusion equation for the infinite interval:

$$G(x-\xi,\, t) = \frac{1}{\sqrt{4\pi\kappa t}}\,\exp\left[-\frac{(x-\xi)^2}{4\kappa t}\right]. \tag{3.11.29}$$

Therefore, in terms of $G(x-\xi,t)$, solution (3.11.28) can be written as

$$u(x,t) = \int_{-\infty}^{\infty} f(\xi)\; G(x-\xi,t)\; d\xi, \tag{3.11.30}$$

so, in the limit as $t \to 0+$, this formally becomes

$$u(x,0) = f(x) = \int_{-\infty}^{\infty} f(\xi) \lim_{t\to 0+} G(x-\xi,t)\; d\xi.$$

The limit of $G(x-\xi,t)$, as $t \to 0+$, represents the Dirac delta function

$$\delta(x-\xi) = \lim_{t\to 0+} \frac{1}{2\sqrt{\pi\kappa t}} \exp\left[-\frac{(x-\xi)^2}{4\kappa t}\right]. \tag{3.11.31}$$

It is important to point out that the integrand in (3.11.30) consists of the initial temperature distribution $f(x)$ and the Green's function $G(x-\xi,t)$, which represents the temperature response along the rod at time t due to an initial unit impulse of heat at $x=\xi$. The physical meaning of the solution (3.11.30) is that the initial temperature distribution $f(x)$ is decomposed into a spectrum of impulses of magnitude $f(\xi)$ at each point $x=\xi$ to form the resulting temperature $f(\xi)\,G(x-\xi,t)$. According to the *principle of superposition*, the resulting temperature is integrated to find the solution (3.11.30).

We make the change of variable

$$\frac{\xi-x}{2\sqrt{\kappa t}} = \zeta, \quad d\zeta = \frac{d\xi}{2\sqrt{\kappa t}}$$

to express solution (3.11.28) in the form

$$u(x,t) = \frac{1}{\sqrt{\pi}} \int_{-\infty}^{\infty} f\left(x+2\sqrt{\kappa t}\,\zeta\right) \exp\left(-\zeta^2\right) d\zeta. \tag{3.11.32}$$

The integral solution (3.11.32) or (3.11.28) is called the *Poisson integral representation* of the temperature distribution. This integral is convergent for all time $t>0$, and the integrals obtained from (3.11.32) by differentiation under the integral sign with respect to x and t are uniformly convergent in the neighborhood of the point (x,t). Hence, the solution $u(x,t)$ and its derivatives of all orders exist for $t>0$.

Finally, we consider two special cases:

(a) $f(x)=\delta(x)$, and (b) $f(x)=T_0\,H(x)$, where T_0 is a constant.

For case (a), the solution (3.11.28) reduces to

$$u(x,t) = \frac{1}{\sqrt{4\pi\kappa t}} \int_{-\infty}^{\infty} \delta(\xi) \exp\left[-\frac{(x-\xi)^2}{4\kappa t}\right] d\xi$$

$$= \frac{1}{\sqrt{4\pi\kappa t}} \exp\left(-\frac{x^2}{4\kappa t}\right). \tag{3.11.33}$$

This is usually called the *Green's function* or the *fundamental solution* of the diffusion equation and is drawn in Figure 3.15 for different values of $\tau = 2\sqrt{\kappa t}$.

At any time t, the solution $u(x,t)$ is Gaussian. The peak height of $u(x,t)$ decreases inversely with $\sqrt{\kappa t}$, whereas the width of the solution $\left(x \approx \sqrt{\kappa t}\right)$ increases with $\sqrt{\kappa t}$. In fact, the initially sharply peaked profile is gradually smoothed out as $t \to \infty$ under the action of diffusion. These are remarkable features for diffusion phenomena.

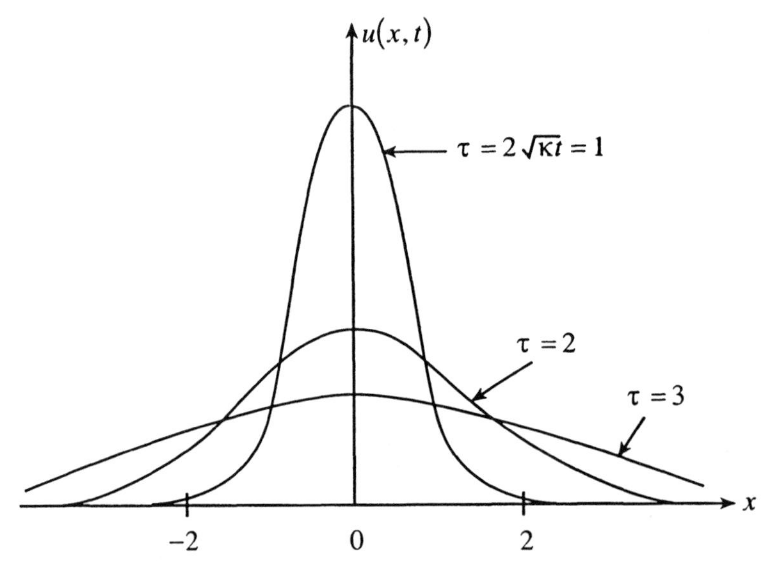

Figure 3.15. The temperature distribution $u(x,t)$.

For case (b), the initial condition is discontinuous. In this case, the solution is

$$u(x,t) = \frac{T_0}{2\sqrt{\pi \kappa t}} \int_0^\infty \exp\left[-\frac{(x-\xi)^2}{4\kappa t}\right] d\xi. \tag{3.11.34}$$

Introducing the change of variable $\eta = \dfrac{\xi - x}{2\sqrt{\kappa t}}$, we can express solution (3.11.34) in the form

$$u(x,t) = \frac{T_0}{\sqrt{\pi}} \int_{-x/2\sqrt{\kappa t}}^\infty e^{-\eta^2}\, d\eta = \frac{T_0}{2} \operatorname{erfc}\left(-\frac{x}{2\sqrt{\kappa t}}\right)$$

$$= \frac{T_0}{2}\left[1 + \operatorname{erf}\left(\frac{x}{2\sqrt{\kappa t}}\right)\right]. \tag{3.11.35}$$

This shows that, at $t = 0$, the solution coincides with the initial data $u(x,0) = T_0$. The graph of $\frac{1}{T_0}\,u(x,t)$ against x is shown in Figure 3.16. As t increases, the discontinuity is gradually smoothed out, whereas the width of the transition zone increases as $\sqrt{\kappa t}$.

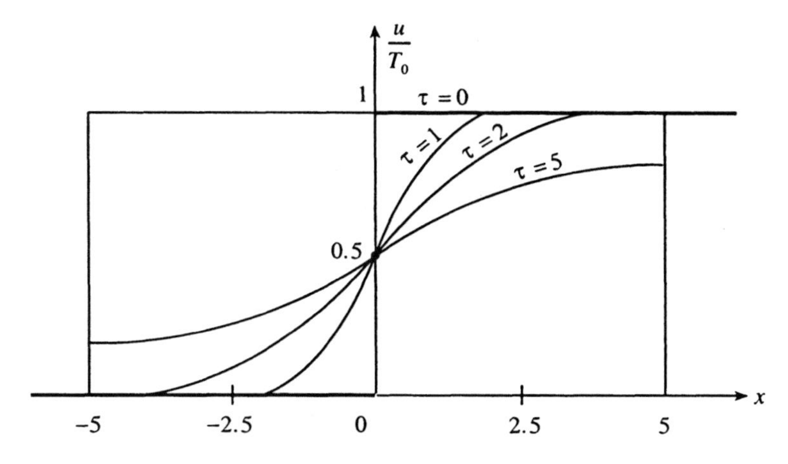

Figure 3.16. The temperature distribution due to discontinuous initial data for different values of $\tau = 2\sqrt{\kappa t} = 0, 1, 2, 5$.

***Example 3.11.4* (*The Inhomogeneous Cauchy Problem for the Wave Equation*).** We use the joint Laplace and Fourier transform method to solve the inhomogeneous Cauchy problem

$$u_{tt} - c^2 u_{xx} = q(x,t), \quad x \in R, \quad t > 0, \tag{3.11.36}$$

$$u(x,0) = f(x), \quad u_t(x,0) = g(x) \quad \text{for all} \quad x \in R, \tag{3.11.37}$$

where $q(x,t)$ is a given function representing a source term.

We define the joint Laplace and Fourier transform of $u(x,t)$ by

$$\bar{\hat{u}}(k,s) = \mathscr{L}\left[\mathscr{F}\{u(x,t)\}\right] = \frac{1}{\sqrt{2\pi}} \int_{-\infty}^{\infty} e^{-ikx}\,dx \int_{0}^{\infty} e^{-st} u(x,t)\,dt. \tag{3.11.38}$$

Application of the joint transform leads to the solution of the transformed problem in the form

$$\bar{\hat{u}}(k,s)=\frac{s\hat{f}(k)+\hat{g}(k)+\bar{\hat{q}}(k,s)}{\left(s^2+c^2k^2\right)}. \tag{3.11.39}$$

The inverse Laplace transform of (3.11.39) gives

$$\hat{u}(k,t)=\hat{f}(k)\cos(ckt)+\frac{1}{ck}\hat{g}(k)\sin(ckt)+\frac{1}{ck}\mathscr{L}^{-1}\left\{\frac{ck}{s^2+c^2k^2}\cdot\bar{\hat{q}}(k,s)\right\}$$

$$=\hat{f}(k)\cos(ckt)+\frac{\hat{g}(k)}{ck}\sin(ckt)+\frac{1}{ck}\int_0^t \sin ck(t-\tau)\,\hat{q}(k,\tau)\,d\tau. \tag{3.11.40}$$

The inverse Fourier transform leads to the solution

$$u(x,t)=\frac{1}{4\pi}\int_{-\infty}^{\infty}\left(e^{ickt}+e^{-ickt}\right)e^{ikx}\hat{f}(k)\,dk+\frac{1}{4\pi}\int_{-\infty}^{\infty}\left(e^{ickt}-e^{-ickt}\right)e^{ikx}\cdot\frac{\hat{g}(k)}{ick}\,dk$$

$$+\frac{1}{2\pi}\cdot\frac{1}{2c}\int_0^t d\tau\int_{-\infty}^{\infty}\frac{\hat{q}(k,\tau)}{ik}\left[e^{ick(t-\tau)}-e^{-ick(t-\tau)}\right]e^{ikx}\,dk.$$

We next use the following results

$$f(x)=\mathscr{F}^{-1}\left\{\hat{f}(k)\right\}=\frac{1}{2\pi}\int_{-\infty}^{\infty}e^{ikx}\hat{f}(k)\,dk$$

and

$$g(x)=\mathscr{F}^{-1}\left\{\hat{g}(k)\right\}=\frac{1}{2\pi}\int_{-\infty}^{\infty}e^{ikx}\hat{g}(k)\,dk$$

to obtain the final form of the solution

$$u(x,t)=\frac{1}{2}\left[f(x+ct)+f(x-ct)\right]+\frac{1}{2c}\int_{x-ct}^{x+ct}g(\xi)\,d\xi$$

$$+\frac{1}{2c}\int_0^t d\tau\,\frac{1}{2\pi}\int_{-\infty}^{\infty}\hat{q}(k,\tau)\,dk\int_{x-c(t-\tau)}^{x+c(t-\tau)}e^{ik\xi}\,d\xi$$

$$=\frac{1}{2}\left[f(x-ct)+f(x+ct)\right]+\frac{1}{2c}\int_{x-ct}^{x+ct}g(\xi)\,d\xi+\frac{1}{2c}\int_0^t d\tau\int_{x-c(t-\tau)}^{x+c(t-\tau)}q(\xi,\tau)\,d\xi.$$

(3.11.41)

In the case of the homogeneous Cauchy problem, $q(x,t) \equiv 0$, the solution reduces to the famous d'Alembert solution

$$u(x,t) = \frac{1}{2}\left[f(x-ct) + f(x+ct)\right] + \frac{1}{2c}\int_{x-ct}^{x+ct} g(\tau)\,d\tau. \qquad (3.11.42)$$

It can be verified by direct substitution that $u(x,t)$ satisfies the homogeneous wave equation, provided f is twice differentiable and g is differentiable. Further, the d'Alembert solution (3.11.42) can be used to show that this problem is *well-posed.* The solution (3.11.42) consists of terms involving $\frac{1}{2}\, f(x \pm ct)$ and the term involving the integral of g. Both terms combined together suggest that the value of the solution at position x and time t depends only on the initial values of $f(x)$ at points $x \pm ct$ and the value of the integral of g between these points. The interval $(x-ct,\, x+ct)$ is called the *domain of dependence* of (x,t). The terms involving $f(x \pm ct)$ in (3.11.42) show that equal waves are propagated along the characteristics with constant velocity c.

In particular, if $g(x) = 0$, the solution is represented by the first two terms in (3.11.42), that is,

$$u(x,t) = \frac{1}{2}\left[f(x-ct) + f(x+ct)\right]. \qquad (3.11.43)$$

Physically, this solution shows that the initial data are split into two equal waves similar in shape to the initial displacement but of half the amplitude. These waves propagate in the opposite direction with the same constant speed c as shown in Figure 3.17.

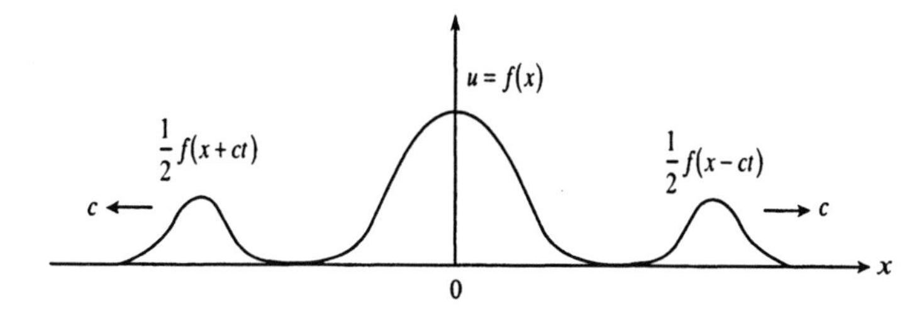

Figure 3.17. Splitting of initial data into equal waves.

To investigate the physical significance of the d'Alembert solution (3.11.42), it is convenient to rewrite the solution in the form

$$u(x,t)=\frac{1}{2}f(x-ct)-\frac{1}{2c}\int_0^{x-ct}g(\tau)d\tau+\frac{1}{2}f(x+ct)+\frac{1}{2c}\int_0^{x+ct}g(\tau)\,d\tau, \tag{3.11.44}$$

$$=\Phi(x-ct)+\Psi(x+ct), \tag{3.11.45}$$

where, in terms of the characteristic variables $\xi=x-ct$ and $\eta=x+ct$,

$$\Phi(\xi)=\frac{1}{2}f(\xi)-\frac{1}{2c}\int_0^{\xi}g(\tau)d\tau,\quad \Psi(\eta)=\frac{1}{2}f(\eta)+\frac{1}{2c}\int_0^{\eta}g(\tau)\,d\tau. \tag{3.11.46a,b}$$

Physically, $\Phi(x-ct)$ represents a progressive wave propagating in the positive x-direction with constant speed c without change of shape, as shown in Figure 3.18. Similarly, $\Psi(x+ct)$ also represents a progressive wave traveling in the negative x-direction with the same speed without change of shape.

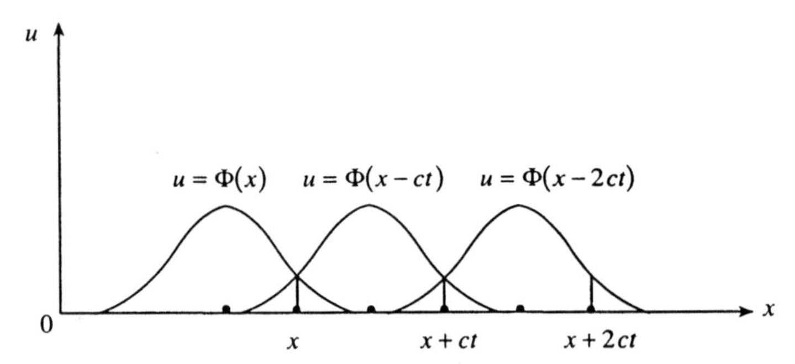

Figure 3.18. Graphical representation of solution.

In particular, if $f(x)=\exp(-x^2)$ and $g(x)=0$ then, the d'Alembert solution (3.11.42) reduces to

$$u(x,t)=\frac{1}{2}\left[\exp\left\{-(x-ct)^2\right\}+\exp\left\{-(x+ct)^2\right\}\right]. \tag{3.11.47}$$

This shows that the initial wave profile breaks up into two identical traveling waves of half the amplitude moving in opposite directions with speed c.

On the other hand, if $f(x)=0$ and $g(x)=\delta(x)$, the d'Alembert solution (3.11.42) becomes

$$u(x,t)=\frac{1}{2c}\int_{x-ct}^{x+ct}\delta(\xi)\,d\xi=\frac{1}{2c}\int_{x-ct}^{x+ct}H'(\xi)\,d\xi$$

$$=\frac{1}{2c}\left[H(x+ct)-H(x-ct)\right]$$

$$=\frac{1}{2c}\begin{bmatrix}1, & |x|<ct \\ 0, & |x|>ct>0\end{bmatrix} \tag{3.11.48}$$

$$=\frac{1}{2c}\,H\left(c^2t^2-x^2\right). \tag{3.11.49}$$

Example 3.11.5 **(*The Linearized Korteweg–de Vries Equation*).** The linearized Korteweg-de Vries (KdV) equation for the free surface elevation $\eta(x,t)$ in an inviscid water of constant depth h is

$$\eta_t+c\eta_x+\frac{ch^2}{6}\eta_{xxx}=0,\quad -\infty<x<\infty,\quad t>0, \tag{3.11.50}$$

where $c=\sqrt{gh}$ is the shallow water speed.

We solve equation (3.11.50) with the initial condition

$$\eta(x,0)=f(x),\quad -\infty<x<\infty. \tag{3.11.51}$$

Application of the Fourier transform $\mathscr{F}\{\eta(x,t)\}=\hat{\eta}(k,t)$ to the KdV system gives the solution for $\hat{\eta}(k,t)$ in the form

$$\hat{\eta}(k,t)=\hat{f}(k)\exp\left[ikct\left(\frac{k^2h^2}{6}-1\right)\right].$$

The inverse transform gives

$$\eta(x,t)=\frac{1}{2\pi}\int_{-\infty}^{\infty}\hat{f}(k)\exp\left[ik\left\{(x-ct)+\left(\frac{cth^2}{6}\right)k^2\right\}\right]dk. \tag{3.11.52}$$

In particular, if $f(x)=\delta(x)$, then (3.11.52) reduces to the Airy integral

$$\eta(x,t)=\frac{1}{\pi}\int_0^{\infty}\cos\left[k(x-ct)+\left(\frac{cth^2}{6}\right)k^3\right]dk \tag{3.11.53}$$

which is, in terms of the Airy function,

$$\eta(x,t) = \left(\frac{cth^2}{2}\right)^{-\frac{1}{3}} \mathrm{Ai}\left[\left(\frac{cth^2}{2}\right)^{-\frac{1}{3}}(x-ct)\right], \qquad (3.11.54)$$

where the Airy function $\mathrm{Ai}(z)$ is defined by

$$\mathrm{Ai}(z) = \frac{1}{\pi}\int_0^\infty \cos\left(kz + \frac{1}{3}k^3\right) dk. \qquad (3.11.55)$$

3.12 Applications of Multiple Fourier Transforms to Partial Differential Equations

The theory of the Fourier transform in $L^1(\mathbb{R}^n)$ is somewhat similar to the one-dimensional case. Moreover, the extension to $L^2(\mathbb{R}^n)$ is also possible and it has similar properties, including the inversion formula and the Plancherel theorem. In this section, we simply include some examples of applications of multiple Fourier transforms to partial differential equations.

Definition 3.12.1 Under the assumptions on $f(x) \in L^1(\mathbb{R}^n)$ similar to those made for the one dimensional case, the *multiple Fourier transform* of $f(\boldsymbol{x})$, where $\boldsymbol{x} = (x_1, x_2, \ldots, x_n)$ is the n-dimensional vector, is defined by

$$\mathscr{F}\{f(\boldsymbol{x})\} = \hat{f}(\boldsymbol{\kappa}) = \frac{1}{(2\pi)^{n/2}} \int_{-\infty}^{\infty} \cdots \int_{-\infty}^{\infty} \exp\{-i(\boldsymbol{\kappa}\cdot\boldsymbol{x})\} f(\boldsymbol{x})\, d\boldsymbol{x}, \qquad (3.12.1)$$

where $\boldsymbol{\kappa} = (k_1, k_2, \ldots, k_n)$ is the n-dimensional transform vector and $\boldsymbol{\kappa}\cdot\boldsymbol{x} = (k_1x_1 + k_2x_2 + \cdots + k_nx_n)$.

The inverse Fourier transform is similarly defined by

$$\mathscr{F}^{-1}\{\hat{f}(\boldsymbol{\kappa})\} = f(\boldsymbol{x}) = \frac{1}{(2\pi)^{n/2}} \int_{-\infty}^{\infty} \cdots \int_{-\infty}^{\infty} \exp\{i(\boldsymbol{\kappa}\cdot\boldsymbol{x})\} \hat{f}(\boldsymbol{\kappa})\, d\boldsymbol{\kappa}. \qquad (3.12.2)$$

In particular, the *double Fourier transform* is defined by

$$\mathscr{F}\{f(x,y)\} = \hat{f}(k,\ell) = \frac{1}{2\pi} \int_{-\infty}^{\infty} \int_{-\infty}^{\infty} \exp\{-i(\boldsymbol{\kappa}\cdot\boldsymbol{r})\} f(x,y)\, dx\, dy, \qquad (3.12.3)$$

where $\boldsymbol{r} = (x, y)$ and $\boldsymbol{\kappa} = (k, \ell)$.

The inverse Fourier transform is given by

$$\mathscr{F}^{-1}\left\{\hat{f}(k,\ell)\right\} = f(x,y) = \frac{1}{2\pi}\int_{-\infty}^{\infty}\int_{-\infty}^{\infty}\exp\{i(\boldsymbol{\kappa}\cdot\boldsymbol{r})\}\hat{f}(k,\ell)\,dk\,d\ell. \quad (3.12.4)$$

Similarly, the three-dimensional Fourier transform and its inverse are defined by the integrals

$$\mathscr{F}\{f(x,y,z)\} = \hat{f}(k,\ell,m)$$

$$= \frac{1}{(2\pi)^{3/2}}\int_{-\infty}^{\infty}\int_{-\infty}^{\infty}\int_{-\infty}^{\infty}\exp\{-i(\boldsymbol{\kappa}\cdot\boldsymbol{r})\}\,f(x,y,z)\,dx\,dy\,dz, \quad (3.12.5)$$

and

$$\mathscr{F}^{-1}\left\{\hat{f}(k,\ell,m)\right\} = f(x,y,z)$$

$$= \frac{1}{(2\pi)^{3/2}}\int_{-\infty}^{\infty}\int_{-\infty}^{\infty}\int_{-\infty}^{\infty}\exp\{i(\boldsymbol{\kappa}\cdot\boldsymbol{r})\}\,\hat{f}(k,\ell,m)\,dk\,d\ell\,dm. \quad (3.12.6)$$

The operational properties of these multiple Fourier transforms are similar to those of the one-dimensional case. In particular, results (3.3.16) and (3.3.17) relating the Fourier transforms of partial derivatives to the Fourier transforms of given functions are also valid for the higher dimensional case. In higher dimensions, they are applied to the transforms of partial derivatives of $f(\boldsymbol{x})$ under the assumptions that $f(x_1, x_2, \ldots, x_n)$ and its partial derivatives vanish at infinity.

We illustrate the multiple Fourier transform method by the following examples of applications.

Example 3.12.1 (The Dirichlet Problem for the Three-Dimensional Laplace Equation in the Half-Space). The boundary-value problem for $u(x,y,z)$ satisfies the following equation and boundary conditions:

$$\nabla^2 u \equiv u_{xx} + u_{yy} + u_{zz} = 0, \quad -\infty < x, y < \infty, \quad z > 0, \quad (3.12.7)$$

$$u(x,y,0) = f(x,y), \qquad -\infty < x, y < \infty \quad (3.12.8)$$

$$u(x,y,z) \to 0, \quad \text{as } r = \sqrt{x^2 + y^2 + z^2} \to \infty. \quad (3.12.9)$$

We apply the double Fourier transform defined by (3.12.3) to the system (3.12.7)-(3.12.9) which reduces to

$$\frac{d^2\hat{u}}{dz^2} - \kappa^2\,\hat{u} = 0 \qquad \text{for } z > 0,$$

$$\hat{u}(k,\ell,0) = \hat{f}(k,\ell).$$

Thus, the solution of this transformed problem is

$$\hat{u}(k,\ell,z) = \hat{f}(k,\ell)\exp(-|\kappa|z) = \hat{f}(k,\ell)\,\hat{g}(k,\ell), \tag{3.12.10}$$

where $\boldsymbol{\kappa} = (k,\ell)$ and $\hat{g}(k,\ell) = \exp(-|\kappa|z)$, so that

$$g(x,y) = \mathscr{F}^{-1}\left\{\exp(-|\kappa|z)\right\} = \frac{z}{\left(x^2+y^2+z^2\right)^{3/2}}. \tag{3.12.11}$$

Applying the convolution theorem to (3.12.10), we obtain the formal solution

$$u(x,y,z) = \frac{1}{2\pi}\int_{-\infty}^{\infty}\int_{-\infty}^{\infty} f(\xi,\eta)\,g(x-\xi,y-\eta,z)\,d\xi\,d\eta$$

$$= \frac{z}{2\pi}\int_{-\infty}^{\infty}\int_{-\infty}^{\infty}\frac{f(\xi,\eta)\,d\xi\,d\eta}{\left[(x-\xi)^2+(y-\eta)^2+z^2\right]^{3/2}}. \tag{3.12.12}$$

Example 3.12.2 (The Two-Dimensional Diffusion Equation). We solve the two-dimensional diffusion equation

$$u_t = K\nabla^2 u, \quad -\infty < x,y < \infty, \quad t > 0, \tag{3.12.13}$$

with the initial conditions

$$u(x,y,0) = f(x,y) \quad -\infty < x,y < \infty, \tag{3.12.14}$$

$$u(x,y,t) \to 0, \;\text{ as } r = \sqrt{x^2+y^2} \to \infty, \tag{3.12.15}$$

where K is the diffusivity constant.

The double Fourier transform of $u(x,y,t)$, defined by (3.12.3), is used to reduce the system (3.12.13)-(3.12.14) into the form

$$\frac{d\hat{u}}{dt} = -\kappa^2 K\hat{u}, \qquad t > 0,$$

$$\hat{u}(k,\ell,0) = \hat{f}(k,\ell).$$

The solution of this differential system is

$$\hat{u}(k,\ell,t) = \hat{f}(k,\ell)\exp(-tK\kappa^2) = \hat{f}(k,\ell)\hat{g}(k,\ell), \tag{3.12.16}$$

where

$$\hat{g}(k,\ell) = \exp(-K\kappa^2 t)$$

and hence,

$$g(x,y) = \mathscr{F}^{-1}\left\{\exp(-tK\kappa^2)\right\} = \frac{1}{2Kt}\exp\left[-\frac{x^2+y^2}{4Kt}\right]. \tag{3.12.17}$$

Finally, using the convolution theorem to (3.12.16) gives the formal solution

$$u(x,y,t) = \frac{1}{4\pi Kt}\int_{-\infty}^{\infty}\int_{-\infty}^{\infty} f(\xi,\eta)\exp\left[-\frac{(x-\xi)^2+(y-\eta)^2}{4Kt}\right]d\xi\,d\eta. \tag{3.12.18}$$

Or, equivalently,

$$u(x,y,t) = \frac{1}{4\pi Kt}\int_{-\infty}^{\infty}\int_{-\infty}^{\infty} f(\boldsymbol{r}')\exp\left\{-\frac{|\boldsymbol{r}-\boldsymbol{r}'|^2}{4Kt}\right\}d\boldsymbol{r}', \tag{3.12.19}$$

where $\mathbf{r}' = (\xi,\eta)$.

We make the change of variable $(\boldsymbol{r}'-\boldsymbol{r}) = \sqrt{4Kt}\,\boldsymbol{R}$ to reduce (3.12.19) into the form

$$u(x,y,t) = \frac{1}{\pi\sqrt{4Kt}}\int_{-\infty}^{\infty}\int_{-\infty}^{\infty} f\left(\boldsymbol{r}+\sqrt{4Kt}\,\boldsymbol{R}\right)\exp(-R^2)\,d\boldsymbol{R}. \tag{3.12.20}$$

Similarly, the formal solution of the initial-value problem for the three-dimensional diffusion equation

$$u_t = K(u_{xx}+u_{yy}+u_{zz}), \quad -\infty < x,y,z < \infty, \quad t>0 \tag{3.12.21}$$

$$u(x,y,z,0) = f(x,y,z), \quad -\infty < x,y,z < \infty \tag{3.12.22}$$

is given by

$$u(x,y,z,t) = \frac{1}{(4\pi Kt)^{3/2}}\int\int_{-\infty}^{\infty}\int f(\xi,\eta,\zeta)\,\exp\left(-\frac{r^2}{4Kt}\right)d\xi\,d\eta\,d\zeta, \tag{3.12.23}$$

where

$$r^2 = (x-\xi)^2+(y-\eta)^2+(z-\zeta)^2.$$

Or, equivalently,

$$u(x,y,z,t)=\frac{1}{(4\pi Kt)^{3/2}}\int\int_{-\infty}^{\infty}\int f(\mathbf{r}')\exp\left\{-\frac{|\mathbf{r}-\mathbf{r}'|^2}{4Kt}\right\}d\xi\,d\eta\,d\zeta, \quad (3.12.24)$$

where $\mathbf{r}=(x,y,z)$ and $\mathbf{r}'=(\xi,\eta,\zeta)$.

Making the change of variable $\boldsymbol{r}'-\boldsymbol{r}=\sqrt{4tK}\;\boldsymbol{R}$, solution (1.8.24) reduces to

$$u(x,y,z,t)=\frac{1}{\pi^{3/2}4Kt}\int\int_{-\infty}^{\infty}\int f\left(\boldsymbol{r}+\sqrt{4Kt}\;\boldsymbol{R}\right)\exp\left(-R^2\right)d\boldsymbol{R}. \quad (3.12.25)$$

This is known as the *Fourier solution.*

Example 3.12.3 **(*The Cauchy Problem for the Two–Dimensional Wave Equation*).** The initial value problem for the wave equation in two dimensions is governed by

$$u_{tt}=c^2\left(u_{xx}+u_{yy}\right), \quad -\infty<x,y<\infty, \quad t>0 \quad (3.12.26)$$

with the initial data

$$u(x,y,0)=0, \quad u_t(x,y,0)=f(x,y), \quad -\infty<x,y<\infty, \quad (3.12.27a,b)$$

where c is a constant. We assume that $u(x,y,z)$ and its first partial derivatives vanish at infinity.

We apply the two-dimensional Fourier transform defined by (3.12.3) to the system (3.12.26)-(3.12.27a,b), which becomes

$$\frac{d^2\hat{u}}{dt^2}+c^2\kappa^2\hat{u}=0, \qquad \kappa^2=k^2+\ell^2,$$

$$\hat{u}(k,\ell,0)=0, \quad \left(\frac{d\hat{u}}{dt}\right)_{t=0}=\hat{f}(k,\ell).$$

The solution of this transformed system is

$$\hat{u}(k,\ell,t)=\hat{f}(k,\ell)\;\frac{\sin(c\kappa t)}{c\kappa}. \quad (3.12.28)$$

The inverse Fourier transform gives the formal solution

$$u(x,y,t)=\frac{1}{2\pi c}\int_{-\infty}^{\infty}\int\exp(i\boldsymbol{\kappa}\cdot\boldsymbol{r})\;\frac{\sin(c\kappa t)}{\kappa}\;\hat{f}(\boldsymbol{\kappa})\,d\boldsymbol{\kappa} \quad (3.12.29)$$

$$=\frac{1}{4i\pi c}\int_{-\infty}^{\infty}\int_{-\infty}^{\infty}\frac{\hat{f}(\boldsymbol{\kappa})}{\kappa}\left[\exp\left\{i\kappa\left(\frac{\boldsymbol{\kappa}\cdot\boldsymbol{r}}{\kappa}+ct\right)\right\}-\exp\left\{i\kappa\left(\frac{\boldsymbol{\kappa}\cdot\boldsymbol{r}}{\kappa}-ct\right)\right\}\right]d\boldsymbol{\kappa}. \quad (3.12.30)$$

The form of this solution reveals some interesting features of the wave equation. The exponential terms $\exp\left\{i\kappa\left(ct \pm \frac{\boldsymbol{\kappa}\cdot\boldsymbol{r}}{\kappa}\right)\right\}$ involved in the integral solution (3.12.30) represent plane wave solutions of the wave equation (3.12.26). Thus, the solutions remain constant on the planes $\boldsymbol{\kappa}\cdot\boldsymbol{r}$ = constant that move parallel to themselves with velocity c. Evidently, solution (3.12.30) represents a superposition of the plane wave solutions traveling in all possible directions.

Similarly, the solution of the Cauchy problem for the three-dimensional wave equation

$$u_{tt} = c^2\left(u_{xx} + u_{yy} + u_{zz}\right), \quad -\infty < x, y, z < \infty, \quad t > 0, \tag{3.12.31}$$

$$u(x,y,z,0) = 0, \quad u_t(x,y,z,0) = f(x,y,z), \quad -\infty < x,y,z < \infty, \tag{3.12.32a,b}$$

is given by

$$u(\mathbf{r},t) = \frac{1}{2ic(2\pi)^{3/2}} \int\!\!\int\!\!\int_{-\infty}^{\infty} \frac{\hat{f}(\boldsymbol{\kappa})}{\kappa}\left[\exp\left\{i\kappa\left(\frac{\boldsymbol{\kappa}\cdot\boldsymbol{r}}{\kappa} + ct\right)\right\} - \exp\left\{i\kappa\left(\frac{\boldsymbol{\kappa}\cdot\boldsymbol{r}}{\kappa} - ct\right)\right\}\right] d\boldsymbol{\kappa}, \tag{3.12.33}$$

where $\boldsymbol{r} = (x,y,z)$ and $\boldsymbol{\kappa} = (k,\ell,m)$.

In particular, when $f(x,y,z) = \delta(x)\delta(y)\delta(z)$, so that $\hat{f}(\boldsymbol{\kappa}) = (2\pi)^{-3/2}$, solution (3.12.33) becomes

$$u(\mathbf{r},t) = \frac{1}{(2\pi)^3} \int\!\!\int\!\!\int_{-\infty}^{\infty} \frac{\sin c\kappa t}{c\kappa} \exp\{i(\boldsymbol{\kappa}\cdot\boldsymbol{r})\}\, d\boldsymbol{\kappa}. \tag{3.12.34}$$

In terms of the spherical polar coordinates (κ, θ, ϕ), where the polar axis (the z-axis) is taken along the $\boldsymbol{r}$ direction with $\boldsymbol{\kappa}\cdot\boldsymbol{r} = \kappa r\cos\theta$, we write (3.12.34) in the form

$$\begin{aligned}
u(r,t) &= \frac{1}{(2\pi)^3} \int_0^{2\pi} d\phi \int_0^{\pi} d\theta \int_0^{\infty} \exp(i\kappa r\cos\theta)\, \frac{\sin c\kappa t}{c\kappa} \cdot \kappa^2 \sin\theta\, d\kappa \\
&= \frac{1}{2\pi^2 cr} \int_0^{\infty} \sin(c\kappa t)\, \sin(\kappa r)\, d\kappa \\
&= \frac{1}{8\pi^2 cr} \int_{-\infty}^{\infty} \left[e^{i\kappa(ct-r)} - e^{i\kappa(ct+r)}\right] d\kappa .
\end{aligned}$$

Or,

$$u(r,t) = \frac{1}{4\pi cr}\left[\delta(ct - r) - \delta(ct + r)\right]. \tag{3.12.35}$$

For $t > 0$, $ct + r > 0$, so that $\delta(ct + r) = 0$ and hence the solution is

$$u(r,t) = \frac{1}{4\pi cr}\,\delta(ct - r) = \frac{1}{4\pi rc^2}\,\delta\left(t - \frac{r}{c}\right). \tag{3.12.36}$$

3.13 Construction of Green's Functions by the Fourier Transform Method

Many physical problems are described by second-order nonhomogeneous differential equations with homogeneous boundary conditions or by second-order homogeneous equations with nonhomogeneous boundary conditions. Such problems can be solved by a powerful method based on a device known as Green's functions.

We consider a nonhomogeneous partial differential equation of the form

$$L_x u(\mathbf{x}) = f(\mathbf{x}), \tag{3.13.1}$$

where $\mathbf{x} = (x, y, z)$ is a vector in three (or higher) dimensions, L_x is a linear partial differential operator in three or more independent variables with constant coefficients, and $u(\mathbf{x})$ and $f(\mathbf{x})$ are functions of three or more independent variables. The Green's function $G(\mathbf{x}, \boldsymbol{\xi})$ of this problem satisfies the equation

$$L_x G(\mathbf{x}, \boldsymbol{\xi}) = \delta(\mathbf{x} - \boldsymbol{\xi}) \tag{3.13.2}$$

and represents the effect at the point $\mathbf{x}$ of the Dirac delta function source at the point $\boldsymbol{\xi} = (\xi, \eta, \zeta)$.

Multiplying (3.13.2) by $f(\boldsymbol{\xi})$ and integrating over the volume V of the $\boldsymbol{\xi}$ space, so that $dV = d\xi\, d\eta\, d\zeta$, we obtain

$$\int_V L_x G(\mathbf{x}, \boldsymbol{\xi}) f(\boldsymbol{\xi})\, d\boldsymbol{\xi} = \int_V \delta(\mathbf{x} - \boldsymbol{\xi}) f(\boldsymbol{\xi})\, d\boldsymbol{\xi} = f(\mathbf{x}). \tag{3.13.3}$$

Interchanging the order of the operator L_x and integral sign in (3.13.3) gives

$$L_x\left[\int_V G(x,\xi)f(\xi)d\xi\right] = f(x). \tag{3.13.4}$$

A simple comparison of (3.13.4) with (3.13.1) leads to solution of (3.13.1) in the form

$$u(x) = \int_V G(x,\xi)f(\xi)d\xi. \tag{3.13.5}$$

Clearly, (3.13.5) is valid for any finite number of components of x. Accordingly, the Green's function method can be applied, in general, to any linear, constant coefficient, inhomogeneous partial differential equations in any number of independent variables.

Another way to approach the problem is by looking for the inverse operator L_x^{-1}. If it is possible to find L_x^{-1}, then the solution of (3.13.1) can be obtained as $u(x) = L_x^{-1}(f(x))$. It turns out that, in many important cases, it is possible, and the inverse operator can be expressed as an integral operator of the form

$$u(x) = L_x^{-1}(f(\xi)) = \int_V G(x,\xi)f(\xi)d\xi. \tag{3.13.6}$$

The kernel $G(x,\xi)$ is called the *Green's function* which is, in fact, the characteristic of the operator L_x for any finite number of independent variables.

The main goal of this section is to develop a general method of Green's functions for several examples of applications.

***Example 3.13.1* (*Green's Function for the One-Dimensional Diffusion Equation*).** We consider the inhomogeneous one-dimensional diffusion equation

$$u_t - \kappa u_{xx} = f(x)\delta(t), \quad x \in R, \quad t > 0 \tag{3.13.7}$$

with the initial and boundary conditions

$$u(x,0) = 0 \quad \text{for} \quad x \in R \quad \text{and} \quad u(x,t) \to 0 \quad \text{as} \quad |x| \to \infty. \tag{3.13.8a,b}$$

We take the Laplace transform with respect to t and the Fourier transform with respect to x to (3.13.7)-(3.13.12a,b), so that

$$\hat{\bar{u}}(k,s) = \frac{\hat{f}(k)}{\left(s + \kappa k^2\right)}, \tag{3.13.9}$$

$$\hat{u}(k,t) \to 0 \quad \text{as} \quad |k| \to \infty. \tag{3.13.10}$$

The inverse Laplace transform of (3.13.9) gives

$$\hat{u}(k,t)=\hat{f}(k)\exp\left(-\kappa k^2 t\right)=\hat{f}(k)\,\hat{g}(k), \tag{3.13.11}$$

where $\hat{g}(k)=\exp\left(-\kappa k^2 t\right)$ and hence (3.2.4) gives

$$g(x)=\mathscr{F}^{-1}\left\{\exp\left(-\kappa k^2 t\right)\right\}=\frac{1}{\sqrt{4\pi\kappa t}}\exp\left(-\frac{x^2}{4\kappa t}\right). \tag{3.13.12}$$

Application of the inverse Fourier transform combined with Convolution Theorem 3.3.7 gives

$$u(x,t)=\frac{1}{\sqrt{2\pi}}\int_{-\infty}^{\infty}e^{ikx}\hat{f}(k)\,\hat{g}(k)\,d\kappa=\frac{1}{\sqrt{2\pi}}\int_{-\infty}^{\infty}f(\xi)\,g(x-\xi)\,d\xi$$

$$=\frac{1}{\sqrt{4\kappa\pi t}}\int_{-\infty}^{\infty}f(\xi)\exp\left[-\frac{(x-\xi)^2}{4\kappa t}\right]d\xi=\int_{-\infty}^{\infty}f(\xi)\,G(x,t;\xi)\,d\xi, \tag{3.13.13}$$

where the Green's function $G(x,t;\xi)$ is given by

$$G(x,t;\xi)=\frac{1}{\sqrt{4\pi\kappa t}}\exp\left[-\frac{(x-\xi)^2}{4\kappa t}\right]. \tag{3.13.14}$$

Evidently, $G(x,t)=G(x,t;0)$ is an even function of x, and at any time t, the spatial distribution of $G(x,t)$ is Gaussian. The amplitude (or peak height) of $G(x,t)$ decreases inversely with $\sqrt{\kappa t}$, whereas the width of the peak increases with $\sqrt{\kappa t}$. The evolution of $G(x,t)=u(x,t)$ has already been plotted against x for different values of $\tau=2\sqrt{\kappa t}$ in Figure 3.15.

Example 3.13.2 **(*Green's Function for the Two-Dimensional Diffusion Equation*).** We consider the two-dimensional diffusion equation

$$u_t-K\nabla^2 u=f(x,y)\,\delta(t),\qquad -\infty<x,y<\infty,\quad t>0 \tag{3.13.15}$$

with the initial and boundary conditions

$$u(x,y,0)=0,\qquad \text{for all}\quad (x,y)\in R^2, \tag{3.13.16}$$

$$u(x,y,t)\to 0\quad \text{as}\quad r=\sqrt{x^2+y^2}\to\infty, \tag{3.13.17}$$

where K is the diffusivity constant.

Application of the Laplace transform and the double Fourier transform (3.12.3) to the preceding differential system gives

$$\bar{\hat{u}}(\kappa, s) = \frac{\hat{f}(k,l)}{\left(s + K\kappa^2\right)}, \tag{3.13.18}$$

where $\boldsymbol{\kappa} = (k,l)$.

The inverse Laplace transform of (3.13.18) gives

$$\hat{u}(\kappa, t) = \hat{f}(k,l)\exp\left(-K\kappa^2 t\right) = \hat{f}(k,l)\hat{g}(k,l), \tag{3.13.19}$$

where $\hat{g}(k,l) = \exp\left(-K\kappa^2 t\right)$, so that

$$g(x,y) = \mathscr{F}^{-1}\left\{\exp\left(-K\kappa^2 t\right)\right\} = \frac{1}{2Kt}\exp\left[-\frac{\left(x^2+y^2\right)}{4Kt}\right]. \tag{3.13.20}$$

Finally, the convolution theorem of the Fourier transform gives the formal solution

$$u(x,y,t) = \frac{1}{4\pi Kt}\int_{-\infty}^{\infty}\int f(\xi,\eta)\exp\left[-\frac{(x-\xi)^2+(y-\eta)^2}{4Kt}\right]d\xi\, d\eta, \tag{3.13.21}$$

$$= \int_{-\infty}^{\infty}\int_{-\infty}^{\infty} f(\xi)G(\boldsymbol{r}, \boldsymbol{\xi})d\boldsymbol{\xi}, \tag{3.13.22}$$

where $\boldsymbol{r} = (x,y)$ and $\boldsymbol{\xi} = (\xi, \eta)$, and the Green's function $G(\boldsymbol{r}, \boldsymbol{\xi})$ is given by

$$G(\boldsymbol{r}, \boldsymbol{\xi}) = \frac{1}{(4\pi Kt)}\exp\left[-\frac{|\boldsymbol{r}-\boldsymbol{\xi}|^2}{4Kt}\right], \tag{3.13.23}$$

Similarly, we can construct the Green's function for the three-dimensional diffusion equation

$$u_t - K\nabla^2 u = f(\boldsymbol{r})\delta(t) \quad \text{for} \quad -\infty < x,y,z < \infty,\ t > 0 \tag{3.13.24}$$

with the initial and boundary data

$$u(\boldsymbol{r}, 0) = 0 \quad \text{for} \quad -\infty < x,y,z < \infty, \tag{3.13.25}$$

$$u(\boldsymbol{r}, t) \to 0 \quad \text{as} \quad r = \left(x^2+y^2+z^2\right)^{\frac{1}{2}} \to \infty, \tag{3.13.26}$$

where $\boldsymbol{r} = (x,y,z)$.

Application of the Laplace transform of $u(\boldsymbol{r},t)$ with respect to t and the three-dimensional Fourier transform (3.12.5) with respect to x, y, z gives the solution

$$u(\boldsymbol{r},t)=\frac{1}{(4\pi Kt)^{\frac{3}{2}}}\iiint_{-\infty}^{\infty} f(\boldsymbol{\xi})\,\exp\left[-\frac{|\boldsymbol{r}-\boldsymbol{\xi}|^2}{4Kt}\right]d\boldsymbol{\xi}, \tag{3.13.27}$$

$$=\iiint_{-\infty}^{\infty} f(\boldsymbol{\xi})\,G(\boldsymbol{r},\boldsymbol{\xi})\,d\boldsymbol{\xi}, \tag{3.13.28}$$

where $\boldsymbol{\xi}=(\xi,\eta,\zeta)$, and the Green's function is given by

$$G(\boldsymbol{r},\boldsymbol{\xi})=\frac{1}{(4\pi Kt)^{\frac{3}{2}}}\exp\left[-\frac{|\boldsymbol{r}-\boldsymbol{\xi}|^2}{4Kt}\right]. \tag{3.13.29}$$

In fact, the same method of construction can be used to find the Green's function for the n-dimensional diffusion equation

$$u_t - K\nabla_n^2 u = f(\boldsymbol{r})\,\delta(t), \quad \boldsymbol{r}\in R^n, \quad t>0, \tag{3.13.30}$$

$$u(\boldsymbol{r},0)=0 \qquad \text{for all} \quad \boldsymbol{r}\in R^n, \tag{3.13.31}$$

$$u(\boldsymbol{r},t)\to 0 \quad \text{as} \quad r=\left(x_1^2+x_2^2+\cdots+x_n^2\right)^{\frac{1}{2}}\to\infty, \tag{3.13.32}$$

where $\boldsymbol{r}=(x_1,x_2,\ldots,x_n)$ and ∇_n^2 is the n-dimensional Laplacian given by

$$\nabla_n^2=\frac{\partial^2}{\partial x_1^2}+\frac{\partial^2}{\partial x_2^2}+\cdots+\frac{\partial^2}{\partial x_n^2}. \tag{3.13.33}$$

The solution of this problem is given by

$$u(\boldsymbol{r},t)=\frac{1}{(4\pi Kt)^{\frac{n}{2}}}\int_{-\infty}^{\infty} f(\boldsymbol{\xi})\,G(\boldsymbol{r},\boldsymbol{\xi})\,d\boldsymbol{\xi}, \tag{3.13.34}$$

where $\boldsymbol{\xi}=(\xi_1,\xi_2,\ldots,\xi_n)$ and the n-dimensional Green's function $G(\boldsymbol{r},\boldsymbol{\xi})$ is given by

$$G(\boldsymbol{r},\boldsymbol{\xi})=\frac{1}{(4\pi Kt)^{\frac{n}{2}}}\exp\left[-\frac{|\boldsymbol{r}-\boldsymbol{\xi}|^2}{4Kt}\right]. \tag{3.13.35}$$

Example 3.13.3 (The Three-Dimensional Poisson Equation). The solution of the Poisson equation

$$-\nabla^2 u = f(\boldsymbol{r}), \tag{3.13.36}$$

where $\boldsymbol{r} = (x, y, z)$, is given by

$$u(\boldsymbol{r}) = \iiint_{-\infty}^{\infty} G(\boldsymbol{r}, \boldsymbol{\xi}) f(\boldsymbol{\xi}) d\boldsymbol{\xi}, \tag{3.13.37}$$

where the *Green's function* $G(\boldsymbol{r}, \boldsymbol{\xi})$ of the operator, $-\nabla^2$, is given by

$$G(\boldsymbol{r}, \boldsymbol{\xi}) = \frac{1}{4\pi} \frac{1}{|\boldsymbol{r} - \boldsymbol{\xi}|}. \tag{3.13.38}$$

To obtain the fundamental solution, we have to solve the equation

$$-\nabla^2 G(\boldsymbol{r}, \boldsymbol{\xi}) = \delta(x - \xi)\delta(y - \eta)\delta(z - \zeta), \qquad \boldsymbol{r} \neq \boldsymbol{\xi}. \tag{3.13.39}$$

Application of the three-dimensional Fourier transform defined by (3.12.5) to equation (3.13.39) gives

$$\kappa^2 \hat{G}(\boldsymbol{\kappa}, \boldsymbol{\xi}) = \frac{1}{(2\pi)^{3/2}} \exp(-i\boldsymbol{\kappa} \cdot \boldsymbol{\xi}), \tag{3.13.40}$$

where $\hat{G}(\boldsymbol{\kappa}, \boldsymbol{\xi}) = \mathscr{F}\{G(\boldsymbol{r}, \boldsymbol{\xi})\}$ and $\boldsymbol{\kappa} = (k, \ell, m)$.

The inverse Fourier transform gives the formal solution

$$\begin{aligned} G(\boldsymbol{r}, \boldsymbol{\xi}) &= \frac{1}{(2\pi)^3} \iiint_{-\infty}^{\infty} \exp\{i\boldsymbol{\kappa} \cdot (\boldsymbol{r} - \boldsymbol{\xi})\} \frac{d\boldsymbol{\kappa}}{\kappa^2} \\ &= \frac{1}{(2\pi)^3} \iiint_{-\infty}^{\infty} \exp(i\boldsymbol{\kappa} \cdot \boldsymbol{R}) \frac{d\boldsymbol{\kappa}}{\kappa^2}, \end{aligned} \tag{3.13.41}$$

where $\boldsymbol{R} = \boldsymbol{r} - \boldsymbol{\xi}$.

We evaluate this integral using the spherical polar coordinates in the $\boldsymbol{\kappa}$-space with the axis along the R-axis. In terms of spherical polar coordinates (κ, θ, ϕ), so that $\boldsymbol{\kappa} \cdot \boldsymbol{R} = \kappa R \cos\theta$, where $R = |\boldsymbol{r} - \boldsymbol{\xi}|$. Thus, solution (3.13.41) becomes

$$\begin{aligned} G(\boldsymbol{r}, \boldsymbol{\xi}) &= \frac{1}{(2\pi)^3} \int_0^{2\pi} d\phi \int_0^{\pi} d\theta \int_0^{\infty} \exp(i\kappa R \cos\theta)\, \kappa^2 \sin\theta \cdot \frac{d\kappa}{\kappa^2} \\ &= \frac{1}{(2\pi)^2} \int_0^{\infty} 2 \frac{\sin(\kappa R)}{\kappa R} d\kappa = \frac{1}{4\pi R} = \frac{1}{4\pi |\boldsymbol{r} - \boldsymbol{\xi}|}, \end{aligned} \tag{3.13.42}$$

provided that $R > 0$.

In electrodynamics, the fundamental solution (3.13.42) has a well-known interpretation. Physically, it represents the potential at point $\boldsymbol{r}$ generated by the unit point-charge distribution at point $\boldsymbol{\xi}$. This is what can be expected because $\delta(\boldsymbol{r}-\boldsymbol{\xi})$ is the charge density corresponding to a unit point charge at $\boldsymbol{\xi}$.

The solution of (3.13.36) is then given by

$$u(\boldsymbol{r})=\int\int_{-\infty}^{\infty}\int G(\boldsymbol{r},\boldsymbol{\xi})\,f(\boldsymbol{\xi})\,d\boldsymbol{\xi}=\frac{1}{4\pi}\int\int_{-\infty}^{\infty}\int\frac{f(\boldsymbol{\xi})\,d\boldsymbol{\xi}}{|\boldsymbol{r}-\boldsymbol{\xi}|}. \tag{3.13.43}$$

The integrand in (3.13.43) consists of the given charge distribution $f(\boldsymbol{r})$ at $\boldsymbol{r}=\boldsymbol{\xi}$ and the Green's function $G(\boldsymbol{r},\boldsymbol{\xi})$. Physically, $G(\boldsymbol{r},\boldsymbol{\xi})\,f(\boldsymbol{\xi})$ represents the resulting potentials due to elementary point charges, and the total potential due to a given charge distribution $f(\boldsymbol{r})$ is then obtained by the integral superposition of the resulting potentials. This is called the *principle of superposition.*

Example 3.13.4 **(*The Two-Dimensional Helmholtz Equation*).** Find the fundamental solution of the two-dimensional Helmholtz equation

$$-\nabla^2 G+\alpha^2 G=\delta(x-\xi)\,\delta(y-\eta), \quad -\infty<x,y<\infty. \tag{3.13.44}$$

It is convenient to change variables $x-\xi=x^*$, $y-\eta=y^*$. Consequently, dropping the asterisks, equation (3.13.44) reduces to the form

$$G_{xx}+G_{yy}-\alpha^2 G=-\delta(x)\,\delta(y). \tag{3.13.45}$$

Application of the double Fourier transform $\hat{G}(\boldsymbol{\kappa})=\mathscr{F}\{G(x,y)\}$ defined by (3.12.3) to equation (3.13.45) gives the solution

$$\hat{G}(\boldsymbol{\kappa})=\frac{1}{2\pi}\frac{1}{(\kappa^2+\alpha^2)}, \tag{3.13.46}$$

where $\boldsymbol{\kappa}=(k,\ell)$ and $\kappa^2=k^2+\ell^2$.

The inverse Fourier transform (3.12.4) yields the solution

$$G(x,y)=\frac{1}{4\pi^2}\int_{-\infty}^{\infty}\int_{-\infty}^{\infty}\exp(i\boldsymbol{\kappa}\cdot\boldsymbol{x})\left(\kappa^2+\alpha^2\right)^{-1}dk\,d\ell. \tag{3.13.47}$$

In terms of polar coordinates $(x,y)=r(\cos\theta,\sin\theta)$, $(k,\ell)=\rho(\cos\phi,\sin\phi)$, the integral solution (3.13.47) becomes

$$G(x,y) = \frac{1}{4\pi^2}\int_0^\infty \frac{\rho\, d\rho}{(\rho^2+\alpha^2)}\int_0^{2\pi}\exp\{ir\rho\cos(\phi-\theta)\}\,d\phi,$$

which is, replacing the second integral by $2\pi J_0(r\rho)$,

$$= \frac{1}{2\pi}\int_0^\infty \frac{\rho\, J_0(r\rho)\,d\rho}{(\rho^2+\alpha^2)}. \tag{3.13.48}$$

In terms of the original coordinates, the fundamental solution of equation (3.13.44) is given by

$$G(\mathbf{r},\boldsymbol{\xi}) = \frac{1}{2\pi}\int_0^\infty \rho(\rho^2+\alpha^2)^{-1}\, J_0\left[\rho\left\{(x-\xi)^2+(y-\eta)^2\right\}^{\frac{1}{2}}\right]d\rho. \tag{3.13.49}$$

Accordingly, the solution of the inhomogeneous equation

$$(\nabla^2+\alpha^2)u = -f(x,y) \tag{3.13.50}$$

is given by

$$u(x,y) = \int_{-\infty}^{\infty}\int G(\mathbf{r},\boldsymbol{\xi})\, f(\boldsymbol{\xi})\,d\boldsymbol{\xi}, \tag{3.13.51}$$

where $G(\mathbf{r},\boldsymbol{\xi})$ is given by (3.13.49).

Since the integral solution (3.13.48) does not exist for $\alpha = 0$, the Green's function for the two-dimensional Poisson equation (3.13.44) cannot be derived from (3.13.48). Instead, we differentiate (3.13.48) with respect to r to obtain

$$\frac{\partial G}{\partial r} = \frac{1}{2\pi}\int_0^\infty \frac{\rho^2\, J_0'(r\rho)\,d\rho}{(\rho^2+\alpha^2)},$$

which is, for $\alpha = 0$,

$$\frac{\partial G}{\partial r} = \frac{1}{2\pi}\int_0^\infty J_0'(r\rho)\,d\rho = -\frac{1}{2\pi r}.$$

Integrating this result gives the Green's function

$$G(r,\theta) = -\frac{1}{2\pi}\log r.$$

In terms of the original coordinates, the Green's function becomes

$$G(\mathbf{r},\boldsymbol{\xi}) = -\frac{1}{4\pi}\log\left[(x-\xi)^2+(y-\eta)^2\right]. \tag{3.13.52}$$

This is the Green's function for the two-dimensional Poisson equation $\nabla^2 = -f(x,y)$. Thus, the solution of the Poisson equation is

$$u(x,y) = \int_{-\infty}^{\infty}\int G(\boldsymbol{r},\xi)\, f(\xi)\, d\xi, \tag{3.13.53}$$

where $G(\boldsymbol{r},\xi)$ is given by (3.13.52).

***Example 3.13.5** (Green's function for the Three-Dimensional Helmholtz Equation)*. We consider the three-dimensional wave equation

$$-\left[u_{tt} - c^2\nabla^2 u\right] = q(\boldsymbol{r},t), \tag{3.13.54}$$

where $q(\boldsymbol{r},t)$ is a source. If $q(\boldsymbol{r},t) = q(\boldsymbol{r})\exp(-i\omega t)$ represents a source oscillating with a single frequency ω, then, as expected, at least after an initial transient period, the entire motion reduces to a wave motion with the same frequency ω so that we can write $u(\boldsymbol{r},t) = u(\boldsymbol{r})\exp(-i\omega t)$. Consequently, the wave equation (3.13.54) reduces to the three-dimensional Helmholtz equation

$$-\left(\nabla^2 + k^2\right)u(\boldsymbol{r}) = f(\boldsymbol{r}), \tag{3.13.55}$$

where $k = \left(\dfrac{\omega}{c}\right)$ and $f(\boldsymbol{r}) = c^{-2}q(\boldsymbol{r})$. The function $u(\boldsymbol{r})$ satisfies this equation in some domain $D \subset R$ with boundary ∂D, and it also satisfies some prescribed boundary conditions. We also assume that $u(\boldsymbol{r})$ satisfies the Sommerfeld radiation condition which simply states that the solution behaves like outgoing waves generated by the source. In the limit as $\omega \to 0$ or $k \to 0$ and $f(\boldsymbol{r})$ can be interpreted as a heat source, equation (3.13.55) results into a three-dimensional Poisson equation. The solution $u(\boldsymbol{r})$ would represent the steady temperature distribution in region D due to the heat source $f(\boldsymbol{r})$. However, in general, $u(\boldsymbol{r})$ can be interpreted as a function of physical interest.

We construct a Green's function $G(\boldsymbol{r},\xi)$ for equation (3.13.55) so that $G(\boldsymbol{r},\xi)$ satisfies the equation

$$-\left(\nabla^2 + k^2\right)G = \delta(x)\,\delta(y)\,\delta(z). \tag{3.13.56}$$

Using the spherical polar coordinates, the three-dimensional Laplacian can be expressed in terms of radial coordinate r only so that (3.13.56) assumes the form

$$-\left[\frac{1}{r^2}\frac{\partial}{\partial r}\left(r^2\frac{\partial G}{\partial r}\right)+k^2G\right]=\frac{\delta(r)}{4\pi r^2},\quad 0<r<\infty, \tag{3.13.57}$$

with the radiation condition

$$\lim_{r\to\infty} r\left(G_r+ikG\right)=0. \tag{3.13.58}$$

For $r>0$, the Green's function G satisfies the homogeneous equation

$$\frac{1}{r^2}\frac{\partial}{\partial r}\left(r^2\frac{\partial G}{\partial r}\right)+k^2G=0. \tag{3.13.59}$$

Or, equivalently,

$$\frac{\partial^2}{\partial r^2}(rG)+k^2(rG)=0. \tag{3.13.60}$$

This equation admits a solution of the form

$$rG(r)=Ae^{ikr}+Be^{-ikr} \tag{3.13.61}$$

or

$$G(r)=A\frac{e^{ikr}}{r}+B\frac{e^{-ikr}}{r}, \tag{3.13.62}$$

where A and B are arbitrary constants. In order to satisfy the radiation condition, we must set $A=0$, and hence the solution (3.13.62) becomes

$$G(r)=B\frac{e^{-ikr}}{r}. \tag{3.13.63}$$

To determine B, we use the spherical surface S_ε of radius ε, so that

$$\lim_{\varepsilon\to0}\int_{S_\varepsilon}\frac{\partial G}{\partial r}\,dS=-\lim_{\varepsilon\to0}\int_{S_\varepsilon}\frac{B}{r}\,e^{-ikr}\left(\frac{1}{r}+ik\right)\,dS=1, \tag{3.13.64}$$

from which we find $B=-\dfrac{1}{4\pi}$ as $\varepsilon\to0$. Consequently, the Green's function takes the form

$$G(r)=\frac{e^{-ikr}}{4\pi r}. \tag{3.13.65}$$

Physically, this represents outgoing spherical waves radiating away from the source at the origin. With a point source at a point ξ, the Green's function is represented by

$$G(\boldsymbol{r},\xi)=\frac{\exp\{-ik|\boldsymbol{r}-\boldsymbol{\xi}|\}}{4\pi|\boldsymbol{r}-\boldsymbol{\xi}|}, \tag{3.13.66}$$

where $\boldsymbol{r}$ and $\boldsymbol{\xi}$ are position vectors in R^3.

Finally, when $k=0$, this result reduces exactly to the Green's function for the three-dimensional Poisson equation.

Example 3.13.6 (One-Dimensional Inhomogeneous Wave Equation). We first consider the one-dimensional inhomogeneous wave equation

$$-\left[u_{xx}-\frac{1}{c^2}u_{tt}\right]=q(x,t), \quad x\in R, \quad t>0, \tag{3.13.67}$$

with the initial and boundary conditions

$$u(x,0)=0, \quad u_t(x,0)=0 \quad \text{for} \quad x\in R, \tag{3.13.68a,b}$$

$$u(x,t)\to 0 \qquad \text{as} \quad |x|\to\infty. \tag{3.13.69}$$

The Green's function $G(x,t)$ for this problem satisfies the equation

$$-\left[G_{xx}-\frac{1}{c^2}G_{tt}\right]=\delta(x)\,\delta(t) \tag{3.13.70}$$

and the same initial and boundary conditions (3.13.68a,b)-(3.13.69) satisfied by $G(x,t)$.

We apply the joint Laplace transform with respect to t and the Fourier transform (3.2.3) with respect to x to equation (3.13.70) so that

$$\hat{\bar{G}}(k,s)=\left(k^2+\frac{s^2}{c^2}\right)^{-1}, \tag{3.13.71}$$

where k and s represent the Fourier and Laplace transform variables respectively.

The inverse Fourier transform of (3.13.71) gives

$$\bar{G}(x,s)=\mathscr{F}^{-1}\left\{\left(k^2+\frac{s^2}{c^2}\right)^{-1}\right\}=\frac{c}{2s}\exp\left(-\frac{s}{c}|x|\right). \tag{3.13.72}$$

Finally, the inverse Laplace transform yields the Green's function with a source at the origin

$$G(x,t) = \frac{c}{2}\mathcal{L}^{-1}\left\{\frac{1}{s}\exp\left(-\frac{s}{c}|x|\right)\right\} = \frac{c}{2}H\left(t - \frac{|x|}{c}\right), \tag{3.13.73}$$

where $H(t)$ is the Heaviside unit step function.

With a point source at (ξ, τ), the Green's function takes the form

$$G(x,t;\xi,\tau) = \frac{c}{2}H\left(t - \tau - \frac{|x-\xi|}{c}\right). \tag{3.13.74}$$

This function is also called the *Riemann function* for the wave equation. The result (3.13.74) shows that $G = 0$ unless the point (x,t) lies within the *characteristic cone* defined by the inequality $c(t-\tau) > |x - \xi|$.

The solution of equation (3.13.67) is

$$\begin{aligned} u(x,t) &= \int_{-\infty}^{\infty} d\xi \int_0^t G(x,t;\xi,\tau)\,q(\xi,\tau)\,d\tau \\ &= \frac{c}{2}\int_{-\infty}^{\infty} d\xi \int_0^t H\left(t - \tau - \frac{|x-\xi|}{c}\right) q(\xi,\tau)\,d\tau \end{aligned} \tag{3.13.75}$$

which is, since $H = 1$ for $x - c(t-\tau) < \xi < x + c(t-\tau)$ and zero outside,

$$= \frac{c}{2}\int_0^t d\tau \int_{x-c(t-\tau)}^{x+c(t-\tau)} q(\xi,\tau)\,d\xi = \frac{c}{2}\iint_D q(\xi,\tau)\,d\tau\,d\xi, \tag{3.13.76}$$

where D is the triangular domain (characteristic triangle) made up of two points $(x \mp ct, 0)$ on the x-axis and vertex (x,t) off the x-axis in the (x,t)-plane.

Thus, the solution of the general Cauchy problem described in Example 3.11.4 can be obtained by adding (3.13.75) to the d'Alembert solution (3.11.42), and hence it reduces to (3.11.41).

Example 3.13.7 **(*Green's Function for the Three-Dimensional Inhomogeneous Wave Equation*).** The three-dimensional inhomogeneous wave equation is given by

$$-\left[\nabla^2 u - \frac{1}{c^2}u_{tt}\right] = f(\mathbf{r},t), \quad -\infty < x,y,z < \infty, \quad t > 0, \tag{3.13.77}$$

where $\mathbf{r} = (x,y,z)$, and the Laplacian is given by

$$\nabla^2 = \frac{\partial^2}{\partial x^2} + \frac{\partial^2}{\partial y^2} + \frac{\partial^2}{\partial z^2}. \tag{3.13.78}$$

The initial and boundary conditions are

$$u(\mathbf{r},0) = 0, \quad u_t(\mathbf{r},t) = 0, \tag{3.13.79a,b}$$

$$u(\mathbf{r},t) \to 0, \quad \text{as} \quad r \to \infty. \tag{3.13.80}$$

The Green's function $G(\boldsymbol{x},t)$ for this problem satisfies the equation

$$-\left[\nabla^2 G - \frac{1}{c^2} G_{tt}\right] = \delta(x)\,\delta(y)\,\delta(z)\,\delta(t), \quad -\infty < x, y, z < \infty, \quad t > 0, \tag{3.13.81}$$

with the same initial and boundary data (3.13.79a,b)-(3.13.80).

Application of the joint Laplace and Fourier transform (3.12.5) gives

$$\hat{\bar{G}}(\boldsymbol{\kappa},s) = \frac{c^2}{(2\pi)^{3/2}} \cdot \frac{1}{(s^2 + c^2\kappa^2)}, \quad \boldsymbol{\kappa} = (k,\ell,m). \tag{3.13.82}$$

The joint inverse transform yields the integral solution

$$G(\boldsymbol{x},t) = \frac{c}{(2\pi)^3} \int\!\!\int\!\!\int_{-\infty}^{\infty} \frac{\sin(c\kappa t)}{\kappa} \exp(i\boldsymbol{\kappa}\cdot\boldsymbol{x})\, d\boldsymbol{\kappa}. \tag{3.13.83}$$

In terms of the spherical polar coordinates with the polar axis along the vector $\boldsymbol{x}$, so that $\boldsymbol{\kappa}\cdot\boldsymbol{x} = \kappa r\cos\theta$, $r = |\boldsymbol{x}|$ and $d\boldsymbol{\kappa} = \kappa^2 d\kappa \sin\theta\, d\theta\, d\phi$, integral (3.13.83) assumes the form

$$G(\boldsymbol{x},t) = \frac{c}{(2\pi)^3} \int_0^{2\pi} d\phi \int_0^{\infty} \kappa \sin(c\kappa t)\, d\kappa \int_0^{\pi} \exp(i\kappa r\cos\theta)\, \sin\theta\, d\theta, \tag{3.13.84}$$

$$= \frac{c}{4\pi^2 ri} \int_0^{\infty} \left(e^{i\kappa r} - e^{-i\kappa r}\right) \sin(c\kappa t)\, d\kappa = -\frac{c}{8\pi^2 r} \int_0^{\infty} \left(e^{i\kappa r} - e^{-i\kappa r}\right)\left(e^{ic\kappa t} - e^{-ic\kappa t}\right) d\kappa$$

$$= \frac{c}{8\pi^2 r}\left[\int_0^{\infty} \left\{e^{i\kappa(ct-r)} + e^{-i\kappa(ct-r)}\right\} d\kappa - \int_0^{\infty} \left\{e^{i\kappa(ct+r)} + e^{-i\kappa(ct+r)}\right\} d\kappa\right]$$

$$= \frac{c}{8\pi^2 r}\left[\int_{-\infty}^{\infty} e^{i\kappa(ct-r)} d\kappa - \int_{-\infty}^{\infty} e^{i\kappa(ct+r)} d\kappa\right]$$

$$= \frac{2\pi c}{8\pi^2 r}\left[\delta(ct-r) - \delta(ct+r)\right]. \tag{3.13.85}$$

For $t > 0$, $ct + r > 0$ and hence $\delta(ct + r) = 0$. Thus,

$$G(x,t) = \frac{1}{4\pi r}\,\delta\left(t - \frac{r}{c}\right) \tag{3.13.86}$$

in which the formula $\delta(ax) = \frac{1}{a}\delta(x)$ is used.

If the source is located at $(\xi, \eta, \zeta, \tau) = (\boldsymbol{\xi}, \tau)$, the desired Green's function is given by

$$G(x,t;\boldsymbol{\xi},\tau) = \frac{c}{4\pi|x-\boldsymbol{\xi}|}\left[\delta\left(t - \tau - \frac{|x-\boldsymbol{\xi}|}{c}\right) - \delta\left(t - \tau + \frac{|x-\boldsymbol{\xi}|}{c}\right)\right]. \tag{3.13.87}$$

It should be noted that the Green's function (3.13.86) for the hyperbolic equation is a generalized function, whereas in the other examples of Green's functions, it was always a piecewise analytic function. In general, the Green's function for an elliptic function is always analytic, whereas the Green's function for a hyperbolic equation is a generalized function.

3.14 Exercises

1. Find the Fourier transforms of each of the following functions:

   (a) $f(t) = t\,\exp(-a|t|),\ a > 0,$ (b) $f(t) = t\,\exp(-at^2),\ a > 0,$

   (c) $f(x) = e^t\,\exp(-e^t),$ (d) $f(t) = \begin{Bmatrix} 1-|x|, & |x| \le 1, \\ 0, & |x| > 1, \end{Bmatrix},$

   (e) $f(t) = t^2\,\exp\left(-\frac{1}{2}t^2\right),$ (f) $f(t) = \exp(-at^2 + bt),$

   (g) $f(t) = \delta^{(n)}(t),$ (h) $f(t) = |t|^{a-1},$

   (i) $f(t) = \dfrac{\sin^2 at}{\pi a t^2},$ (j) $f(t) = \chi_\tau(t)\cos\omega_0 t,$

   (k) $f(t) = \dfrac{1}{\sqrt{t}}\,J_{n+\frac{1}{2}}(t),$ (l) $f(t) = \exp(iat),$

   where $J_n(t)$ is the Bessel function.

2. Use the Fourier transform with respect to x to show that

   (a) $\mathscr{F}\{H(cf - |x|)\} = \left(\frac{2}{k}\right)\sin(ckt),$

(b) $\mathscr{F}\{\delta(x-ct)\}+\delta(x+ct)=2\cos(ckt)$,

where k is the Fourier transform variable.

3. If $p(t)=\int_{-\infty}^{\infty}(t-\xi)\, f(t-\xi)\, f'(\xi)\, d\xi$, show that

$$\hat{p}(\omega)=-\frac{1}{2}\frac{d}{d\omega}\left\{\hat{f}^2(\omega)\right\}.$$

4. If $f(t)$ has a finite discontinuity at a point $t=a$, prove that

$$\mathscr{F}\{f'(t)\}=(i\omega)\,\hat{f}(\omega)-e^{-ia\omega}[f]_a,$$

where $[f]_a=f(a+0)-f(a-0)$.

Generalize this result for $\mathscr{F}\left\{f^{(n)}(t)\right\}$.

5. Prove the following results:

(a) $\mathscr{F}\left\{\left(a^2-t^2\right)^{-\frac{1}{2}}\, H(a-|t|)\right\}=\pi\, J_0(ak),\quad a>0$,

(b) $\mathscr{F}\{\rho_n(t)\, H(1-|t|)\}=\sqrt{\frac{2\pi}{\omega}}\,(-i)^n\, J_{n+\frac{1}{2}}(\omega)$.

6. Use result (3.3.11) to find

(a) $\mathscr{F}\left\{x^n\exp\left(-\frac{x^2}{2}\right)\right\}$ and (b) $\mathscr{F}\left\{x^2\exp\left(-ax^2\right)\right\}$.

7. If $h(t)=f(t)*g(t)$, show that

(a) $h(t-t_0)=f(t-t_0)*g(t)=\int_{-\infty}^{\infty}f(\tau-t_0)\, g(t-\tau)\, d\tau$,

(b) $|a|\, h\left(\frac{t}{a}\right)=f\left(\frac{t}{a}\right)*g\left(\frac{t}{a}\right)$.

8. Prove the following results for the convolution of the Fourier transform:

(a) $\delta(x)*f(x)=f(x)$, (b) $\delta'(x)*f(x)=f'(x)$,

(c) $\frac{d}{dx}\{f(x)*g(x)\}=f'(x)*g(x)=f(x)*g'(x)$.

9. If $G(x,t)=\frac{1}{\sqrt{4\pi\kappa t}}\exp\left(-\frac{x^2}{4\kappa t}\right)$, show that

$$G(x,t) * G(x,t) = G(x,2t).$$

10. Show that

$$\mathscr{F}\left\{\exp\left(-a^2t^2\right)\right\} * \mathscr{F}\left\{\exp\left(-b^2t^2\right)\right\} = \left(\frac{\pi}{ab}\right) \exp\left(-\frac{\omega^2}{4c^2}\right),$$

where $\dfrac{1}{c^2} = \left(\dfrac{1}{a^2} + \dfrac{1}{b^2}\right)$.

11. If $f \in L^1(\mathbb{R})$ and $F \in L^1(\mathbb{R})$ is defined by

$$F(t) = \int_{-\infty}^{t} f(x)\,dx = \int_{-\infty}^{\infty} f(x)\, \chi_{(-\infty,t)}(x)\, dx \quad \text{for } t \in \mathbb{R},$$

show that

$$\hat{F}(\omega) = (i\omega)^{-1}\, \hat{f}(\omega) \qquad \text{for all } \omega \neq 0.$$

12. Prove that the Gaussian kernel $G_\lambda(t)$ is a summability kernel.

Examine the nature of $G_\lambda(t)$ and $\hat{G}_\lambda(\omega)$ as $\lambda \to \infty$.

13. Show that, for the Gaussian kernel G_λ,

(a) $G_\lambda^{(t)} * f(t) = \displaystyle\int_{-\infty}^{\infty} \exp\left(-\frac{\omega^2}{4\lambda^2}\right) \hat{f}(\omega)\, e^{i\omega t} d\omega,$

(b) $\displaystyle\lim_{\lambda \to \infty} \left(G_\lambda * f\right)(t) = f(t).$

14. If the Poisson kernel $P_\lambda(x) = \lambda\, P(x)$, $P(x) = \dfrac{1}{\pi}\left(1 + x^2\right)^{-1}$, show that

(a) $\hat{P}_\lambda(\omega) = \exp\left(-\left|\dfrac{\omega}{\lambda}\right|\right),$

(b) $\left(P_\lambda * f\right)(t) = \displaystyle\int_{-\infty}^{\infty} \exp\left(-\left|\frac{\omega}{\lambda}\right|\right) \hat{f}(\omega) \exp(i\omega t)\, d\omega,$

(c) $\displaystyle\lim_{\lambda \to \infty} \left(P_\lambda * f\right)(t) = f(t).$

15. Find the normalized autocorrelation function $\gamma(t)$ for functions

(a) $f(t) = \begin{cases} 0, & t < 0, \\ 1 - t, & 0 < t < 1, \\ 0, & t > 1. \end{cases}$

(b) $f(t) = e^{-at}\, H(t).$

16. In terms of Fejér kernel $F_\lambda(x)$, the Vallée-Poussin kernel is defined by

$$V_\lambda(x) = 2\ F_{2\lambda}(x) - F_\lambda(x).$$

Show that

$$\hat{V}_\lambda(\omega) = \begin{Bmatrix} 1, & |\omega| \le \lambda, \\ 2 - \dfrac{|\omega|}{\lambda}, & \lambda \le |\omega| \le 2\lambda, \\ 0, & |\omega| \ge 2\lambda \end{Bmatrix}.$$

17. If $\Delta(t)$ is a triangular function defined by (3.2.10), show that

(a) $\hat{\Delta}_1(\omega) = F_1(\omega)$,

(b) $\mathscr{F}\{\Delta_\lambda(t)\} = \hat{\Delta}_\lambda(\omega) = F_\lambda(\omega)$, where $F_\lambda(x)$ is a Fejér kernel.

18. For the Fejér kernel $F_n(t)$ defined by

$$F_n(t) = \sum_{k=-n}^{n} \left(1 - \frac{|k|}{n+1}\right) t^n,$$

show that

(a) $\displaystyle\sum_{n=-\infty}^{\infty} f(t + 2n\pi) = \lim_{N\to\infty} \frac{1}{2\pi} \sum_{n=-N}^{N} \left(1 - \frac{|n|}{N+1}\right) \hat{f}(n)\, e^{int}$,

(b) $\displaystyle F_n\left(e^{it}\right) = \frac{1}{n+1} \left(\frac{\sin\left(\frac{n+1}{2}\right) t}{\sin \frac{t}{2}} \right)^2.$

19. Verify the equality of the uncertainty principle for the Gaussian function

$$f(t) = \exp\left(-\alpha t^2\right).$$

20. Find the system impulse function of the Gaussian filter

$$\hat{\phi}(\omega) = A_0 \exp\left(-a\omega^2 - i\omega t_0\right).$$

21. Find the transfer function of the cosine filter whose amplitude is defined by

$$\hat{A}(\omega) = \begin{Bmatrix} a + b \cos\left(\dfrac{nT\omega}{\omega_0}\right), & |\omega| < \omega_0, \\ 0, & |\omega| > \omega_0 \end{Bmatrix}.$$

22. Use the Fourier transform method to solve the following ordinary differential equations for $x \in R$:

(a) $y''(x) + x\,y'(x) + y(x) = 0$,

(b) $y''(x) + x\,y'(x) + x\,y(x) = 0$,

(c) $y''(x) + 2a y'(x) + \sigma^2 y(x) = f(x)$,

(d) $y''(x) - y(x) = -2\,f(x)$,

where $f(x) = 0$ for $x < -a$ and for $x > a$, and $f(x)$ and its derivatives vanish as $|x| \to \infty$.

23. Let f be a continuous non-negative function defined on $[-\pi, \pi]$ such that $\operatorname{supp} f \subseteq [-\pi + \varepsilon, \pi - \varepsilon]$, for some $0 < \varepsilon < \pi$, and $\int_{-\pi}^{\pi} f(x)\,dx = 2\pi$.

Let g be a 2π-periodic extension of f onto the entire line $\mathbb{R}$. Define

$$k_n(x) = n g(nx) \quad \text{for } n = 1, 2, \ldots.$$

Show that $\{k_n(x)\}$ is a summability kernel.

24. If $\hat{f}(\omega) = \mathscr{F}\{\exp(-t^2)\}$, show that $\hat{f}(\omega)$ satisfies the differential equation

$$2\frac{d}{d\omega}\hat{f}(\omega) + \omega\,\hat{f}(\omega) = 0.$$

Hence, show that

$$\hat{f}(\omega) = \sqrt{\pi}\,\exp\left(-\frac{\omega^2}{4}\right).$$

25. Show that (a) the Fourier transform of the Dirac comb

$$D(t) = \sum_{n=-\infty}^{\infty} \delta(t - nT)$$

is

$$\hat{D}(\omega) = \sum_{n=-\infty}^{\infty} \exp(-i n \omega T).$$

(b) Derive the Poisson summation formula for the Dirac comb

$$\sum_{n=-\infty}^{\infty} \exp(-i n \omega t) = \left(\frac{2\pi}{T}\right) \sum_{r=-\infty}^{\infty} \delta\left(\omega - \frac{2\pi r}{T}\right).$$

26. Prove that

$$\mathscr{F}\left\{\exp\left[-(a-ib)\,t^2\right]\right\}=\left(\frac{\pi}{a-ib}\right)^{\frac{1}{2}}\exp\left[-\frac{(a+ib)\omega^2}{4\left(a^2+b^2\right)}\right].$$

27. Show that the two-dimensional convolution

$$h(x,y)=(f*g)(x,y)=\int_{-\infty}^{\infty} f(\boldsymbol{\xi})\,g(\boldsymbol{r}-\boldsymbol{\xi})\,d\boldsymbol{\xi}$$

has the Fourier transform given by

$$\hat{h}(\boldsymbol{\omega})=\hat{f}(\boldsymbol{\omega})\,\hat{g}(\boldsymbol{\omega}),$$

where $\boldsymbol{\xi}=(\xi,\eta)$, $\boldsymbol{r}=(x,y)$ and $\boldsymbol{\omega}=(\omega,\sigma)$.

28. For the two-dimensional Fourier transform, prove the following results:

(a) $\displaystyle\int_{-\infty}^{\infty} f(\boldsymbol{r})\,\overline{g(\boldsymbol{r})}\,d\boldsymbol{r}=\frac{1}{2\pi}\int_{-\infty}^{\infty}\hat{f}(\boldsymbol{\omega})\,\overline{\hat{g}(\boldsymbol{\omega})}\,d\boldsymbol{\omega}$ (Poisson's formula),

(b) $\displaystyle\int_{-\infty}^{\infty}\left|f(\boldsymbol{r})\right|^2 d\boldsymbol{r}=\frac{1}{2\pi}\int_{-\infty}^{\infty}\left|\hat{f}(\boldsymbol{\omega})\right|^2 d\boldsymbol{\omega}$ (Plancherel's formula),

where $\boldsymbol{r}=(x,y)$ and $\boldsymbol{\omega}=(\omega,\sigma)$.

29. (a) If $f,g,h\in L^2(\mathbb{R})$ and $f(x,y)=g(x)\,h(y)$, show that

$$\hat{f}(\omega,\sigma)=\hat{g}(\omega)\,\hat{h}(\sigma),$$

where $\hat{g}$ and $\hat{h}$ are the one-dimensional Fourier transforms of g and h respectively.

(b) Apply the result in Exercise 29(a) to calculate the Fourier transform of the two-dimensional characteristic function

$$f(x,y)=\begin{cases}1, & |x|\le a \text{ and } |y|\le a\\ 0, & \text{otherwise}\end{cases}.$$

30. If F is the autocorrelation function of $f\in L^2(\mathbb{R})$, show that

(a) $|F(t)|\le \|f\|_2^2$ for all $t\in\mathbb{R}$,

(b) F is uniformly continuous on $\mathbb{R}$.

31. Solve the Bernoulli-Euler beam equation

$$EI\,\frac{d^4u}{dx^4}+\kappa u=W(x), \qquad -\infty<x<\infty,$$

where EI is the flexural rigidity, κ is the foundation modulus of the beam, $W(x)$ has a compact support, and u,u',u'',u''' all tend to zero as $|x|\to\infty$.

32. Use the Fourier transform to solve the following ordinary differential equations

(a) $\ddot{y}(t)+2\alpha\,\dot{y}(t)+\omega^2\,y(t)=f(t), \quad \dot{y}(t)=\dfrac{dy}{dt},$

(b) $y''(x)+x\,y'(x)+x\,y(x)=0$.

33. Solve the following equations for a function f

(a) $\displaystyle\int_{-\infty}^{\infty}\exp\left(-at^2\right)\,f(x-t)\,dt=\exp\left(-bx^2\right), \quad a,b>0,$

(b) $\displaystyle\int_{-\infty}^{\infty}\phi(x-t)\,f(t)\,dt=\psi(x).$

34. Solve the inhomogeneous diffusion problem

$$u_t-\kappa\,u_{xx}=q(x,t), \quad x\in\mathbb{R}, \quad t>0,$$

$$u(x,0)=f(x) \qquad \text{for all } x\in\mathbb{R}.$$

35. (a) Find the Green's function for the one-dimensional Klein-Gordon equation

$$u_{tt}-c^2u_{xx}+d^2u=p(x,t), \quad x\in\mathbb{R}, \quad t>0$$

with the initial and boundary conditions

$$u(x,0)=0=u_t(x,0) \quad \text{for all } x\in\mathbb{R},$$

$$u(x,t)\to 0 \qquad \text{as } |x|\to\infty, \quad t>0,$$

where c and d are constants and $p(x,t)$ is a given function.

(b) Derive the Green's function for both two-dimensional and three-dimensional Klein-Gordon equations.

36. Solve the biharmonic equation

$$u_{tt}+u_{xxxx}=0, \quad x\in\mathbb{R}, \quad t>0,$$

$$u(x,0)=f(x) \quad \text{and} \quad u_t(x,0)=0 \quad \text{for } x\in\mathbb{R}.$$

37. Find the solution of the telegraph equation

$$u_{tt} - c^2 u_{xx} + 2a u_t = 0, \quad x \in \mathbb{R}, \quad t > 0,$$
$$u(x,0) = 0, \qquad u_t(x,0) = g(x),$$

where c and a are constants.

38. Solve the initial value problem for the dissipative wave equation

$$u_{tt} - c^2 u_{xx} + \alpha u_t = 0, \quad x \in \mathbb{R}, \quad t > 0,$$
$$u(x,0) = f(x), \qquad u_t(x,0) = g(x) \quad \text{for } x \in \mathbb{R},$$

where $\alpha > 0$ is the dissipation parameter.

39. Solve the Cauchy problem

$$u_t = \kappa u_{xx}, \quad x \in \mathbb{R}, \quad t > 0,$$
$$u(x,0) = \exp\left(-ax^2\right), \quad u \to 0 \quad \text{as } |x| \to \infty.$$

40. Find the Green's function $G(\mathbf{x},t)$ which satisfies the differential system

$$G_{tt} - c^2\nabla^2 G + d^2 G = \delta(x)\,\delta(y)\,\delta(z)\,\delta(t), \; \mathbf{x} \in \mathbb{R}^3, \; t > 0,$$
$$G(\mathbf{x},0) = 0 = G_t(\mathbf{x},0) \quad \text{for all } \mathbf{x} \in \mathbb{R}^3,$$
$$G(\mathbf{x},t) \to 0 \quad \text{as } |\mathbf{x}| \to \infty,$$

where $x = (x, y, z)$, c and d are constants, and $\nabla^2 \equiv \dfrac{\partial^2}{\partial x^2} + \dfrac{\partial^2}{\partial y^2} + \dfrac{\partial^2}{\partial z^2}$.

41. For a signal f with the Gaussian amplitude modulation and a cubic phase modulation

$$f(t) = \left(\frac{\alpha}{\pi}\right)^{\frac{1}{4}} \exp\left(-\frac{1}{2}\alpha t^2 + \frac{1}{3} i \gamma t^3 + 2\pi i \nu_0 t\right),$$

show that the average frequency and the bandwidth square are

$$\langle \nu \rangle = \left(\frac{\gamma}{8\pi\alpha} + \nu_0\right) \quad \text{and} \quad B^2 = \frac{1}{8\pi^2}\left(\alpha + \frac{\gamma^2}{\alpha^2}\right).$$

42. For a signal with the Gaussian amplitude modulation and a sinusoidal modulated frequency

$$f(t) = \left(\frac{\alpha}{\pi}\right)^{\frac{1}{4}} \exp\left(-\frac{1}{2}\alpha t^2 + i m \sin 2\pi \nu_m t + 2\pi i \nu_0 t\right),$$

find the average frequency and the bandwidth of the signal.

Chapter 4

The Gabor Transform and Time-Frequency Signal Analysis

"What is clear and easy to grasp attracts us; complications deter."

David Hilbert

"Motivated by 'quantum mechanics', in 1946 the physicist Gabor defined elementary time-frequency atoms as waveforms that have a minimal spread in a time-frequency plane. To measure time-frequency 'information' content, he proposed decomposing signals over these elementary atomic waveforms. By showing that such decompositions are closely related to our sensitivity to sounds, and that they exhibit important structures in speech and music recordings, Gabor demonstrated the importance of localized time-frequency signal processing."

Stéphane Mallat

4.1 Introduction

Signals are, in general, nonstationary. A complete representation of nonstationary signals requires frequency analysis that is local in time, resulting in the time-frequency analysis of signals. The Fourier transform analysis has long been recognized as the great tool for the study of stationary signals and processes where the properties are statistically invariant over time. However, it cannot be used for the frequency analysis that is local in time. In recent years, several useful methods have been developed for the time-frequency signal analysis. They include the Gabor transform, the Zak transform, and the wavelet transform.

It has already been stated in section 1.2 that decomposition of a signal into a small number of elementary waveforms that are localized in time and frequency plays a remarkable role in signal processing. Such a decomposition reveals important structures in analyzing nonstationary signals such as speech and music. In order to measure localized frequency components of sounds, Gabor (1946) first introduced the windowed Fourier transform (or the local time-frequency transform), which may be called the Gabor transform, and suggested the representation of a signal in a joint time-frequency domain. Subsequently, the Gabor analysis has effectively been applied in many fields of science and engineering, such as image analysis and image compression, object and pattern recognition, computer vision, optics, and filter banks. Since medical signal analysis and medical signal processing play a crucial role in medical diagnostics, the Gabor transform has also been used for the study of brain functions, ECC signals, and other medical signals.

This chapter deals with classification of signals, joint time-frequency analysis of signals, and the Gabor transform and its basic properties, including the inversion formula. Special attention is given to the discrete Gabor transform and the Gabor representation problem. Included are the Zak transform, its basic properties, and applications for studying the orthogonality and completeness of the Gabor frames in the critical case.

4.2 Classification of Signals and the Joint Time-Frequency Analysis of Signals

Many physical quantities including pressure, sound waves, electric fields, voltage, electric current, and electromagnetic fields vary with time t. These quantities are called *signals* or *waveforms*. Example of signals include speech signals, optical signals, acoustic signals, biomedical signals, radar, and sonar. Indeed, signals are very common in the real world.

In general, there are two kinds of signals: (i) *deterministic* and (ii) *random* (or *stochastic*). A signal is called *deterministic* if it can be determined explicitly, under identical conditions, in terms of a mathematical relationship. A deterministic signal is referred to as *periodic* or *transient* if the signal repeats continuously at regular intervals of time or decays to zero after a finite time interval. Periodic and transient signals are shown in Figures 4.1(a), 4.1(b) and Figure 4.2.

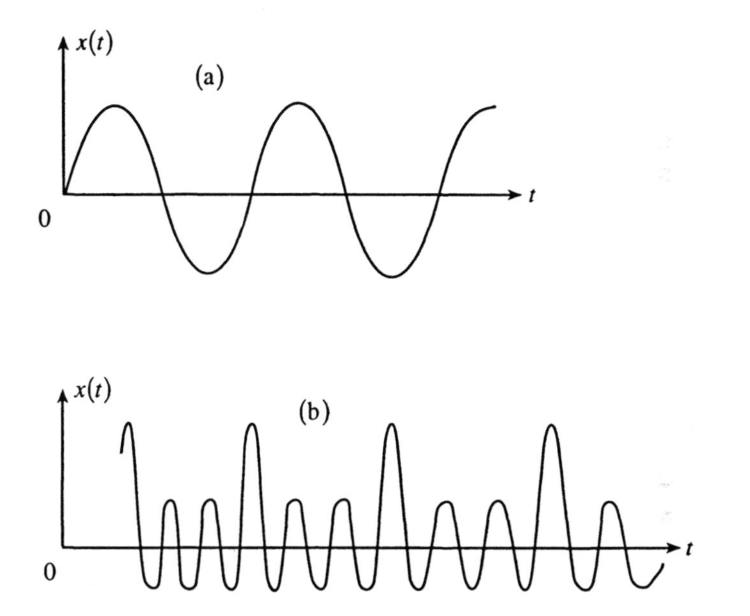

Figure 4.1. (a) Sinusoidal periodic signal; (b) Nonsinusoidal periodic signal.

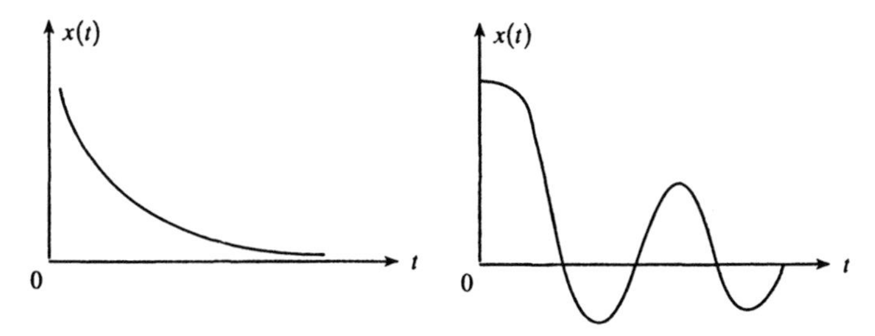

Figure 4.2. Transient signals.

On the other hand, signals are, in general, random or stochastic in nature in the sense that they cannot be determined precisely at any given instant of time even under identical conditions. Obviously, probabilistic and statistical information is required for a description of random signals. It is necessary to consider a particular random process that can produce a set of time-histories, known as an *ensemble*. This can represent an experiment producing random data, which is repeated n times to give an ensemble of n separate records (See Figure 4.3).

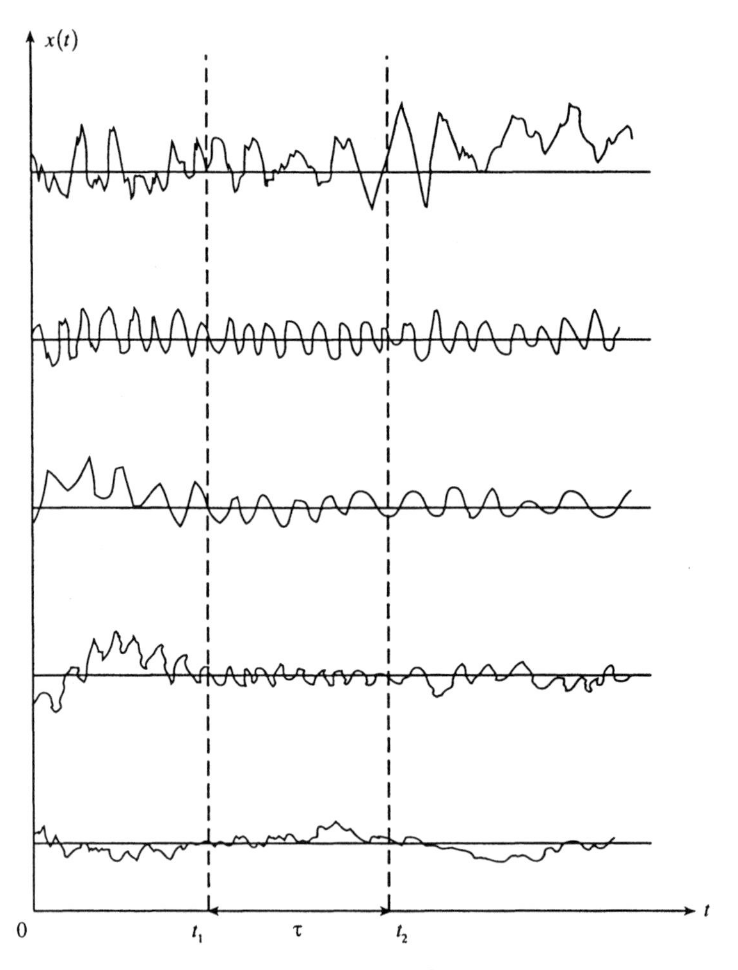

Figure 4.3. Ensemble of n records.

The *average value* at time t over the ensemble x is defined by

$$\langle x(t) \rangle = \lim_{n\to\infty} \frac{1}{n} \sum_{k=1}^{n} x_k(t), \tag{4.2.1}$$

where x takes any one of a set of values x_k, and $k = 1,2,\ldots,n$.

The *average value* of the product of two samples taken at two separate times t_1 and t_2 is called the *autocorrelation function* R, for each separate record, defined by

$$R(\tau) = \lim_{n\to\infty} \frac{1}{n} \sum_{k=1}^{n} x_k(t_1)\, x_k(t_2), \tag{4.2.2}$$

where $\tau = t_1 - t_2$. The process of finding these values is referred to as *ensemble averaging* and may be continued over the entire record length to provide statistical information on the complex set of records.

A signal is called *stationary* if the values of $\langle x(t) \rangle$ and $R(t)$ remain constant for all possible values of t and $R(\tau)$ depends only on the time displacement $\tau = t_1 - t_2$ (see Figure 4.4(a)). In most practical situations, a signal is called stationary if $\langle x(t) \rangle$ and $R(\tau)$ are constant over the finite record length T.

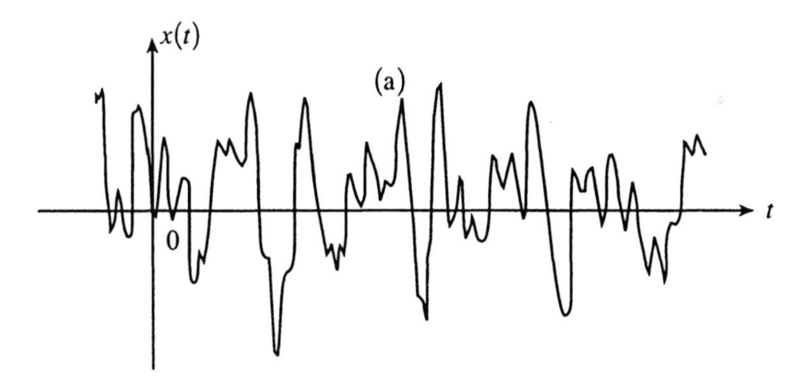

Figure 4.4. (a) Stationary random signal.

A signal is called *nonstationary* if the values of $\langle x(t) \rangle$ and $R(\tau)$ vary with time (see Figure 4.4(b)). However, in many practical situations, the change of time is very slow, so the signal can be regarded as stationary. Under certain conditions, we regard a signal as stationary by considering the statistical characteristic of a single long record. The average value of a signal $x(t)$ over a time length T is defined by

$$\bar{x} = \lim_{T\to\infty} \frac{1}{T} \int_0^T x(t)\, dt, \tag{4.2.3}$$

where $\bar{x}$ is used to represent a single time-history average to distinguish it from the ensemble average $\langle x \rangle$.

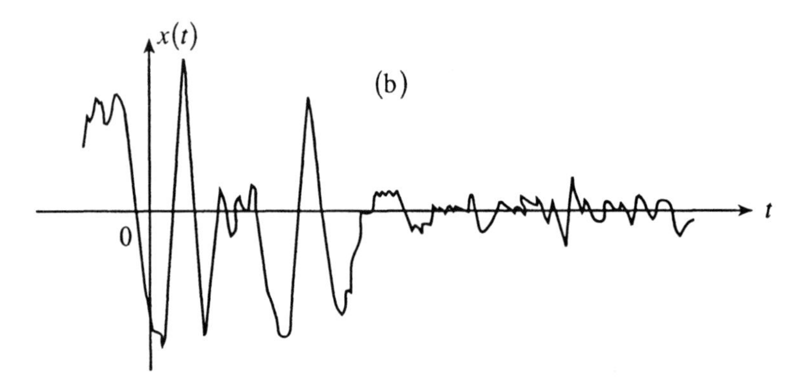

Figure 4.4. (b) Nonstationary random signal.

Similarly, the *autocorrelation* function over a single time length T is defined by

$$R(\tau) = \lim_{T\to\infty} \frac{1}{T} \int_0^T x(t)\, x(t+\tau)\, dt. \tag{4.2.4}$$

Under certain circumstances, the ensemble average can be obtained from computing the time average so

$$\langle x \rangle = \bar{x} \tag{4.2.5}$$

for all values of time t. Then, this process is called an *ergodic random process*. By definition, this must be a stationary process. However, the converse is not necessarily true, that is, a stationary random process need not be ergodic.

Finally, we can introduce various ensemble averages of $x(t)$ which take any one of the values $x_k(t)$, $k = 1, 2, \ldots, n$ at time t in terms of probability $P_x(x_k(t))$. The ensemble average of x is then defined by

$$\langle x \rangle = \sum_{k=1}^{n} P_x(x_k)\, x_k . \tag{4.2.6}$$

We now consider two random variables $x_i(t)$ and $x_k(s)$ which are values of a random process x at times t and s with the joint probability distribution $P_x^{t,s}(x_i, x_k)$. Then, the *autocorrelation function*, $R(t,s)$ of the random process x is defined by

$$R(t,s) = \langle x(t)\, x(s) \rangle = \sum_{i,k} P_x^{t,s}(x_i, x_k)\, x_i\, x_k . \tag{4.2.7}$$

This function provides a great deal of information about the random process and arises often in signal analysis. For a random stationary process, $P_x^{(t,s)}$ is a function of $\tau = t - s$ only, so that

$$R(t,s) = R(t-s) = R(\tau) \tag{4.2.8}$$

and hence, $R(-\tau) = R(\tau)$ and R is an even function.

Signals can be described in a time domain or in a frequency domain by the traditional method of Fourier transform analysis. The frequency description of signals is known as the *frequency* (or *spectral*) analysis. It was recognized long ago that a global Fourier transform of a long time signal is of little practical value in analyzing the frequency spectrum of the signal. From the Fourier spectrum (or spectral function) $\hat{f}(\omega)$ of a signal $f(t)$, it is always possible to determine which frequencies were present in the signal. However, there is absolutely no indication as to when those frequencies existed. So, the Fourier transform analysis cannot provide any information regarding either a time evolution of spectral characteristics or a possible localization with respect to the time variable. Transient signals such as a speech signals or ECG signals (see Figure 4.5) require the idea of frequency analysis that is local in time.

In general, the frequency of a signal varies with time, so there is a need for a joint time-frequency representation of a signal in order to describe fully the characteristics of the signal. Thus, both the analysis and processing of non-stationary signals require specific mathematical methods which go beyond the classical Fourier transform analysis. Gabor (1946) was the first to introduce the joint time-frequency representation of a signal. Almost simultaneously, Ville (1948) first introduced the Wigner distribution into time-frequency signal analysis to unfold the signal in the time-frequency plane in such a way that this development led to a joint representation in time-frequency atoms.

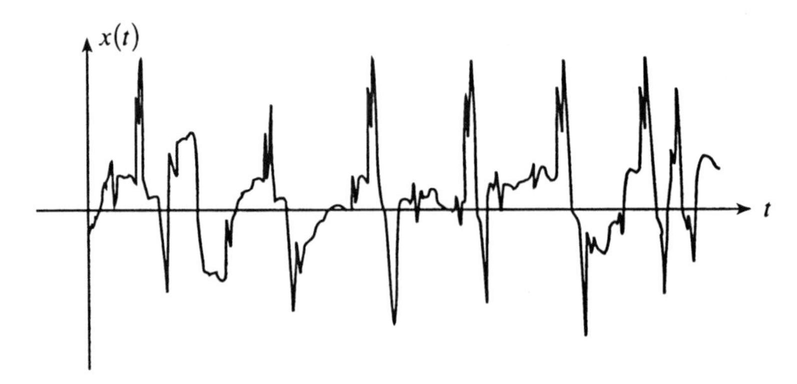

Figure 4.5. ECG signal of a human heart.

4.3 Definition and Examples of the Gabor Transform

Gabor (1946) first introduced a time-localization window function $g_a(t-b)$ for extracting local information from a Fourier transform of a signal, where the parameter a measures the width of the window, and the parameter b is used to represent translation of the window to cover the whole time domain. The idea is to use this window function in order to localize the Fourier transform, then shift the window to another position, and so on. This remarkable property of the Gabor transform provides the local aspect of the Fourier transform with time resolution equal to the size of the window. Thus, the Gabor transform is often called the *windowed Fourier transform*. Gabor first introduced

$$g_{t,\omega}(\tau) = \exp(i\omega\tau)\, g(\tau - t) = M_\omega T_t\, g(\tau), \tag{4.3.1}$$

as the window function by first translating in time and then modulating the function $g(t) = \pi^{-\frac{1}{4}} \exp\left(-2^{-1} t^2\right)$ which is called the *canonical coherent state* in quantum physics. The energy associated with the function $g_{t,\omega}$ is localized in the neighborhood of t in an interval of size σ_t, measured by the standard deviation of $|g|^2$. Evidently, the Fourier transform of $g_{t,\omega}(\tau)$ with respect to τ is given by

$$\hat{g}_{t,\omega}(\nu) = \hat{g}(\nu - \omega) \exp\{-it(\nu - \omega)\}. \tag{4.3.2}$$

Obviously, the energy of $\hat{g}_{t,\omega}$ is concentrated near the frequency ω in an interval of size σ_ω which measures the frequency dispersion (or bandwidth) of $\hat{g}_{t,\omega}$. In a time-frequency (t,ω) plane, the energy spread of the Gabor atom $\hat{g}_{t,\omega}$ can be represented by the rectangle with center at $(\langle t\rangle, \langle\omega\rangle)$ and sides σ_t (along the time axis) and σ_ω (along the frequency axis). According to the Heisenberg uncertainty principle, the area of the rectangle is at least $\frac{1}{2}$; that is, $\sigma_t\sigma_\omega \geq \frac{1}{2}$. This area is minimum when g is a Gaussian function, and the corresponding $g_{t,\omega}$ is called the *Gabor function* (or *Gabor wavelet*).

Definition 4.3.1 (The Continuous Gabor Transform). The continuous Gabor transform of a function $f \in L^2(\mathbb{R})$ with respect to a window function $g \in L^2(\mathbb{R})$ is denoted by $\mathcal{G}[f](t,\omega) = \tilde{f}_g(t,\omega)$ and defined by

$$\mathcal{G}[f](t,\omega) = \tilde{f}_g(t,\omega) = \int_{-\infty}^{\infty} f(\tau)\, g(\tau - t)\, e^{-i\omega\tau} d\tau = \left(f, \overline{g_{t,\omega}}\right), \tag{4.3.3}$$

where $g_{t,\omega}(\tau) = \bar{g}(\tau - t)\exp(i\omega\tau)$, so, $\|g_{t,\omega}\| = \|g\|$ and hence, $g_{t,\omega} \in L^2(\mathbb{R})$.

Clearly, the Gabor transform $\tilde{f}_g(t,\omega)$ of a given signal f depends on both time t and frequency ω. For any fixed t, $\tilde{f}_g(t,\omega)$ represents the frequency distribution at time t. Usually, only values of $f(\tau)$ for $\tau \leq t$ can be used in computing $\tilde{f}_g(t,\omega)$. In a system of finite memory, there exists a time interval $T > 0$ such that only the values $f(\tau)$ for $\tau > t - T$ can affect the output at time t. Thus, the transform function $\tilde{f}_g(t,\omega)$ depends only on $f(\tau)$ for $t - T \leq \tau \leq t$. Mathematically, if $g_{t,\omega}(\tau)$ vanishes outside $[-T, 0]$ such that supp $g \subset [-T, 0]$, then $g_{t,\omega}(\tau)$ can be used to localize the signal in time. For any $t \in \mathbb{R}$, we can define $f_t(\tau) = g(\tau - t)\, f(\tau)$ so that supp $f_t \subset [t - T, t]$. Therefore, $f_t(\tau)$ can be regarded as a localized version of f that depends only on the values of $f(\tau)$ in $t - T \leq \tau \leq t$. If g is continuous, then the values of $f_t(\tau)$ with $\tau \approx t - T$ and

$\tau \approx t$ are small. This means that the localization is smooth, and this particular feature plays an important role in signal processing.

In physical applications, f and g represent signals with finite energy. In quantum physics, $\tilde{f}_g(t,\omega)$ is referred to as the *canonical coherent state* representation of f. The term *coherent state* was first used by Glauber (1964) in quantum optics.

We next discuss the following consequences of the preceding definition.

1. For a fixed t, the Fourier transform of $f_t(\tau)$ with respect to τ is given by

$$\tilde{f}_g(t,\omega) = \mathscr{F}\{f_t(\tau)\} = \hat{f}_t(\nu), \tag{4.3.4}$$

where $f_t(\tau) = f(\tau)\, g(\tau - t)$.

2. If the window g is real and symmetric with $g(\tau) = g(-\tau)$ and if g is normalized so that $\|g\| = 1$ and $\|g_{t,\omega}\| = \|g(\tau - t)\| = 1$ for any $(t,\omega) \in \mathbb{R}^2$, then the Gabor transform of $f \in L^2(\mathbb{R})$ becomes

$$\tilde{f}_g(t,\omega) = (f, g_{t,\omega}) = \int_{-\infty}^{\infty} f(\tau)\, g(\tau - t)\, e^{-i\omega\tau} d\tau. \tag{4.3.5}$$

This can be interpreted as the *short-time Fourier transform* because the multiplication by $g(\tau - t)$ induces localization of the Fourier integral in the neighborhood of $\tau = t$.

Application of the Schwarz inequality (2.6.1) to (4.3.5) gives

$$\left|\tilde{f}_g(t,\omega)\right| = \left|(f, g_{t,\omega})\right| \le \|f\| \|g_{t,\omega}\| = \|f\| \|g\|.$$

This shows that the Gabor transform $\tilde{f}_g(t,\omega)$ is *bounded*.

3. The energy density defined by

$$\left|\tilde{f}_g(t,\omega)\right|^2 = \left|\int_{-\infty}^{\infty} f(\tau)\, g(\tau - t)\, e^{-i\omega\tau} d\tau\right|^2 \tag{4.3.6}$$

measures the energy of a signal in the time-frequency plane in the neighborhood of the point (t,ω).

4. It follows from definition (4.3.3) with a fixed ω that

$$\tilde{f}_g(t,\omega) = e^{-i\omega t} \int_{-\infty}^{\infty} f(\tau)\, g(\tau - t)\, e^{i\omega(t-\tau)} d\tau = e^{-i\omega t} (f * g_\omega)(t), \tag{4.3.7}$$

where $g_\omega(\tau) = e^{i\omega\tau} g(\tau)$ and $g(-\tau) = g(\tau)$. Furthermore, by the Parseval relation (3.4.34) of the Fourier transform, we find

$$\tilde{f}_g(t,\omega) = \left(f, \bar{g}_{t,\omega}\right) = \left(\hat{f}, \hat{\bar{g}}_{t,\omega}\right) = e^{i\omega t} \int_{-\infty}^{\infty} \hat{f}(v)\, \hat{g}(v-\omega)\, e^{-ivt}\, dv. \tag{4.3.8}$$

Except for the factor $\exp(i\omega t)$, result (4.3.8) is almost identical with (4.3.3), but the time variable t is replaced by the frequency variable ω, and the time window $g(\tau - t)$ is replaced by the frequency window $\hat{g}(v-\omega)$. The extra factor $\exp(i\omega t)$ in (4.3.8) is associated with the Weyl commutation relations of the Weyl-Heisenberg group which describe translations in time and frequency. If the window is well-localized in frequency and in time, that is, if $\hat{g}(v-\omega)$ is small outside a small frequency band in addition to $g(\tau)$ being small outside a small time interval, then (4.3.8) reveals that the Gabor transform gives a local time-frequency analysis of the signal f in the sense that it provides accurate information of f simultaneously in both time and frequency domains. However, all functions, including the window function, satisfy the Heisenberg uncertainty principle, that is, the joint resolution $\sigma_t \sigma_\omega$ of a signal cannot be arbitrarily small and has always greater than the minimum value $\frac{1}{2}$ which is attained only for the Gaussian window function $g(t) = \exp\left(-at^2\right)$.

5. For a fixed ω, the Fourier transform of $\tilde{f}_g(t,\omega)$ with respect to t is given by the following:

$$\mathcal{F}\left\{\tilde{f}_g(t,\omega)\right\} = \hat{\tilde{f}}_g(v,\omega) = \hat{f}(v+\omega)\, \hat{g}(v). \tag{4.3.9}$$

This follows from the Fourier transform of (4.3.7) with respect to t

$$\mathcal{F}\left\{\tilde{f}_g(t,\omega)\right\} = \mathcal{F}\left\{e^{-i\omega t}\left(f * g_\omega\right)(t)\right\} = \hat{f}(\omega+v)\, \hat{g}(v).$$

6. If $g(t) = \exp\left(-\frac{1}{4}t^2\right)$, then

$$\tilde{f}_g(t,\omega) = \sqrt{2}\, \exp\left(i\omega t - \omega^2\right)(Wf)(t + 2i\omega), \tag{4.3.10}$$

where W represents the *Weierstrass transformation* of $f(x)$ defined by

$$W\left[f(x)\right] = \frac{1}{2\sqrt{\pi}} \int_{-\infty}^{\infty} f(x)\, \exp\left[-\frac{1}{4}(t-x)^2\right] dx. \tag{4.3.11}$$

7. The time width σ_t around t and the frequency spread σ_ω around ω are independent of t and ω.

We have, by definition, and the Gabor window function (4.3.1),

$$\sigma_t^2 = \int_{-\infty}^{\infty} (\tau - t)^2 \left|g_{t,\omega}(\tau)\right|^2 d\tau = \int_{-\infty}^{\infty} (\tau - t)^2 \left|g(\tau - t)\right|^2 d\tau = \int_{-\infty}^{\infty} \tau^2 \left|g(\tau)\right|^2 d\tau.$$

Similarly, we obtain, by (4.3.2),

$$\sigma_\omega^2 = \frac{1}{2\pi} \int_{-\infty}^{\infty} (v - \omega)^2 \left|\hat{g}_{t,\omega}(v)\right|^2 dv = \frac{1}{2\pi} \int_{-\infty}^{\infty} (v - \omega)^2 \left|\hat{g}(v)\right|^2 dv = \frac{1}{2\pi} \int_{-\infty}^{\infty} v^2 \left|\hat{g}(v)\right|^2 dv.$$

Thus, both σ_t and σ_ω are independent of t and ω. The energy spread of $g_{t,\omega}(\tau)$ can be represented by the Heisenberg rectangle centered at (t,ω) with the area $\sigma_t \sigma_\omega$ which is independent of t and ω. This means that the Gabor transform has the same resolution in the time-frequency plane.

Example 4.3.1 Obtain the Gabor transform of functions

(a) $f(\tau) = 1$, (b) $f(\tau) = \exp(-i\sigma\tau)$.

We obtain

(a) $$\tilde{f}_g(t,\omega) = \int_{-\infty}^{\infty} g(\tau - t)\, e^{-i\omega\tau} d\tau = e^{-i\omega t}\hat{g}(\omega).$$

(b) $$\tilde{f}_g(t,\omega) = \int_{-\infty}^{\infty} e^{-i\tau(\omega+\sigma)}\, g(\tau - t)\, d\tau = \exp\{-it(\sigma + \omega)\}\, \hat{g}(\sigma + \omega).$$

Example 4.3.2 Find the Gabor transform of functions

(a) $f(\tau) = \delta(\tau)$, (b) $f(\tau) = \delta(\tau - t_0)$.

We have

(a) $$\tilde{f}_g(t,\omega) = \int_{-\infty}^{\infty} \delta(\tau)\, g(\tau - t)\, e^{-i\omega\tau} d\tau = g(-t).$$

(b) $$\tilde{f}_g(t,\omega) = \int_{-\infty}^{\infty} \delta(\tau - t_0)\, g(\tau - t)\, e^{-i\omega\tau} d\tau = e^{-i\omega t_0} g(t_0 - t).$$

Example 4.3.3 Find the Gabor transform of the function $f(\tau)=\exp\left(-a^2\tau^2\right)$ with $g(\tau)=1$.

We have

$$\tilde{f}_g(t,\omega)=\int_{-\infty}^{\infty}\exp\left\{-\left(a^2\tau^2+i\omega\tau\right)\right\}d\tau=\hat{f}(\omega)=\frac{\sqrt{\pi}}{a}\exp\left(-\frac{\omega^2}{4a^2}\right).$$

4.4 Basic Properties of Gabor Transforms

Theorem 4.4.1 (Linearity). If the Gabor transforms of two functions f_1 and f_2 exist with respect to a window function g, then

$$\mathcal{G}\left[a\,f_1+b\,f_2\right](t,\omega)=a\,\mathcal{G}\left[f_1\right](t,\omega)+b\,\mathcal{G}\left[f_2\right](t,\omega), \tag{4.4.1}$$

where a and b are two arbitrary constants.

The proof easily follows from the definition of the Gabor transform and is left as an exercise.

Theorem 4.4.2 If f and $g\in L^2(\mathbb{R})$, then the following results hold:

(a) (*Translation*): $\mathcal{G}\left[T_a f\right](t,\omega)=e^{ia\omega}\,\mathcal{G}[f](t-a,\omega)$, (4.4.2)

(b) (*Modulation*): $\mathcal{G}\left[M_a f\right](t,\omega)=\mathcal{G}[f](t,\omega-a)$, (4.4.3)

(c) (*Conjugation*): $\mathcal{G}\left[\bar{f}\right](t,\omega)=\overline{\mathcal{G}[f](t,-\omega)}$. (4.4.4)

Proof. (a) We have, by definition,

$$\begin{aligned}\mathcal{G}\left[T_a f\right](t,\omega)=\mathcal{G}\left[f(\tau-a)\right](t,\omega)&=\int_{-\infty}^{\infty}f(\tau-a)\,g(\tau-t)\,e^{-i\omega\tau}\,d\tau\\&=e^{-i\omega a}\int_{-\infty}^{\infty}f(x)\,g\left(x-\overline{t-a}\right)e^{-i\omega x}\,dx\\&=e^{-i\omega a}\,\mathcal{G}[f](t-a,\,\omega).\end{aligned}$$

(b) We have

$$\begin{aligned}\mathcal{G}[M_a f](t,\omega) &= \mathcal{G}\left[e^{ia\tau} f(\tau)\right](t,\omega)\\ &= \int_{-\infty}^{\infty} f(\tau)\, g(\tau-t)\, e^{-i\tau(\omega-a)}\, d\tau\\ &= \mathcal{G}[f](t,\omega-a).\end{aligned}$$

(c) It follows from definition (4.3.3) with a real window function g that

$$\begin{aligned}\mathcal{G}\left[\bar{f}\right](t,\omega) &= \int_{-\infty}^{\infty} \overline{f(\tau)}\, g(\tau-t)\, e^{-i\omega\tau} d\tau = \overline{\int_{-\infty}^{\infty} f(\tau)\, g(\tau-t)\, e^{i\omega\tau}\, d\tau}\\ &= \overline{\mathcal{G}[f](\tau,-\omega)}.\end{aligned}$$

Theorem 4.4.3 If two signals $f,g \in L^2(\mathbb{R})$, then

$$\int_{-\infty}^{\infty}\int_{-\infty}^{\infty} \left|\tilde{f}_g(t,\omega)\right|^2 dt\, d\omega = \|f\|_2^2\, \|g\|_2^2.$$

Proof. The left-hand side of the above result is equal to

$$\begin{aligned}\int_{-\infty}^{\infty}\int_{-\infty}^{\infty} \left|\tilde{f}_g(t,\omega)\right|^2 dt\, d\omega &= \int_{-\infty}^{\infty}\int_{-\infty}^{\infty} \left|\int_{-\infty}^{\infty} f(\tau)\, g(\tau-t)\, e^{-i\omega\tau}\, d\tau\right|^2 dt\, d\omega\\ &= \int_{-\infty}^{\infty}\int_{-\infty}^{\infty} \left|\int_{-\infty}^{\infty} h_t(\tau)\, e^{-i\omega\tau}\, d\tau\right|^2 dt\, d\omega, \quad h_t(\tau) = f(\tau)\, g(\tau-t)\\ &= \int_{-\infty}^{\infty} dt \int_{-\infty}^{\infty} \left|\hat{h}_t(\omega)\right|^2 d\omega\\ &= \int_{-\infty}^{\infty} \left\|\hat{h}_t(\omega)\right\|^2 dt\\ &= \int_{-\infty}^{\infty} \left\|h_t(\tau)\right\|^2 dt, \qquad \text{by Plancherel's theorem}\\ &= \int_{-\infty}^{\infty} dt \int_{-\infty}^{\infty} |f(\tau)|^2\, |g(\tau-t)|^2\, d\tau\\ &= \int_{-\infty}^{\infty} |f(\tau)|^2\, d\tau \int_{-\infty}^{\infty} |g(x)|^2\, dx = \|f\|_2^2\, \|g\|_2^2.\end{aligned}$$

This completes the proof.

Theorem 4.4.4 (Parseval's Formula). If $\tilde{f}_g(t,\omega) = \mathcal{G}[f](t,\omega)$ and $\tilde{h}_g(t,\omega) = \mathcal{G}[h](t,\omega)$, then the Parseval formula for the Gabor transform is given by

$$\left(\tilde{f},\tilde{h}\right) = \|g\|^2 (f,h), \tag{4.4.5}$$

where

$$\left(\tilde{f},\tilde{h}\right) = \left(\tilde{f},\tilde{h}\right)_{L^2(\mathbb{R}^2)} = \int_{-\infty}^{\infty}\int_{-\infty}^{\infty} \tilde{f}_g(t,\omega)\, \overline{\tilde{h}_g}(t,\omega)\, dt\, d\omega. \tag{4.4.6}$$

In particular, if $\|g\| = 1$, then the Gabor transformation is an isometry from $L^2(\mathbb{R})$ into $L^2(\mathbb{R}^2)$.

Proof. We first note that, for a fixed t,

$$\tilde{f}_g(t,\omega) = \mathcal{F}\{f_t(\tau)\} = \mathcal{F}\{f(\tau)\, g_t(\tau)\},$$

where $g_t(\tau) = g(\tau - t)$.

Thus, the Parseval formula (3.4.34) for the Fourier transform gives

$$\begin{aligned}
\int_{-\infty}^{\infty} \tilde{f}(t,\omega)\, \overline{\tilde{h}}(t,\omega)\, d\omega &= \left(\mathcal{F}\{f\, g_t\},\, \mathcal{F}\{h\, g_t\}\right) \\
&= \left(f\, g_t,\, h\, g_t\right) = \int_{-\infty}^{\infty} f(\tau)\, g(\tau - t)\, \overline{h}(\tau)\, \overline{g(\tau - t)}\, d\tau \\
&= \int_{-\infty}^{\infty} f(\tau)\, \overline{h}(\tau)\, |g(\tau - t)|^2\, d\tau.
\end{aligned}$$

Integrating this result with respect to t from $-\infty$ to $+\infty$ gives

$$\begin{aligned}
\left(\tilde{f},\tilde{h}\right) &= \int_{-\infty}^{\infty}\int_{-\infty}^{\infty} \tilde{f}(t,\omega)\, \overline{\tilde{h}}(t,\omega)\, dt\, d\omega \\
&= \int_{-\infty}^{\infty} f(\tau)\, \overline{h}(\tau)\, d\tau \int_{-\infty}^{\infty} |g(\tau - t)|^2\, dt \\
&= \int_{-\infty}^{\infty} f(\tau)\, \overline{h}(\tau)\, d\tau \int_{-\infty}^{\infty} |g(x)|^2\, dx \qquad (\tau - t = x) \\
&= \|g\|^2 (f,h).
\end{aligned}$$

This proves the result.

If $\|g\| = 1$, then (4.4.5) shows isometry from $L^2(\mathbb{R})$ to $L^2(\mathbb{R}^2)$.

Theorem 4.4.5 (Inversion Theorem). If a function $f \in L^2(\mathbb{R})$, then

$$f(\tau) = \frac{1}{2\pi} \frac{1}{\|g\|^2} \int_{-\infty}^{\infty}\int_{-\infty}^{\infty} \tilde{f}_g(t,\omega)\, g(\tau - t)\, e^{i\omega\tau}\, d\omega\, dt. \tag{4.4.7}$$

First Proof. It follows from the continuous Gabor transform (4.3.3) that

$$\tilde{f}_g(t,\omega) = \mathcal{F}\{f(\tau)\, g(\tau - t)\},$$

where the Fourier transform with respect to τ is taken.

Application of the inverse Fourier transform to this result gives

$$f(\tau)\, g(\tau - t) = \mathcal{F}^{-1}\left\{\tilde{f}_g(t,\omega)\right\} = \frac{1}{2\pi} \int_{-\infty}^{\infty} e^{i\omega\tau}\, \tilde{f}_g(t,\omega)\, d\omega.$$

Multiplying this result by $\bar{g}(\tau - t)$ and integrating with respect to t yields

$$f(\tau) \int_{-\infty}^{\infty} |g(\tau - t)|^2\, dt = \frac{1}{2\pi} \int_{-\infty}^{\infty}\int_{-\infty}^{\infty} e^{i\omega\tau}\, \bar{g}(\tau - t)\, \tilde{f}_g(\omega, t)\, d\omega\, dt.$$

Or, equivalently,

$$f(\tau)\, \|g\|^2 = \frac{1}{2\pi} \int_{-\infty}^{\infty}\int_{-\infty}^{\infty} e^{i\omega\tau}\, \bar{g}(\tau - t)\, \tilde{f}_g(t,\omega)\, d\omega\, dt.$$

This proves the inversion theorem.

Second Proof. We apply the inverse Fourier transform of $f(\tau)$ and the Parseval formula to replace $\|g\|^2$ by $\frac{1}{2\pi}\|\hat{g}\|^2$ so that

$$f(\tau)\, \|g\|^2 = \frac{1}{2\pi} \int_{-\infty}^{\infty} e^{i\omega\tau}\, \hat{f}(\omega)\, d\omega \cdot \frac{1}{2\pi} \|\hat{g}\|^2$$

$$= \frac{1}{2\pi} \int_{-\infty}^{\infty} e^{i\omega\tau}\, \hat{f}(\omega)\, d\omega \cdot \frac{1}{2\pi} \int_{-\infty}^{\infty} |\hat{g}(v)|^2\, dv.$$

Since the integral is true for any arbitrary ω, we replace ω by $\omega + v$ to obtain

$$
\begin{aligned}
f(\tau)\,\|g\|^2 &= \frac{1}{2\pi}\int_{-\infty}^{\infty} e^{i\tau(\omega+v)}\,\hat{f}(\omega+v)\,d\omega \cdot \frac{1}{2\pi}\int_{-\infty}^{\infty}\hat{g}(v)\,\hat{g}(v)\,dv \\
&= \frac{1}{2\pi}\int_{-\infty}^{\infty} e^{i\omega\tau}\,d\omega \cdot \frac{1}{2\pi}\int_{-\infty}^{\infty} e^{i\tau v}\left[\hat{f}(\omega+v)\,\hat{g}(v)\right]\hat{g}(v)\,dv \\
&= \frac{1}{2\pi}\int_{-\infty}^{\infty} e^{i\omega\tau}\,d\omega \cdot \left[\frac{1}{2\pi}\int_{-\infty}^{\infty} e^{iv\tau}\,\hat{\tilde{f}}_g(v+\omega)\,\hat{g}(v)\,dv\right], \qquad \text{by (4.3.9)} \\
&= \frac{1}{2\pi}\int_{-\infty}^{\infty} e^{i\omega\tau}\,d\omega \cdot \left[\tilde{f}_g(\tau,\omega) * g(\tau)\right], \qquad \text{by (3.3.23)} \\
&= \frac{1}{2\pi}\int_{-\infty}^{\infty} e^{i\omega\tau}\,d\omega \int_{-\infty}^{\infty} \tilde{f}_g(t,\omega)\,g(\tau-t)\,dt \\
&= \frac{1}{2\pi}\int_{-\infty}^{\infty}\int_{-\infty}^{\infty} e^{i\omega\tau}\,\tilde{f}_g(t,\omega)\,g(\tau-t)\,dt\,d\omega.
\end{aligned}
$$

This proves the inversion theorem.

Theorem 4.4.6 (Conservation of Energy). If $f \in L^2(\mathbb{R})$, then

$$
\|f\|_2^2 = \frac{1}{2\pi}\int_{-\infty}^{\infty}\int_{-\infty}^{\infty}\left|\tilde{f}_g(t,\omega)\right|^2 dt\,d\omega, \tag{4.4.8}
$$

where g is a normalized window function $\left(\|g\| = 1\right)$.

Proof. Using (4.3.9) dealing with the Fourier transform of $\tilde{f}_g(t,\omega)$ with respect to t, we apply the Plancherel formula to the right-hand side of (4.4.8) to obtain

$$
\begin{aligned}
\frac{1}{2\pi}\int_{-\infty}^{\infty}\int_{-\infty}^{\infty}\left|\tilde{f}_g(t,\omega)\right|^2 dt\,d\omega &= \frac{1}{2\pi}\int_{-\infty}^{\infty} d\omega\,\frac{1}{2\pi}\int_{-\infty}^{\infty}\left|\mathscr{F}\left\{\tilde{f}_g(t,\omega)\right\}\right|^2 dv \\
&= \frac{1}{2\pi}\int_{-\infty}^{\infty} d\omega\,\frac{1}{2\pi}\int_{-\infty}^{\infty}\left|\hat{f}(\omega+v)\right|^2\left|\hat{g}(v)\right|^2 dv \\
&= \frac{1}{2\pi}\int_{-\infty}^{\infty} d\omega\,\frac{1}{2\pi}\int_{-\infty}^{\infty}\left|\hat{f}(\omega)\right|^2\left|\hat{g}(v)\right|^2 dv \\
&= \frac{1}{2\pi}\int_{-\infty}^{\infty}\left|\hat{f}(\omega)\right|^2 d\omega, \qquad \text{since } \|\hat{g}\| = \frac{1}{2\pi}\int_{-\infty}^{\infty}\left|\hat{g}(v)\right|^2 dv = 1 \\
&= \int_{-\infty}^{\infty}\left|f(\tau)\right|^2 d\tau = \|f\|_2^2.
\end{aligned}
$$

This completes the proof.

Physically, the Gabor transformation transforms a signal f of one variable τ to a function $\tilde{f}$ of two variables t and ω *without* changing its total energy.

4.5 Frames and Frame Operators

The concept of frames in a Hilbert space was originally introduced by Duffin and Schaeffer (1952) in the context of nonharmonic Fourier series only six years after Gabor (1946) published his famous work. In signal processing, this concept has become useful in analyzing the completeness and stability of linear discrete signal representations. A frame is a set of vectors $\{\phi_n\}_{n\in\Gamma}$ that characterizes any signal f from its inner products $\{(f,\phi_n)\}_{n\in\Gamma}$, where Γ is the index set, which may be finite or infinite.

Definition 4.5.1 (Basis). A sequence of vectors $\{x_n\}$ in a Hilbert space H is called a *basis* (*Schauder basis*) of H if to each $x\in H$, there corresponds a unique sequence of scalars $\{a_n\}_{n=1}^{\infty}$ such that

$$x=\sum_{n=1}^{\infty}a_n x_n, \tag{4.5.1}$$

where the convergence is defined by the norm.

Definition 4.5.2 (Orthogonal Basis and Orthonormal Basis). A basis $\{x_n\}_{n=1}^{\infty}$ of H is called *orthogonal* if $(x_n,x_m)=0$ for $n\neq m$.

An orthogonal basis is called *orthonormal* if $(x_n,x_n)=1$ for all n.

An orthogonal basis $\{x_n\}_{n=1}^{\infty}$ is complete in the sense that if $(x,x_n)=0$ for all n, then $x=0$ (see Theorem 2.10.4).

Every separable Hilbert space has an orthonormal basis, and for an orthonormal basis the expansion (4.5.1) has the form

$$x=\sum_{n=1}^{\infty}(x,x_n)\,x_n, \tag{4.5.2}$$

with

$$\|x\|^2 = \sum_{n=1}^{\infty} |(x, x_n)|^2 . \tag{4.5.3}$$

More generally, for any $x, y \in H$,

$$(x, y) = \sum_{n=1}^{\infty} (x, x_n)\overline{(y, x_n)}. \tag{4.5.4}$$

It can be proved that every basis $\{x_n\}_{n=1}^{\infty}$ of a Hilbert space H possesses a unique *biorthogonal* basis $\{x_n^*\}_{n=1}^{\infty}$ which implies that

$$(x_m, x_n^*) = \delta_{m,n},$$

and, for every $x \in H$, we have

$$x = \sum_{n=1}^{\infty} (x, x_n^*) x_n = \sum_{n=1}^{\infty} (x, x_n) x_n^* .$$

If $(x_m, x_n^*) = 0$ for $m \neq n$, but (x_n, x_n^*) is not necessarily equal to one, $\{x_n^*\}_{n=1}^{\infty}$ is called a *biorthogonal basis* of $\{x_n\}_{n=1}^{\infty}$. In this case, we have, for any $x \in H$,

$$x = \sum_{n=1}^{\infty} (\bar{a}_n)^{-1} (x, x_n^*) x_n = \sum_{n=1}^{\infty} (a_n)^{-1} (x, x_n)\, x_n^* , \tag{4.5.5}$$

where $a_n = (x_n^*, x_n) \neq 0$. It can be shown that $a_n \neq 0$ for all n.

Definition 4.5.3 (Bounded Basis, Unconditional Basis, and Riesz Basis). If $\{x_n\}$ is a basis in a separable Hilbert space H, then

(i) $\{x_n\}$ is called a *bounded basis* if there exist two nonnegative numbers A and B such that

$$A \leq \|x_n\| \leq B \qquad \text{for all } n.$$

(ii) $\{x_n\}$ is called an *unconditional basis* in a separable Hilbert space H if

$$\sum a_n x_n \in H \quad \text{implies that} \quad \sum |a_n| x_n \in H .$$

(iii) $\{x_n\}$ is called a *Riesz basis* if there exist a topological isomorphism $T: H \to H$ and an orthonormal basis $\{y_n\}$ of H such that $T x_n = y_n$ for every n.

Remark. In a Hilbert space, all bounded unconditional bases are equivalent to an orthonormal basis. In other words, if $\{x_n\}$ is a bounded unconditional basis, then there exists an orthonormal basis $\{e_n\}$ and a topological isomorphism $T: H \to H$ such that $Te_n = x_n$ for all n.

Definition 4.5.4 (Frame). A sequence $\{x_n\}$ in a separable Hilbert space H (not necessarily a basis of H) is called a *frame* if there exist two numbers A and B with $0 < A \le B < \infty$ such that

$$A\|x\|^2 \le \sum_n |(x, x_n)|^2 \le B\|x\|^2. \tag{4.5.6}$$

The numbers A and B are called the *frame bounds*. If $A = B$, the frame is said to be *tight*. The frame is called *exact* if it ceases to be a frame whenever any single element is deleted from the frame.

Definition 4.5.5 (Frame Operator). To each frame $\{x_n\}$ there corresponds an operator T, called the *frame operator*, from H into itself defined by

$$Tx = \sum_n (x, x_n)\, x_n \qquad \text{for all} \quad x \in H. \tag{4.5.7}$$

Remark. The x_n's are not necessarily linearly independent. Since $\sum_n |(x, x_n)|^2$ is a series of positive real numbers, it converges absolutely and hence, unconditionally.

The following example shows that tightness and exactness are not related.

Example 4.5.1 If $\{e_n\}_{n=1}^{\infty}$ is an orthonormal basis of H, then

(i) $\{e_1, e_1, e_2, e_2, \cdots\}$ is a tight frame with frame bounds $A = 2 = B$, but it is not exact.

(ii) $\{\sqrt{2}\, e_1, e_2, e_3, \cdots\}$ is an exact frame but not tight since the frame bounds are easily seen as $A = 1$ and $B = 2$.

(iii) $\left\{e_1, \frac{e_2}{\sqrt{2}}, \frac{e_2}{\sqrt{2}}, \frac{e_3}{\sqrt{3}}, \frac{e_3}{\sqrt{3}}, \frac{e_3}{\sqrt{3}}, \cdots\right\}$ is a tight frame with the frame bound $A = 1$ but not an orthonormal basis.

(iv) $\left\{e_1, \frac{e_2}{2}, \frac{e_3}{3}, \cdots\right\}$ is a complete orthogonal sequence but is not a frame.

If $\{x_n\}$ is an orthonormal basis of H, then the Parseval formula holds, that is, for any $x \in H$,

$$\|x\|^2 = \sum_n |(x, x_n)|^2 .$$

It follows from the definition of frame that $\{x_n\}$ is a tight frame with frame bounds $A = B = 1$.

But the converse is not necessarily true. That is, tight frames are not necessarily orthonormal. For example, $H = \mathbb{C}^2$ and

$$e_1 = (0,1),\ e_2 = \left(\frac{\sqrt{3}}{2}, -\frac{1}{2}\right),\ e_3 = \left(-\frac{\sqrt{3}}{2}, -\frac{1}{2}\right).$$

For any $x = (x_1, x_2) \in H$, we have

$$\begin{aligned}\sum_{i=1}^{3} |(x, e_i)|^2 &= |x_2|^2 + \left|\frac{\sqrt{3}}{2}x_1 - \frac{1}{2}x_2\right|^2 + \left|-\frac{\sqrt{3}}{2}x_1 - \frac{1}{2}x_2\right|^2 \\ &= |x_2|^2 + \frac{1}{2}\left(3|x_1|^2 + |x_2|^2\right) \\ &= \frac{3}{2}\left(|x_1|^2 + |x_2|^2\right) = \frac{3}{2}\|x\|^2 .\end{aligned}$$

Thus, three vectors (e_1, e_2, e_3) define a tight frame with the frame bounds $A = B = \frac{3}{2}$ but they are not orthonormal since (e_1, e_2, e_3) are not linearly independent.

Theorem 4.5.1 If a sequence $\{x_n\}$ is a tight frame in H with the frame bound $A = 1$, and if $\|x_n\| = 1$ for all n, then $\{x_n\}$ is an orthonormal basis of H.

Proof. It follows from (4.5.6) that

$$\|x_m\|^2 = \sum_n |(x_m, x_n)|^2 = \|x_m\|^4 + \sum_{m \neq n} |(x_m, x_n)|^2 .$$

Since $\|x_m\| = 1$, the above equality implies that

$$(x_m, x_n) = 0 \qquad \text{for} \quad m \neq n.$$

The completeness of $\{x_n\}$ is a consequence of the fact that frames are complete.

To check this, suppose $x \in H$ such that $(x, x_n) = 0$ for all n. Then, the relation

$$A\|x\|^2 \leq \sum_n |(x, x_n)|^2 = 0$$

implies that $x = 0$.

***Theorem** 4.5.2* Suppose a sequence $\{x_n\}$ is a separable Hilbert space H. Then, the following are equivalent.

(a) The frame operator $Tx = \sum_n (x, x_n) x_n$ is a bounded linear operator on it with $AI \leq T \leq BI$, where I is the identity operator on H.

(b) $\{x_n\}_{n=-\infty}^{\infty}$ is a frame with frame bounds A and B.

Proof. If (a) holds, then the relation $AI \leq T \leq BI$ is equivalent to

$$(AIx, x) \leq (Tx, x) \leq (BIx, x) \qquad \text{for all } x \in H. \tag{4.5.8}$$

Since I is an identity operator, $(Ix, x) = \|x\|^2$. Also,

$$(Tx, x) = \left(\sum_n (x, x_n) x_n, x\right) = \sum_n (x, x_n)(x_n, x)$$

$$= \sum_n (x, x_n)\overline{(x, x_n)} = \sum_n |(x, x_n)|^2.$$

Evidently, inequality (4.5.8) gives

$$A\|x\|^2 \leq \sum_n |(x, x_n)|^2 \leq B\|x\|^2.$$

This shows that (a) implies (b).

We next prove that (b) implies (a). Suppose (b) holds, that is, $\{x_n\}$ is a frame with frame bounds A and B. Recall that in any Hilbert space H the norm of any element $x \in H$ is given by

$$\|x\| = \sup_{\|y\|=1} |(x, y)| \qquad \text{for } y \in H.$$

For a fixed $x \in H$, we consider

$$T_N x = \sum_{n=-N}^{N} (x, x_n) x_n .$$

For $0 \le M \le N$, we have, by the Schwarz inequality,

$$\begin{aligned}
\|T_N x - T_M x\|^2 &= \sup_{\|y\|=1} |(T_N x - T_M x, y)|^2 \\
&= \sup_{\|y\|=1} \left| \sum_{M+1 \le |n| < N} (x, x_n)(x_n, y) \right|^2 \\
&\le \sup_{\|y\|=1} \left(\sum_{M+1 \le |n| \le N} |(x, x_n)|^2 \right) \left(\sum_{M+1 \le |n| \le N} |(x_n, y)|^2 \right) \\
&\le \sup_{\|y\|=1} \left(\sum_{M+1 \le |n| \le N} |(x, x_n)|^2 \right) B \|y\|^2, \qquad \text{by (4.5.6)} \\
&= B \left(\sum_{M+1 \le |n| \le N} |(x, x_n)|^2 \right) \to 0 \qquad \text{as } M, N \to \infty.
\end{aligned}$$

Thus, $\{T_N x\}$ is a Cauchy sequence in H and hence it is convergent as $N \to \infty$. Therefore,

$$\lim_{N \to \infty} T_N x = Tx .$$

Next, we use the preceding argument to obtain

$$\begin{aligned}
\|Tx\|^2 &= \sup_{\|y\|=1} |(Tx, y)|^2 = \sup_{\|y\|=1} \left| \left(\sum_n (x, x_n) x_n, x \right) \right|^2 \\
&= \sup_{\|y\|=1} \left| \sum_n (x, x_n)(x_n, x) \right|^2 \\
&\le B \left(\sum_n |(x, x_n)|^2 \right) \le B^2 \|x\|^2 .
\end{aligned}$$

This implies that $\|T\| \le B$, and hence the frame operator T is bounded.

Since $(Ix, x) = \|x\|^2$, it follows from definition (4.5.6) that

$$A(Ix, x) \le (Tx, x) \le B(Ix, x),$$

which is equivalent to the relation

$$AI \le T \le BI.$$

This complete the proof.

Theorem 4.5.3 Suppose $\{x_n\}_{n=-\infty}^{\infty}$ is a frame on a separable Hilbert space with frame bounds A and B, and T is the corresponding frame operator. Then,
(a) T is invertible and $B^{-1}I \le T^{-1} \le A^{-1}I$. Furthermore, T^{-1} is a positive operator and hence it is self-adjoint.
(b) $\{T^{-1}x_n\}$ is a frame with frame bounds B^{-1} and A^{-1} with $A^{-1} \ge B^{-1} > 0$, and it is called the *dual frame* of $\{x_n\}$.
(c) Every $x \in H$ can be expressed in the form

$$x = \sum_n \left(x, T^{-1}x_n\right) x_n = \sum_n \left(x, x_n\right) T^{-1} x_n. \tag{4.5.9}$$

The frame $\{T^{-1}x_n\} = \{\tilde{x}_n\}$ is called the *dual frame* of $\{x_n\}$. It is easy to verify that the dual frame of $\{\tilde{x}_n\}$ is the original frame $\{x_n\}$. According to formula (4.5.9), the reconstruction formula for x has the form

$$x = \sum_n \left(x, \tilde{x}_n\right) x_n = \sum_n \left(x, x_n\right) \tilde{x}_n. \tag{4.5.10}$$

Proof *(a)*. Since the frame operator T satisfies the relation

$$AI \le T \le BI, \tag{4.5.11}$$

it follows that

$$\left(I - B^{-1}T\right) \le \left(I - B^{-1}AI\right) = \left(1 - \frac{A}{B}\right) I$$

and hence

$$\left\|I - B^{-1}T\right\| \le \left\|\left(1 - \frac{A}{B}\right)\right\| < 1.$$

Thus, $B^{-1}T$ is invertible and consequently so is T. We next multiply (4.5.11) by T^{-1} and use the fact that T^{-1} commutes with I and T to obtain

$$B^{-1}I \le T \le A^{-1}I.$$

In view of the fact that

$$\left(T^{-1}x,\, x\right) = \left(T^{-1}x,\, T\left(T^{-1}x\right)\right) \ge A\left(T^{-1}x,\, T^{-1}x\right) = A\left\|T^{-1}x\right\|^2 \ge 0,$$

we conclude that T^{-1} is a positive operator and hence it is self-adjoint.

(b) Since T^{-1} is self-adjoint, we have

$$\sum_n \left(x,\, T^{-1}x_n\right) T^{-1}x_n = T^{-1}\left(\sum_n \left(T^{-1}x,\, x_n\right) x_n\right) = T^{-1}\left(T\left(T^{-1}x\right)\right) = T^{-1}x. \quad (4.5.12)$$

This gives

$$\left(\sum_n \left(x,\, T^{-1}x_n\right) T^{-1}x_n,\, x\right) = \left(T^{-1}x,\, x\right).$$

Or,

$$\sum_n \left(x,\, T^{-1}x_n\right)\left(T^{-1}x_n,\, x\right) = \left(T^{-1}x,\, x\right).$$

Hence,

$$\sum_n \left(x,\, T^{-1}x_n\right)\overline{\left(x,\, T^{-1}x_n\right)} = \left(T^{-1}x,\, x\right).$$

Or,

$$\sum_n \left|\left(x,\, T^{-1}x_n\right)\right|^2 = \left(T^{-1}x,\, x\right).$$

Using the result from (a), that is, $B^{-1}I \le T^{-1} \le A^{-1}I$, it turns out that

$$B^{-1}(Ix,x) \le \left(T^{-1}x,x\right) \le A^{-1}(Ix,x)$$

and hence

$$B^{-1}\|x\|^2 \le \sum_n \left|\left(x,\, T^{-1}x_n\right)\right|^2 \le A^{-1}\|x\|^2. \quad (4.5.13)$$

This shows that $\left\{T^{-1}x_n\right\}$ is a frame with frame bounds B^{-1} and A^{-1}.

(c) We replace x by $T^{-1}x$ in (4.5.7) to derive

$$x = \sum_n \left(T^{-1}x,\, x_n\right) x_n = \sum_n \left(x,\, T^{-1}x_n\right) x_n.$$

Similarly, replacing x by Tx in (4.5.12) gives

$$x = \sum_n \left(Tx,\, T^{-1}x_n\right) T^{-1}x_n = \sum \left(T^{-1}Tx,\, x_n\right) T^{-1}x_n = \sum_n \left(x,\, x_n\right) T^{-1}x_n.$$

This completes the proof.

Theorem 4.5.4 Suppose $\{x_n\}_{n=-\infty}^{\infty}$ is a frame on a separable Hilbert space H with frame bounds A and B. If there exists a sequence of scalars $\{c_n\}$ such that $x = \sum_n c_n x_n$, then

$$\sum_n |c_n|^2 = \sum_n |a_n|^2 + \sum_n |a_n - c_n|^2, \tag{4.5.14}$$

where $a_n = (x, T^{-1}x_n)$ so that $x = \sum a_n x_n$.

Proof. Note that $(x_n, T^{-1}x) = (T^{-1}x_n, x) = \bar{a}_n$. Substituting $x = \sum a_n x_n$ into the first term in the inner product $(x, T^{-1}x)$ gives

$$(x, T^{-1}x) = \left(\sum_n a_n x_n, T^{-1}x\right) = \sum_n a_n (x_n, T^{-1}x) = \sum_n |a_n|^2.$$

Similarly, substituting $x = \sum_n c_n x_n$ into the first term in $(x, T^{-1}x)$ yields

$$(x, T^{-1}x) = \left(\sum_n c_n x_n, T^{-1}x\right) = \sum_n c_n (x_n, T^{-1}x) = \sum_n c_n \bar{a}_n.$$

Consequently,

$$\sum_n |a_n|^2 = \sum_n c_n \bar{a}_n. \tag{4.5.15}$$

Finally, we obtain, by using (4.5.15),

$$\sum_n |a_n|^2 + \sum_n |a_n - c_n|^2 = \sum_n |a_n|^2 + \sum_n \left(|a_n|^2 - a_n\bar{c}_n - \bar{a}_n c_n + |c_n|^2\right) = \sum_n |c_n|^2.$$

This completes the proof.

Theorem 4.5.5 A necessary and sufficient condition for a sequence $\{x_n\}$ on a Hilbert space H to be an exact frame is that the sequence $\{x_n\}$ be a bounded unconditional basis of H.

Proof. The condition is necessary.

We assume that $\{x_n\}$ is an exact frame with frame bounds A and B. Then, $\{x_n\}$ and $\{T^{-1}x_n\}$ are biorthonormal. For a fixed m, we have

$$A\left\|T^{-1}x_m\right\|^2 \le \sum_n \left|\left(T^{-1}x_m,\, x_n\right)\right|^2 = \left|\left(T^{-1}x_m,\, x_m\right)\right|^2 \le \left\|T^{-1}x_m\right\|^2 \left\|x_m\right\|^2$$

and

$$\left\|x_m\right\|^4 = \left|\left(x_m, x_m\right)\right|^2 \le \sum_n \left|\left(x_m, x_n\right)\right|^2 \le B\left\|x_m\right\|^2.$$

Consequently,

$$A \le \left\|x_m\right\|^2 \le B.$$

Hence, the sequence $\{x_n\}$ is bounded in norm, and $x \in H$ can be represented as

$$x = \sum_n \left(x,\, T^{-1}x_n\right) x_n\,.$$

It remains to show that this representation is unique. If $x = \sum_n c_n x_n$, then

$$\left(x,\, T^{-1}x_m\right) = \left(\sum_n c_n x_n,\; T^{-1}x_m\right) = \sum_n c_n \left(x_n,\, T^{-1}x_m\right) = c_m\,.$$

Thus, the sequence $\{x_n\}$ is a basis. Since the series converges unconditionally, the basis is unconditional.

The condition is sufficient.

We assume that $\{x_n\}$ is a bounded unconditional basis of *H*. Then, there exists an orthonormal basis $\{e_n\}$ and a topological isomorphism $T\colon H \to H$ such that $Te_n = x_n$ for all *n*. For $x \in H$, we have

$$\sum_n \left|\left(x, x_n\right)\right|^2 = \sum_n \left|\left(x, Te_n\right)\right|^2 = \sum_n \left|\left(T^*x, e_n\right)\right|^2 = \left\|T^*x\right\|^2,$$

where T^* is the adjoint of *T*. But

$$\left\|T^{*-1}\right\|^{-1}\|x\| \le \left\|T^*x\right\| \le \left\|T^*\right\|\|x\|.$$

Hence, the sequence $\{x_n\}$ is a frame which is obviously an exact frame because it ceases to be a frame whenever any element is deleted from the sequence. This completes the proof.

4.6 Discrete Gabor Transforms and the Gabor Representation Problem

In many applications to physical and engineering problems, it is more important, at least from a computational viewpoint, to work with discrete transforms rather than continuous ones. In sampling theory, the sample points are defined by $v = m\omega_0$ and $\tau = n t_0$, where m,n are integers and t_0 and ω_0 are positive quantities. The *discrete Gabor functions* are defined by

$$g_{m,n}(t) = \exp\left(2\pi m\,\omega_0 t\right)\, g\left(t - n\,t_0\right) = M_{2\pi m\omega_0}\, T_{nt_0}\, g(t), \tag{4.6.1}$$

where $g \in L^2(\mathbb{R})$ is a fixed function and t_0 and ω_0 are the time shift and the frequency shift parameters, respectively. A typical set of Gabor functions is shown in Figure 4.6.

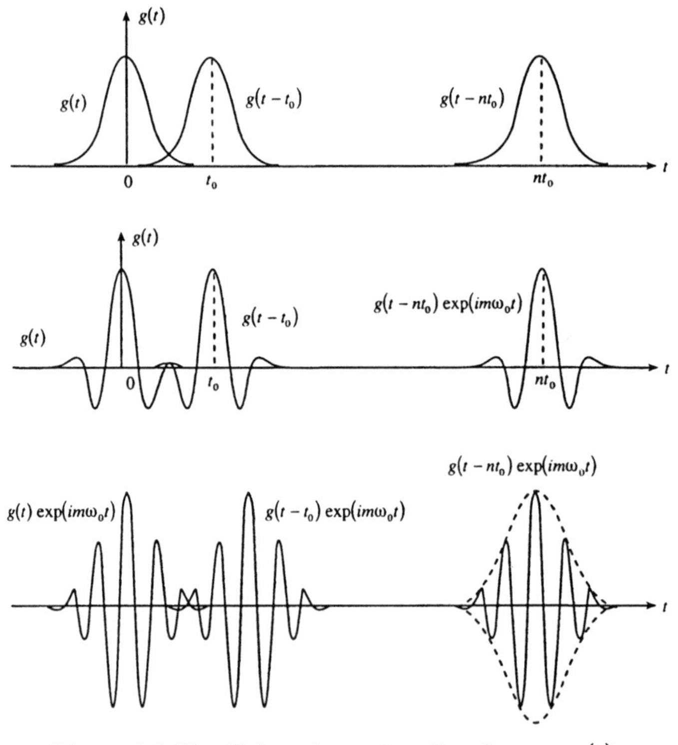

Figure 4.6. The Gabor elementary functions $g_{m,n}(t)$.

These functions are also called the *Weyl-Heisenberg coherent states* which arise from translations and modulations of the Gabor window function. From a physical point of view, these coherent states are of great interest and have several important applications in quantum mechanics. Following Gabor's analysis, various other functions have been introduced as window functions instead of the Gaussian function which was originally used by Gabor. In order to expand general functions (quantum mechanical states) with respect to states with minimum uncertainty, von Neumann (1945) introduced a set of coherent states on lattice constants $\omega_0 t_0 = h$ in the phase space with position and momentum as coordinates where h is the Planck constant. These states, associated with the *Weyl-Heisenberg group*, are in fact the same as used by Gabor. The time-frequency lattice with lattice constants $\omega_0 t_0 = 1$ is also called the *von Neumann lattice*.

Definition 4.6.1 (Discrete Gabor Transform). The discrete Gabor transform is defined by

$$\tilde{f}(m,n) = \int_{-\infty}^{\infty} f(t)\, \overline{g}_{m,n}(t)\, dt = \left(f, g_{m,n}\right). \tag{4.6.2}$$

The double series

$$\sum_{m,n=-\infty}^{\infty} \tilde{f}(m,n)\, g_{m,n}(t) = \sum_{m,n=-\infty}^{\infty} \left(f, g_{m,n}\right) g_{m,n}(t) \tag{4.6.3}$$

is called the *Gabor series* of f.

It is of special interest to find the inverse of the discrete Gabor transform so that $f \in L^2(\mathbb{R})$ can be determined by the formula

$$\tilde{f}(mt_0, n\omega_0) = \int_{-\infty}^{\infty} f(t)\, g_{m,n}(t)\, dt = \left(f, \overline{g}_{m,n}\right). \tag{4.6.4}$$

The set of sample points $\left\{(mt_0, n\omega_0)\right\}_{m,n=-\infty}^{\infty}$ is called the *Gabor lattice*. The answer to the question of finding the inverse is in the affirmative if the set of functions $\left\{g_{m,n}(t)\right\}$ forms an orthonormal basis, or more generally, if the set is a frame for $L^2(\mathbb{R})$. A system $\left\{g_{m,n}(t)\right\} = \left\{M_{2\pi m\omega_0} T_{nt_0}\, g(t)\right\}$ is called a *Gabor frame* or *Weyl-Heisenberg frame* in $L^2(\mathbb{R})$ if there exist two constants $A, B > 0$ such that

$$A\|f\|^2 \leq \sum_{m,n=-\infty}^{\infty} \left|\left(f, g_{m,n}\right)\right|^2 \leq B\|f\|^2 \tag{4.6.5}$$

holds for all $f \in L^2(\mathbb{R})$. For a Gabor frame $\{g_{m,n}(t)\}$, the *analysis operator* T_g is defined by

$$T_g f = \left\{\left(f, g_{m,n}\right)\right\}_{m,n}, \tag{4.6.6}$$

and its *synthesis operator* T_g^* is defined by

$$T_g^* c_{m,n} = \sum_{m,n=-\infty}^{\infty} c_{m,n} \, g_{m,n}, \tag{4.6.7}$$

where $c_{m,n} \in \ell^2(\mathbb{Z})$. Both T_g and T_g^* are bounded linear operators and in fact are adjoint operators with respect to the inner product $(\,,\,)$. The *Gabor frame* operator S_g is defined by $S_g = T_g^* T_g$. More explicitly,

$$S_g f = \sum_{m,n=-\infty}^{\infty} \left(f, g_{m,n}\right) g_{m,n}. \tag{4.6.8}$$

If $\{g_{m,n}\}$ constitute a Gabor frame for $L^2(\mathbb{R})$, any function $f \in L^2(\mathbb{R})$ can be expressed as

$$f(t) = \sum_{m,n=-\infty}^{\infty} \left(f, g_{m,n}\right) g_{m,n}^* = \sum_{m,n=-\infty}^{\infty} \left(f, g_{m,n}^*\right) g_{m,n}, \tag{4.6.9}$$

where $\{g_{m,n}\}$ is called the *dual frame* given by $g_{m,n}^* = S_g^{-1} g_{m,n}$. Equation (4.6.9) provides an answer for constructing *f* from its Gabor transform $\left(f, g_{m,n}\right)$ for a given window function *g*.

Finding the conditions on t_0, ω_0, and g under which the Gabor series of *f* determines *f* or converges to it is known as the *Gabor representation problem*. For an appropriate function g, the answer is positive provided that $0 < \omega_0 t_0 < 1$. If $0 < \omega_0 t_0 < 1$, the reconstruction is *stable* and g can have a good time and frequency localization. This is in contrast with the case when $\omega_0 t_0 = 1$, where the construction is *unstable* and g cannot have a good time and frequency localization. For the case when $\omega_0 t_0 > 1$, the reconstruction of *f* is, in general, impossible no matter how g is selected.

4.7 The Zak Transform and Time-Frequency Signal Analysis

Historically, the Zak transform (ZT), known as the *Weil-Brezin* transform in harmonic analysis, was introduced by Gelfand (1950) in his famous paper on eigenfunction expansions associated with Schrödinger operators with periodic potentials. This transform was also known as the *Gelfand mapping* in the Russian mathematical literature. However, Zak (1967, 1968) independently rediscovered it as the $k-q$ transform in solid state physics to study a quantum-mechanical representation of the motion of electrons in the presence of an electric or magnetic field. Although the Gelfand-Weil-Brezin-Zak transform seems to be a more appropriate name for this transform, there is a general consensus among scientists to name it as the Zak transform since Zak himself first recognized its deep significance and usefulness in a more general setting. In recent years, the Zak transform has been widely used in time-frequency signal analysis, in the coherent states representation in quantum field theory, and also in mathematical analysis of Gabor systems. In particular, the Zak transform has also been useful for a study of the Gabor representation problem, where this transform has successfully been utilized to investigate the orthogonality and completeness of the Gabor frames in the critical case.

Definition 4.7.1 (The Zak Transform). The Zak transform, $(\mathfrak{Z}_a f)(t,\omega)$, of a function $f \in L^2(\mathbb{R})$ is defined by the series

$$(\mathfrak{Z}_a f)(t,\omega) = \sqrt{a}\sum_{n=-\infty}^{\infty} f(at+an)\exp(-2\pi i n\omega), \tag{4.7.1}$$

where $a(>0)$ is a fixed parameter and t and ω are real.

If f represents a signal, then its Zak transform can be treated as the joint time-frequency representation of the signal f. It can also be considered as the discrete Fourier transform of f in which an infinite set of samples in the form $f(at+an)$ is used for $n = 0, \pm 1, \pm 2, \cdots$. Without loss of generality, we set $a = 1$ so that we can write $(\mathfrak{Z} f)(t,\omega)$ in the explicit form

$$(\mathfrak{Z} f)(t,\omega) = F(t,\omega) = \sum_{n=-\infty}^{\infty} f(t+n)\exp(-2n\pi i\omega). \tag{4.7.2}$$

This transform satisfies the *periodic relation*

$$(\mathfrak{Z} f)(t,\omega+1) = (\mathfrak{Z} f)(t,\omega), \tag{4.7.3}$$

and the following *quasiperiodic relation*

$$(\mathfrak{Z} f)(t+1,\,\omega) = \exp(2\pi i\omega)(\mathfrak{Z} f)(t,\omega), \tag{4.7.4}$$

and therefore the Zak transform $\mathfrak{Z} f$ is completely determined by its values on the unit square $S = [0,1]\times[0,1]$.

It is easy to prove that the Zak transform of f can be expressed in terms of the Zak transform of its Fourier transform $\hat{f}(\nu) = \mathscr{F}\{f(t)\}$ defined by (3.3.19b). More precisely,

$$(\mathfrak{Z} f)(t,\omega) = \exp(2\pi i\,\omega\, t)(\mathfrak{Z} \hat{f})(\omega,\,-t). \tag{4.7.5}$$

To prove this result, we define a function g for fixed t and ω by

$$g(x) = \exp(-2\,i\,\omega\,\pi\,x)\; f(x+t).$$

Then, it follows that

$$\begin{aligned}
\hat{g}(\nu) &= \int_{-\infty}^{\infty} g(x)\,\exp(-2\pi\, i\,x\,\nu)\;dx \\
&= \int_{-\infty}^{\infty} f(x+t)\,\exp\{-2\pi\, i\,x(\nu+\omega)\}\;dx \\
&= e^{2\pi i(\omega+\nu)t}\int_{-\infty}^{\infty} f(u)\,\exp\{-2\pi\, i(\nu+\omega)u\}\;du \\
&= \exp\{2\pi\, i(\omega+\nu)t\}\;\hat{f}(\nu+\omega).
\end{aligned}$$

We next use the Poisson summation formula (3.5.7) in the form

$$\sum_{n=-\infty}^{\infty} g(n) = \sum_{n=-\infty}^{\infty} \hat{g}(2\pi n).$$

Or, equivalently,

$$\begin{aligned}
\sum_{n=-\infty}^{\infty} f(t+n)\,\exp(-2\pi\, i\,\omega\, n) &= \exp(2\pi\, i\,\omega\, t)\sum_{n=-\infty}^{\infty} \exp[2\pi\, i(2n\pi)t]\;\hat{f}(\omega+2\pi\, n) \\
&= \exp(2\pi\, i\,\omega\, t)\sum_{m=-\infty}^{\infty} \hat{f}(\omega+m)\,\exp(2\pi\, i\,m\,t).
\end{aligned}$$

This gives the desired result (4.7.5).

The following results can be easily verified:

$$\left(\mathfrak{Z}\,\mathcal{F} f\right)(\omega,t)=\exp(2\pi i\omega t)\left(\mathfrak{Z} f\right)(-t,\,\omega), \tag{4.7.6}$$

$$\left(\mathfrak{Z}\,\mathcal{F}^{-1} f\right)(\omega,t)=\exp(2\pi i\omega t)\left(\mathfrak{Z} f\right)(t,-\omega). \tag{4.7.7}$$

If $g_{m,n}(t)=\exp(-2\pi i m t)\, g(t-n)$, then

$$\left(\mathfrak{Z} g_{mn}\right)(t,\omega)=\exp\left[-2\pi i(mt+n\omega)\right]\left(\mathfrak{Z} g(t,\omega)\right). \tag{4.7.8}$$

We next observe that $L^2(S)$ is the set of all square integrable complex-valued functions F on the unit square S, that is,

$$\int_0^1\int_0^1 |F(t,\omega)|^2\, dt\, d\omega<\infty .$$

It is easy to check that $L^2(S)$ is a Hilbert space with the inner product

$$(F,G)=\int_0^1\int_0^1 F(t,\omega)\,\overline{G}(t,\omega)\, dt\, d\omega \tag{4.7.9}$$

and the norm

$$\|F\|=\left[\int_0^1\int_0^1 |F(t,\omega)|^2\, dt\, d\omega\right]^{\frac{1}{2}}. \tag{4.7.10}$$

The set

$$\left\{M_{m,n}=M_{2\pi m,2\pi n}(t,\omega)=\exp\left[2\pi i(mt+n\omega)\right]\right\}_{m,n=-\infty}^{\infty} \tag{4.7.11}$$

forms an orthonormal basis of $L^2(S)$.

Example 4.7.1 If

$$\phi_{m,n;a}(x)=\frac{1}{\sqrt{a}}\, T_{na}\, M_{2\pi m/a}\, \chi_{[0,a]}(x), \tag{4.7.12}$$

where $a>0$, then

$$\left(\mathfrak{Z}_a\, \phi_{m,n;a}\right)(t,\omega)=e_m(t)\, e_n(\omega), \tag{4.7.13}$$

where $e_k(t)=\exp(2\pi i k t)$.

We have

$$\phi_{m,n;a}(x) = \frac{1}{\sqrt{a}} \exp\left[2\pi i m\left(\frac{x-na}{a}\right)\right] \chi_{[0,a]}(x-na)$$
$$= \frac{1}{\sqrt{a}} \exp\left(\frac{2\pi i m x}{a}\right) \chi_{[na,(n+1)a]}(x).$$

Thus, we obtain

$$\left(\mathfrak{Z}_a \phi_{m;n,a}\right)(t,\omega) = \sum_{k=-\infty}^{\infty} \exp\left[\frac{2\pi i m}{a}(at+ak)\right] \chi_{[na,na+a]}(at+ak)$$
$$= \sum_{k=-\infty}^{\infty} e_m(t)\, e^{-2\pi i k\omega}\, \chi_{[n-k,n+1-k]}(t)$$
$$= e_m(t)\, e_n(\omega).$$

4.8 Basic Properties of Zak Transforms

1. (*Linearity*). The Zak transform is linear, that is, for any two constants a, b,

$$\left[\mathfrak{Z}(af+bg)\right](t,\omega) = a\left(\mathfrak{Z} f\right)(t,\omega) + b\left(\mathfrak{Z} f\right)(t,\omega). \tag{4.8.1}$$

2. (*Translation*). For any real a and integer m,

$$\left[\mathfrak{Z}(T_a f)\right](t,\omega) = \left(\mathfrak{Z} f\right)(t-a,\omega), \tag{4.8.2}$$

$$\left[\mathfrak{Z}(T_{-m} f)\right](t,\omega) = \exp(2\pi i m\omega)\left(\mathfrak{Z} f\right)(t,\omega). \tag{4.8.3}$$

3. (*Modulation*).

$$\left[\mathfrak{Z}(M_b f)\right](t,\omega) = e^{ibt}\left(\mathfrak{Z} f\right)\left(t,\, \omega - \frac{b}{2\pi}\right), \tag{4.8.4}$$

$$\left[\mathfrak{Z}\left(M_{2\pi b} f\right)\right](t,\omega) = \exp(2\pi i b t)\left(\mathfrak{Z} f\right)(t,\, \omega - b). \tag{4.8.5}$$

4. (*Translation and Modulation*).

$$\mathfrak{Z}\left[M_{2\pi m}\, T_n f\right](t,\omega) = \exp\left[2\pi i(mt - n\omega)\right]\left(\mathfrak{Z} f\right)(t,\omega). \tag{4.8.6}$$

5. (*Conjugation*).

$$\left(\mathfrak{Z} \bar{f}\right)(t,\omega) = \overline{\left(\mathfrak{Z} f\right)}(t,\, -\omega). \tag{4.8.7}$$

6. (*Symmetry*).

(a) If f is an even function, then

$$\left(\mathfrak{Z} f\right)(t,\omega) = \left(\mathfrak{Z} f\right)(-t,\, -\omega). \tag{4.8.8}$$

(b) If f is an odd function, then

$$(\mathfrak{Z} f)(t,\omega) = -(\mathfrak{Z} f)(-t, -\omega). \tag{4.8.9}$$

If f is a real and even function, it follows from (4.8.7) that

$$(\mathfrak{Z} f)(t,\omega) = \overline{(\mathfrak{Z} f)}(t, -\omega) = (\mathfrak{Z} f)(-t, -\omega). \tag{4.8.10}$$

7. (*Inversion*). For $t,\omega \in \mathbb{R}$,

$$f(t) = \int_0^1 (\mathfrak{Z} f)(t,\omega)\, d\omega, \tag{4.8.11}$$

$$\hat{f}(\omega) = \int_0^1 \exp(-2\pi i \omega t)(\mathfrak{Z} f)(t,\omega)\, dt, \tag{4.8.12}$$

$$f(x) = \int_0^1 \exp(-2\pi i x t)(\mathfrak{Z} \hat{f})(t,x)\, dt. \tag{4.8.13}$$

8. (*Dilation*).

$$\left(\mathfrak{Z} D_{\frac{1}{a}} f\right)(t,\omega) = (\mathfrak{Z}_a f)\left(at, \frac{\omega}{a}\right). \tag{4.8.14}$$

9. (*Product and Convolution of Zak Transforms*).

Results (4.7.3) and (4.7.4) show that the Zak transform is not periodic in the two variables t and ω. The product of two Zak transforms is periodic in t and ω.

Proof. We consider the product

$$F(t,\omega) = (\mathfrak{Z} f)(t,\omega)\overline{(\mathfrak{Z} g)}(t,\omega) \tag{4.8.15}$$

and find from (4.7.4) that

$$\overline{(\mathfrak{Z} g)}(t,\omega) = \exp(-2\pi i \omega)(\mathfrak{Z} g)(t,\omega).$$

Therefore, it follows that

$$F(t+1,\omega) = (\mathfrak{Z} f)(t,\omega)\overline{(\mathfrak{Z} g)}(t,\omega) = F(t,\omega),$$

$$F(t,\omega+1) = (\mathfrak{Z} f)(t,\omega)\overline{(\mathfrak{Z} g)}(t,\omega) = F(t,\omega).$$

These show that F is periodic in t and ω. Consequently, it can be expanded in a Fourier series on a unit square

$$F(t,\omega) = \sum_{m,n=-\infty}^{\infty} c_{m,n} \exp(2\pi i m t) \exp(2\pi i n \omega), \tag{4.8.16}$$

where

$$c_{m,n} = \int_0^1 \int_0^1 F(t,\omega) \exp(-2\pi i m t) \exp(-2\pi i n \omega)\, dt\, d\omega .$$

If we assume that the series involved are uniformly convergent, we can interchange the summation and integration to obtain

$$\begin{aligned}
c_{m,n} &= \int_0^1 \int_0^1 \left[\sum_{r=-\infty}^{\infty} f(t+r)\exp(-2\pi i r \omega)\right]\left[\sum_{s=-\infty}^{\infty} \bar{g}(t+s)\exp(2\pi i s \omega)\right] \\
&\qquad\qquad \times \exp\{-2\pi i (mt + n\omega)\}\, dt\, d\omega \\
&= \int_0^1 \left[\sum_{r=-\infty}^{\infty} f(t+r)\right]\left[\sum_{s=-\infty}^{\infty} \bar{g}(t+s)\right] \exp(-2\pi i m t)\, dt \\
&\qquad\qquad \times \int_0^1 \exp\{2\pi i \omega (s-n-r)\}\, d\omega \\
&= \int_0^1 \left[\sum_{r=-\infty}^{\infty} f(t+r)\, \bar{g}(t+n+r)\right] \exp(-2\pi i m t)\, dt \\
&= \sum_{r=-\infty}^{\infty} \int_r^{r+1} f(x)\, \bar{g}(x+n) \exp\{-2\pi i m (x-r)\}\, dx \\
&= \int_{-\infty}^{\infty} f(x)\, \bar{g}(x+n) \exp(-2\pi i m x)\, dx \\
&= \left(f(x),\ e^{2\pi i m x}\, g(x+n)\right) \\
&= \left(f,\ M_{2\pi m} T_{-n}\, g\right).
\end{aligned}$$

Consequently, (4.8.16) becomes

$$(\mathfrak{Z} f)(t,\omega)\overline{(\mathfrak{Z} g)}(t,\omega) = \sum_{m,n=-\infty}^{\infty} \left(f, M_{2\pi m} T_{-n}\, g\right) \exp\{2\pi i (mt + n\omega)\}. \tag{4.8.17}$$

This completes the proof.

Theorem 4.8.1 Suppose H is a function of two real variables t and s satisfying the condition

$$H(t+1, s+1) = H(t,s), \qquad s,t \in \mathbb{R}, \tag{4.8.18}$$

and

$$h(t) = \int_{-\infty}^{\infty} H(t,s)\, f(s)\, ds, \tag{4.8.19}$$

where the integral is absolutely and uniformly convergent.

Then,

$$(\mathfrak{Z} f)(t,\omega) = \int_{0}^{1} (\mathfrak{Z} f)(s,\omega)\, \Phi(t,s,\omega)\, ds, \tag{4.8.20}$$

where Φ is given by

$$\Phi(t,s,\omega) = \sum_{n=-\infty}^{\infty} H(t+n,s)\, \exp(-2\pi i n\omega), \quad 0 \le t,\, s,\, \omega \le 1. \tag{4.8.21}$$

Proof. It follows from the definition of the Zak transform of $h(t)$ that

$$(\mathfrak{Z} h)(t,\omega) = \sum_{k=-\infty}^{\infty} h(t+k)\, e^{-2\pi i k\omega} = \sum_{k=-\infty}^{\infty} e^{-2\pi i k\omega} \int_{-\infty}^{\infty} H(t+k,s)\, f(s)\, ds$$

$$= \sum_{k=-\infty}^{\infty} e^{-2\pi i k\omega} \sum_{m=-\infty}^{\infty} \int_{m}^{m+1} H(t+k,s)\, f(s)\, ds$$

$$= \sum_{k=-\infty}^{\infty} e^{-2\pi i k\omega} \sum_{m=-\infty}^{\infty} \int_{0}^{1} H(t+k,s+m)\, f(s+m)\, ds$$

$$= \int_{0}^{1} \left[\sum_{k,m=-\infty}^{\infty} H(t+k,s+m)\, f(s+m)\, \exp(-2\pi i k\omega) \right] ds,$$

which is, due to (4.8.17),

$$= \int_{0}^{1} \left[\sum_{k,m=-\infty}^{\infty} H(t+k-m,s)\, f(s+m)\, \exp(-2\pi i k\omega) \right] ds$$

$$= \int_{0}^{1} \left[\sum_{m,n=-\infty}^{\infty} H(t+n,s)\, f(s+m)\, \exp\{-2\pi i (m+n)\omega\} \right] ds$$

$$= \int_{0}^{1} (\mathfrak{Z} f)(s,\omega)\, \Phi(t,s,\omega)\, ds. \tag{4.8.22}$$

This completes the proof.

In particular, if $H(t,s) = H(t-s)$,

$$\Phi(t,s,\omega) = \sum_{n=-\infty}^{\infty} H(t-s+n)\exp(-2\pi i n\omega) = (\mathfrak{Z} H)(t-s,\,\omega).$$

Consequently, Theorem 4.8.1 leads to the following convolution theorem.

Theorem 4.8.2 (Convolution Theorem). If

$$h(t) = \int_{-\infty}^{\infty} H(t-s)\, f(s)\, ds = (H * f)(t),$$

then (4.8.20) reduces to the form

$$(\mathfrak{Z} h)(t,\omega) = \int_0^1 (\mathfrak{Z} H)(t-s)(\mathfrak{Z} f)(s,\omega)\,ds = \mathfrak{Z}(H * f)(t,\,\omega). \tag{4.8.23}$$

Example 4.8.1 If $H(t) = \sum_{k=-\infty}^{\infty} a_k\, \delta(t-k)$, then

$$\mathfrak{Z}(H * f)(t,\omega) = A(\omega)(\mathfrak{Z} f)(t,\omega), \tag{4.8.24}$$

where

$$A(\omega) = \sum_{k=-\infty}^{\infty} a_k \exp(-2\pi i k\omega).$$

Clearly,

$$\begin{aligned}
\mathfrak{Z}(H * f)(t,\omega) &= \mathfrak{Z}\left[\int_{-\infty}^{\infty} H(t-s)\, f(s)\, ds\right](t,\omega) \\
&= \mathfrak{Z}\left[\sum_{k=-\infty}^{\infty} a_k \int_{-\infty}^{\infty} \delta(t-s-k)\, f(s)\, ds\right](t,\omega) \\
&= \mathfrak{Z}\left[\sum_{k=-\infty}^{\infty} a_k\, f(t-k)\right](t,\omega) \\
&= \sum_{k=-\infty}^{\infty} a_k \sum_{n=-\infty}^{\infty} f(t+n-k)\exp(-2\pi i n\omega) \\
&= \sum_{k=-\infty}^{\infty} a_k \sum_{m=-\infty}^{\infty} f(t+m)\exp\{-2\pi i\omega(m+k)\} \\
&= A(\omega)\,(\mathfrak{Z} f)(t,\omega).
\end{aligned}$$

Theorem 4.8.3 The Zak transformation is a unitary mapping from $L^2(\mathbb{R})$ to $L^2(S)$.

Proof. It follows from the definition of the inner product (4.7.9) in $L^2(S)$ that

$$\begin{aligned}(\mathfrak{Z}_a f, \mathfrak{Z}_a g) &= a\int_0^1\int_0^1\left[\sum_{n=-\infty}^{\infty} f(at+an)e^{-2\pi in\omega}\right]\left[\sum_{m=-\infty}^{\infty} \bar{g}(at+am)e^{2\pi im\omega}\right]dt\,d\omega \\ &= a\int_0^1\left[\sum_{n=-\infty}^{\infty} f(at+an)\,\bar{g}(at+an)\right]dt \\ &= \sum_{n=-\infty}^{\infty}\int_{na}^{(n+1)a} f(x)\,\bar{g}(x)\,dx \\ &= \int_{-\infty}^{\infty} f(x)\,\bar{g}(x)\,dx = (f,g). \end{aligned} \tag{4.8.25}$$

In particular, if $f = g$, we obtain from (4.8.25) that

$$\|\mathfrak{Z}_a f\|^2 = \|f\|^2. \tag{4.8.26}$$

This means that the Zak transform is an isometry from $L^2(\mathbb{R})$ into $L^2(S)$.

Further, Example 4.7.1 shows that $\{\phi_{m,na}(x)\}_{m,n=-\infty}^{\infty}$ is an orthonormal basis of $L^2(\mathbb{R})$. Hence, the Zak transform is a one-to-one mapping of an orthonormal basis of $L^2(\mathbb{R})$ onto an orthonormal basis of $L^2(S)$. This proves the theorem.

4.9 Applications of Zak Transforms and the Balian-Low Theorem

It has already been mentioned that the Zak transform plays a major role in the study of the Gabor representation problem in signal analysis and the coherent states representation in quantum physics. Furthermore, the Zak transform is particularly useful in proving the Balian-Low theorem (BLT) which is also a fundamental result in time-frequency analysis. For a detailed investigation of these problems, we need the following results.

If $t_0, \omega_0 > 0$, $g \in L^2(\mathbb{R})$, and

$$g_{m,n}(t) = g_{m\omega_0, nt_0}(t) = M_{2\pi m\omega_0} T_{nt_0} g(t) = \exp(2\pi i m \omega_0 t)\, g(t - nt_0) \tag{4.9.1}$$

is a Gabor system (or Weyl-Heisenberg system), then it is easy to verify that, if $\omega_0 t_0 = 1$,

$$\begin{aligned} \mathfrak{Z}_{t_0}\left[g_{m,n}(t)\right](t,\omega) &= \exp\{2\pi i(mt - n\omega)\}\left(\mathfrak{Z}_{t_0} g\right)(t,\omega) \\ &= e_m(t)\, e_{-n}(\omega)\left(\mathfrak{Z}_{t_0} g\right)(t,\omega), \end{aligned} \tag{4.9.2}$$

where $e_k(t) = \exp(2\pi i k t)$.

Furthermore, if $\{g_{m,n}(t)\}$ is a frame in $L^2(\mathbb{R})$, then the frame operator S is given by

$$Sf = \sum_{m,n=-\infty}^{\infty} \left(f, g_{m,n}\right) g_{m,n}, \tag{4.9.3}$$

where $f \in L^2(\mathbb{R})$.

Theorem 4.9.1 If $t_0, \omega_0 > 0$, $g \in L^2(\mathbb{R})$, and $\{g_{m,n}\}_{m,n=-\infty}^{\infty}$ is a frame in $L^2(\mathbb{R})$, then its dual frame $\{S^{-1} g_{m,n}\}_{m,n=-\infty}^{\infty}$ is also generated by one single function. More precisely,

$$S^{-1} g_{m,n} = g_{m,n}^{*}, \tag{4.9.4}$$

where $g^* = S^{-1} g$.

Proof. For any $f \in L^2(\mathbb{R})$ and fixed integer k, we have

$$
\begin{aligned}
S\left(T_{kt_0} f\right)(t) &= \sum_{m,n=-\infty}^{\infty} \left(T_{kt_0} f, g_{m,n}\right) g_{m,n}(t) \\
&= \sum_{m,n=-\infty}^{\infty} \exp\left(-2\pi i m \omega_0 k t_0\right)\left(f, g_{m,n-k}\right) g_{m,n}(t) \\
&= \sum_{m,n=-\infty}^{\infty} \exp\left(-2\pi i m \omega_0 k t_0\right)\left(f, g_{m,n}\right) g_{m,k+n}(t) \\
&= \sum_{m,n=-\infty}^{\infty} \left(f, g_{m,n}\right) \exp\left\{2\pi i m \omega_0 \left(t - k t_0\right)\right\} g\left(t - n t_0 - k t_0\right) \\
&= \sum_{m,n=-\infty}^{\infty} \left(f, g_{m,n}\right) T_{kt_0}\left[\exp\left(2\pi i m \omega_0 t\right) g\left(t - n t_0\right)\right] \\
&= T_{kt_0}\left(S f(t)\right), \qquad (4.9.5)
\end{aligned}
$$

in which $T_{kt_0} \exp\left(2\pi i m \omega_0 t\right) = \exp\left(2\pi i m \omega_0 t\right)$ is used. This shows that S commutes with T_{kt_0}.

Similarly, S commutes with modulation operator $M_{2\pi k \omega_0}$ and hence

$$
S\left(M_{2\pi k \omega_0} T_{st_0} f\right) = M_{2\pi k \omega_0} T_{st_0} (Sf). \qquad (4.9.6)
$$

Consequently,

$$
S^{-1}\left(M_{2\pi k \omega_0} T_{st_0} f\right) = M_{2\pi k \omega_0} T_{st_0} f^*, \qquad (4.9.7)
$$

where $f^* = S^{-1} f$. Putting $f = g$ in (4.9.7) gives

$$
S^{-1}\left(g_{m,n}\right) = S^{-1}\left(M_{2\pi m \omega_0} T_{nt_0} g\right) = M_{2\pi m \omega_0} T_{nt_0} S^{-1} g = M_{2\pi m \omega_0} T_{nt_0} g^* = g^*_{m,n}.
$$

This completes the proof.

Remark. The elements of the dual frame $\left\{g^*_{m,n}\right\}$ are generated by a single function g^*, analogously to $g_{m,n}$. To compute the dual system, it is necessary to find the *dual atom* $g^* = S^{-1} g$ and compute all other elements $g^*_{m,n}$ of the dual frame by modulation and translation.

Some important properties of the Gabor system $\left\{g_{m,n}\right\}$ for $\omega_0 t_0 = 1$ are given by the following:

Theorem 4.9.2 If $t_0, \omega_0 > 0$ such that $\omega_0 t_0 = 1$ and $g \in L^2(\mathbb{R})$, then the following statements are equivalent:

(i) There exist two constants A and B such that

$$0 < A \le \left|\left(\mathfrak{Z}_{t_0} g\right)(t,\omega)\right|^2 \le B < \infty .$$

(ii) The Gabor system $\left\{g_{m,n}(t) = \exp\left(2\pi i m \omega_0 t\right) g\left(t - n t_0\right)\right\}_{m,n=\infty}^{\infty}$ is a frame in $L^2(\mathbb{R})$ with the frame bounds A and B.

(iii) The system $\left\{g_{m,n}(t)\right\}_{m,n=-\infty}^{\infty}$ is an exact frame in $L^2(\mathbb{R})$ with the frame bounds A and B.

If any of the above statements are satisfied, then there exists a unique representation of any $f \in L^2(\mathbb{R})$ in the form

$$f(t) = \sum_{m,n=-\infty}^{\infty} a_{m,n}\, g_{m,n}(t) = \sum_{m,n=-\infty}^{\infty} \left(f, g_{m,n}^{*}\right) g_{m,n}(t), \tag{4.9.8}$$

where

$$a_{m,n} = \left(f,\, g_{m,n}^{*}\right) = \int_0^1 \int_0^1 \frac{\left(\mathfrak{Z}_{t_0} f\right)(t,\omega)}{\left(\mathfrak{Z}_{t_0} g\right)(t,\omega)}\, e_{-m}(t)\, e_n(\omega)\, dt\, d\omega . \tag{4.9.9}$$

Proof. We first show that (i) implies (ii). Since Theorem 4.8.3 asserts that the Zak transformation is a unitary mapping from $L^2(\mathbb{R})$ onto $L^2(S)$, it suffices to prove that $\left\{\left(\mathfrak{Z}_{t_0} g_{m,n}\right)(t,\omega)\right\}_{m,n=-\infty}^{\infty}$ is a frame in $L^2(S)$. Let $h \in L^2(S)$. Since $\left(\mathfrak{Z}_{t_0} g\right)$ is bounded, $h\overline{\left(\mathfrak{Z}_{t_0} g\right)} \in L^2(S)$, and hence, it follows from (4.9.2) that

$$\left(h,\, \mathfrak{Z}_{t_0} g_{m,n}\right) = \left(h,\, e_m(t) e_{-n}(\omega)\, \mathfrak{Z}_{t_0} g\right) = \left(h\overline{\left(\mathfrak{Z}_{t_0} g\right)},\, e_m(t) e_{-n}(\omega)\right). \tag{4.9.10}$$

Since $\left\{e_{m\omega_0}\, e_{-nt_0}\right\}$ is an orthonormal basis of $L^2(S)$, the Parseval relation implies that

$$\sum_{m,n=-\infty}^{\infty} \left|\left(h,\, \mathfrak{Z}_{t_0} g_{m,n}\right)\right|^2 = \left\|h\, \overline{\left(\mathfrak{Z}_{t_0} g\right)}\right\|^2 . \tag{4.9.11}$$

Combining this equality with the inequalities

$$A\|h\|^2 \le \left\| h\overline{\left(\mathfrak{Z}_{t_0} g\right)} \right\|^2 \le B\|h\|^2$$

leads to the result

$$A\|h\|^2 \le \sum_{m,n=-\infty}^{\infty} \left|\left(h, \mathcal{Z}_{t_0} g_{m,n}\right)\right|^2 \le B\|h\|^2.$$

This shows that $\left(\mathcal{Z}_{t_0} g_{m,n}\right)(t,\omega)$ is a frame in $L^2(S)$,

We next show that (ii) implies (i). If (ii) holds, then $\left\{e_m(t)e_{-n}(\omega)\left(\mathcal{Z}_{t_0} g\right)\right\}$ is a frame in $L^2(S)$ with frame bounds A and B. Hence, for any $h \in L^2(S)$, we must have

$$A\|h\|^2 \le \sum_{m,n=-\infty}^{\infty} \left|\left(h, e_m(t)e_{-n}(\omega)\left(\mathcal{Z}_{t_0} g\right)\right)\right|^2 \le B\|h\|^2. \tag{4.9.12}$$

It follows from (4.9.10) and (4.9.11) that

$$\begin{aligned}\sum_{m,n=-\infty}^{\infty} \left|\left(h, e_m(t)e_{-n}(\omega)\left(\mathcal{Z}_{t_0} g\right)\right)\right|^2 &= \sum_{m,n=-\infty}^{\infty} \left|\left(h\overline{\left(\mathcal{Z}_{t_0} g\right)}, e_m(t)e_{-n}(\omega)\right)\right|^2 \\ &= \left\|h\overline{\left(\mathcal{Z}_{t_0} g\right)}\right\|^2.\end{aligned} \tag{4.9.13}$$

Combining (4.9.12) and (4.9.13) together gives

$$A\|h\|^2 \le \left\|h\overline{\left(\mathcal{Z}_{t_0} g\right)}\right\|^2 \le B\|h\|^2$$

which implies (i).

Next, we prove that (ii) implies (iii). Suppose (ii) is satisfied. Then $\left\{e_{m\omega_0}\, e_{-nt_0}\left(\mathcal{Z}_{t_0} g\right)\right\}$ represents a frame in $L^2(S)$. But (i) implies $\left(\mathcal{Z}_{t_0} g\right)$ is bounded. Hence the mapping $F: L^2(S) \to L^2(S)$ defined by

$$F(h) = F\left(\mathcal{Z}_{t_0} g\right), \qquad h \in L^2(S) \tag{4.9.14}$$

is a topological isomorphism that maps the orthonormal basis $\left\{e_m\, e_{-n}\right\}$ onto $\left\{\left(\mathcal{Z}_{t_0} g_{m,n}\right)(t,\omega)\right\}$. Thus, $\left\{\left(\mathcal{Z}_{t_0} g_{m,n}\right)(t,\omega)\right\}$ is a Riesz basis on $L^2(S)$ and hence so is $\left\{g_{m,n}(t,\omega)\right\}$ in $L^2(\mathbb{R})$. In view of the fact that $\left\{g_{m,n}(t,\omega)\right\}$ is a Riesz basis in $L^2(\mathbb{R})$, $\left\{g_{m,n}\right\}$ is an exact frame for $L^2(\mathbb{R})$.

Finally, that (iii) implies (ii) is obvious. To prove (4.9.9), we first prove that

$$\mathcal{Z}_{t_0}(Sf) = \left(\mathcal{Z}_{t_0} f\right)\left|\left(\mathcal{Z}_{t_0} g\right)\right|^2, \tag{4.9.15}$$

where S is the frame operator associated with the frame $\{g_{m,n}(x)\}$. Since $\{e_m(t)e_{-n}(\omega)\}$ is an orthonormal basis for $L^2(S)$, it follows from (4.8.24) and (4.9.2) that

$$\begin{aligned}
\mathfrak{Z}_{t_0}(Sf) &= \mathfrak{Z}_{t_0}\left(\sum_{m,n=-\infty}^{\infty}(f, g_{m,n})\, g_{m,n}\right) \\
&= (\mathfrak{Z}_{t_0} g) \sum_{m,n=-\infty}^{\infty}(f, g_{m,n})\, e_m(t)e_{-n}(\omega) \\
&= (\mathfrak{Z}_{t_0} g) \sum_{m,n=-\infty}^{\infty}(\mathfrak{Z}_{t_0} f, \mathfrak{Z}_{t_0} g_{m,n})\, e_m(t)e_{-n}(\omega), \qquad \text{by (4.8.25)} \\
&= (\mathfrak{Z}_{t_0} g) \sum_{m,n=-\infty}^{\infty}(\mathfrak{Z}_{t_0} f, \mathfrak{Z}_{t_0} g\; e_m(t)e_{-n}(\omega))\, e_m(t)e_{-n}(\omega) \\
&= (\mathfrak{Z}_{t_0} g) \sum_{m,n=-\infty}^{\infty}(\mathfrak{Z}_{t_0} f\; \overline{\mathfrak{Z}_{t_0} g},\; e_m(t)e_{-n}(\omega))\, e_m(t)e_{-n}(\omega) \\
&= (\mathfrak{Z}_{t_0} f)\left|(\mathfrak{Z}_{t_0} g)\right|^2.
\end{aligned}$$

This proves the result (4.9.15).

If we replace f by $S^{-1}f$ in (4.9.15), we obtain

$$\mathfrak{Z}_{t_0}(S^{-1}f) = \frac{(\mathfrak{Z}_{t_0} f)}{\left|(\mathfrak{Z}_{t_0} g)\right|^2}, \tag{4.9.16}$$

which is, by putting $f = g$,

$$\mathfrak{Z}_{t_0} g^* = \frac{1}{\overline{(\mathfrak{Z}_{t_0} g)}}, \quad g^* = S^{-1}g. \tag{4.9.17}$$

In view of (4.8.25), (4.9.2), (4.9.17), and Theorem 4.9.1, it turns out that

$$\begin{aligned}
a_{m,n} &= (f, S^{-1}g_{m,n}) = (f, g^*_{m,n}) = (\mathfrak{Z}_{t_0} f, \mathfrak{Z}_{t_0} g^*_{m,n}) \\
&= (\mathfrak{Z}_{t_0} f, e_m(t)e_{-n}(\omega)\, \mathfrak{Z}_{t_0} g^*) \\
&= \left(\mathfrak{Z}_{t_0} f, \frac{e_m(t)e_{-n}(\omega)}{\overline{(\mathfrak{Z}_{t_0} g)}}\right) = \left(\frac{\mathfrak{Z}_{t_0} f}{\mathfrak{Z}_{t_0} g}, e_m(t)e_{-n}(\omega)\right)
\end{aligned}$$

which gives (4.9.9).

The *Gabor representation problem* can be stated as follows.

Given $g \in L^2(\mathbb{R})$ and two real numbers t_0 and ω_0 different from zero, is it possible to represent any $f \in L^2(\mathbb{R})$ in the series form

$$f(t) = \sum_{m,n=-\infty}^{\infty} a_{m,n}\, g_{m,n}(t), \tag{4.9.18}$$

where $g_{m,n}$ is the Gabor system defined by (4.9.1) and $a_{m,n}$ are constants? Under what conditions is this representation unique?

Evidently, the above representation is possible, if the Gabor system $\{g_{m,n}\}$ forms an orthonormal basis or a frame in $L^2(\mathbb{R})$, and the uniqueness of the representation depends on whether the Gabor functions form a complete set in $L^2(\mathbb{R})$. The Zak transform is used to study this representation problem with two positive real numbers t_0 and ω_0 with $\omega_0 t_0 = 1$. We also use the result (4.9.2).

Theorem 4.9.4 If t_0 and ω_0 are two positive real numbers with $\omega_0 t_0 = 1$ and $g \in L^2(\mathbb{R})$, then

(i) the Gabor system $\{g_{m,n}\}$ is an orthonormal basis of $L^2(\mathbb{R})$ if and only if $\left|\left(\mathfrak{Z}_{t_0} g\right)\right| = 1$ almost everywhere.

(ii) the Gabor system $\{g_{m,n}\}$ is complete in $L^2(\mathbb{R})$ if and only if $\left|\left(\mathfrak{Z}_{t_0} g\right)\right| > 0$ almost everywhere.

Proof. (i) It follows from (4.8.25), (4.9.2), and Theorem 4.8.3 that

$$\left(g_{k,\ell}, g_{m,n}\right) = \left(\mathfrak{Z}_{t_0} g_{k,\ell}, \mathfrak{Z}_{t_0} g_{m,n}\right) = \int_0^1\int_0^1 e_k(t)\, e_{-\ell}(\omega)\, \bar{e}_m(t)\, \bar{e}_{-n}(\omega) \left|\left(\mathfrak{Z}_{t_0} g\right)\right|^2 dt\, d\omega .$$

This shows that the set $\{\mathfrak{Z}_{t_0} g_{m,n}\}$ is an orthonormal basis in $L^2(\mathbb{R})$ if and only if $\left|\left(\mathfrak{Z}_{t_0} g\right)\right| = 1$ almost everywhere.

An argument similar to above gives

$$\begin{aligned}\left(f, g_{m,n}\right) &= \left(\mathfrak{Z}_{t_0} f, \mathfrak{Z}_{t_0} g_{m,n}\right) = \left(\mathfrak{Z}_{t_0} f, e_m(t)\, e_{-n}(\omega)\, \mathfrak{Z}_{t_0} g\right) \\ &= \left(\mathfrak{Z}_{t_0} f\, \overline{\mathfrak{Z}_{t_0} g}\,,\, e_m(t)\, e_{-n}(\omega)\right).\end{aligned} \tag{4.9.19}$$

This implies that $\{g_{m,n}\}$ is complete in $L^2(\mathbb{R})$ if and only if $\mathfrak{Z}_{t_0} g \neq 0$ almost everywhere.

The answer to the Gabor representation problem can be summarized as follows.

The properties of the Gabor system $\{g_{m,n}\}$ are related to the density of the rectangular lattice $\Lambda = \{n t_0, m \omega_0\} = n\mathbb{Z} \times m\mathbb{Z}$ in the time-frequency plane. Small values of t_0, ω_0 correspond to a high density for Λ, whereas large values of t_0, ω_0 correspond to low density. Thus, it is natural to classify Gabor systems according to the following sampling density of the time-frequency lattice.

Case (i) (Oversampling). A Gabor system $\{g_{m,n}\}$ can be a frame where $0 < \omega_0 t_0 < 1$. In this case, frames exist with excellent time-frequency localization.

Case (ii) (Critical Sampling). This critical case corresponds to $\omega_0 t_0 = 1$, and there is a frame, and orthonormal basis exist, but g has bad localization properties either in time or in the frequency domain. More precisely, this case leads to the celebrated result in the time-frequency analysis which is known as the *Balian-Low Theorem* (BLT), originally and independently stated by Balian (1981) and Low (1985) as follows.

Theorem 4.9.5 (Balian-Low). If a Gabor system $\{g_{m,n}\}$ defined by (4.6.1) with $\omega_0 t_0 = 1$ forms an orthonormal basis in $L^2(\mathbb{R})$, then either $\int_{-\infty}^{\infty} |t\, g(t)|^2$ or $\int_{-\infty}^{\infty} |\omega\, \hat{g}(\omega)|^2 \, d\omega$ must diverge, or equivalently,

$$\int_{-\infty}^{\infty} |t\, g(t)|^2 \, dt \int_{-\infty}^{\infty} |\omega\, \hat{g}(\omega)|^2 \, d\omega = \infty. \tag{4.9.20}$$

The condition $\omega_0 t_0 = 1$ associated with the density $\Lambda = 1$ can be interpreted as a Nyquist phenomenon for the Gabor system. In this critical situation, the time-frequency shift operators that are used to build a coherent frame commute with each other.

For an elegant proof of the Balian-Low theorem using the Zak transform, we refer the reader to Daubechies (1992) or Beneditto and Frazier (1994).

Case (iii) (Undersampling). In this case, $\omega_0 t_0 > 1$. There is no frame of the form $\{g_{m,n}\}$ for any choice of the Gabor window function g. In fact, $\{g_{m,n}\}$ is incomplete in the sense that there exist $f \in L^2(\mathbb{R})$ such that $(f, g_{m,n}) = 0$ for all m, n but $f \neq 0$.

These three cases can be represented by three distinct regions in the $t_0 - \omega_0$ plane, where the critical curve $\omega_0 t_0 = 1$ represents a *hyperbola* which separates the region $\omega_0 t_0 < 1$, where an exact frame exists with an excellent time-frequency localization from the region $\omega_0 t_0 > 1$ with no frames.

There exist many examples for g so that $\{g_{m,n}\}$ is a frame or even an orthonormal basis for $L^2(\mathbb{R})$. We give two examples of functions for which the family $\{M_{m\omega_0} T_{nt_0} g\}$ represents an orthonormal basis.

Example 4.9.1 (Characteristic Function). This function $g(t) = \chi_{[0,1]}(t)$ is defined by

$$g(t) = \begin{cases} 1, & 0 \le t \le 1 \\ 0, & \text{otherwise} \end{cases}.$$

Clearly,

$$\int_{-\infty}^{\infty} |\omega \hat{g}(\omega)|^2 \, d\omega = \infty.$$

Example 4.9.2 (Sinc Function). In this case,

$$g(t) = \text{sinc}(t) = \frac{\sin \pi t}{\pi t}.$$

Evidently, $\int_{-\infty}^{\infty} |t\, g(t)|^2 \, dt = \infty.$

Thus, these examples lead to systems with bad localization properties in either time or frequency. Even if the orthogonality requirement is dropped, we cannot construct Riesz bases with good time-frequency localization properties for the critical case $\omega_0 t_0 = 1$. This constitutes the contents of the Balian-Low theorem which describes one of the fundamental features of Gabor wavelet analysis.

4.10 Exercises

1. If $g(x) = \frac{1}{\sqrt{4\pi a}} \exp\left(-\frac{x^2}{4a}\right)$ is a Gaussian window, show that

   (a) $$\int_{-\infty}^{\infty} \tilde{f}_g(t,\omega)\, dt = \hat{f}(\omega), \qquad \omega \in \mathbb{R}.$$

   Give a significance of result 1(a).

   (b) $$\hat{g}(\nu) = \exp\left(-a\nu^2\right).$$

2. Suppose $g_{t,\omega}(\tau) = g(\tau - t)\exp(i\omega\tau)$ where g is a Gaussian window defined in Exercise 1, show that

   (a) $$\hat{g}_{t,\omega}(\nu) = \exp\left[-i(\nu-\omega)\,t - a(\nu-\omega)^2\right].$$

   (b) $$\tilde{f}_g(t,\omega) = \frac{1}{2\pi}\left(\hat{f}, \hat{g}_{t,\omega}\right) = \frac{1}{2\pi}\, e^{i\omega t}\, \tilde{\hat{f}}_{\hat{g}}(t,\omega).$$

3. For the Gaussian window defined in Exercise 1, introduce

   $$\sigma_t^2 = \frac{1}{\|g\|_2}\left\{\int_{-\infty}^{\infty} \tau^2 g^2(\tau)\, d\tau\right\}^{\frac{1}{2}}.$$

   Show that the radius of the window function is $\sqrt{a}$ and the width of the window is twice the radius.

4. If $e_1 = (1,0)$, $e_2 = \left(-\frac{1}{2}, \frac{\sqrt{3}}{2}\right)$, $e_3 = \left(-\frac{1}{2}, -\frac{\sqrt{3}}{2}\right)$ represent a set of vectors, show that, for any vector $x = (x_1, x_2)$,

   $$\sum_{n=1}^{3} \left|(x, e_n)\right|^2 = \frac{3}{2}\|x\|^2.$$

   Hence, show that $\{e_i\}$ is a tight frame and $e_n^* = \frac{2}{3} e_n$.

5. If $e_1 = (1,0)$, $e_2 = (0,1)$, $e_3 = (-1,0)$, $e_4 = (0,-1)$ form a set of vectors, show that, for any vector $x = (x_1, x_2)$,

$$\sum_{n=1}^{4} |(x, e_n)|^2 = 2\,\|x\|^2,$$

and

$$x = \sum_{k=1}^{4} \frac{1}{2}\,(x, x_k)\,x_k\,.$$

6. If $e_1 = (0,1)$, $e_2 = \left(-\frac{1}{2}, \frac{\sqrt{3}}{2}\right)$ and $e_3 = \left(-\frac{1}{2}, -\frac{3}{2}\right)$ represent a set of vectors and $x = [x_1, x_2]^T$, show that

$$\sum_{n=1}^{3} |(x, e_n)|^2 = \frac{1}{2}\left(x_1^2 + 5x_2^2\right),$$

and

$$\frac{1}{2}\left(x_1^2 + x_2^2\right) \le \sum_{n=1}^{3} |(x, e_n)|^2 \le \frac{5}{2}\left(x_1^2 + x_2^2\right).$$

7. Show that the set of elements $\{e_n\}$ in a Hilbert space $\mathbb{C}^2$ forms a tight frame.

8. If g is a continuous function on $\mathbb{R}$ and if there exists an $\varepsilon > 0$ such that $|g(x)| \le A(1+|x|)^{-1-\varepsilon}$, show that

$$g_{m,n}(x) = \exp(2\pi i m x)\; g(x-n)$$

cannot be a frame for $L^2(\mathbb{R})$.

9. Show that the marginals of the Zak transform are given by

$$\int_0^1 (\mathfrak{Z} f)(t, \omega)\, d\omega = f(t),$$

$$\int_0^1 \exp(-2\pi i \omega t)(\mathfrak{Z} f)(t, \omega)\, dt = \hat{f}(\omega).$$

10. If $f(t)$ is time-limited to $-a \le t \le a$ and band-limited to $-b \le \omega \le b$, where $0 \le a,\, b \le \frac{1}{2}$, then the following results hold:

$$(\mathfrak{Z} f)(\tau,\omega) = f(\tau), \quad |\tau| \le \frac{1}{2}, \quad \omega \in \mathbb{R},$$

$$(\mathfrak{Z} f)(\tau,\omega) = \exp(2\pi i \omega \tau)\, \hat{f}(\omega), \quad |\omega| \le \frac{1}{2}, \quad \tau \in \mathbb{R}.$$

Show that the second of the above results gives the Shannon's sampling formula

$$f(t) = \sum_{n=-\infty}^{\infty} \frac{\sin 2\pi b(n-t)}{\pi(n-t)}, \qquad t \in \mathbb{R}.$$

11. If $g = \chi_{[0,1]}$, $g_{m,n}(x) = \exp(2\pi i m x)\, g(x-n)$, where $m,n \in \mathbb{Z}$ is an orthonormal basis of $L^2(\mathbb{R})$, show that the first integral

$$\int_{-\infty}^{\infty} t\, |g(t)|^2\, dt$$

in the Balian-Low theorem is finite, whereas the second integral

$$\int_{-\infty}^{\infty} \omega\, |\hat{g}(\omega)|^2\, d\omega = \infty.$$

12. If $g(x) = \text{sinc}\,(x) = \dfrac{\sin \pi x}{\pi x}$, $g_{m,n}(x) = \exp(2\pi i m x)\, g(x-n)$ is an orthonormal basis of $L^2(\mathbb{R})$, show that the first integral in the Balian-Low theorem

$$\int_{-\infty}^{\infty} t\, |g(t)|^2\, dt = \infty,$$

and the second integral in the Balian-Low theorem is finite.

Chapter 5

The Wigner-Ville Distribution and Time-Frequency Signal Analysis

"As long as a branch of knowledge offers an abundance of problems, it is full of vitality."

David Hilbert

"Besides linear time-frequency representations like the short-time Fourier transform, the Gabor transform, and the wavelet transform, an important contribution to this development has undoubtedly been the *Wigner distribution* (WD) which holds an exceptional position within the field of bilinear/quadratic time-frequency representations."

W. Mecklenbräuker and F. Hlawatsch

5.1 Introduction

Although time-frequency analysis of signals had its origin almost fifty years ago, there has been major development of the time-frequency distributions approach in the last two decades. The basic idea of the method is to develop a joint function of time and frequency, known as a time-frequency distribution, that can describe the energy density of a signal simultaneously in both time and frequency. In principle, the time-frequency distributions characterize phenomena in a two-dimensional time-frequency plane. Basically, there are two kinds of time-frequency representations. One is the quadratic method covering the time-frequency distributions, and the other is the linear approach including the Gabor transform, the Zak transform, and the wavelet transform analysis. So, the time-frequency signal analysis deals with time-frequency representations of signals and with problems related to their definition, estimation and interpretation, and it has evolved into a widely recognized applied discipline of signal processing. From theoretical and application points of view, the Wigner-Ville distribution (WVD) or the Wigner-Ville transform (WVT) plays a major role in the time-frequency signal analysis for the following reasons. First, it provides a high-resolution representation in both time and frequency for non-stationary signals. Second, it has the special properties of satisfying the time and frequency marginals in terms of the instantaneous power in time and energy spectrum in frequency and the total energy of the signal in the time and frequency plane. Third, the first conditional moment of frequency at a given time is the derivative of the phase of the signal at that time. Fourth, the theory of the Wigner-Ville distribution was reformulated in the context of sonar and radar signal analysis, and a new function, the so-called Woodward ambiguity function, was introduced by Woodward in 1953 for the mathematical analysis of sonar and radar systems. In analogy with the Heisenberg uncertainty principle in quantum mechanics, Woodward introduced the radar uncertainty principle which says that the range and velocity of a target cannot be measured precisely and simultaneously.

This chapter is devoted to the Wigner-Ville distribution (or the Wigner-Ville transform) and the ambiguity function and their basic structures and properties. Special attention is given to fairly exact mathematical treatment with examples and applications to the time-frequency signal analysis in general and the radar signal analysis in particular. The relationship between the Wigner-Ville distribution and the ambiguity function is discussed. A comparison of some of the major properties of these transformations is made. In the end, recent generalizations of the Wigner-Ville distribution are briefly discussed.

5.2 Definition and Examples of the Wigner-Ville Distribution

Definition 5.2.1 (The Cross Wigner-Ville Distribution). If $f, g \in L^2(\mathbb{R})$, the *cross Wigner-Ville distribution* of f and g is defined by

$$W_{f,g}(t,\omega)=\int_{-\infty}^{\infty} f\left(t+\frac{\tau}{2}\right)\bar{g}\left(t-\frac{\tau}{2}\right)e^{-i\omega\tau}\,d\tau. \tag{5.2.1}$$

Introducing a change of variable $t+\frac{\tau}{2}=x$ gives an equivalent definition of $W_{f,g}(t,\omega)$ in the form

$$W_{f,g}(t,\omega)=2\exp(2i\omega t)\int_{-\infty}^{\infty} f(x)\,\bar{g}(2t-x)\exp(-2i\omega x)\,dx \tag{5.2.2}$$

$$=2\exp(2i\omega t)\,\tilde{f}_h(2t,2\omega), \tag{5.2.3}$$

where $h(x)=\bar{g}(-x)$.

It follows from definition (5.2.1) that the cross Wigner-Ville distribution is the Fourier transform of the function

$$h_t(\tau)=f\left(t+\frac{\tau}{2}\right)\bar{g}\left(t-\frac{\tau}{2}\right)$$

with respect to τ. Hence, $W_{f,g}(t,\omega)$ is a complex-valued function in the time-frequency plane. In other words,

$$W_{f,g}(t,\omega)=\mathscr{F}\{h_t(\tau)\}=\hat{h}_t(\omega). \tag{5.2.4}$$

On the other hand, the Fourier transform of the cross Wigner-Ville distribution with respect to ω is given by

$$\begin{aligned}\hat{W}_{f,g}(t,\sigma)&=\int_{-\infty}^{\infty} e^{-i\omega\sigma}W_{f,g}(t,\omega)\,d\omega\\&=\int_{-\infty}^{\infty} e^{-i\omega\sigma}d\omega\int_{-\infty}^{\infty} h_t(\tau)\,e^{-i\omega\tau}d\tau\\&=\int_{-\infty}^{\infty} h_t(\tau)\,d\tau\int_{-\infty}^{\infty} e^{-i\omega(\tau+\sigma)}d\omega\\&=2\pi\int_{-\infty}^{\infty} h_t(\tau)\,\delta(\tau+\sigma)\,d\tau=2\pi\,h_t(-\sigma)\\&=2\pi\,f\left(t-\frac{\sigma}{2}\right)\bar{g}\left(t+\frac{\sigma}{2}\right).\end{aligned} \tag{5.2.5}$$

Or, equivalently,

$$\hat{W}_{f,g}(t,-\sigma)=2\pi\,f\left(t+\frac{\sigma}{2}\right)\bar{g}\left(t-\frac{\sigma}{2}\right). \tag{5.2.6}$$

Definition 5.2.2 (The Auto Wigner-Ville Distribution). If $f = g$ in (5.2.1)-(5.2.3), then $W_{f,f}(t,\omega) = W_f(t,\omega)$ is called the *auto Wigner-Ville distribution* and defined by

$$W_f(t,\omega) = \int_{-\infty}^{\infty} f\left(t+\frac{\tau}{2}\right) \bar{f}\left(t-\frac{\tau}{2}\right) e^{-i\omega\tau} d\tau \tag{5.2.7}$$

$$= 2\exp(2i\omega t) \int_{-\infty}^{\infty} f(x)\, \bar{f}(2t-x) \exp(-2i\omega x)\, dx \tag{5.2.8}$$

$$= 2\exp(2i\omega t)\, \tilde{f}_h(2t, 2\omega), \tag{5.2.9}$$

where $h(x) = \bar{f}(-x)$.

Obviously, results (5.2.4)-(5.2.6) hold for the auto Wigner-Ville distribution.

Furthermore, the Wigner-Ville distribution of a real signal is an even function of the frequency. More precisely,

$$W_f(t,\omega) = \int_{-\infty}^{\infty} f\left(t+\frac{\tau}{2}\right) \bar{f}\left(t-\frac{\tau}{2}\right) e^{-i\omega\tau} d\tau, \qquad (\tau = -x)$$

$$= \int_{-\infty}^{\infty} \bar{f}\left(t+\frac{x}{2}\right) \bar{\bar{f}}\left(t-\frac{x}{2}\right) e^{i\omega x} dx$$

$$= W_{\bar{f}}(t, -\omega). \tag{5.2.10}$$

Often both the cross Wigner-Ville distribution and the auto Wigner-Ville distribution are usually referred to simply as the *Wigner-Ville distribution* or *Wigner-Ville transform*.

The formula (5.2.5) can also be written as

$$\frac{1}{2\pi} \int_{-\infty}^{\infty} e^{-i\sigma\omega}\, W_{f,g}(t,\omega)\, d\omega = f\left(t-\frac{\sigma}{2}\right) \bar{g}\left(t+\frac{\sigma}{2}\right),$$

which is, by putting $t + \dfrac{\sigma}{2} = t_1$ and $t - \dfrac{\sigma}{2} = t_2$,

$$\frac{1}{2\pi} \int_{-\infty}^{\infty} e^{-i(t_1-t_2)\omega}\, W_{f,g}\left(\frac{t_1+t_2}{2}, \omega\right) d\omega = f(t_2)\, \bar{g}(t_1). \tag{5.2.11}$$

Putting $t_1 = 0$ and $t_2 = t$ in (5.2.11) gives a representation of $f(t)$ in terms of $W_{f,g}$ in the form

$$f(t)\,\overline{g}(0)=\frac{1}{2\pi}\int_{-\infty}^{\infty} e^{it\omega}\,W_{f,g}\left(\frac{t}{2},\,\omega\right)\,d\omega\,, \tag{5.2.12}$$

provided $\overline{g}(0)\neq 0$. This is the *inversion formula* for the Wigner-Ville distribution.

In particular, if we substitute $t_1=t_2=t$ in (5.2.11), we find the *inversion formula*

$$\frac{1}{2\pi}\int_{-\infty}^{\infty} W_{f,g}(t,\omega)\,d\omega = f(t)\,\overline{g}(t), \tag{5.2.13}$$

and when $f=g$, we obtain the marginal integral over all time

$$\frac{1}{2\pi}\int_{-\infty}^{\infty} W_{f}(t,\omega)\,d\omega = |f(t)|^{2}. \tag{5.2.14}$$

This implies that the integral of the Wigner-Ville distribution over the frequency at any time t is equal to the *time energy density* (*instantaneous power*) of a signal f.

Integrating (5.2.13) with respect to time gives

$$\frac{1}{2\pi}\int_{-\infty}^{\infty}\int_{-\infty}^{\infty} W_{f,g}(t,\omega)\,dt\,d\omega = \int_{-\infty}^{\infty} f(t)\,\overline{g}(t)\,dt = (f,g). \tag{5.2.15}$$

Similarly, integrating (5.2.14) with respect to time t yields the total energy over the whole time-frequency plane (t,ω),

$$\frac{1}{2\pi}\int_{-\infty}^{\infty}\int_{-\infty}^{\infty} W_{f}(t,\omega)\,dt\,d\omega = \int_{-\infty}^{\infty} |f(t)|^{2}\,dt = \|f\|_{2}^{2}. \tag{5.2.16}$$

We can also define the Wigner-Ville distribution of the Fourier spectrum $\hat{f}$ and $\hat{g}$ by

$$\begin{aligned} W_{\hat{f},\hat{g}}(\omega,t) &= \frac{1}{2\pi}\int_{-\infty}^{\infty} \hat{f}\left(\omega+\frac{\tau}{2}\right)\overline{\hat{g}}\left(\omega-\frac{\tau}{2}\right)e^{it\tau}\,d\tau \\ &= \frac{1}{2\pi}\int_{-\infty}^{\infty} f(x)\,e^{-i\omega x}\,dx\int_{-\infty}^{\infty}\overline{\hat{g}}\left(\omega-\frac{\tau}{2}\right)\exp\left[i\tau\left(t-\frac{x}{2}\right)\right]d\tau \end{aligned} \tag{5.2.17}$$

$$= 2\exp(2i\omega t)\int_{-\infty}^{\infty} f(x)\, e^{-2i\omega x}\, dx \cdot \frac{1}{2\pi}\int_{-\infty}^{\infty} \overline{\hat{g}}\,(u)\exp\left[iu(x-2t)\right] du, \;\left(\omega - \frac{\tau}{2} = u\right)$$

$$= 2\exp(2i\omega t)\int_{-\infty}^{\infty} f(x)\,\overline{g}\,(2t-x)\exp(-2i\omega x)\, dx$$

$$= W_{f,g}(t,\omega). \tag{5.2.18}$$

Thus, (5.2.17) can be used as another equivalent definition of the Wigner-Ville distribution due to symmetry between time and frequency, as expressed by the important relation (5.2.5).

It also follows from (5.2.17) and (5.2.18) that the Fourier transform of $W_{f,g}(t,\omega)$ with respect to t is

$$\int_{-\infty}^{\infty} e^{-it\tau}\, W_{f,g}(t,\omega)\, dt = \hat{f}\left(\omega + \frac{\tau}{2}\right)\overline{\hat{g}}\left(\omega - \frac{\tau}{2}\right). \tag{5.2.19}$$

Putting $\tau = 0$ in (5.2.19) gives

$$\int_{-\infty}^{\infty} W_{f,g}(t,\omega)\, dt = \hat{f}(\omega)\,\overline{\hat{g}}\,(\omega). \tag{5.2.20}$$

When $\hat{f} = \hat{g}$, this leads to the marginal integral over all time t giving the signal frequency energy density, that is, the *energy density spectrum* (or *spectral energy density*),

$$\int_{-\infty}^{\infty} W_f(t,\omega)\, dt = \left|\hat{f}(\omega)\right|^2. \tag{5.2.21}$$

Integrating (5.2.20) with respect to frequency ω yields

$$\int_{-\infty}^{\infty}\int_{-\infty}^{\infty} W_{f,g}(t,\omega)\, dt\, d\omega = \int_{-\infty}^{\infty} \hat{f}(\omega)\,\overline{\hat{g}}\,(\omega)\, d\omega = \left(\hat{f},\hat{g}\right). \tag{5.2.22}$$

Thus, we obtain the total energy of the signal f over the whole (t,ω) plane as

$$\int_{-\infty}^{\infty}\int_{-\infty}^{\infty} W_f(t,\omega)\, dt\, d\omega = \int_{-\infty}^{\infty}\left|\hat{f}(\omega)\right|^2 d\omega = \left\|\hat{f}\right\|_2^2. \tag{5.2.23}$$

Thus, we obtain a fundamental theorem from the above analysis in the form:

Theorem 5.2.1 (Time and Frequency Energy Densities, and the Total Energy). If $f \in L^2(\mathbb{R})$, then the Wigner-Ville distribution satisfies the time and frequency marginal integrals (5.2.14) and (5.2.21) respectively, and the integral of the Wigner-Ville distribution over the entire time-frequency plane yields the total energy of the signal, that is,

$$\frac{1}{2\pi}\int_{-\infty}^{\infty}\int_{-\infty}^{\infty} W_f(t,\omega)\,dt\,d\omega = \frac{1}{2\pi}\int_{-\infty}^{\infty}\left|\hat{f}(\omega)\right|^2 d\omega = \int_{-\infty}^{\infty}|f(t)|^2\,dt. \tag{5.2.24}$$

Physically, the Wigner-Ville distribution can be interpreted as the time-frequency energy distribution.

Finally, it may be noted from (5.2.1) that

$$W_{f,g}(t,0) = \int_{-\infty}^{\infty} f\left(t+\frac{\tau}{2}\right)\bar{g}\left(t-\frac{\tau}{2}\right)d\tau = 2\int_{-\infty}^{\infty} f(x)\,\bar{g}\,(2t-x)\,dx. \tag{5.2.25}$$

$$W_{f,g}(0,0) = 2\int_{-\infty}^{\infty} f(x)\,\bar{g}(-x)\,dx = 2\int_{-\infty}^{\infty}\hat{f}(x)\,\bar{\hat{g}}(-x)\,dx = 2\pi\,W_{\hat{f},\hat{g}}(0,0). \tag{5.2.26}$$

In particular, if $f = g$ and $f(-x) = f(x)$, then (5.2.26) becomes

$$W_f(0,0) = 2\int_{-\infty}^{\infty}|f(x)|^2\,dx = 2\pi\;W_{\hat{f},\hat{g}}(0,0). \tag{5.2.27}$$

Moreover, it follows from (5.2.17) that

$$\begin{aligned} W_{\hat{f},\hat{g}}(\omega,0) &= \frac{1}{2\pi}\int_{-\infty}^{\infty}\hat{f}\left(\omega+\frac{\tau}{2}\right)\bar{\hat{g}}\left(\omega-\frac{\tau}{2}\right)d\tau \\ &= \frac{1}{2\pi}\cdot 2\int_{-\infty}^{\infty} f(x)\,\bar{g}(-x)\,e^{-2i\omega x}dx \\ &= \frac{1}{2\pi}\,W_{f,g}(0,\omega). \end{aligned} \tag{5.2.28}$$

Application of Schwarz's inequality (see Debnath and Mikusinski, 1999) gives

$$\begin{aligned} |W_f(t,\omega)|^2 &\le \left[\int_{-\infty}^{\infty}\left|f\left(t+\frac{\tau}{2}\right)\right|\left|f\left(t-\frac{\tau}{2}\right)\right|d\tau\right]^2 \\ &\le \int_{-\infty}^{\infty}\left|f\left(t+\frac{\tau}{2}\right)\right|^2 d\tau\int_{-\infty}^{\infty}\left|\bar{f}\left(t-\frac{\tau}{2}\right)\right|^2 d\tau \\ &= 4\int_{-\infty}^{\infty}|f(x)|^2\,dx\int_{-\infty}^{\infty}|\bar{f}(x)|^2\,dx \\ &= 4\,\|f\|_2^2\,\|\bar{f}\|_2^2 = 4\,\|f\|_2^4. \end{aligned} \tag{5.2.29}$$

Clearly, for all t and ω,

$$\left|W_f(t,\omega)\right| \le 2\,\|f\|_2^2 = 2\int_{-\infty}^{\infty} |f(x)|^2\,dx = W_f(0,0). \tag{5.2.30}$$

Example 5.2.1 Find the Wigner-Ville transform of a Gaussian signal

$$f(t) = \left(\pi\sigma^2\right)^{-\frac{1}{4}} \exp\left(-\frac{t^2}{2\sigma^2}\right). \tag{5.2.31}$$

We have, by definition (5.2.7),

$$\begin{aligned} W_f(t,\omega) &= \int_{-\infty}^{\infty} f\left(t+\frac{\tau}{2}\right)\bar{f}\left(t-\frac{\tau}{2}\right)\exp(-i\omega\tau)\,d\tau \\ &= \frac{1}{\sigma\sqrt{\pi}}\exp\left(-\frac{t^2}{\sigma^2}\right)\int_{-\infty}^{\infty}\exp\left[-\left(i\omega\tau+\frac{\tau^2}{4\sigma^2}\right)\right]d\tau \\ &= 2\exp\left[-\left(\frac{t^2}{\sigma^2}+\omega^2\sigma^2\right)\right] = |f(t)|^2\left|\hat{f}(\omega)\right|^2. \end{aligned} \tag{5.2.32}$$

In particular, when $\sigma = 1$, then (5.2.32) becomes

$$W_f(t,\omega) = 2\exp\left[-\left(t^2+\omega^2\right)\right]. \tag{5.2.33}$$

This shows that the Wigner-Ville distribution of a Gaussian signal is also Gaussian in both time t and frequency ω.

Example 5.2.2 Find the Wigner-Ville transform of a *harmonic* signal (or the so-called *plane wave*)

$$f(t) = A\exp\left(i\omega_0 t\right), \tag{5.2.34}$$

where the constant frequency ω_0 is the derivative of the phase of the signal, that is, $\omega_0 = \frac{d}{dt}\left(\omega_0 t\right)$.

We have, by definition,

$$W_f(t,\omega) = A\bar{A} \int_{-\infty}^{\infty} f\left(t+\frac{\tau}{2}\right) \bar{f}\left(t-\frac{\tau}{2}\right) \exp(-i\omega\tau)\, d\tau$$

$$= |A|^2 \int_{-\infty}^{\infty} \exp\left[i\omega_0\left\{\left(t+\frac{\tau}{2}\right)-\left(t-\frac{\tau}{2}\right)\right\}\right] \exp(-i\omega\tau)\, d\tau$$

$$= |A|^2 \int_{-\infty}^{\infty} \exp\left[i\tau(\omega_0-\omega)\right] d\tau = 2\pi|A|^2\,\delta(\omega-\omega_0). \tag{5.2.35}$$

Physically, this means that only one frequency $\omega = \omega_0$ manifests itself, that is, what is expected as the local frequency of a plane wave.

Example 5.2.3 Find the Wigner-Ville distribution of a *quadratic-phase signal* (or *chirp*)

$$f(t) = A\exp\left(\frac{1}{2}\, i a t^2\right), \tag{5.2.36}$$

where the instantaneous frequency of a time-varying signal is defined as the derivative of the phase of that signal, that is, $\omega_f(t) = \frac{d}{dt}\arg f(t) = \frac{d}{dt}\left(\frac{1}{2}at^2\right) = at$ which is a linear function in time t.

We have

$$W_f(t,\omega) = A\bar{A} \int_{-\infty}^{\infty} f\left(t+\frac{\tau}{2}\right) \bar{f}\left(t-\frac{\tau}{2}\right) \exp(-i\omega\tau)\, d\tau$$

$$= |A|^2 \int_{-\infty}^{\infty} \exp\left[\frac{ia}{2}\left\{\left(t+\frac{\tau}{2}\right)^2-\left(t-\frac{\tau}{2}\right)^2\right\}\right] \exp(-i\omega\tau)\, d\tau$$

$$= |A|^2 \int_{-\infty}^{\infty} \exp\left[i\tau(at-\omega)\right] d\tau = |A|^2\, 2\pi\,\delta(\omega-at). \tag{5.2.37}$$

The quadratic-phase signal $f(t)$ represents, at least for small time t, (that is, in the paraxial approximation), a spherical wave whose curvature is equal to a. The Wigner-Ville distribution of this signal shows that, at any time t, only one frequency $\omega = at$ manifests itself. This corresponds to a ray picture of a spherical wave.

Example 5.2.4 Find the Wigner-Ville distribution of a point source at $t = t_0$ described by the *impulse signal*

$$f(t) = \delta(t - t_0). \tag{5.2.38}$$

Its Wigner-Ville distribution is given by

$$\begin{aligned}
W_f(t,\omega) &= \int_{-\infty}^{\infty} \delta\left(t - t_0 + \frac{\tau}{2}\right) \delta\left(t - t_0 - \frac{\tau}{2}\right) e^{-i\omega\tau}\, d\tau \\
&= 2\int_{-\infty}^{\infty} \delta(x + t - t_0)\, \delta(x - t + t_0)\, e^{-2i\omega x}\, dx \\
&= 2\, e^{-2i\omega(t-t_0)}\, \delta[2(t - t_0)] \\
&= e^{-2i\omega(t-t_0)}\, \delta(t - t_0).
\end{aligned} \tag{5.2.39}$$

At a particular time $t = t_0$, all frequencies are present, and there is no contribution at other points. This is exactly what is expected as the local frequency spectrum of a point source.

Example 5.2.5 If $f(t) = g(t)\exp\left(\frac{ia}{2}t^2\right)$, show that

$$W_f(t,\omega) = W_g(t,\omega - at) = W_g(t,\omega) * \delta(\omega - at), \tag{5.2.40}$$

where $*$ is the convolution with respect to frequency ω.

We have, by definition,

$$\begin{aligned}
W_f(t,\omega) &= \int_{-\infty}^{\infty} f\left(t + \frac{\tau}{2}\right) \bar{f}\left(t - \frac{\tau}{2}\right) \exp(-i\omega\tau)\, d\tau \\
&= \int_{-\infty}^{\infty} g\left(t + \frac{\tau}{2}\right) \bar{g}\left(t - \frac{\tau}{2}\right) \left[\exp\left\{\frac{ia}{2}\left(t + \frac{\tau}{2}\right)^2 - \frac{ia}{2}\left(t - \frac{\tau}{2}\right)^2\right\}\right] e^{-i\omega\tau} d\tau \\
&= \int_{-\infty}^{\infty} g\left(t + \frac{\tau}{2}\right) \bar{g}\left(t - \frac{\tau}{2}\right) \exp[-i\tau(\omega - at)]\, d\tau \\
&= W_g(t,\omega - at).
\end{aligned}$$

Example 5.2.6 If $f(t) = A_1 e^{i\omega_1 t}$ and $g(t) = A_2 e^{i\omega_2 t}$ represent two plane waves, then

$$W_{f,g}(t,\omega) = 2\pi A_1 \overline{A}_2 \exp\left[i(\omega_1 - \omega_2)t\right] \delta\left(\omega - \frac{\omega_1 + \omega_2}{2}\right). \tag{5.2.41}$$

We have, by definition,

$$\begin{aligned} W_{f,g}(t,\omega) &= A_1 \overline{A}_2 \int_{-\infty}^{\infty} \exp\left[i\omega_1\left(t + \frac{\tau}{2}\right) - i\omega_2\left(t - \frac{\tau}{2}\right)\right] e^{-i\omega\tau} d\tau \\ &= A_1 \overline{A}_2 e^{i(\omega_1 - \omega_2)t} \int_{-\infty}^{\infty} \exp\left[-i\tau\left(\omega - \frac{\omega_1 + \omega_2}{2}\right)\right] d\tau \\ &= 2\pi A_1 \overline{A}_2 \exp\left[i(\omega_1 - \omega_2)t\right] \delta\left(\omega - \frac{\omega_1 + \omega_2}{2}\right). \end{aligned}$$

***Example** 5.2.7* If $f(t) = \chi_{[-T,T]}(t)$, then

$$W_f(t,\omega) = \left(\frac{2}{\omega}\right) \sin\left\{2\omega\left(T - |t|\right)\right\}. \tag{5.2.42}$$

The solution is left as an exercise.

***Example** 5.2.8* Show that the Wigner-Ville distribution of a smooth-phase signal

$$f(t) = \exp\left[i\gamma(t)\right] \tag{5.2.43}$$

is given by

$$W_f(t,\omega) \approx 2\pi\delta\left(\omega - \frac{d\gamma}{dt}\right), \tag{5.2.44}$$

where $\gamma(t)$ is a smooth function of time.

It follows from the definition (5.2.1) that

$$\begin{aligned} W_f(t,\omega) &= \int_{-\infty}^{\infty} \exp\left[i\gamma\left(t + \frac{\tau}{2}\right) - i\gamma\left(t - \frac{\tau}{2}\right)\right] e^{-i\omega\tau} d\tau \\ &= \int_{-\infty}^{\infty} \exp\left[-i\tau\left(\omega - \frac{d\gamma}{dt}\right)\right] d\tau \\ &= 2\pi\delta\left(\omega - \frac{d\gamma}{dt}\right). \end{aligned}$$

This shows that, at a particular time t, only one frequency $\omega = \frac{d\gamma}{dt}$ manifests itself.

Example 5.2.9 Find the Wigner-Ville distributions for Gaussian signals

(a) $$f(t) = \left(\pi\sigma^2\right)^{-\frac{1}{4}} \exp\left(i\omega_0 t - \frac{t^2}{2\sigma^2}\right), \quad \sigma > 0; \tag{5.2.45}$$

(b) $$f(t) = \left(\pi\sigma^2\right)^{-\frac{1}{4}} \exp\left[i\omega_0 t - \frac{1}{2\sigma^2}(t - t_0)^2\right]. \tag{5.2.46}$$

(a) It follows readily from the definition that

$$W_f(t,\omega) = 2\exp\left[-\left\{\frac{t^2}{\sigma^2} + \sigma^2(\omega - \omega_0)^2\right\}\right]. \tag{5.2.47}$$

(b) It is easy to check that

$$W_f(t,\omega) = 2\exp\left[-\left\{\frac{1}{\sigma^2}(t - t_0)^2 + \sigma^2(\omega - \omega_0)^2\right\}\right]. \tag{5.2.48}$$

This shows that the Wigner-Ville distribution of a Gaussian signal is also Gaussian in both time t and frequency ω with center at (t_0, ω_0).

Example 5.2.10 For a Gaussian beam $f(t)$, which is a Gaussian signal multiplied by a quadratic-phase signal, that is,

$$f(t) = \left(\pi\sigma^2\right)^{-\frac{1}{4}} \exp\left(\frac{1}{2} i a t^2 - \frac{t^2}{\sigma^2}\right). \tag{5.2.49}$$

The Wigner-Ville distribution of the Gaussian beam is given by

$$W_f(t,\omega) = 2\exp\left[-\left\{\frac{t^2}{\sigma^2} + \sigma^2(\omega - at)^2\right\}\right]. \tag{5.2.50}$$

This also follows from the definition (5.2.7) or from Example 5.2.5.

5.3 Basic Properties of the Wigner-Ville Distribution

(a) (*Nonlinearity*). The Wigner-Ville distribution is nonlinear. This means that the Wigner-Ville distribution of the sum of two signals is not simply the sum of the Wigner-Ville distributions of the signals. It readily follows from the definition that

$$W_{f_1+f_2,g_1+g_2}(t,\omega) = W_{f_1,g_1}(t,\omega) + W_{f_1,g_2}(t,\omega) + W_{f_2,g_1}(t,\omega) + W_{f_2,g_2}(t,\omega). \quad (5.3.1)$$

In particular,

$$W_{af+bg}(t,\omega) = |a|^2 W_f(t,\omega) + a\bar{b}\, W_{f,g}(t,\omega) + \bar{a}\, b\, W_{g,f}(t,\omega) + |b|^2 W_g(t,\omega), \quad (5.3.2)$$

where a and b are two constants, and

$$W_{f+g}(t,\omega) = W_f(t,\omega) + W_g(t,\omega) + 2\,\mathrm{Re}\, W_{f,g}(t,\omega). \quad (5.3.3)$$

To prove (5.3.2), we write

$$W_{af+bg}(t,\omega) = \int_{-\infty}^{\infty} \left[a\, f\left(t+\frac{\tau}{2}\right) + b\, g\left(t+\frac{\tau}{2}\right)\right]\left[\bar{a}\,\bar{f}\left(t-\frac{\tau}{2}\right) + \bar{b}\,\bar{g}\left(t-\frac{\tau}{2}\right)\right] e^{-i\omega\tau} d\tau$$

$$= |a|^2 W_f(t,\omega) + a\bar{b}\, W_{f,g}(t,\omega) + \bar{a}\, b\, W_{g,f}(t,\omega) + |b|^2 W_g(t,\omega).$$

(b) (*Translation*). $W_{T_a f, T_a g}(t,\omega) = W_{f,g}(t-a,\omega),$ (5.3.4)

In particular,

$$W_{T_a f}(t,\omega) = W_f(t-a,\omega). \quad (5.3.5)$$

This means that the time shift of signals corresponds to a time shift of the Wigner-Ville distribution.

Proof. We have, by definition,

$$W_{T_a f, T_a g}(t,\omega) = \int_{-\infty}^{\infty} f\left(t-a+\frac{\tau}{2}\right) \bar{g}\left(t-a-\frac{\tau}{2}\right) e^{-i\omega\tau}\, d\tau$$

$$= W_{f,g}(t-a,u).$$

(c) (*Complex Conjugation*).

$$\overline{W}_{f,g}(t,\omega) = W_{g,f}(t,\omega). \quad (5.3.6)$$

From this hermiticity property of the Wigner-Ville distribution, it follows that the auto WVD is a real-valued even function for complex signals and continuous in both variables t and ω, which can be represented graphically as a surface over the time-frequency plane.

We have, by definition,

$$\overline{W}_{f,g}(t,\omega)=\int_{-\infty}^{\infty}\bar{f}\left(t+\frac{\tau}{2}\right)g\left(t-\frac{\tau}{2}\right)e^{i\omega\tau}\,d\tau$$

$$=\int_{-\infty}^{\infty}g\left(t+\frac{x}{2}\right)\bar{f}\left(t-\frac{x}{2}\right)e^{-i\omega x}\,dx=W_{g,f}(t,\omega).$$

(d) (*Modulation*).

$$W_{M_b f, M_b g}(t,\omega)=W_{f,g}(t,\omega-b), \tag{5.3.7}$$

$$W_{M_b f, g}(t,\omega)=e^{ibt}\,W_{f,g}\left(t,\omega-\frac{b}{2}\right), \tag{5.3.8}$$

$$W_{f, M_b g}(t,\omega)=e^{-ibt}\,W_{f,g}\left(t,\omega-\frac{b}{2}\right). \tag{5.3.9}$$

In particular,

$$W_{M_b f}(t,\omega)=W_f(t,\omega-b). \tag{5.3.10}$$

We have, by definition,

$$W_{M_b f, M_b g}(t,\omega)=\int_{-\infty}^{\infty}\exp\left\{ib\left(t+\frac{\tau}{2}\right)\right\}f\left(t+\frac{\tau}{2}\right)\exp\left\{-ib\left(t-\frac{\tau}{2}\right)\right\}\bar{g}\left(t-\frac{\tau}{2}\right)\times$$
$$\times e^{-i\omega\tau}\,d\tau$$

$$=\int_{-\infty}^{\infty}f\left(t+\frac{\tau}{2}\right)\bar{g}\left(t-\frac{\tau}{2}\right)\exp\left[-i\tau(\omega-b)\right]d\tau=W_{f,g}(t,\omega-b).$$

Similarly, we obtain (5.3.8) and (5.3.9).

(e) (*Translation and Modulation*).

$$W_{M_b T_a f, M_b T_a g}(t,\omega)=W_{T_a M_b f, T_a M_b g}(t,\omega)=W_{f,g}(t-a,\omega-b). \tag{5.3.11}$$

This follows from the joint application of (5.3.4) and (5.3.7).

In particular,

$$W_{M_b T_a f}(t,\omega)=W_{T_a M_b f}(t,\omega)=W_{f,g}(t-a,\omega-b). \tag{5.3.12}$$

Proof. Set $u(t) = M_b T_a f = e^{ibt} f(t-a)$, $v(t) = M_b T_a g = e^{ibt} g(t-a)$. Thus,

$$W_{u,v}(t,\omega) = \int_{-\infty}^{\infty} u\left(t+\frac{\tau}{2}\right) \bar{v}\left(t-\frac{\tau}{2}\right) e^{-i\omega t}\, dt$$

$$= \int_{-\infty}^{\infty} e^{ib\left(t+\frac{\tau}{2}\right)} f\left(t-a+\frac{\tau}{2}\right) e^{-ib\left(t-\frac{\tau}{2}\right)} \bar{g}\left(t-a-\frac{\tau}{2}\right) e^{-i\omega\tau}\, d\tau$$

$$= W_{f,g}(t-a,\omega-b).$$

This completes the proof.

(f) The Wigner-Ville distribution of the convolution of two signals is the convolution in time of their corresponding Wigner-Ville distributions. More precisely, for any two signals f and g, the following result holds:

$$W_{f*g}(t,\omega) = \int_{-\infty}^{\infty} W_f(u,\omega)\, W_g(t-u,\omega)\, du, \tag{5.3.13}$$

where $(f*g)(t)$ is the convolution of f and g.

Proof. We have, by definition,

$$W_{f*g}(t,\omega) = \int_{-\infty}^{\infty} (f*g)\left(t+\frac{\tau}{2}\right) \overline{(f*g)}\left(t-\frac{\tau}{2}\right) e^{-i\omega\tau}\, d\tau$$

$$= \int_{-\infty}^{\infty} \left[\int_{-\infty}^{\infty} f(x)\, g\left(t+\frac{\tau}{2}-x\right) dx\right]\left[\int_{-\infty}^{\infty} \bar{f}(y)\, \bar{g}\left(t-\frac{\tau}{2}-y\right) dy\right] e^{-i\omega\tau}\, d\tau$$

which is, by putting, $x = u + \frac{p}{2}$, $y = u - \frac{p}{2}$, $\tau = p + q$,

$$= \int_{-\infty}^{\infty} \left[\int_{-\infty}^{\infty}\int_{-\infty}^{\infty} f\left(u+\frac{p}{2}\right) \bar{f}\left(u-\frac{p}{2}\right) g\left(t-u+\frac{q}{2}\right) \bar{g}\left(t-u-\frac{q}{2}\right) dp\, dq\right]$$
$$\times \exp\left[-i(p+q)\omega\right] du$$

$$= \int_{-\infty}^{\infty} W_f(u,\omega)\, W_g(t-u,\omega)\, du.$$

This completes the proof.

(g) (*General Modulation*). The Wigner-Ville distribution of the modulated signal $f(t)m(t)$ is the convolution of $W_f(t,\omega)$ and $W_m(t,\omega)$ in the frequency variable, that is,

$$W_{fm}(t,\omega)=\frac{1}{2\pi}\int_{-\infty}^{\infty}W_f(t,u)\,W_m(t,\omega-u)\,du. \tag{5.3.14}$$

Proof. We have

$$\begin{aligned}
W_{fm}(t,\omega) &= \int_{-\infty}^{\infty} f\left(t+\frac{x}{2}\right) m\left(t+\frac{x}{2}\right) \bar{f}\left(t-\frac{x}{2}\right) \overline{m}\left(t-\frac{x}{2}\right) e^{-i\omega x}\,dx \\
&= \int_{-\infty}^{\infty} f\left(t+\frac{x}{2}\right) \bar{f}\left(t-\frac{x}{2}\right) dx \int_{-\infty}^{\infty} m\left(t+\frac{y}{2}\right) \overline{m}\left(t-\frac{y}{2}\right) e^{-i\omega y}\,\delta(y-x)\,dy \\
&= \frac{1}{2\pi}\int_{-\infty}^{\infty} du \left[\int_{-\infty}^{\infty} f\left(t+\frac{x}{2}\right) \bar{f}\left(t-\frac{x}{2}\right) e^{-ixu}\,dx \right. \\
&\qquad \left. \times \int_{-\infty}^{\infty} m\left(t+\frac{y}{2}\right) \overline{m}\left(t-\frac{y}{2}\right) e^{-i(\omega-u)y}\,dy\right] \\
&= \frac{1}{2\pi}\int_{-\infty}^{\infty} W_f(t,u)\,W_m(t,\omega-u)\,du.
\end{aligned}$$

(h) (*The Pseudo Wigner-Ville Distribution*). We consider a family of signals f_t and g_t defined by

$$f_t(\tau)=f(\tau)\,w_f(\tau-t),\qquad g_t(\tau)=g(\tau)\,w_g(\tau-t),$$

where w_f and w_g are called the *window functions*.

For a fixed t, we can evaluate the Wigner-Ville distribution of f_t and g_t so that, by (5.3.14),

$$W_{f_t,g_t}(\tau,\omega)=\frac{1}{2\pi}\int_{-\infty}^{\infty}W_{f,g}(\tau,u)\,W_{w_f,w_g}(\tau-t,\omega-u)\,du, \tag{5.3.15}$$

where t represents the position of the window as it moves along the time axis. Obviously, (5.3.15) is a family of the WVD, and a particular member of this family is obtained by putting $\tau=t$ so that

$$W_{f_t,g_t}(t,\omega)=\frac{1}{2\pi}\int_{-\infty}^{\infty}W_{f,g}(t,u)\,W_{w_f,w_g}(0,\omega-u)\,du. \tag{5.3.16}$$

We next define a *pseudo Wigner-Ville distribution* (PWVD) of f and g by (5.3.16) and write

$$PW_{f,g}(t,\omega)=\left[W_{f_t,g_t}(\tau,\omega)\right]_{\tau=t}. \tag{5.3.17}$$

This is similar to the Wigner-Ville distribution, but, in general, is not a Wigner-Ville distribution. Even though the notation does not indicate explicit dependence on the window functions, the PWVD of two functions actually depends on the window functions. It follows from (5.3.16) that

$$PW_{f,g}(t,\omega) = \frac{1}{2\pi} W_{f,g}(t,\omega) * W_{w_f,w_g}(t,\omega), \tag{5.3.18}$$

where the convolution is taken with respect to the frequency variable ω. In particular,

$$PW_f(t,\omega) = \frac{1}{2\pi} W_f(t,\omega) * W_{w_f}(t,\omega) = \frac{1}{2\pi} \int_{-\infty}^{\infty} W(t,u)\, W_{w_f}(t,\omega - u)\, du. \tag{5.3.19}$$

This can be interpreted that the pseudo Wigner-Ville distribution of a signal is a smoothed version of the original WVD with respect to the frequency variable.

(i) (*Dilation*). If $D_c f(t) = \frac{1}{\sqrt{|c|}} f\left(\frac{t}{c}\right)$, $c \neq 0$, then

$$W_{D_c f, D_c g}(t,\omega) = W_{f,g}\left(\frac{t}{c}, c\omega\right). \tag{5.3.20}$$

In particular,

$$W_{D_c f}(t,\omega) = W_f\left(\frac{t}{c}, c\omega\right). \tag{5.3.21}$$

Proof. We have, by definition,

$$\begin{aligned} W_{D_c f, D_c g}(t,\omega) &= \frac{1}{|c|} \int_{-\infty}^{\infty} f\left(\frac{t}{c} + \frac{\tau}{2c}\right) \bar{g}\left(\frac{t}{c} - \frac{\tau}{2c}\right) e^{-i\omega\tau}\, d\tau \\ &= \int_{-\infty}^{\infty} f\left(\frac{t}{c} + \frac{x}{2}\right) \bar{g}\left(\frac{t}{c} - \frac{x}{2}\right) e^{-i(c\omega)x}\, dx \\ &= W_{f,g}\left(\frac{t}{c}, c\omega\right). \end{aligned}$$

(j) (*Multiplication*). If $M f(t) = t\, f(t)$, then

$$2t\, W_{f,g}(t,\omega) = W_{Mf,g}(t,\omega) + W_{f,Mg}(t,\omega). \tag{5.3.22}$$

Proof. We have, by definition,

$$2t\, W_{f,g}(t,\omega) = \int_{-\infty}^{\infty} \left(t + \frac{\tau}{2} + t - \frac{\tau}{2}\right) f\left(t + \frac{\tau}{2}\right) \bar{g}\left(t - \frac{\tau}{2}\right) e^{-i\omega\tau}\, d\tau$$

$$= W_{Mf,g}(t,\omega) + W_{f,Mg}(t,\omega).$$

(k) (*Differentiation*).

$$W_{Df,g}(t,\omega) + W_{f,Dg}(t,\omega) = 2i\omega\, W_{f,g}(t,\omega). \tag{5.3.23}$$

In particular,

$$W_{Df,f}(t,\omega) + W_{f,Df}(t,\omega) = 2i\omega\, W_f(t,\omega). \tag{5.3.24}$$

Proof. We apply the Fourier transform of the left-hand side of (5.3.24) with respect to t to obtain

$$\mathcal{F}\left\{W_{Df,g}(t,\omega)\right\} + \mathcal{F}\left\{W_{f,Dg}(t,\omega)\right\}$$

$$= i\left(\omega + \frac{\tau}{2}\right) \hat{f}\left(\omega + \frac{\tau}{2}\right) \bar{\hat{g}}\left(\omega - \frac{\tau}{2}\right) + i\left(\omega - \frac{\tau}{2}\right) \hat{f}\left(\omega + \frac{\tau}{2}\right) \bar{\hat{g}}\left(\omega - \frac{\tau}{2}\right),$$

by (5.2.8)

$$= 2i\omega\, \hat{f}\left(\omega + \frac{\tau}{2}\right) \bar{\hat{g}}\left(\omega - \frac{\tau}{2}\right)$$

$$= 2i\omega\, \mathcal{F}\left\{W_{f,g}(t,\omega)\right\} = \mathcal{F}\left\{2i\omega\, W_{f,g}(t,\omega)\right\}.$$

Application of the inverse Fourier transform completes the proof of (5.3.23).

(l) (*Time and Frequency Moments*).

$$\frac{1}{2\pi} \int_{-\infty}^{\infty} \int_{-\infty}^{\infty} t^n\, W_{f,g}(t,\omega)\, dt\, d\omega = \int_{-\infty}^{\infty} t^n\, f(t)\, \bar{g}(t)\, dt, \tag{5.3.25}$$

$$\int_{-\infty}^{\infty} \int_{-\infty}^{\infty} \omega^n\, W_{f,g}(t,\omega)\, dt\, d\omega = \int_{-\infty}^{\infty} \omega^n\, \hat{f}(\omega)\, \bar{\hat{g}}(\omega)\, d\omega. \tag{5.3.26}$$

In particular,

$$\frac{1}{2\pi} \int_{-\infty}^{\infty} \int_{-\infty}^{\infty} t^n\, W_f(t,\omega)\, dt\, d\omega = \int_{-\infty}^{\infty} t^n\, |f(t)|^2\, dt, \tag{5.3.27}$$

$$\int_{-\infty}^{\infty} \int_{-\infty}^{\infty} \omega^n\, W_f(t,\omega)\, dt\, d\omega = \int_{-\infty}^{\infty} \omega^n\, |\hat{f}(\omega)|^2\, d\omega. \tag{5.3.28}$$

Proof. We have, by definition,

$$\frac{1}{2\pi}\int_{-\infty}^{\infty}\int_{-\infty}^{\infty} t^n\, W_{f,g}(t,\omega)\, dt\, d\omega$$

$$= \frac{1}{2\pi}\int_{-\infty}^{\infty}\int_{-\infty}^{\infty} t^n\, dt\, d\omega \int_{-\infty}^{\infty} f\left(t+\frac{\tau}{2}\right)\bar{g}\left(t-\frac{\tau}{2}\right) e^{-i\omega\tau}\, d\tau$$

$$= \int_{-\infty}^{\infty} t^n\, dt \int_{-\infty}^{\infty} f\left(t+\frac{\tau}{2}\right)\bar{g}\left(t-\frac{\tau}{2}\right)\delta(\tau)\, d\tau$$

$$= \int_{-\infty}^{\infty} t^n\, f(t)\,\bar{g}(t)\, dt.$$

Similarly, from (5.2.20), we obtain

$$\int_{-\infty}^{\infty}\int_{-\infty}^{\infty} \omega^n\, W_{f,g}(t,\omega)\, dt\, d\omega = \int_{-\infty}^{\infty} \omega^n\, \hat{f}(\omega)\,\bar{\hat{g}}(\omega)\, d\omega.$$

Theorem 5.3.1 (Moyal's Formulas). If f_1, g_1, f_2 and g_2 belong to $L^2(\mathbb{R})$, then the following Moyal's formulas hold:

$$\frac{1}{2\pi}\int_{-\infty}^{\infty}\int_{-\infty}^{\infty} W_{f_1,g_1}(t,\omega)\,\overline{W}_{f_2,g_2}(t,\omega)\, dt\, d\omega = (f_1, f_2)\overline{(g_1, g_2)}, \tag{5.3.29}$$

$$\frac{1}{2\pi}\int_{-\infty}^{\infty}\int_{-\infty}^{\infty} \left|W_{f,g}(t,\omega)\right|^2\, dt\, d\omega = \|f\|^2\,\|g\|^2, \tag{5.3.30}$$

$$\frac{1}{2\pi}\int_{-\infty}^{\infty}\int_{-\infty}^{\infty} W_f(t,\omega)\,\overline{W}_g(t,\omega)\, dt\, d\omega = (f,g)\overline{(f,g)} = \left|(f,g)\right|^2. \tag{5.3.31}$$

Proof. It is clear from (5.2.6) that, for fixed t, the Fourier transform of $W_{f,g}(t,\omega)$ with respect to ω is

$$\hat{W}_{f,g}(t,\sigma) = 2\pi\, f\left(t-\frac{\sigma}{2}\right)\bar{g}\left(t+\frac{\sigma}{2}\right). \tag{5.3.32}$$

Thus, it follows from the Parseval formula for the Fourier transform that

$$\frac{1}{2\pi}\int_{-\infty}^{\infty} W_{f_1,g_1}(t,\omega)\,\overline{W}_{f_2,g_2}(t,\omega)\,d\omega$$

$$=\frac{1}{(2\pi)^2}\int_{-\infty}^{\infty} \hat{W}_{f_1,g_1}(t,\sigma)\,\overline{\hat{W}}_{f_2,g_2}(t,\sigma)\,d\sigma$$

$$=\int_{-\infty}^{\infty} f_1\left(t-\frac{\sigma}{2}\right)\bar{g}_1\left(t+\frac{\sigma}{2}\right)\bar{f}_2\left(t-\frac{\sigma}{2}\right)g_2\left(t+\frac{\sigma}{2}\right)d\sigma, \quad \text{by (5.3.32).}$$

Integrating both sides with respect to t over $\mathbb{R}$ gives

$$\frac{1}{2\pi}\int_{-\infty}^{\infty}\int_{-\infty}^{\infty} W_{f_1,g_1}(t,\omega)\,\overline{W}_{f_2,g_2}(t,\omega)\,d\omega\,dt$$

$$=\int_{-\infty}^{\infty}\int_{-\infty}^{\infty} f_1\left(t-\frac{\sigma}{2}\right)\bar{g}_1\left(t+\frac{\sigma}{2}\right)\bar{f}_2\left(t-\frac{\sigma}{2}\right)g_2\left(t+\frac{\sigma}{2}\right)d\sigma\,dt$$

which is, putting $t-\frac{\sigma}{2}=x$ and $t+\frac{\sigma}{2}=y$,

$$=\int_{-\infty}^{\infty} f_1(x)\,\bar{f}_2(x)\,dx\;\overline{\int_{-\infty}^{\infty} g_1(y)\,\bar{g}_2(y)\,dy}$$

$$=(f_1,f_2)\overline{(g_1,g_2)}.$$

This completes the proof of (5.3.29).

In particular, if $f_1=f_2=f$ and $g_1=g_2=g$, then (5.3.29) reduces to (5.3.30).

However, we give another proof of (5.3.31) as follows:

We use the definition (5.2.1) to replace $W_f(t,\omega)$ and $W_g(t,\omega)$ on the left hand side of (5.3.31) so that

$$\frac{1}{2\pi}\int_{-\infty}^{\infty}\int_{-\infty}^{\infty} W_f(t,\omega)\,\overline{W}_g(t,\omega)\,dt\,d\omega$$

$$=\frac{1}{2\pi}\int_{-\infty}^{\infty}\int_{-\infty}^{\infty}\int_{-\infty}^{\infty}\int_{-\infty}^{\infty} f\left(t+\frac{r}{2}\right)\bar{f}\left(t-\frac{r}{2}\right)\bar{g}\left(t+\frac{s}{2}\right)g\left(t-\frac{s}{2}\right)$$

$$\times\exp\left[i(s-r)\omega\right]dr\,ds\,dt\,d\omega$$

which is, by replacing the ω-integral with the delta function,

$$= \int_{-\infty}^{\infty}\int_{-\infty}^{\infty}\int_{-\infty}^{\infty} f\left(t+\frac{r}{2}\right) \bar{f}\left(t-\frac{r}{2}\right) \bar{g}\left(t+\frac{s}{2}\right) g\left(t-\frac{s}{2}\right) \delta(s-r)\, dr\, ds\, dt$$

$$= \int_{-\infty}^{\infty}\int_{-\infty}^{\infty} f\left(t+\frac{r}{2}\right) \bar{f}\left(t-\frac{r}{2}\right) \bar{g}\left(t+\frac{r}{2}\right) g\left(t-\frac{r}{2}\right) dr\, dt$$

which is, due to change of variables $t+\frac{r}{2}=x$ and $t-\frac{r}{2}=y$,

$$= \int_{-\infty}^{\infty} f(x)\, \bar{g}(x)\, dx \; \overline{\int_{-\infty}^{\infty} f(y)\, \bar{g}(y)\, dy}$$

$$= (f,g)\,\overline{(f,g)} = |(f,g)|^2 .$$

Theorem 5.3.2 (Convolution with Respect to Both Variables). If two signals f and g belong to $L^2(\mathbb{R})$, then

$$\left(W_f * W_g\right)(a,b) = \int_{-\infty}^{\infty}\int_{-\infty}^{\infty} W_f(t,\omega)\, W_g(a-t, b-\omega)\, dt\, d\omega$$

$$= 2\pi \left|\left(T_a M_b h, f\right)\right|^2 = 2\pi \left|A_{f,h}(a,b)\right|^2, \tag{5.3.33}$$

where $h(x)=g(-x)$ and $A_{f,h}$ is the cross ambiguity function defined in Section 5.5 by (5.5.1).

If g is even, then

$$\left(W_f * W_g\right)(a,b) = 2\pi \left|\left(T_a M_b g, f\right)\right|^2 = 2\pi \left|A_{f,g}(a,b)\right|^2 . \tag{5.3.34}$$

Proof. It readily follows from the convolution theorem of the Fourier transform that

$$\int_{-\infty}^{\infty} W_f(t,\omega)\, W_g(a-t,b-\omega)\, d\omega = \frac{1}{2\pi} \int_{-\infty}^{\infty} \hat{W}_f(t,\sigma)\, \hat{W}_g(a-t,\sigma)\, e^{ib\sigma}\, d\sigma$$

which is, by (5.3.32),

$$= 2\pi \int_{-\infty}^{\infty} f\left(t-\frac{\sigma}{2}\right) \bar{f}\left(t+\frac{\sigma}{2}\right) g\left(a-t-\frac{\sigma}{2}\right) \bar{g}\left(a-t-\frac{\sigma}{2}\right) \exp(ib\sigma)\, d\sigma$$

which is, due to the change of variables $t+\frac{\sigma}{2}=x$,

$$= 4\pi \int_{-\infty}^{\infty} \bar{f}(x)\, f(2t - x)\, g(a - x)\, \bar{g}(a + x - 2t)\, \exp\left[2ib(x - t)\right] dx.$$

Integrating this result with respect to t yields

$$\int_{-\infty}^{\infty}\int_{-\infty}^{\infty} W_f(t,\omega)\, W_g(a - t, b - \omega)\, d\omega\, dt = 4\pi \int_{-\infty}^{\infty} \bar{f}(x)\, g(a - x)\, \exp(2ibx)\, dx$$

$$\times \int_{-\infty}^{\infty} f(2t - x)\, \bar{g}(a + x - 2t)\, \exp(-2ibt)\, dt$$

which is, by substitution of $2t - x = u$,

$$= 2\pi \int_{-\infty}^{\infty} \bar{f}(x)\, g(a - x)\, \exp(ibx)\, dx \int_{-\infty}^{\infty} f(u)\, \bar{g}(a - u)\, \exp(-ibu)\, du.$$

This leads to (5.3.33) and hence to (5.3.34).

5.4 The Wigner-Ville Distribution of Analytic Signals and Band-Limited Signals

Gabor first used the Hilbert transform to define a complex signal $f(t)$ of time t by

$$f(t) = u(t) + iv(t) = u(t) + i(Hu)(t), \tag{5.4.1}$$

where $v(t)$ is the Hilbert transform of $u(t)$ defined by

$$(Hu)(t) = \frac{1}{\pi} \int_{-\infty}^{\infty} \frac{u(x)\, dx}{(x - t)}, \tag{5.4.2}$$

and the integral is treated as the Cauchy principle value. This signal $f(t)$ is called an *analytic signal* or the *analytic part of the signal* $u(t)$. The imaginary part of $f(t)$ is called the *quadrature* function of $u(t)$. In electrical systems, the output, $v(t) = (Hu)(t)$, for a given input $u(t)$ is known as the *quadrature filter*. Obviously, the quadrature filter of $\sin\omega t$ is $\cos\omega t$ and that of $\chi_{a,b}(t)$ is $\pi^{-1}\ell n\left|(b - t)(a - t)^{-1}\right|$ (see Debnath, 1995).

For an analytic signal in the form

$$f(t) = a(t)\exp\{i\theta(t)\},$$

the instantaneous frequency (IF), $f_i(t)$ is defined by

$$f_i(t) = \frac{1}{2\pi}\frac{d}{dt}\theta(t). \tag{5.4.3}$$

The complex spectrum $\hat{f}(\omega)$ of $f(t)$ can be expressed in the form

$$\hat{f}(\omega) = \hat{a}(\omega)\, e^{i\hat{\theta}(\omega)}, \tag{5.4.4}$$

where $a(t)$ and $\hat{a}(\omega)$ are positive functions.

Another quantity of interest in the time-frequency analysis of a signal is the group delay (GD) of a signal defined by

$$\tau_g(\omega) = -\frac{1}{2\pi}\frac{d}{d\omega}\hat{\theta}(\omega). \tag{5.4.5}$$

In many applications, the group delay is used to characterize the time-frequency law of a signal. Therefore, it is natural to relate the two quantities IF and GD. In order to achieve this, the Fourier transform of the signal is used. For signals of the form $f(t) = a(t)\exp\{i\theta(t)\}$ with a large BT product, where B is a finite bandwidth and T is a finite duration of a signal, and a monotonic instantaneous frequency law, the Fourier transform can be approximated by the stationary phase approximation method (see Myint-U and Debnath, 1987) as follows:

$$\mathscr{F}\left\{a(t)\, e^{i\theta(t)}\right\} = \int_{-\infty}^{\infty} a(t)\exp\left[i\{\theta(t) - \omega t\}\right] dt$$

$$\sim \left\{\frac{2\pi}{|\theta''(\sigma)|}\right\}^{\frac{1}{2}} a(\sigma)\exp\left[i\left\{\theta(\sigma) - \omega\sigma \pm \frac{\pi}{4}\right\}\right], \tag{5.4.6}$$

where σ is a stationary point given by the roots of the equation

$$\frac{d}{dt}\theta(t) = \omega.$$

For signals for large BT, the IF and GD are approximately inverses of each other, that is,

$$f_i(t) = \frac{1}{\tau_g(\omega)}. \tag{5.4.7}$$

Ville utilized the same definition (5.2.7) for an analytic signal $f(t)$ and then introduced the instantaneous frequency $f_i(t)$ by

$$f_i(t) = \int_{-\infty}^{\infty} \omega\, W(t,\omega)\, d\omega \div \int_{-\infty}^{\infty} W(t,\omega)\, d\omega. \tag{5.4.8}$$

Before we compute the Wigner-Ville distribution of analytic signals, it should be noted that the analytic signal is almost universally used in the time-frequency signal analysis. Using the analytic form eliminates cross-terms between positive and negative frequency components of the signal. However, for certain low-frequency signals, there may be undue smoothing of the low-frequency time components of the time-frequency representation due to the frequency domain window implied by using the analytic signals. In that case, the original real signal should be more appropriate.

The Fourier spectrum of the analytic signal f is given by

$$\hat{f}(\omega) = \begin{Bmatrix} 2\hat{u}(\omega), & \omega > 0 \\ \hat{u}(0), & \omega = 0 \\ 0, & \omega < 0 \end{Bmatrix}. \tag{5.4.9}$$

For band limited signal f, $\hat{f}(\omega) = 0$ for $|\omega| > \omega_0$. Then, it follows from (5.2.18) that

$$W_f(t, \omega) = 0 \qquad \text{for } |\omega| > \omega_0 \text{ and all } t.$$

This result is also true for an analytic signal f and hence,

$$W_f(t, \omega) = 0 \qquad \text{for } \omega < 0.$$

The relation between $W_f(t, \omega)$ and $W_u(t, \omega)$ can be determined by using (5.2.18) and (5.3.6). It follows from definition (5.2.17) that

$$W_{\hat{f}}(\omega, t) = \frac{1}{2\pi} \int_{-\infty}^{\infty} \hat{f}\left(\omega + \frac{\tau}{2}\right) \bar{\hat{f}}\left(\omega - \frac{\tau}{2}\right) e^{it\tau} d\tau$$

which is, by (5.4.9),

$$= \frac{2}{\pi} \int_{-2\omega}^{2\omega} \hat{u}\left(\omega + \frac{\tau}{2}\right) \bar{\hat{u}}\left(\omega - \frac{\tau}{2}\right) e^{it\tau} d\tau, \qquad \omega > 0.$$

It is also clear from the definition (5.2.17) that the Fourier transform of $W_{\hat{f}}(\omega, t)$ with respect to t is $\hat{f}\left(\omega + \frac{\tau}{2}\right) \bar{\hat{f}}\left(\omega - \frac{\tau}{2}\right)$. In view of this result, (5.2.18) takes the form

$$\begin{aligned} W_{\hat{f}}(\omega, t) &= \frac{2}{\pi} \int_{-2\omega}^{2\omega} e^{it\tau} d\tau \int_{-\infty}^{\infty} e^{-i\tau x}\, W_{\hat{u}}(\omega, x)\, dx \\ &= \frac{2}{\pi} \int_{-\infty}^{\infty} W_{\hat{u}}(\omega, x)\, dx \int_{-2\omega}^{2\omega} e^{i\tau(t-x)}\, d\tau \\ &= \frac{4}{\pi} \int_{-\infty}^{\infty} W_{\hat{u}}(\omega, t - \xi)\, \frac{\sin 2\omega\xi}{\xi}\, d\xi, \qquad \omega > 0. \end{aligned} \tag{5.4.10}$$

Similarly, we obtain

$$W_f(t, \omega) = \frac{4}{\pi} H(\omega) \int_{-\infty}^{\infty} W_u(t - \xi, \omega)\, \frac{\sin 2\omega\xi}{\xi}\, d\xi. \tag{5.4.11}$$

Thus, (5.4.11) shows that the Wigner-Ville distribution of an analytic signal f

exists only for positive frequencies, and it has no contributions for any negative frequency.

Finally, we calculate the Wigner-Ville distribution of band-limited functions. If $f(t)$ and $g(t)$ are signals band limited to $[-\omega_0, \omega_0]$, that is, their Fourier transforms $\hat{f}(\omega)$ and $\hat{g}(\omega)$ vanish for $|\omega| > \omega_0$. Consequently, their Wigner-Ville distribution is also band-limited in ω, that is,

$$W_{\hat{f},\hat{g}}(\omega, t) = 0 \qquad \text{for } |\omega| > \omega_0. \tag{5.4.12}$$

Using (5.2.18), it turns out that

$$W_{f,g}(t, \omega) = 0 \qquad \text{for } |\omega| > \omega_0 \text{ and all } t. \tag{5.4.13}$$

Under these conditions, the Shannon sampling formula (3.6.19) asserts that

$$f(t) = \sum_{n=-\infty}^{\infty} f(t_n) \frac{\sin \omega_0 (t - t_n)}{\omega_0 (t - t_n)}$$

and

$$g(-t) = \sum_{m=-\infty}^{\infty} g(-t_m) \frac{\sin \omega_0 (t - t_m)}{\omega_0 (t - t_m)},$$

when $t_s = \dfrac{\pi s}{\omega_0}$, $s = n$ or m.

We multiply the above two series together and use the fact that the sequence

$$\left\{ \frac{\sin \pi(t - n)}{\pi(t - n)} \right\}_{n=-\infty}^{\infty}$$

is orthonormal in $(-\infty, \infty)$ so that $W_{f,g}(0, 0)$ in (5.2.25) can be expressed as

$$W_{f,g}(0, 0) = 2 \int_{-\infty}^{\infty} f(x)\, \bar{g}(-x)\, dx$$

$$= \left(\frac{2\pi}{\omega_0} \right) \sum_{n=-\infty}^{\infty} f(t_n)\, g(-t_n). \tag{5.4.14}$$

This confirms that the Wigner-Ville distribution of two band-limited signals can be expressed in terms of their samples taken at $t_n = \dfrac{n\pi}{\omega_0}$.

5.5 Definitions and Examples of the Woodward Ambiguity Functions

During the 1950s, the theory of the Wigner-Ville distribution was reformulated in the context of sonar and radar signal analysis, where the echo from a transmitted signal is used to find the position and velocity of a target. A new function, the so-called ambiguity function, was introduced by Woodward (1953) for the mathematical analysis of sonar and radar signals.

Definition 5.5.1 (The Cross Ambiguity Function). The cross ambiguity function of two signals $f, g \in L^2(\mathbb{R})$ is denoted by $A_{f,g}(t,\omega)$ and defined by

$$A_{f,g}(t,\omega) = \int_{-\infty}^{\infty} f\left(\tau + \frac{t}{2}\right) \bar{g}\left(\tau - \frac{t}{2}\right) e^{-i\omega\tau} \, d\tau. \tag{5.5.1}$$

In radar technology, the ambiguity function is interpreted as a time-frequency correlation of signals f and g.

However, if $f, g \in L^1(\mathbb{R})$, then $A_{f,g}(t,\omega)$ exists for all $t, \omega \in \mathbb{R}$. For a fixed ω, we set

$$F(\tau,t) = f\left(\tau + \frac{t}{2}\right) \bar{g}\left(\tau - \frac{t}{2}\right) \exp(-i\omega\tau).$$

Then, it follows from the translation invariant property of the Lebesgue measure that

$$\int_{-\infty}^{\infty} \left|A_{f,g}(t,\omega)\right| dt \le \int_{-\infty}^{\infty}\int_{-\infty}^{\infty} \left|F(\tau,t)\right| d\tau \, dt \le \|f\|_1 \|g\|_1 < \infty.$$

On the other hand, the existence of (5.5.1) follows from the Schwarz inequality. Putting $\tau + \frac{1}{2} = x$, definition (5.5.1) is equivalent to

$$A_{f,g}(t,\omega) = \exp\left(\frac{1}{2} i\omega t\right) \int_{-\infty}^{\infty} f(x) \, \bar{g}(x-t) \, e^{-i\omega x} \, dx \tag{5.5.2}$$

$$= \exp\left(\frac{1}{2} i\omega t\right) \tilde{f}_{\bar{g}}(t,\omega). \tag{5.5.3}$$

Result (5.5.3) shows that the cross ambiguity function is related to the Gabor transform of the function f with respect to the window function $\bar{g}$.

In the context of radar technology, the cross ambiguity function of two radar signals reflected by a moving target plays an important role, where t denotes the time delay and ω is the Doppler frequency shift.

The definition (5.5.1) also reveals that $A_{f,g}(t,\omega)$ is the Fourier transform of the function

$$k_t(\tau) = f\left(\tau + \frac{t}{2}\right) \bar{g}\left(\tau - \frac{t}{2}\right) \tag{5.5.4}$$

with respect to the variable τ, that is,

$$A_{f,g}(t,\omega)=\hat{k}_t(\omega). \tag{5.5.5}$$

It also follows from the definition (5.5.1) that

$$A_{f,g}(t,0)=\int_{-\infty}^{\infty} f\left(\tau+\frac{t}{2}\right)\bar{g}\left(\tau-\frac{t}{2}\right)d\tau, \quad \left(\tau-\frac{t}{2}=x\right)$$

$$=\int_{-\infty}^{\infty} f(x+t)\,\bar{g}(x)\,dx\equiv R_{f,g}(t), \tag{5.5.6}$$

where $R_{f,g}(t)$ is called the *cross-correlation* function of f and g. In particular, if $f=g$, then $R_f(t)\equiv A_f(t,0)$ is the autocorrelation function of f defined already by (3.4.9).

Definition 5.5.2 (Autoambiguity Function). If $f=g$ in (5.5.1), then $A_{f,f}(t,\omega)\equiv A_f(t,\omega)$ is called the *autoambiguity function* of f defined by

$$A_f(t,\omega)=\int_{-\infty}^{\infty} f\left(\tau+\frac{t}{2}\right)\bar{f}\left(\tau-\frac{t}{2}\right)e^{-i\omega\tau}\,d\tau, \tag{5.5.7}$$

$$=\exp\left(\frac{1}{2}\,i\omega t\right)\int_{-\infty}^{\infty} f(x)\,\bar{f}(x-t)\,e^{-i\omega x}\,dx. \tag{5.5.8}$$

Both cross ambiguity and autoambiguity functions are simply referred to as *ambiguity functions*.

It is easy to see that the cross Wigner-Ville distribution is closely related to the cross ambiguity function. Making a change of variable $t+\frac{\tau}{2}=x$ in (5.2.1) gives

$$W_{f,g}(t,\omega)=2\exp(2i\omega t)\int_{-\infty}^{\infty} f(x)\,\bar{g}(2t-x)\exp(-2i\omega x)\,dx$$

$$=2\,A_{f,h}(2t,2\omega), \tag{5.5.9}$$

where $h(x)=g(-x)$.

On the other hand, the Fourier transform of $A_{f,g}(t,\omega)$ with respect to ω is given by

$$\hat{A}_{f,g}(t,\sigma) = \int_{-\infty}^{\infty} e^{-i\omega\sigma} A_{f,g}(t,\omega)\, d\omega$$

$$= \int_{-\infty}^{\infty} e^{-i\omega\sigma}\, d\omega \int_{-\infty}^{\infty} k_t(\tau)\, e^{-i\omega\tau}\, d\tau$$

$$= \int_{-\infty}^{\infty} k_t(\tau)\, d\tau \int_{-\infty}^{\infty} e^{-i\omega(\tau+\sigma)}\, d\omega$$

$$= 2\pi\, k_t(-\sigma)$$

$$= 2\pi\, f\left(-\sigma + \frac{t}{2}\right) \bar{g}\left(-\sigma - \frac{t}{2}\right). \tag{5.5.10}$$

Or, equivalently,

$$\hat{A}_{f,g}(t, -\sigma) = 2\pi\, f\left(\sigma + \frac{t}{2}\right) \bar{g}\left(\sigma - \frac{t}{2}\right). \tag{5.5.11}$$

The double Fourier transform of $A_f(t,\omega)$ with respect to t and ω gives

$$\hat{\hat{A}}(\tau,\sigma) = \int_{-\infty}^{\infty}\int_{-\infty}^{\infty} \exp[-i(t\tau + \omega\sigma)]\, A(t,\omega)\, dt\, d\omega$$

$$= \int_{-\infty}^{\infty} e^{-it\tau}\, dt \int_{-\infty}^{\infty} e^{-i\omega\sigma} A(t,\omega)\, d\omega$$

$$= 2\pi \int_{-\infty}^{\infty} f\left(-\sigma + \frac{t}{2}\right) \bar{f}\left(-\sigma - \frac{t}{2}\right) e^{-it\tau}\, dt$$

$$= 2\pi\, W_f(-\sigma,\tau). \tag{5.5.12}$$

Or, equivalently,

$$\frac{1}{2\pi}\int_{-\infty}^{\infty}\int_{-\infty}^{\infty} \exp[-i(t\tau - \omega\sigma)]\, A(\tau,\sigma)\, d\tau\, d\sigma = W_f(\omega,t). \tag{5.5.13}$$

Similarly, the double Fourier transform of $W_f(t,\omega)$ with respect to t and ω is

$$\hat{\hat{W}}_f(\tau,\sigma) = \int_{-\infty}^{\infty}\int_{-\infty}^{\infty} \exp[-i(t\tau + \omega\sigma)]\, W_f(t,\omega)\, dt\, d\omega$$

$$= 2\pi \int_{-\infty}^{\infty} f\left(t - \frac{\sigma}{2}\right) \bar{f}\left(t + \frac{\sigma}{2}\right) e^{-it\tau}\, dt$$

$$= 2\pi\, A_f(-\sigma,\tau). \tag{5.5.14}$$

Or, equivalently,

$$\frac{1}{2\pi}\int_{-\infty}^{\infty}\int_{-\infty}^{\infty}\exp[-i(t\tau-\omega\sigma)]\, W_f(\tau,\sigma)\, d\tau\, d\sigma = A_f(\omega,t). \qquad (5.5.15)$$

Substituting $\frac{t}{2}-\sigma = t_1$ and $-\left(\frac{t}{2}+\sigma\right) = t_2$ in (5.5.10) gives the inversion formula

$$\frac{1}{2\pi}\int_{-\infty}^{\infty}\exp\left[\frac{1}{2}i(t_1+t_2)\omega\right]A_{f,g}(t_1-t_2,\omega)\, d\omega = f(t_1)\,\overline{g}(t_2). \qquad (5.5.16)$$

In particular, if $t_1 = t_2 = t$, we find

$$\frac{1}{2\pi}\int_{-\infty}^{\infty} e^{i\omega t}\, A_{f,g}(0,\omega)\, d\omega = f(t)\,\overline{g}(t), \qquad (5.5.17)$$

and if $f = g$, then

$$\frac{1}{2\pi}\int_{-\infty}^{\infty} e^{i\omega t}\, A_f(0,\omega)\, d\omega = |f(t)|^2. \qquad (5.5.18)$$

Integrating (5.5.17) and (5.5.18) with respect to t yields the following results:

$$\frac{1}{2\pi}\int_{-\infty}^{\infty}\int_{-\infty}^{\infty} e^{i\omega t}\, A_{f,g}(0,\omega)\, d\omega\, dt = \int_{-\infty}^{\infty} f(t)\,\overline{g}(t)\, dt = (f,g), \qquad (5.5.19)$$

$$\frac{1}{2\pi}\int_{-\infty}^{\infty}\int_{-\infty}^{\infty} e^{i\omega t}\, A_f(0,\omega)\, d\omega\, dt = \int_{-\infty}^{\infty} |f(t)|^2\, dt = \|f\|_2. \qquad (5.5.20)$$

Putting $t_1 = t$ and $t_2 = 0$ in (5.5.16) gives $f(t)$ in terms of $A_{f,g}(t,\omega)$ in the form

$$f(t)\,\overline{g}(0) = \frac{1}{2\pi}\int_{-\infty}^{\infty}\exp\left(\frac{1}{2}i\omega t\right)A_{f,g}(t,\omega)\, d\omega, \qquad (5.5.21)$$

provided $\overline{g}(0) \neq 0$. Result (5.5.21) is also called an *inversion formula*.

We can also define the ambiguity function of the Fourier spectrum $\hat{f}$ and $\hat{g}$ by

$$A_{\hat{f},\hat{g}}(t,\omega) = \int_{-\infty}^{\infty}\hat{f}\left(\tau+\frac{t}{2}\right)\,\overline{\hat{g}}\left(\tau-\frac{t}{2}\right)\, e^{-i\omega\tau}\, d\tau \qquad (5.5.22)$$

$$= \int_{-\infty}^{\infty} \bar{g}\left(\tau - \frac{t}{2}\right) e^{-i\omega\tau}\, d\tau \int_{-\infty}^{\infty} f(x) \exp\left\{-i\left(\tau + \frac{t}{2}\right)x\right\} dx$$

$$= \int_{-\infty}^{\infty} f(x)\, e^{-\frac{itx}{2}}\, dx \int_{-\infty}^{\infty} \bar{g}\left(\tau - \frac{t}{2}\right) e^{-i\tau(\omega+x)}\, d\tau$$

$$= \exp\left(-\frac{1}{2}\, i\omega t\right) \int_{-\infty}^{\infty} f(x)\, \bar{g}(x+\omega)\, e^{-itx}\, dx, \qquad \left(\tau - \frac{t}{2} = u\right)$$

$$= A_{f,g}(-\omega, t). \tag{5.5.23}$$

Or, equivalently,

$$A_{f,g}(t,\omega) = A_{\hat{f},\hat{g}}(\omega, -t) = \int_{-\infty}^{\infty} \hat{f}\left(\tau + \frac{\omega}{2}\right) \bar{\hat{g}}\left(\tau - \frac{\omega}{2}\right) e^{it\tau}\, d\tau. \tag{5.5.24}$$

In particular,

$$A_{f,g}(0,\omega) = \int_{-\infty}^{\infty} \hat{f}\left(\tau - \frac{\omega}{2}\right) \bar{\hat{g}}\left(\tau + \frac{\omega}{2}\right) d\tau. \tag{5.5.25}$$

It follows from (5.5.24) that

$$\frac{1}{2\pi} \int_{-\infty}^{\infty} A_{f,g}(t,\omega)\, e^{it\tau}\, dt = \hat{f}\left(\tau - \frac{\omega}{2}\right) \bar{\hat{g}}\left(\tau + \frac{\omega}{2}\right). \tag{5.5.26}$$

Putting $\tau - \frac{\omega}{2} = t_1$ and $\tau + \frac{\omega}{2} = t_2$ in (5.5.26) gives

$$\frac{1}{2\pi} \int_{-\infty}^{\infty} A_{f,g}(t, t_2 - t_1) \exp\left[\frac{1}{2}\, i(t_1 + t_2)t\right] dt = \hat{f}(t_1)\, \bar{\hat{g}}(t_2). \tag{5.5.27}$$

In particular, if $t_1 = t_2 = x$, we obtain

$$\frac{1}{2\pi} \int_{-\infty}^{\infty} A_{f,g}(t,0)\, e^{ixt}\, dt = \hat{f}(x)\, \bar{\hat{g}}(x), \tag{5.5.28}$$

$$\frac{1}{2\pi} \int_{-\infty}^{\infty}\int_{-\infty}^{\infty} A_{f,g}(t,0)\, e^{ixt}\, dt\, dx = \int_{-\infty}^{\infty} \hat{f}(x)\, \bar{\hat{g}}(x)\, dx = \left(\hat{f}, \hat{g}\right). \tag{5.5.29}$$

$$\frac{1}{2\pi} \int_{-\infty}^{\infty}\int_{-\infty}^{\infty} A_{f}(t,0)\, e^{ixt}\, dt\, dx = \int_{-\infty}^{\infty} \hat{f}(x)\, \bar{\hat{f}}(x)\, dx = \left\|\hat{f}\right\|_2. \tag{5.5.30}$$

Example 5.5.1 For a Gaussian signal, $f(t) = A\exp\left(-at^2\right)$, $a > 0$; the ambiguity function is

$$A_f(t,\omega)=|A|^2\sqrt{\frac{\pi}{2a}}\exp\left[-\left(\frac{at^2}{2}+\frac{\omega^2}{8a}\right)\right]. \tag{5.5.31}$$

It follows from definition (5.5.7) that

$$\begin{aligned}A_f(t,\omega)&=A\bar{A}\int_{-\infty}^{\infty}\exp\left[-a\left(\tau+\frac{t}{2}\right)^2-a\left(\tau-\frac{t}{2}\right)^2\right]e^{-i\omega\tau}\,d\tau\\&=|A|^2\exp\left(-\frac{1}{2}at^2\right)\int_{-\infty}^{\infty}\exp\left(-2a\tau^2\right)e^{-i\omega\tau}\,d\tau\\&=|A|^2\sqrt{\frac{\pi}{2a}}\exp\left[-\left(\frac{at^2}{2}+\frac{\omega^2}{8a}\right)\right].\end{aligned}$$

Example 5.5.2 For a quadratic-phase signal, $f(t)=A\exp\left(\frac{1}{2}iat^2\right)$, the ambiguity function is

$$A_f(t,\omega)=|A|^2\,2\pi\,\delta(\omega-at). \tag{5.5.32}$$

We have, by definition (5.5.7),

$$\begin{aligned}A_f(t,\omega)&=|A|^2\int_{-\infty}^{\infty}\exp\left[\frac{1}{2}ia\left\{\left(\tau+\frac{t}{2}\right)^2-\left(\tau-\frac{t}{2}\right)^2\right\}\right]e^{-i\omega\tau}\,d\tau\\&=|A|^2\int_{-\infty}^{\infty}\exp\left[-i\tau(\omega-at)\right]d\tau\\&=|A|^2\cdot 2\pi\,\delta(\omega-at).\end{aligned}$$

Example 5.5.3 If $f(t)=\chi_{(-T,T)}(t)$ is the characteristic function, then

$$A_f(t,\omega)=\left(\frac{2}{\omega}\right)\sin\left(\omega T-\frac{1}{2}|t|\omega\right)\chi_{[-2T,2T]}(t). \tag{5.5.33}$$

It follows from definition (5.5.8) that

$$\begin{aligned}A_f(t,\omega)&=\exp\left(\frac{1}{2}i\omega t\right)\int_{-\infty}^{\infty}f(x)\,f(x-t)\,e^{-i\omega x}\,dx\\&=\exp\left(\frac{1}{2}i\omega t\right)\int_{-T}^{T}f(x-t)\,e^{-i\omega x}\,dx.\end{aligned}$$

If $t>0$, then

$$A_f(t,\omega) = \exp\left(\frac{1}{2}\, i\,\omega\, t\right) \int_{-T}^{T} \chi_{(-T+t,T+t)}(x)\; e^{-i\omega x}\, dx$$

$$= \exp\left(\frac{1}{2}\, i\,\omega\, t\right) \chi_{[0,2T]}(t) \int_{-T+t}^{T} e^{-i\omega x}\, dx$$

$$= \left(\frac{2}{\omega}\right) \sin\left(\omega T - \frac{\omega t}{2}\right) \chi_{[0,2T]}(t).$$

Similarly, if $t < 0$, then

$$A_f(t,\omega) = \exp\left(\frac{1}{2}\, i\,\omega\, t\right) \chi_{[-2T,0]}(t) \int_{-T}^{T+t} \exp(-i\,\omega\, x)\, dx$$

$$= \left(\frac{2}{\omega}\right) \sin\left(\omega T + \frac{\omega t}{2}\right) \chi_{[-2T,0]}(t).$$

Thus,

$$A_f(t,\omega) = \left(\frac{2}{\omega}\right) \sin\left(\omega T - \frac{\omega}{2}|t|\right) \chi_{[-2T,2T]}(t).$$

Example 5.5.4 For a harmonic signal (or a plane wave) $f(t) = A\exp(i\,a\,t)$, the ambiguity function is

$$A_f(\omega,t) = 2\pi\, |A|^2\, \delta(\omega - a). \tag{5.5.34}$$

We have, by definition,

$$A_f(t,\omega) = A\,\bar{A} \int_{-\infty}^{\infty} \exp[-i\,\tau(\omega - a)]\; d\tau$$

$$= |A|^2\; 2\pi\,\delta(\omega - a).$$

Example 5.5.5 If $f(t) = g(t)\, \exp\left(\frac{ia}{2}\, t^2\right)$, then

$$A_f(t,\omega) = A_g(t,\omega - at). \tag{5.5.35}$$

We have, by definition,

$$A_f(t,\omega) = \int_{-\infty}^{\infty} g\left(\tau + \frac{t}{2}\right) \bar{g}\left(\tau - \frac{t}{2}\right) \exp[-i\,\tau(\omega - at)]\; d\tau$$

$$= A_g(t,\omega - at).$$

5.6 Basic Properties of Ambiguity Functions

(a) (*Nonlinearity*). For any four signals f_1, f_2, g_1 and g_2,

$$A_{f_1+f_2,g_1+g_2}(t,\omega)=A_{f_1,g_1}(t,\omega)+A_{f_1,g_2}(t,\omega)+A_{f_2,g_1}(t,\omega)+A_{f_2,g_2}(t,\omega). \tag{5.6.1}$$

In particular,

$$A_{af+bg}(t,\omega)=|a|^2 A_f(t,\omega)+a\bar{b}\,A_{f,g}(t,\omega)+\bar{a}b\,A_{g,f}(t,\omega)+|b|^2 A_g(t,\omega). \tag{5.6.2}$$

where a and b are two constants and

$$A_{f+g}(t,\omega)=A_f(t,\omega)+A_g(t,\omega)+2\operatorname{Re}A_{f,g}(t,\omega). \tag{5.6.3}$$

To prove (5.6.2), we write

$$\begin{aligned}A_{af+bg}(t,\omega)&=\int_{-\infty}^{\infty}\left[a\,f\left(\tau+\frac{t}{2}\right)+b\,g\left(\tau+\frac{t}{2}\right)\right]\left[\bar{a}\,\bar{f}\left(\tau-\frac{t}{2}\right)+\bar{b}\,\bar{g}\left(\tau-\frac{t}{2}\right)\right]e^{-i\omega\tau}d\tau\\&=|a|^2 A_f(t,\omega)+a\,\bar{b}\,A_{f,g}(t,\omega)+\bar{a}\,b\,A_{g,f}(t,\omega)+|b|^2 A_g(t,\omega).\end{aligned}$$

(b) (*Translation*).

$$A_{T_af,g}(t,\omega)=e^{-i\omega a}A_{f,g}(t,\omega). \tag{5.6.4}$$

In particular,

$$A_{T_af}(t,\omega)=e^{-i\omega a}A_f(t,\omega). \tag{5.6.5}$$

(c) (*Complex Conjugation*).

$$\bar{A}_{f,g}(t,\omega)=A_{g,f}(-t,-\omega). \tag{5.6.6}$$

(d) (*Modulation*).

$$A_{M_bf,M_bg}(t,\omega)=e^{ibt}A_{f,g}(t,\omega), \tag{5.6.7}$$

$$A_{M_bf,g}(t,\omega)=\exp\left(\frac{1}{2}i\,b\,t\right)A_{f,g}(t,\omega), \tag{5.6.8}$$

$$A_{f,M_bg}(t,\omega)=\exp\left(-\frac{1}{2}\,i\,b\,t\right)A_{f,g}(t,\omega). \tag{5.6.9}$$

In particular,

$$A_{M_bf}(t,\omega)=e^{ibt}A_f(t,\omega). \tag{5.6.10}$$

In general, a more general modulation property holds:

$$A_{Mf}(t,\omega)=\frac{1}{2\pi}\int_{-\infty}^{\infty}A_f(t,u)\,A_m(t,\omega-u)\,du, \tag{5.6.11}$$

where $M f(t)=f(t)\,m(t)$ which represents a signal modified by $m(t)$.

(e) (*Dilation*). If $D_c f(t)=\frac{1}{\sqrt{|c|}}\,f\left(\frac{t}{c}\right)$, $c\neq 0$, then

$$A_{D_c f,D_c g}(t,\omega)=A_{f,g}\left(\frac{t}{c},\,\omega c\right). \tag{5.6.12}$$

In particular,

$$A_{D_c f}(t,\omega)=A_f\left(\frac{t}{c},\,\omega c\right). \tag{5.6.13}$$

Proof. We have, by definition,

$$\begin{aligned}A_{D_c f,D_c g}(t,\omega)&=\frac{1}{|c|}\int_{-\infty}^{\infty}f\left(\frac{\tau}{c}+\frac{t}{2c}\right)\overline{g}\left(\frac{\tau}{c}-\frac{t}{2c}\right)e^{-i\omega\tau}\,d\tau\\&=\int_{-\infty}^{\infty}f\left(x+\frac{t}{2c}\right)\overline{g}\left(x-\frac{t}{2c}\right)e^{-i(\omega c)x}\,dx\\&=A_{f,g}\left(\frac{t}{c},\,\omega c\right).\end{aligned}$$

(f) (*Translation and Modulation*).

$$A_{T_aM_bf,g}(t,\omega)=A_{M_bT_af,g}(t,\omega)=\exp\left[i\left(\frac{1}{2}\,bt+ab-a\omega\right)\right]A_{f,T_{-a}g}(t,\omega-b). \tag{5.6.14}$$

$$A_{T_aM_bf,T_aM_bg}(t,\omega)=A_{M_bT_af,M_bT_ag}(t,\omega)=\exp\left[i(bt-\omega a)\right]A_{f,g}(t,\omega). \tag{5.6.15}$$

(g) (*Convolution*).

$$A_{f*g}(t,\omega)=\int_{-\infty}^{\infty}A_f(u,\omega)\,A_g(t-u,\omega)\,du. \tag{5.6.16}$$

(h) (*Differentiation*).

$$A_{Df,g}(t,\omega)+A_{f,Dg}(t,\omega)=i\omega\,A_{f,g}(t,\omega). \tag{5.6.17}$$

In particular,

$$A_{Df,f}(t,\omega)+A_{f,Df}(t,\omega)=i\omega\,A_f(t,\omega). \tag{5.6.18}$$

(i) (*Coordinate Transformations*). Let $SL(2,\mathbb{R})$ represent the group of all 2×2 real matrices S of determinant one acting on $\mathbb{R}$ by

$$Su = \begin{pmatrix} au + bv \\ cu + dv \end{pmatrix},$$

where $S = \begin{pmatrix} a & b \\ c & d \end{pmatrix}$, $ad - bc = 1$ and $\boldsymbol{u} = \begin{pmatrix} u \\ v \end{pmatrix}$.

We define the matrices $P, Q,$ and R by

$$P(\alpha) = \begin{pmatrix} 1 & -\alpha \\ 0 & 1 \end{pmatrix}, \quad Q(\beta) = \begin{pmatrix} 1 & 0 \\ -\beta & 1 \end{pmatrix}, \quad R(\gamma) = \begin{pmatrix} \gamma & 0 \\ 0 & \frac{1}{\gamma} \end{pmatrix}.$$

In particular, when $b = c = 0$, then $a = \frac{1}{d} = \gamma$ and $S = R(\gamma)$.

We then calculate

$$\begin{aligned}
A_f\left[\left(R(\gamma)\boldsymbol{u}\right)^T\right] &= A_f\left(\gamma\, u, \frac{v}{\gamma}\right) \\
&= \exp\left(\frac{1}{2} i\, u\, v\right) \int_{-\infty}^{\infty} f(x)\, \bar{f}(x - \gamma\, u)\, \exp\left(-i\, \frac{v}{\gamma}\, x\right) dx \\
&= \gamma \exp\left(\frac{1}{2} i\, u\, v\right) \int_{-\infty}^{\infty} f(\gamma\, y)\, \bar{f}(\gamma\, y - \gamma\, u)\, e^{-ivy} dy \\
&= \gamma \exp\left(\frac{1}{2} i\, u\, v\right) \int_{-\infty}^{\infty} g(x)\, \bar{g}(x - u)\, e^{-ivx} dx \\
&= \gamma\, A_f(u, v),
\end{aligned} \tag{5.6.19}$$

where the superscript T stands for the transpose of the matrix and $g(x) = f(\gamma\, x)$.

In general,

$$A_f\left[\left(S\boldsymbol{u}\right)^T\right] = A_f(au + bv,\, cu + dv).$$

$$A_f\left[\left(P(\alpha)\boldsymbol{u}\right)^T\right] = A_f(u - \alpha v,\, v).$$

$$A_f\left[\left(Q(\beta)\boldsymbol{u}\right)^T\right] = A_f(u,\, v - \beta u).$$

We evaluate (5.6.21) by using (5.5.23) so that

$$A_f(u-\alpha v,\, v) = A_{\hat{f}}(-v,\, u-\alpha v)$$

$$= \exp\left[\frac{i}{2}\left(\alpha v^2 - u v\right)\right] \int_{-\infty}^{\infty} \hat{f}(\omega)\, \bar{\hat{f}}(\omega+v) \exp\left[-i\omega\left(u-\alpha v\right)\right] d\omega.$$

We define $\hat{g}(\omega) = c\, \hat{f}(\omega) \exp\left(\frac{-i\alpha\omega^2}{2}\right)$ with $|c| = 1$. It is easy to check that $A_{\hat{g}}(-v,u) = A_{\hat{f}}(-v,\, u-\alpha v)$. Thus, we find by (5.5.23) that

$$A_f(u-\alpha v,\, v) = A_g(u,v), \tag{5.6.20}$$

where g is the inverse Fourier transform of $\hat{g}$.

If we define $g(t) = \exp\left(\frac{1}{2}\, i\beta t^2\right) f(t)$, it follows from direct calculation that

$$A_f(u,\, v-\beta u) = A_g(u,v). \tag{5.6.21}$$

Theorem 5.6.1 (Parseval's Formulas). If f_1, g_1, f_2, and g_2 belong to $L^2(\mathbb{R})$, then

$$\frac{1}{2\pi} \int_{-\infty}^{\infty}\int_{-\infty}^{\infty} A_{f_1,g_1}(t,\omega)\, \bar{A}_{f_2,g_2}(t,\omega)\, dt\, d\omega = \left(f_1, f_2\right)\overline{\left(g_1, g_2\right)}. \tag{5.6.22}$$

In particular,

$$\frac{1}{2\pi} \int_{-\infty}^{\infty}\int_{-\infty}^{\infty} \left|A_{f,g}(t,\omega)\right|^2 dt\, d\omega = \|f\|^2 \|g\|^2, \tag{5.6.23}$$

$$\frac{1}{2\pi} \int_{-\infty}^{\infty}\int_{-\infty}^{\infty} A_f(t,\omega)\, \bar{A}_g(t,\omega)\, dt\, d\omega = \left|(f,g)\right|^2. \tag{5.6.24}$$

Proof. We know from (5.5.10) that the Fourier transform of $A_{f,g}(t,\omega)$ with respect to ω is

$$\hat{A}_{f,g}(t,\sigma) = 2\pi\, f\left(-\sigma + \frac{t}{2}\right) \bar{g}\left(-\sigma - \frac{t}{2}\right). \tag{5.6.25}$$

It follows from the Parseval relation (3.4.32) for the Fourier transform that

$$\frac{1}{2\pi}\int_{-\infty}^{\infty} A_{f_1,g_1}(t,\omega)\,\bar{A}_{f_2,g_2}(t,\omega)\,d\omega$$

$$=\frac{1}{(2\pi)^2}\int_{-\infty}^{\infty}\hat{A}_{f_1,g_1}(t,\sigma)\,\bar{\hat{A}}_{f_1,g_1}(t,\sigma)\,d\sigma$$

$$=\int_{-\infty}^{\infty} f_1\left(-\sigma+\frac{t}{2}\right)\bar{g}_1\left(-\sigma-\frac{t}{2}\right)\bar{f}_2\left(-\sigma+\frac{t}{2}\right)g_2\left(-\sigma-\frac{t}{2}\right)d\sigma.$$

Integrating both sides with respect to t gives

$$\frac{1}{2\pi}\int_{-\infty}^{\infty}\int_{-\infty}^{\infty} A_{f,g}(t,\omega)\,\bar{A}_{f,g}(t,\omega)\,dt\,d\omega$$

$$=\int_{-\infty}^{\infty}\int_{-\infty}^{\infty} f_1\left(-\sigma+\frac{t}{2}\right)\bar{f}_2\left(-\sigma+\frac{t}{2}\right)\bar{g}_1\left(-\sigma-\frac{t}{2}\right)g_2\left(-\sigma-\frac{t}{2}\right)dt\,d\sigma$$

which is, by putting $\frac{t}{2}-\sigma=x$ and $-\left(\frac{t}{2}+\sigma\right)=y$,

$$=\int_{-\infty}^{\infty} f_1(x)\,\bar{f}_2(x)\,dx\int_{-\infty}^{\infty} g_1(y)\,\bar{g}_2(y)\,dy=(f_1,f_2)\overline{(g_1,g_2)}.$$

Hence, (5.6.23) and (5.6.24) follow readily from (5.6.22). Combining (5.3.30) and (5.6.22) gives the following result:

$$\frac{1}{2\pi}\int_{-\infty}^{\infty}\int_{-\infty}^{\infty}|W_f(t,\omega)|^2\,dt\,d\omega=\frac{1}{2\pi}\int_{-\infty}^{\infty}|A_f(\omega,t)|^2\,dt\,d\omega=\|f\|_2^4=A_f^2(0,0). \tag{5.6.26}$$

This equation is known as the *radar uncertainty principle* for the following reason. Since, for any t and ω,

$$|A_{f,g}(t,\omega)|^2\le\left[\int_{-\infty}^{\infty}\left|f\left(\tau+\frac{t}{2}\right)\bar{g}\left(\tau-\frac{t}{2}\right)\right|d\tau\right]^2$$

$$\le\int_{-\infty}^{\infty}|f(x)|^2\,dx\int_{-\infty}^{\infty}|g(y)|^2\,dy$$

$$=\|f\|^2\,\|g\|^2$$

$$=A_f(0,0)\,A_g(0,0)\quad\text{by definition (5.5.1).}$$

In particular,

$$|A_f(t,\omega)|^2\le A_f^2(0,0). \tag{5.6.27}$$

This implies that the ambiguity surface can nowhere be higher than at the origin. In other words, the graph of the function $|A_f(t,\omega)|^2$ cannot be concentrated arbitrarily close to the origin for any function f.

In the context of radar signal analysis, Woodward (1953) pointed out the physical significance of the radar uncertainty principle in the sense that there are limits on resolution performance in range and velocity (or range rate) of a radar signal. In analogy with the Heisenberg uncertainty principle in quantum mechanics, the radar uncertainty principle states that resolution can be high either in range or velocity but not in both parameters at the same time. In other words, the range and velocity of a target cannot be determined exactly and simultaneously.

In order to establish an important inequality involving the second partial derivatives of A_f at the origin, for any signal $f(t)$, the quantities

$$\sigma_t^2 = \frac{1}{\|f\|^2} \int_{-\infty}^{\infty} x^2 |f(x)|^2 \, dx, \quad \sigma_\omega^2 = \frac{1}{\|\hat{f}\|^2} \int_{-\infty}^{\infty} y^2 |\hat{f}(y)|^2 \, dy, \tag{5.6.28a,b}$$

are used as a measure of the signal duration in both the time and frequency domains. From the Heisenberg inequality (3.7.3), we have

$$\sigma_t^2 \, \sigma_\omega^2 \geq \frac{1}{4}. \tag{5.6.29}$$

It follows from (5.5.1) and (5.5.27) that

$$-\frac{\partial^2 A_f(0,0)}{\partial t^2} = \int_{-\infty}^{\infty} x^2 |f(x)|^2 \, dx, \quad -\frac{\partial^2 A_f(0,0)}{\partial \omega^2} = \int_{-\infty}^{\infty} y^2 |\hat{f}(y)|^2 \, dy. \tag{5.6.30a,b}$$

It turns out from (5.6.25a,b)-(5.6.27a,b) that

$$\left[\frac{\partial^2 A_f(0,0)}{\partial t^2} \cdot \frac{\partial^2 A_f(0,0)}{\partial \omega^2} \right]^{\frac{1}{2}} \geq \frac{1}{2} \|f\|^2. \tag{5.6.31}$$

We close this section by including the relationship between the Zak transform and the ambiguity function. We use the product formula for the Zak transform in the form

$$\mathfrak{Z}_f(t,\omega) \, \overline{\mathfrak{Z}}_g(t,\omega) = \sum_{m,n=-\infty}^{\infty} a_{m,n} \exp[2\pi i (mt + n\omega)], \quad 0 \leq t, \omega \leq 1, \tag{5.6.32}$$

where

$$a_{m,n} = \left(f,\, M_{2\pi m} T_{-n} g\right) = \int_{-\infty}^{\infty} f(t)\, \bar{g}(t+n) \exp(-2\pi i m t). \tag{5.6.33}$$

It is convenient to define $\tilde{\bar{A}}_{f,g}(t,\omega) = A_{f,g}(-t,\, 2\pi\omega)$ so that we can write

$$\tilde{\bar{A}}_{f,g}(t,\omega) = \int_{-\infty}^{\infty} f\left(\tau - \frac{t}{2}\right) \bar{g}\left(\tau + \frac{t}{2}\right) e^{-2\pi i \omega \tau} d\tau$$

$$= e^{-\pi i \omega t} \int_{-\infty}^{\infty} f(x)\, \bar{g}(x+t)\, e^{-2\pi i \omega x} dx.$$

Consequently,

$$a_{m,n} = (-1)^{mn}\, \tilde{\bar{A}}_{f,g}(n,m), \tag{5.6.34}$$

and it turns out from (5.6.32) that

$$\left|\mathfrak{Z}_f(t,\omega)\right|^2 = \sum_{m,n=-\infty}^{\infty} (-1)^{mn}\, \tilde{\bar{A}}_f(n,m) \exp\left[2\pi i (mt + n\omega)\right],$$

$$\left|\mathfrak{Z}_g(t,\omega)\right|^2 = \sum_{p,q=-\infty}^{\infty} (-1)^{pq}\, \overline{\tilde{\bar{A}}_g}(q,p) \exp\left[-2\pi i (pt + q\omega)\right].$$

Integrating the product of these last two series over the unit square $(0 \le t \le 1,\ 0 \le \omega \le 1)$ gives

$$\int_0^1\int_0^1 \left|\mathfrak{Z}_f(t,\omega)\right|^2 \left|\mathfrak{Z}_g(t,\omega)\right|^2 dt\, d\omega = \sum_{m,n=-\infty}^{\infty} \tilde{\bar{A}}_f(n,m)\, \overline{\tilde{\bar{A}}}(n,m).$$

On the other hand, result (5.6.32) combined with the Parseval formula for the Fourier series leads to the result

$$\int_0^1\int_0^1 \left|\mathfrak{Z}_f(t,\omega)\right|^2 \left|\mathfrak{Z}_g(t,\omega)\right|^2 dt\, d\omega = \sum_{m,n=-\infty}^{\infty} \left|a_{m,n}\right|^2 = \sum_{m,n=-\infty}^{\infty} \left|\tilde{\bar{A}}_{f,g}(n,m)\right|^2.$$

Evidently, the following interesting relation is obtained from the above result

$$\sum_{m,n=-\infty}^{\infty} \left|\tilde{\bar{A}}_{f,g}(n,m)\right|^2 = \sum_{n,m=-\infty}^{\infty} \tilde{\bar{A}}_f(n,m)\, \overline{\tilde{\bar{A}}_g}(n,m) = \int_0^1\int_0^1 \left|\mathfrak{Z}_f(t,\omega)\, \mathfrak{Z}_g(t,\omega)\right|^2 dt\, d\omega. \tag{5.6.35}$$

5.7 The Ambiguity Transformation and Its Properties

The cross ambiguity function $A_{f,g}(t,\omega)$ is closely related to a bilinear transformation $\mathcal{B}: L^2(\mathbb{R}) \times L^2(\mathbb{R}) \to L^2(\mathbb{R}^2)$ defined by

$$\mathcal{B}(f,g) = \mathcal{B}_{f,g}(t,\omega) = \int_{-\infty}^{\infty} f(x)\,\overline{g}(x-t)\,e^{-i\omega x}dx \tag{5.7.1}$$

$$= \exp\left(-\frac{1}{2}i\omega t\right) A_{f,g}(t,\omega), \tag{5.7.2}$$

where $f, g \in L^2(\mathbb{R})$. The function $\mathcal{B}_{f,g}(t,\omega)$ is often called the *cross ambiguity function* and has the following properties:

(a)
$$\mathcal{B}_{f,g}(t,\omega) = e^{-i\omega t}\,B_{\hat{f},\hat{g}}(-\omega,t). \tag{5.7.3}$$

(b) (*Complex Conjugation*).

$$\overline{\mathcal{B}}_{f,g}(t,\omega) = e^{i\omega t}\,\mathcal{B}_{g,f}(-t,-\omega). \tag{5.7.4}$$

(c) (*Inversion Formula*).

$$\frac{1}{2\pi}\int_{-\infty}^{\infty} \mathcal{B}_{f,g}(x-t,\omega)\,e^{i\omega x}\,d\omega = f(x)\,\overline{g}(t). \tag{5.7.5}$$

Proof. To prove (5.7.3), we use (5.7.2) and (5.5.23) so that

$$\begin{aligned}
\mathcal{B}_{f,g}(t,\omega) &= \exp\left(-\frac{1}{2}i\omega t\right) A_{f,g}(t,\omega) \\
&= \exp(-i\omega t)\exp\left(\frac{1}{2}i\omega t\right) A_{\hat{f},\hat{g}}(\omega,-t) \qquad \text{by (5.5.23)} \\
&= \exp(-i\omega t)\exp\left(\frac{1}{2}i\omega t\right) A_{\hat{f},\hat{g}}(-\omega,t) \\
&= \exp(-i\omega t)\,\mathcal{B}_{\hat{f},\hat{g}}(-\omega,t).
\end{aligned}$$

Taking the complex conjugate of (5.7.2) gives

$$\begin{aligned}
\overline{\mathcal{B}}_{f,g}(t,\omega) &= \exp\left(\frac{1}{2}i\omega t\right) \overline{A}_{f,g}(t,\omega) \\
&= \exp\left(\frac{1}{2}i\omega t\right) A_{g,f}(-t,-\omega), \qquad \text{by (5.6.6)} \\
&= e^{i\omega t}\,\mathcal{B}_{g,f}(-t,-\omega), \qquad \text{by (5.7.2)}.
\end{aligned}$$

To prove the inversion formula, we use (5.7.1) which implies that $\mathcal{B}_{f,g}(t,\omega)$ is the Fourier transform of $f(x)\,\bar{g}(x-t)$ for fixed t. Clearly, the inverse Fourier transform gives

$$\frac{1}{2\pi}\int_{-\infty}^{\infty}\mathcal{B}_{f,g}(t,\omega)\,e^{i\omega x}\,d\omega = f(x)\,\bar{g}(x-t).$$

Replacing t by $x-t$ gives (5.7.5).

(d) (Parseval's Formula). If f_1, f_2, g_1, and g_2 belong to $L^2(\mathbb{R})$, then

$$\frac{1}{2\pi}\int_{-\infty}^{\infty}\int_{-\infty}^{\infty}\mathcal{B}_{f_1,g_1}(t,\omega)\,\overline{\mathcal{B}}_{f_2,g_2}(t,\omega)\,dt\,d\omega = (f_1,f_2)\,\overline{(g_1,g_2)}. \tag{5.7.6}$$

This formula follows from the Parseval formula (5.6.22).

We next put $g=f$ in (5.7.5) with interchanging x and t and set

$$H(t,x) = \frac{1}{2\pi}\int_{-\infty}^{\infty}\mathcal{B}_f(t-x,\omega)\,e^{i\omega t}\,d\omega = f(t)\,\bar{f}(x)\cdot \tag{5.7.7}$$

We consider the mapping $U:\ \mathcal{B}_f(\tau,\omega)\to H(t,x)$ defined by (5.7.7). It is easy to check that

$$\|\mathcal{B}_f\| = \sqrt{2\pi}\ \|H\| = \sqrt{2\pi}\ \|f\|^2, \tag{5.7.8}$$

where the norm is defined in the usual way by

$$\|H\|^2 = \int_{-\infty}^{\infty}\int_{-\infty}^{\infty}|H(t,x)|^2\,dt\,dx. \tag{5.7.9}$$

Furthermore, it is also easy to verify that H satisfies the following functional equation

$$H(t,t)\ge 0, \tag{5.7.10}$$

$$\overline{H}(t,x) = H(x,t), \tag{5.7.11}$$

$$H(t,y)\,H(y,x) = H(y,y)\,H(t,x). \tag{5.7.12}$$

It is clear from definition (5.7.1) that the cross ambiguity transformation $\mathcal{B}$ is a bilinear transformation from $L^2(\mathbb{R})\times L^2(\mathbb{R})\to L^2(\mathbb{R}^2)$ given by

$$\mathcal{B}(f,g) = \mathcal{B}_{f,g}(t,\omega)\cdot$$

We state the following theorems due to Auslander and Tolimieri (1985) without proof.

Theorem 5.7.1. The cross ambiguity transformation $\mathcal{B}$ is continuous, and the image of $\mathcal{B}$ spans a dense subspace of $L^2(\mathbb{R}^2)$.

Theorem 5.7.2. The set of ambiguity functions $\mathcal{B}_f(t, \omega)$ for all $f \in L^2(\mathbb{R})$ is a closed subset of $L^2(\mathbb{R}^2)$.

Theorem 5.7.3. $\mathcal{B}_{f,g}$ is a continuous bounded function which attains its maximum (f, g) at the origin.

Theorem 5.7.4. If f and g belong to $L^2(\mathbb{R})$, $\mathcal{B}_f$ and $\mathcal{B}_g$ are their corresponding ambiguity functions, then $\mathcal{B}_f + \mathcal{B}_g$ is an ambiguity function if and only if $f = ag$, where a is a constant.

The reader is referred to Auslander and Tolimieri (1985) for a complete discussion of proofs of the above theorems. Furthermore, we closely follow Auslander and Tolimieri (1985) without many technical details to show that ambiguity functions represent well-known elements in the theory of unitary representations of the Heisenberg group.

A unitary operation on $L^2(\mathbb{R})$ is a linear mapping U of $L^2(\mathbb{R})$ that satisfies the following

$$(Uf, Ug) = (f, g) \tag{5.7.13}$$

for all f and g that belong to $L^2(\mathbb{R})$. The set of all unitary operators U on $L^2(\mathbb{R})$ forms a group under composition which will be denoted by $\mathcal{U}$. As defined in Section 3.1, T_a and M_b for $a, b \in \mathbb{R}$ are unitary operators of $L^2(\mathbb{R})$ which play an important role in the theory of the ambiguity functions. Also, T_a and M_b are noncommuting operators, and this fact is the mathematical basis for the Heisenberg group in quantum mechanics and is an expression of the Heisenberg uncertainty principle.

We now consider two mappings T and M from R to $\mathcal{U}$ and set

$$\mathcal{T} = T(\mathbb{R}), \quad \mathcal{M} = M(\mathbb{R})$$

so that $\mathcal{T}$ and $\mathcal{M}$ are called the *translation* (or *shift*) and *multiplication* (or *modulation*) operators. Obviously, both $\mathcal{T}$ and $\mathcal{M}$ are subgroups of $\mathcal{U}$. We next introduce the Heisenberg group N consisting of all points $\boldsymbol{x} = (x_1, x_2, x) \in \mathbb{R}^3$. The multiplication law in the group N is given by the formula

$$\boldsymbol{x} \circ \boldsymbol{y} = \left(x_1 + y_1, x_2 + y_2, \; x + y + \frac{1}{2}(x_2 y_1 - x_1 y_2) \right). \tag{5.7.14}$$

It is easy to check that N is a group having centex X consisting of all points

$(0, 0, x)$, where $x \in \mathbb{R}$.

We now define $D : N \to \mathcal{U}$ by setting

$$D_x = C\left(e^{i\lambda(x)}\right) M(x_1)\, T(x_2), \tag{5.7.15}$$

where $C(\lambda) = \lambda I$, $\lambda \in \mathbb{C}$, $|\lambda| = 1$ and I is the identity operator on $L^2(\mathbb{R})$, and $\lambda(x) = x + \frac{1}{2}\, x_1\, x_2$.

Or, equivalently,

$$(D_x f)(t) = C\left(e^{i\lambda(x)}\right) \exp\left[i\, x_1 (t + x_2)\right]\, f(t - x_2). \tag{5.7.16}$$

Then, D: $N \to \mathcal{U}$ is a group homomorphism built in a non-Abelian fashion from the group homomorphisms T and M. The ambiguity function A_f can then be represented in terms of the group homomorphism D, as shown in the next theorem.

Theorem 5.7.5. For $x \in N$ and $f \in L^2(\mathbb{R})$,

$$A_f(x_2, x_1) = e^{ix}\, (f, D_x\, f). \tag{5.7.17}$$

Proof. Since

$$(D_x\, f)(t) = e^{i\lambda(x)} \exp\left[i\, x_1 (t - x_2)\right]\, f(t - x_2),$$

we can find

$$\begin{aligned}
(f, D_x\, f) &= \int_{-\infty}^{\infty} f(t)\, \overline{(D_x\, f)}(t)\, dt \\
&= e^{-i\lambda(x)} \int_{-\infty}^{\infty} f(t) \exp\left[-i\, x_1 (t - x_2)\right]\, \bar{f}(t - x_2)\, dt \\
&= e^{-i\lambda(x)}\, e^{ix_1 x_2} \int_{-\infty}^{\infty} f(t)\, \bar{f}(t - x_2)\, e^{-i x_1 t} dt \\
&= \exp\left(\frac{1}{2} i\, x_1\, x_2 - i\, x\right) \mathcal{B}_f(x_2, x_1), \quad \text{by (5.7.1)} \\
&= e^{-ix}\, A_f(x_2, x_1), \quad \text{by (5.7.2).}
\end{aligned}$$

This completes the proof.

The obvious significance of the result (5.7.17) is that ambiguity functions represent well-defined elements in the theory of unitary representations of the Heisenberg group.

5.8 Discrete Wigner-Ville Distributions

The *cross Wigner-Ville distribution* of two *discrete time signals* $f(n)$ and $g(n)$ is defined by

$$W_{f,g}(n,\theta) = 2 \sum_{m=-\infty}^{\infty} f(n+m)\, \overline{g}(n-m) \exp(-2im\theta). \tag{5.8.1}$$

Thus, $W_{f,g}(n,\theta)$ is a function of the discrete variable n and the continuous variable θ. Moreover, for all n and θ,

$$W_{f,g}(n,\theta+\pi) = W_{f,g}(n,\theta). \tag{5.8.2}$$

This means that the $W_{f,g}(n,\theta)$ is a periodic function of θ with period π.

Clearly,

$$\begin{aligned} W_{f,g}\left(n, \frac{\theta}{2}\right) &= \sum_{m=-\infty}^{\infty} \left[2 f(n+m)\, \overline{g}(n-m)\right] e^{-im\theta} \\ &= \mathscr{F}_d\left[2 f(n+m)\, \overline{g}(n-m)\right], \end{aligned} \tag{5.8.3}$$

where $\mathscr{F}_d$ is the discrete Fourier transformation defined by

$$\mathscr{F}_d\{f(n)\} = \hat{f}(\theta) = \sum_{n=-\infty}^{\infty} f(n)\, e^{-in\theta}. \tag{5.8.4}$$

The inverse transformation, $\mathscr{F}_d^{-1}$ is defined by

$$f(n) = \mathscr{F}_d^{-1}\left\{\hat{f}(\theta)\right\} = \frac{1}{2\pi} \int_{-\pi}^{\pi} \hat{f}(\theta)\, e^{in\theta} d\theta. \tag{5.8.5}$$

The inner products for the signals and their Fourier spectra are defined by

$$(f,g) = \sum_{n=-\infty}^{\infty} f(n)\, \overline{g}(n), \tag{5.8.6}$$

$$\left(\hat{f},\hat{g}\right) = \frac{1}{2\pi} \int_{-\pi}^{\pi} \hat{f}(\theta)\, \overline{\hat{g}}(\theta)\, d\theta. \tag{5.8.7}$$

The *auto Wigner-Ville distribution* of a *discrete signal* is then given by

$$W_f(n,\theta) = 2 \sum_{m=-\infty}^{\infty} f(n+m)\, \bar{f}(n-m) \exp(-2im\theta). \tag{5.8.8}$$

Or, equivalently,

$$W_f\left(n, \frac{\theta}{2}\right) = \sum_{m=\infty}^{\infty} 2\, f(n+m)\, \bar{f}(n-m) \exp(-i m \theta)$$
$$= \mathscr{F}_d\left\{2\, f(n+m)\, \bar{f}(n-m)\right\}. \tag{5.8.9}$$

Both $W_{f,g}(n,\theta)$ and $W_f(n,\theta)$ are usually referred to as the *discrete Wigner-Ville distribution* (DWVD).

In order to obtain a relation similar to (5.2.18), we define the DWVD for the Fourier spectra $\hat{f}(\theta)$ and $\hat{g}(\theta)$ by

$$W_{\hat{f},\hat{g}}(\theta,n) = \frac{1}{\pi} \int_{-\pi}^{\pi} \hat{f}(\theta+\alpha)\, \bar{\hat{g}}(\theta-\alpha) \exp(2 i n \alpha)\, d\alpha, \tag{5.8.10}$$

so that

$$W_{\hat{f},\hat{g}}(\theta,n) = W_{f,g}(n,\theta). \tag{5.8.11}$$

We next discuss the basic properties of discrete Wigner-Ville distributions.

(a) (*Nonlinearity*). The discrete Wigner-Ville distribution is a nonlinear transformation. More precisely,

$$W_{af+bg}(n,\theta) = |a|^2\, W_f(n,\theta) + |b|^2\, W_g(n,\theta) + a\bar{b}\, W_{f,g}(n,\theta) + \bar{a} b\, W_{g,f}(n,\theta), \tag{5.8.12}$$

where a and b are two constants.

More generally, we can prove

$$W_{f_1+g_1, f_2+g_2}(n,\theta) = W_{f_1,g_1}(n,\theta) + W_{f_1,g_2}(n,\theta) + W_{f_2,g_1}(n,\theta) + W_{f_2,g_2}(n,\theta). \tag{5.8.13}$$

These results can easily be proved from the definition.

(b) (*Inversion*). It follows from (5.8.2) that

$$2\, f(n+m)\, \bar{g}(n-m) = \mathscr{F}_d^{-1}\left\{W_{f,g}\left(n, \frac{\theta}{2}\right)\right\} = \frac{1}{2\pi} \int_{-\pi}^{\pi} W_{f,g}\left(n, \frac{\theta}{2}\right) e^{in\theta}\, d\theta.$$

This can be expressed in the form

$$f(n_1)\, \bar{g}(n_2) = \frac{1}{2\pi} \int_{-\pi/2}^{\pi/2} W_{f,g}\left(\frac{n_1+n_2}{2}, \theta\right) \exp\left[i(n_1-n_2)\,\theta\right] d\theta, \tag{5.8.14}$$

where $n_1 = n+m$, $n_2 = n-m$, and $\frac{1}{2}(n_1+n_2)$ is an integer.

In particular, when $n_2 = n_1 = n$, we obtain from (5.8.14)

$$f(n)\,\bar{g}(n)=\frac{1}{2\pi}\int_{-\pi/2}^{\pi/2}W_{f,g}(n,\theta)\,d\theta \tag{5.8.15}$$

$$|f(n)|^2=\frac{1}{2\pi}\int_{-\pi/2}^{\pi/2}W_f(n,\theta)\,d\theta. \tag{5.8.16}$$

They may be referred to as the *inversion formulas*.

Summing (5.8.15) over n gives

$$\sum_{n=-\infty}^{\infty}f(n)\,\bar{g}(n)=\frac{1}{2\pi}\sum_{n=-\infty}^{\infty}\int_{-\pi/2}^{\pi/2}W_{f,g}(n,\theta)\,d\theta=(f,g), \tag{5.8.17}$$

$$\sum_{n=-\infty}^{\infty}f(n)\,\bar{f}(n)=\frac{1}{2\pi}\int_{-\pi/2}^{\pi/2}W_f(n,\theta)\,d\theta=(f,f)=\|f\|^2. \tag{5.8.18}$$

In view of (5.8.10), (5.8.11), and the periodic property of $W_{f,g}(n,\theta)$ with respect to θ, we obtain

$$\frac{1}{2}\,W_{f,g}(n,\theta)=\frac{1}{2\pi}\int_{-\pi}^{\pi}\hat{f}(\theta+\alpha)\,\bar{\hat{g}}(\theta-\alpha)\,\exp(2in\alpha)\,d\alpha,$$

$$\frac{1}{2}W_{f,g}(n,\theta)=\frac{1}{2}W_{f,g}(n,\theta+\pi)=\frac{1}{2\pi}\int_{-\pi}^{\pi}\hat{f}(\theta+n+\alpha)\,\bar{\hat{g}}(\theta+n-\alpha)\,\exp(2in\alpha)\,d\alpha.$$

Adding these two results gives

$$W_{f,g}(n,\theta)=\frac{1}{2\pi}\int_{-\pi}^{\pi}\exp(2in\alpha)\left[\hat{f}(\theta+\alpha)\,\bar{\hat{g}}(\theta-\alpha)+\hat{f}(\theta+\pi+\alpha)\right.$$

$$\left.\times\,\bar{\hat{g}}(\theta+\pi-\alpha)\right]\,d\alpha.$$

This implies that

$$\sum_{n=-\infty}^{\infty}\exp(-2ni\alpha)\,W_{f,g}(n,\theta)=\hat{f}(\theta+\alpha)\,\bar{\hat{g}}(\theta-\alpha)+\hat{f}(\theta+\pi+\alpha)\,\bar{\hat{g}}(\theta+\pi-\alpha).$$

Putting $\theta+\alpha=\theta_1$, $\theta-\alpha=\theta_2$ gives

$$\sum_{n=-\infty}^{\infty}\exp\{(\theta_2-\theta_1)\,in\}\,W_{f,g}\left(n,\,\frac{\theta_1+\theta_2}{2}\right)=\hat{f}(\theta_1)\,\bar{\hat{g}}(\theta_2)+\hat{f}(\theta_1+\pi)\,\bar{\hat{g}}(\theta_2+\pi)$$

whence, by substituting $\theta_1=\theta_2=\theta$,

$$\sum_{n=-\infty}^{\infty}W_{f,g}(n,\theta)=\hat{f}(\theta)\,\bar{\hat{g}}(\theta)+\hat{f}(\theta+\pi)\,\bar{\hat{g}}(\theta+\pi)=2\,\hat{f}(\theta)\,\bar{\hat{g}}(\theta). \tag{5.8.19}$$

In particular, when $f = g$,

$$\sum_{n=-\infty}^{\infty} W_f(n,\theta) = \left|\hat{f}(\theta)\right|^2 + \left|\hat{f}(\theta+\pi)\right|^2. \tag{5.8.20}$$

Finally, integrating (5.8.19) with respect to θ gives

$$\frac{1}{2\pi}\int_{-\pi/2}^{\pi/2}\sum_{n=-\infty}^{\infty} W_{f,g}(n,\theta)\,d\theta = \frac{1}{2\pi}\int_{-\pi}^{\pi}\hat{f}(\theta)\,\bar{\hat{g}}(\theta)\,d\theta = \left(\hat{f},\hat{g}\right). \tag{5.8.21}$$

This is identical with (5.8.17).

(c) (*Conjugation*).

$$\overline{W_{f,g}}(n,\theta) = W_{g,f}(n,\theta). \tag{5.8.22}$$

$$\overline{W}_f(n,\theta) = W_f(n,\theta). \tag{5.8.23}$$

$$W_f(n,\theta) = W_{\bar{f}}(n,-\theta). \tag{5.8.24}$$

(d) (*Translation* or *Time Shift*).

$$W_{T_k f, T_k g}(n,\theta) = W_{f,g}(n-k,\theta), \tag{5.8.25}$$

where $T_k f(n) = f(n-k)$.

(e) (*Modulation*).

$$W_{M_\alpha f, M_\alpha g}(n,\theta) = W_{f,g}(n,\theta-\alpha), \tag{5.8.26}$$

where $(M_\alpha f)(n) = \exp(in\alpha)\,f(n)$.

(f) (*Inner Product*).

$$W_{f,g}(0,0) = 2(f,\operatorname{Re} g). \tag{5.8.27}$$

(g) (*Multiplication*). If $(Mf)(n) = n\,f(n)$, then

$$2n\,W_{f,g}(n,\theta) = W_{Mf,g}(n,\theta) + W_{f,Mg}(n,\theta), \tag{5.8.28}$$

$$\exp(2i\theta)\,W_{f,g}(n,\theta) = W_{T_{-1}f,T_1 g}(n,\theta). \tag{5.8.29}$$

(h) (*Moyal's Formula*).

$$\frac{1}{2\pi}\int_{-\pi/2}^{\pi/2}\sum_{n=-\infty}^{\infty} W_{f_1,g_1}(n,\theta)\,\overline{W}_{f_2,g_2}(n,\theta)\,d\theta$$

$$= (f_1,f_2)\overline{(g_1,g_2)} + (f_1,M_\pi f_2)\overline{(g_1,M_\pi g_2)}, \tag{5.8.30}$$

where $(M_\pi f)(n) = e^{in\pi} f(n)$.

5.9 Cohen's Class of Time-Frequency Distributions

Cohen (1966) has provided a simple method to generate all possible time-frequency distributions. This is known as Cohen's general class of distributions of a signal f defined by

$$C_f(t,\omega) = \int\int\int_{-\infty}^{\infty} \exp\left[i(\upsilon u - \upsilon t - \tau\omega)\right] k(\upsilon,\tau)\, f\left(u + \frac{1}{2}\tau\right) \bar{f}\left(u - \frac{1}{2}\tau\right) d\upsilon\, du\, d\tau, \tag{5.9.1}$$

where $k(\upsilon,\tau)$ is called the *kernel* of the distribution. Different kernels give different time-frequency distributions for the same signal.

For a time-frequency distribution $C_f(t,\omega)$ of a signal f (real or complex) to be interpreted as a joint energy density, it must at least satisfy the following two fundamental properties of nonnegativity and correct marginals for all times and frequencies:

$$C_f(t,\omega) \geq 0, \tag{5.9.2}$$

$$\int_{-\infty}^{\infty} C_f(t,\omega)\, dt = \left|\hat{f}(\omega)\right|^2, \quad \int_{-\infty}^{\infty} C_f(t,\omega)\, d\omega = \left|f(t)\right|^2. \tag{5.9.3a,b}$$

The quantities $|f(t)|^2$ and $\left|\hat{f}(\omega)\right|^2$ are the energy densities of time and frequency, or marginals which are usually interpreted as the *instantaneous power* and the *energy density spectrum*, respectively. Cohen and Posch (1985) referred to joint density functions of time and frequency, $C_f(t,\omega)$ that satisfy the above properties as *proper time-frequency distributions* (TFDs). They have shown that there are an infinite number of TFDs satisfying (5.9.2) and (5.9.3a,b), and it is necessary to consider time-dependent kernels to determine positive TFDs with correct marginals. Therefore, the fundamental question is how to select the kernel of the Cohen-Posch class.

In terms of the Fourier spectrum, Cohen's general class is

$$C_{\hat{f}}(t,\omega) = \int\int\int_{-\infty}^{\infty} \exp\left[i(\upsilon u - \upsilon t - \tau\omega)\right] k(\upsilon,\tau)\, \hat{f}\left(u + \frac{1}{2}\tau\right) \bar{\hat{f}}\left(u - \frac{1}{2}\tau\right) d\upsilon\, du\, d\tau, \tag{5.9.4}$$

as may be verified by expressing the signal in terms of the Fourier spectrum and putting (5.9.1).

It is important to note that the kernel k does not appear in the signal. Thus, we can define the Fourier transform of the kernel by

$$R(t,\tau)=\int_{-\infty}^{\infty} k(\upsilon,\tau)\, e^{-it\upsilon} d\upsilon, \tag{5.9.5}$$

so that Cohen's general class (5.9.1) can be written as

$$C_f(t,\omega)=\int_{-\infty}^{\infty}\int_{-\infty}^{\infty} R(t-u,\tau)\, f\left(u+\frac{1}{2}\tau\right) \bar{f}\left(u-\frac{1}{2}\tau\right) e^{-i\tau\omega} du\, d\tau. \tag{5.9.6}$$

Zhao et al. (1990) introduced a special kernel $\hat{k}(t,\tau)$ to define a new time-frequency distribution that has many remarkable properties. This distribution is now known as the *Zhao-Atlas-Marks distribution* which significantly enhances the time and frequency resolution and eliminates all undesirable cross terms. The kernel $\hat{k}(t,\tau)$ involved in the original work of Zhao et al. (1990) is given by

$$\hat{k}(t,\tau)=g(\tau)\, H(|\tau|-a|t|), \tag{5.9.7}$$

where $H(x)$ is the Heaviside unit step function, $g(\tau)$ is arbitrary and to be specified and the parameter a is assumed to be greater than or equal to 2 so that the finite time support condition is satisfied. The kernel $k(v,\tau)$ can be obtained by inversion so that

$$k(v,\tau)=g(\tau)\cdot\frac{\sin\left(\frac{v|\tau|}{a}\right)}{\pi v}. \tag{5.9.8}$$

Consequently, the Zhao-Atlas-Marks (ZAM) distribution is given by

$$C_f(t,\omega)=\int_{-\infty}^{\infty} g(\tau)\, e^{-i\omega\tau}\, d\tau \int_{-\infty}^{\infty} f\left(u+\frac{1}{2}\tau\right) \bar{f}\left(u-\frac{1}{2}\tau\right) du. \tag{5.9.9}$$

In their original work, Zhao et al. (1990) used $g(\tau)=1$ and $a=2$. The ZAM distribution has been applied to speech signals with remarkable results. For the ZAM distribution to satisfy the condition of finite time support, it is necessary that the nonzero support region of $\hat{k}(t,\tau)$ lie inside the cone-shaped region defined by

$$\Pi_{|\tau|}(t) = \begin{Bmatrix} 1, & -\frac{|\tau|}{2} \le t \le \frac{|\tau|}{2} \\ 0, & \text{otherwise} \end{Bmatrix}. \tag{5.9.10}$$

The infinite-time versions of the Wigner-Ville kernel, $\hat{k}(t,\tau) = \delta(t)$, the Margenau and Hill (1961) kernel, which is the real part of the Kirkwood (1933) and Rihaczek (1968) time-frequency distribution, kernel $\hat{k}(t,\tau) = \frac{1}{2}\left[\delta\left(t+\frac{\tau}{2}\right)+\delta\left(t-\frac{\tau}{2}\right)\right]$, and the Born-Jordan-Cohen kernel (see Cohen, 1995), $\hat{k}(t,\tau) = |\tau|^{-1}\,\Pi_{|\tau|}(t)$, all satisfy the time and frequency marginals.

It was stated earlier that the Wigner-Ville distribution satisfies the marginals but it is not always positive, whereas the spectrogram is manifestly positive, but does not satisfy the marginals. Recently, Loughlin et al. (1994) developed a new general method of construction of positive distributions satisfying marginals of time and frequency. They used the cross-entropy minimization principle to construct TFDs that are members of the Cohen-Posch class. Several examples of these TFDs, including chirps, tones, resonators, speech, and acoustic records of rotating machinery, are given in Loughlin et al. (1994).

On the other hand, if we write

$$C_f(t,\omega) = \int_{-\infty}^{\infty}\int_{-\infty}^{\infty} M_f(\upsilon,\tau)\,\exp\left[-i(\upsilon t + \tau\omega)\right]\,d\upsilon\,d\tau, \tag{5.9.11}$$

where $M(\upsilon,\tau)$ is called the *generalized* ambiguity function defined by

$$\begin{aligned} M_f(\upsilon,t) &= k(\upsilon,\tau)\;A_f(\upsilon,\tau) \\ &= k(\upsilon,\tau)\int_{-\infty}^{\infty}\exp(i\upsilon u)\;f\left(u+\frac{1}{2}\tau\right)\;\bar{f}\left(u-\frac{1}{2}\tau\right)\,du. \end{aligned} \tag{5.9.12}$$

In his book, Cohen (1995) listed several kernels and their corresponding distributions, and we list a few of these distributions. If $k(\upsilon,\tau) = 1$, then (5.9.1) reduces to the Wigner-Ville distribution.

If $k(\upsilon,\tau) = \exp\left(-\frac{\upsilon^2\tau^2}{\sigma}\right)$, (5.9.1) gives the Choi and Williams (1989) distribution in the form

$$C_f(t,\omega) = \sqrt{\pi\sigma} \int_{-\infty}^{\infty}\int_{-\infty}^{\infty} \tau^{-1} \exp\left[-\sigma\left\{(u-t)^2/\tau^2\right\} - i\tau\omega\right] f\left(u+\frac{1}{2}\tau\right)$$

$$\times \bar{f}\left(u-\frac{1}{2}\tau\right) du\, d\tau. \qquad (5.9.13)$$

If the kernel

$$k(\upsilon.\tau) = \int_{-\infty}^{\infty} h\left(u+\frac{1}{2}\tau\right) \bar{h}\left(u-\frac{1}{2}\tau\right) e^{-i\upsilon u} du, \qquad (5.9.14)$$

then (5.9.1) reduces to the spectrogram

$$C_f(t,\omega) = \left|\int_{-\infty}^{\infty} e^{-i\omega\tau} f(\tau)\, h(\tau-t)\, d\tau\right|^2. \qquad (5.9.15)$$

Finally, if the kernel $k(\upsilon,\tau) = \exp\left(\frac{i\upsilon\tau}{2}\right)$, (5.9.1) becomes the Rihaczek distribution. In fact, the generalized Rihaczek distribution of two signals f and g is defined by

$$\mathfrak{R}_{f,g}^{\alpha}(t,\omega) = \int_{-\infty}^{\infty} f\left\{t+\left(\frac{1}{2}-\alpha\right)\tau\right\} \bar{g}\left\{t-\left(\frac{1}{2}+\alpha\right)\tau\right\} e^{-i\omega\tau} d\tau, \qquad (5.9.16)$$

where α is a real constant.

If $\alpha = 0$, (5.9.16) reduces to the cross Wigner-Ville distribution (5.2.1).

If $\alpha = \frac{1}{2}$, (5.9.16) gives the cross Rihaczek distribution $R_{f,g}(t,\omega)$ defined by

$$R_{f,g}(t,\omega) = \mathfrak{R}_{f,g}^{\frac{1}{2}}(t,-\omega) = e^{i\omega t} f(t)\, \hat{\bar{g}}(\omega). \qquad (5.9.17)$$

5.10 Exercises

1. Find the Wigner-Ville transform of the following signals:

(a) $f(t) = \frac{1}{\sqrt{\sigma}} \exp\left(-\frac{t^2}{\sigma^2}\right)$, (b) $f(t) = H(t)$,

(c) $f(t) = \left(\frac{2}{\sigma^2}\right)^{\frac{1}{4}} \exp\left(-\frac{\pi t^2}{\sigma^2} + \frac{1}{2} i\omega_0 t^2\right)$,

(d) $f(t)=\left(\frac{2}{\sigma^2}\right)^{\frac{1}{4}} \exp\left[-\frac{\pi}{\sigma^2}(t-t_0)^2+i\omega_0 t\right].$

2. Find the ambiguity function of the following signals:

(a) $f(t)=\frac{1}{\sqrt{\sigma}} \exp\left(-\frac{t^2}{\sigma^2}\right),$

(b) $f(t)=\frac{1}{\sqrt{\sigma}} \exp\left(-\frac{t^2}{\sigma^2}+iat^2\right).$

3. If $f(t)$ is even, show that

$$A_f(t,\omega)=\frac{1}{2} W_f\left(\frac{t}{2},\frac{\omega}{2}\right).$$

4. If $f(t)=\mathscr{F}\{u(x)\}=\hat{u}(t)$ and $g(t)=\mathscr{F}\{v(x)\}=\hat{v}(t)$, show that

(a) $W_{f,g}(t,\omega)=W_{\hat{u},\hat{v}}(-\omega,t),$

(b) $W_f(t,\omega)=W_{\hat{u}}(-\omega,t).$

5. Use Exercise 4 to prove the following result:

(a) If $f(t)=\chi_a(t)=\begin{Bmatrix}1, & |t|<a\\ 0, & |t|>a\end{Bmatrix}$, then

$$W_f(t,\omega)=\frac{2}{\omega}\sin\{2(a-|t|\omega)\}\ \chi_a(t).$$

(b) If $g(t)=\mathscr{F}\{h(x)\}=\hat{h}(t)=\frac{2}{t}\sin at$, then

$$W_g(t,\omega)=W_h(-\omega,t)=\left(\frac{2}{t}\right)\sin\{2(a-|-\omega|)t\}\ \chi_a(-\omega)$$
$$=\left(\frac{2}{t}\right)\sin\{2(a-|\omega|)t\}\ \chi_a(\omega).$$

6. Use the integral representation of the Dirac delta function to prove the following marginal integrals:

(a) $\frac{1}{2\pi}\int_{-\infty}^{\infty} W_{f,g}(t,\omega)\,d\omega=f(t)\,\bar{g}(t),$

(b) $\frac{1}{2\pi}\int_{-\infty}^{\infty} W_{f,g}(t,\omega)\,dt=\hat{f}(\omega)\,\hat{\bar{g}}(\omega),$

(c) $\frac{1}{2\pi}\int_{-\infty}^{\infty} W_f(t,\omega)\, d\omega = |f(t)|^2$,

(d) $\frac{1}{2\pi}\int_{-\infty}^{\infty} W_f(t,\omega)\, dt = |\hat{f}(\omega)|^2$.

7. If $f(t) = \exp(i\omega_0 t)\, g(t)$, show that $W_f(t,\omega) = W_g(t,\omega-\omega_0)$.

8. Find the Wigner-Ville transform of $f(t) = \exp(i\omega_0 t)\, H(T-|t|)$, where $H(x)$ is the Heaviside unit step function.

9 .Find the Wigner-Ville transform of a sinusiodal signal $f(t) = A\cos(\omega_0 t + \theta)$.

10. If $f,g \in L^2(\mathbb{R})$ and $\mathcal{B}_{f,g}(t,\omega) = 0$, prove that $f = 0$ almost everywhere or $g = 0$ almost everywhere.

11. If $f \in L^2(\mathbb{R})$, show that the set of ambiguity functions A_f is a closed subset of $L^2(\mathbb{R}^2)$.

12. If $f,g \in L^2(\mathbb{R})$, show that $h(t,x) = f(t)\,\bar{g}(x-t)$ is in $L^2(\mathbb{R}^2)$ and $\|h\|_2^2 = \|f\|_2^2\, \|g\|_2^2$.

13. If sequences $f_n \to f$ and $g_n \to g$ in $L^2(\mathbb{R})$, then $\mathcal{B}(f_n,g_n) \to \mathcal{B}(f,g)$.

14. Show that

$\mathcal{B}\left(e^{-t^2}, e^{-(t-a)^2}\right)$ is continuous for $a \in \mathbb{R}$.

15. Show, by direct computation, that $\mathcal{B}_f + \mathcal{B}_{af} = \mathcal{B}\left(\sqrt{1+|a|^2}\, f\right)$.

16. If $f(n) = \begin{Bmatrix} 1, & |n| < N \\ 0, & |n| \ge N \end{Bmatrix}$, show that

$$W_f(n,\theta) = \begin{Bmatrix} \frac{2}{\sin\theta}\cdot\sin\left[2\theta\left(N - |n| + \frac{1}{2}\right)\right], & |n| < N \\ 0, & |n| \ge N \end{Bmatrix}.$$

17. Find the discrete Wigner-Ville transform of a chirp signal

$f(n) = A\exp\left(\frac{i}{2}an^2\right)$.

18. If f and g modulate the carrier signals m_f and m_g, respectively, show that the discrete Wigner-Ville transform is

$$W_{f_m,g_m}(n,\theta)=\frac{1}{2\pi}\int_{-\pi/2}^{\pi/2} W_{f,g}(n,\alpha)\, W_{m_f,m_g}(n,\theta-\alpha)\, d\alpha .$$

19. Prove the following relations for the discrete Wigner-Ville transform:

$$\frac{1}{2\pi}\int_{-\pi/2}^{\pi/2} W_{f,g}(n,\theta)\, d\theta = f(n)\,\overline{g}(n),$$

$$\frac{1}{2\pi}\int_{-\pi/2}^{\pi/2}\left(\sum_n W_{f,g}(n,\theta)\right) d\theta = (f,g).$$

20. (a) Write the definition of Cohen's general class of distributions of signals f and g with $k(\upsilon,\tau)$ as kernel function.

(b) Show that $C_{f,g}(t,\omega)$ is a nonlinear transformation, that is,

$$C_{af+bg}(t,\omega)=|a|^2\, C_f(t,\omega)+|b|^2\, C_g(t,\omega)+a\,\overline{b}\, C_{f,g}(t,\omega)+\overline{a}\, b\, C_{g,f}(t,\omega).$$

(c) For all real t and $f\in L^2(\mathbb{R})$, prove the following results:

$$\int_{-\infty}^{\infty} C_f(t,\omega)\, d\omega=|f(t)|^2, \qquad \int_{-\infty}^{\infty} C_f(t,\omega)\, dt=\left|\hat{f}(\omega)\right|^2,$$

$$\int_{-\infty}^{\infty}\int_{-\infty}^{\infty} C_f(t,\omega)\, dt\, d\omega=\int_{-\infty}^{\infty}|f(t)|^2\, dt=\|f\|_2^2.$$

21. Prove Moyal's relation for Cohen's class of distributions in the form

$$\int_{-\infty}^{\infty}\int_{-\infty}^{\infty} C_f(t,\omega)\,\overline{C}_g(t,\omega)\, dt\, d\omega=\left|\int_{-\infty}^{\infty} f(t)\,\overline{g}(t)\, dt\right|^2=|(f,g)|^2.$$

22. If $\psi_n(x)=2^{\frac{1}{4}}\,2^{-\frac{1}{2}n}\,(n!)^{-\frac{1}{2}}\exp(-\pi x^2)\, H_n\left(\sqrt{2\pi}\,x\right)$ is a Hermite function, where $H_n(x)$ is a Hermite polynomial of degree $n=0,1,2,3\ldots$, show that the Wigner-Ville transform of the signal $\psi_n\left(\frac{x}{\rho}\right)$ is given by

$$W_{\psi_n}(t,\omega)=2(-1)^n\exp\left[-\left(\frac{2\pi t^2}{\rho^2}+\frac{\rho^2\omega^2}{2\pi}\right)\right] L_n\left[2\left(\frac{2\pi t^2}{\rho^2}+\frac{\rho^2\omega^2}{2\pi}\right)\right],$$

where $L_n(x)$ is the Laguerre polynomial of degree $n=0,1,2,\ldots$.

Chapter 6

The Wavelet Transform and Its Basic Properties

"Wavelets are without doubt an exciting and intuitive concept. The concept brings with it a new way of thinking, which is absolutely essential and was entirely missing in previously existing algorithms."

"Today the boundaries between mathematics and signal and image processing have faded, and mathematics has benefitted from the rediscovery of wavelets by experts from other disciplines. The detour through signal and image processing was the most direct path leading from the Haar basis to Daubechies's wavelets."

Yves Meyer

6.1 Introduction

Morlet et al. (1982a,b) modified the Gabor wavelets to study the layering of sediments in a geophysical problem of oil exploration. He recognized certain difficulties of the Gabor wavelets in the sense that the Gabor analyzing function $g_{t,\omega}(\tau) = g(\tau - t)e^{i\omega\tau}$ oscillates more rapidly as the frequency ω tends to infinity. This leads to significant numerical instability in the computation of the coefficients $\left(f, g_{\omega,t}\right)$. On the other hand, $g_{t,\omega}$ oscillates very slowly at low frequencies. These difficulties led to a problem of finding a suitable reconstruction formula. In order to resolve these difficulties, Morlet first made an attempt to use analytic signals $f(t) = a(t)\exp\{i\phi(t)\}$ and then introduced the wavelet ψ defined by its Fourier transform

$$\hat{\psi}(\omega) = \sqrt{2\pi}\, \omega^2 \exp\left(-\frac{1}{2}\omega^2\right), \qquad \omega > 0. \tag{6.1.1}$$

This wavelet corresponds to an analytic signal related to the second derivative $(1-t^2)\exp\left(-\frac{1}{2}t^2\right)$ of the Gaussian function $\exp\left(-\frac{1}{2}t^2\right)$. Thus, the Morlet wavelet turned out to be the modulated Gaussian function. In fact, Morlet's ingenious idea was to filter the signal $f(t)$ with the aid of the filters $\hat{\psi}(a^m\omega)$, $m \in \mathbb{Z}$ so that

$$f(t) \to f_m(t) = \int_{-\infty}^{\infty} f(t-\tau)\, a^{-m}\, \hat{\psi}\left(a^{-m}\tau\right) d\tau. \tag{6.1.2}$$

Morlet's analysis showed that the quantity $\sum_{m\in\mathbb{Z}} \left|\hat{\psi}(a^m\omega)\right|^2$ remained constant for sufficiently small a. It also led to stable and fast reconstruction algorithms of f from f_m even when $a = 2$. Moreover, Morlet suggested sufficiently small mesh sizes so that they allow a good reconstruction algorithm of analytic signals with coefficients

$$c_{mn} = f_m\left(n\,2^m\right) = \left(f(t),\, 2^{-m}\psi\left(2^{-m}t - n\right)\right). \tag{6.1.3}$$

Thus, Morlet's remarkable analysis led to the discovery of the wavelet transform which seems to be an efficient and effective time-frequency representation algorithm. The major difference between the Morlet wavelet representation and the Gabor wavelet is that the former has a more and more acute spatial resolution as the frequency gets higher and higher.

Based on the idea of wavelets as a family of functions constructed from translation and dilation of a single function ψ, called the *mother wavelet* (or *affine coherent states*), we define *wavelets* by

$$\psi_{a,b}(t) = \frac{1}{\sqrt{|a|}}\, \psi\left(\frac{t-b}{a}\right), \qquad a, b \in \mathbb{R}, \quad a \neq 0, \tag{6.1.4}$$

where a is called a *scaling parameter* which measures the degree of compression or scale, and b is a *translation parameter* which determines the time location of the wavelet. Clearly, wavelets $\psi_{a,b}(t)$ generated by the mother wavelet ψ are somewhat similar to the Gabor wavelets $g_{t,\omega}(\tau)$ which can be considered as musical notes that oscillate at the frequency ω inside the envelope defined by $|g(\tau - t)|$ as a function of τ. If $|a| < 1$, the wavelet (6.1.4) is the compressed version (smaller support in time-domain) of the mother wavelet and corresponds mainly to higher frequencies. Thus, wavelets have time-widths adapted to their frequencies. This is the main reason for the success of the Morlet wavelets in signal processing and time-frequency signal analysis. It may be noted that the resolution of wavelets at different scales varies in the time and frequency domains as governed by the Heisenberg uncertainty principle. At large scale, the solution is coarse in the time domain and fine in the frequency domain. On the other hand, as the scale a decreases, the resolution in the time domain decreases (the time resolution becomes finer) while that in the frequency domain increases (the frequency resolution becomes coarser).

We sketch a typical mother wavelet with a compact support $[-T, T]$ in Figure 6.1(a). Different values of the parameter b represent the time localization center, and each $\psi_{a,b}(t)$ is localized around the center $t = b$. As scale parameter a varies, wavelet $\psi_{a,b}(t)$ covers different frequency ranges. Small values of $|a|$ $(0 < |a| << 1)$ result in very narrow windows and correspond to high frequencies or very fine scales $\psi_{a,b}$, as shown in Figure 6.1(b), whereas very large values of $|a|$ $(|a| >> 1)$ result in very wide windows and correspond to small frequencies or very coarse scales $\psi_{a,b}$ as shown in Figure 6.1(c). The wavelet transform (6.2.4) gives a time-frequency description of a signal f. Different shapes of the wavelets are plotted in Figures 6.1(b) and 6.1 (c).

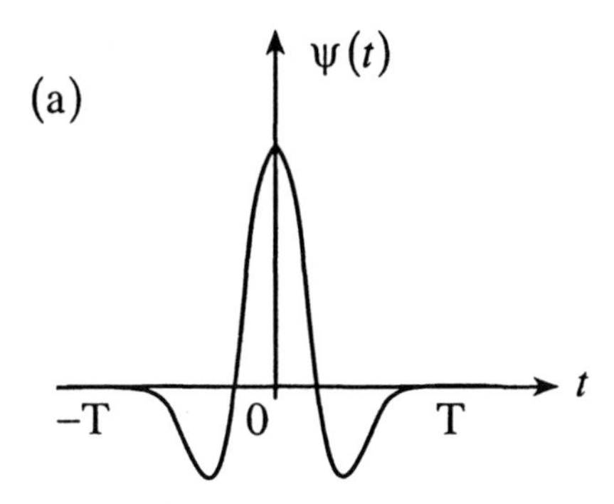

Figure 6.1(a). Typical mother wavelet.

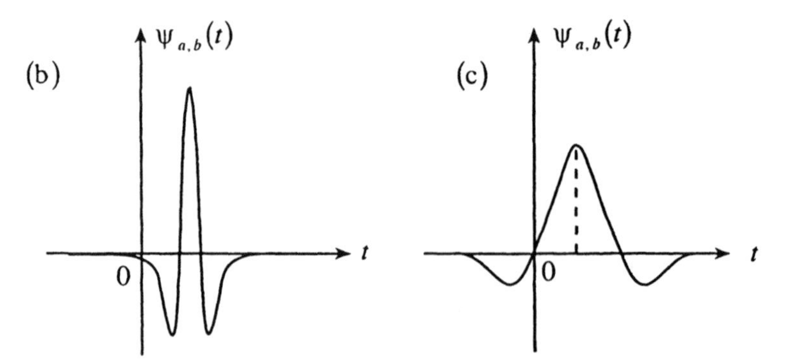

Figure 6.1(b). Compressed and translated wavelet $\psi_{a,b}(t)$ with $0<|a|<<1,\ \ b>0$;

(c) Magnified and translated wavelet $\psi_{a,b}(t)$ with $|a|>>1,\ \ b>0$.

It follows from the preceding discussion that a typical mother wavelet physically appears as a local oscillation (or wave) in which most of the energy is localized to a narrow region in the physical space. It will be shown in Section 6.2 that the time resolution σ_t and the frequency resolution σ_ω are proportional to the scale a and a^{-1}, respectively, and $\sigma_t\sigma_\omega \geq 2^{-1}$. When a decreases or increases, the frequency support of the wavelet atom is shifted toward higher or lower frequencies, respectively. Therefore, at higher frequencies, the time resolution becomes finer (better) and the frequency resolution becomes coarser (worse). On the other hand, the time resolution becomes coarser but the frequency resolution becomes finer at lower frequencies.

Morlet first called his functions "*wavelets of constant shape*" in order to contrast them with the analyzing functions in the short-time Fourier transform which do not have a constant shape. From a group-theoretic point of view, the wavelets $\psi_{a,b}(x)$ are in fact the result of the action of the operators $U(a,b)$ on the function ψ so that

$$[U(a,b)\,\psi](x)=\frac{1}{\sqrt{|a|}}\,\psi\left(\frac{x-b}{a}\right). \tag{6.1.5}$$

These operators are all unitary on the Hilbert space $L^2(\mathbb{R})$ and constitute a representation of the "$ax+b$" group

$$U(a,b)\,U(c,d)=U\,(ac,b+ad), \tag{6.1.6}$$

$$U(1,0) = Id, \tag{6.1.7}$$

$$U(a,b)^{-1} = U\left(\frac{1}{a}, -\frac{b}{a}\right). \tag{6.1.8}$$

This group representation is *irreducible*, that is, for any nonzero $f \in L^2(\mathbb{R})$, there exists no nontrivial g orthogonal to all the $U(a,b)\,f$. In other words, $U(a,b)\,f$ span the entire space. The multiplication of operators defines the product of pairs $(a,b),(c,d) \in \mathbb{R}/\{0\} \times \mathbb{R}$, that is; $(a,b) \circ (c,d) = (ac, b+ad)$. Like the operators $U(a,b)$, the pairs (a,b) together with the operation $\circ$ form a group. The coherent states associated with the $(ax+b)$-group, which are now known as wavelets, were first formulated by Aslaksen and Klauder (1968, 1969). The success of Morlet's numerical algorithms prompted Grossmann to make a more extensive study of the Morlet wavelet transform which led to the recognition that wavelets $\psi_{a,b}(t)$ correspond to a square integrable representation of the affine group.

This chapter is devoted to wavelets and wavelet transforms with examples. The basic ideas and properties of wavelet transforms are discussed with special attention given to the use of different wavelets for resolution and synthesis of signals. This is followed by the definition and properties of discrete wavelet transforms. It is important and useful to consider discrete versions of the continuous wavelet transform due to the fact that, in many applications, especially in signal and image processing, data are represented by a finite number of values.

6.2 Continuous Wavelet Transforms and Examples

An integral transform is an operator T on a space of functions for some X which is defined by

$$(Tf)(y) = \int_X K(x,y)\, f(x)\, dx.$$

The properties of the transform depend on the function K which is called the *kernel* of the transform. For example, in the case of the Fourier transform, we have $K(x,y) = e^{-ixy}$. Note that y can be interpreted as a scaling factor. We take the exponential function $\phi(x) = e^{ix}$ and then generate a family of functions by taking scaled copies of ϕ, that is, $\phi_\alpha(x) = e^{-i\alpha x}$ for all $\alpha \in \mathbb{R}$. The continuous wavelet transform is similar to the Fourier transform in the sense that it is based on a single function ψ and that this function is scaled. But, unlike the Fourier transform, we also shift the function, thus generating a two-parameter family of functions $\psi_{a,b}(t)$ defined by (6.1.4).

We next give formal definitions of a wavelet and a continuous wavelet transform of a function.

Definition 6.2.1 (Wavelet). A wavelet is a function $\psi \in L^2(\mathbb{R})$ which satisfies the condition

$$C_\psi \equiv \int_{-\infty}^{\infty} \frac{|\hat{\psi}(\omega)|^2}{|\omega|}\, d\omega < \infty, \tag{6.2.1}$$

where $\hat{\psi}(\omega)$ is the Fourier transform of $\psi(t)$.

If $\psi \in L^2(\mathbb{R})$, then $\psi_{a,b}(t) \in L^2(\mathbb{R})$ for all a, b. For

$$\|\psi_{a,b}(t)\|^2 = |a|^{-1} \int_{-\infty}^{\infty} \left|\psi\left(\frac{t-b}{a}\right)\right|^2 dt = \int_{-\infty}^{\infty} |\psi(x)|^2 dx = \|\psi\|^2. \tag{6.2.2}$$

The Fourier transform of $\psi_{a,b}(t)$ is given by

$$\hat{\psi}_{a,b}(\omega) = |a|^{-\frac{1}{2}} \int_{-\infty}^{\infty} e^{-i\omega t} \psi\left(\frac{t-b}{a}\right) dt = |a|^{\frac{1}{2}} e^{-ib\omega} \hat{\psi}(a\omega). \tag{6.2.3}$$

Definition 6.2.2 (Continuous Wavelet Transform). If $\psi \in L^2(\mathbb{R})$, and $\psi_{a,b}(t)$ is given by (6.1.4), then the integral transformation $\mathcal{W}_\psi$ defined on $L^2(\mathbb{R})$ by

$$\mathcal{W}_\psi[f](a,b) = (f, \psi_{a,b}) = \int_{-\infty}^{\infty} f(t)\, \overline{\psi_{a,b}(t)}\, dt \tag{6.2.4}$$

is called a *continuous wavelet transform* of $f(t)$. This definition allows us to make the following comments.

First, the kernel $\psi_{a,b}(t)$ in (6.2.2) plays the same role as the kernel $\exp(-i\omega t)$ in the Fourier transform. However, unlike the Fourier transformation, the continuous wavelet transform is not a *single* transform but any transform obtained in this way. Like the Fourier transformation, the continuous wavelet transformation is linear. Second, as a function of b for a fixed scaling parameter a, $\mathcal{W}_\psi[f](a,b)$ represents the detailed information contained in the signal $f(t)$ at the scale a. In fact, this interpretation motivated Morlet et al. (1982a,b) to introduce the translated and scaled versions of a single function for the analysis of seismic waves.

Using the Parseval relation of the Fourier transform, it also follows from (6.2.4) that

$$\mathcal{W}_\psi[f](a,b) = \left(f, \psi_{a,b}\right) = \frac{1}{2\pi}\left(\hat{f}, \hat{\psi}_{a,b}\right)$$

$$= \frac{1}{2\pi}\int_{-\infty}^{\infty}\left\{\sqrt{|a|}\,\hat{f}(\omega)\,\overline{\hat{\psi}}(a\,\omega)\right\} e^{ib\omega}\,d\omega, \qquad \text{by (6.2.3).}$$

This means that

$$\mathcal{F}\left\{\mathcal{W}_\psi[f](a,b)\right\} = \int_{-\infty}^{\infty} e^{-ib\omega}\,\mathcal{W}_\psi[f](a,b)\,db = \sqrt{|a|}\,\hat{f}(\omega)\,\overline{\hat{\psi}}(a\,\omega). \tag{6.2.5}$$

Example 6.2.1 (The Haar Wavelet). The Haar wavelet (Haar, 1910) is one of the classic examples. It is defined by

$$\psi(t) = \begin{Bmatrix} 1, & 0 \le t < \frac{1}{2} \\ -1, & \frac{1}{2} \le t < 1 \\ 0, & \text{otherwise} \end{Bmatrix}. \tag{6.2.6}$$

The Haar wavelet has compact support. It is obvious that

$$\int_{-\infty}^{\infty}\psi(t)\,dt = 0, \qquad \int_{-\infty}^{\infty}|\psi(t)|^2\,dt = 1.$$

This wavelet is very well-localized in the time domain, but it is not continuous. Its Fourier transform $\hat{\psi}(\omega)$ is calculated as follows:

$$\begin{aligned}\hat{\psi}(\omega) &= \int_0^{\frac{1}{2}} e^{-i\omega t}\,dt - \int_{\frac{1}{2}}^{1} e^{-i\omega t}\,dt \\ &= \frac{1}{(-i\omega)}\left\{\left[e^{-i\omega t}\right]_0^{\frac{1}{2}} - \left[e^{-i\omega t}\right]_{\frac{1}{2}}^{1}\right\} \\ &= \left(\frac{i}{\omega}\right)\left(2e^{-\frac{i\omega}{2}} - 1 - e^{-i\omega}\right) \\ &= \frac{\sin^2\frac{\omega}{4}}{\left(\frac{\omega}{4}\right)} \exp\left[\frac{i}{2}(\pi - \omega)\right] \\ &= i\exp\left(-\frac{i\omega}{2}\right)\frac{\sin^2\left(\frac{\omega}{4}\right)}{\left(\frac{\omega}{4}\right)}.\end{aligned} \tag{6.2.7}$$

and

$$\int_{-\infty}^{\infty} \frac{|\hat{\psi}(\omega)|^2}{|\omega|}\,d\omega = 16\int_{-\infty}^{\infty} |\omega|^{-3}\left|\sin\frac{\omega}{4}\right|^4 d\omega < \infty. \tag{6.2.8}$$

Both $\psi(t)$ and $\hat{\psi}(\omega)$ are plotted in Figure 6.2. These figures indicate that the Haar wavelet has good time localization but poor frequency localization. The function $|\hat{\psi}(\omega)|$ is even, attains its maximum at the frequency $\omega_0 \sim 4.662$, and decays slowly as ω^{-1} as $\omega \to \infty$, which means that it does not have compact support in the frequency domain. Indeed, the discontinuity of ψ causes a slow decay of $\hat{\psi}$ as $\omega \to \infty$. Its discontinuous nature is a serious weakness in many applications. However, the Haar wavelet is one of the most fundamental examples that illustrate major features of the general wavelet theory.

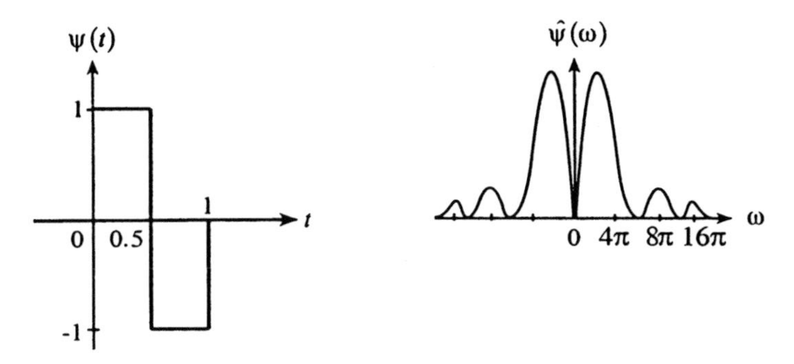

Figure 6.2. The Haar wavelet and its Fourier transform.

Theorem 6.2.1 If ψ is a wavelet and ϕ is a bounded integrable function, then the convolution function $\psi * \phi$ is a wavelet.

Proof. Since

$$\begin{aligned}
\int_{-\infty}^{\infty} |\psi * \phi(x)|^2 dx &= \int_{-\infty}^{\infty} \left| \int_{-\infty}^{\infty} \psi(x-u)\, \phi(u)\, du \right|^2 dx \\
&\le \int_{-\infty}^{\infty} \left(\int_{-\infty}^{\infty} |\psi(x-u)| |\phi(u)|\, du \right)^2 dx \\
&= \int_{-\infty}^{\infty} \left(\int_{-\infty}^{\infty} |\psi(x-u)| |\phi(u)|^{1/2} |\phi(u)|^{1/2}\, du \right)^2 dx \\
&\le \int_{-\infty}^{\infty} \left(\int_{-\infty}^{\infty} |\psi(x-u)^2| |\phi(u)|\, du \int_{-\infty}^{\infty} |\phi(u)|\, du \right) dx \\
&\le \int_{-\infty}^{\infty} |\phi(u)|\, du \int_{-\infty}^{\infty}\int_{-\infty}^{\infty} |\psi(x-u)|^2 |\phi(u)|\, dx\, du \\
&= \left(\int_{-\infty}^{\infty} |\phi(u)|\, du \right)^2 \int_{-\infty}^{\infty} |\psi(x)|^2 dx < \infty,
\end{aligned}$$

we have $\psi * \phi \in L^2(\mathbb{R})$. Moreover,

$$\int_{-\infty}^{\infty} \frac{\left|\mathcal{F}\{\psi * \phi\}\right|^2}{|\omega|}\, d\omega = \int_{-\infty}^{\infty} \frac{\left|\hat{\psi}(\omega)\, \hat{\phi}(\omega)\right|^2}{|\omega|}\, d\omega$$

$$= \int_{-\infty}^{\infty} \frac{\left|\hat{\psi}(\omega)\right|^2}{|\omega|} \left|\hat{\phi}(\omega)\right|^2 d\omega$$

$$\leq \int_{-\infty}^{\infty} \frac{\left|\hat{\psi}(\omega)\right|^2}{|\omega|}\, d\omega \,\sup\left|\hat{\phi}(\omega)\right|^2 < \infty.$$

Thus, the convolution function $\psi * \phi$ is a wavelet.

Example 6.2.2 This example illustrates how to generate other wavelets by using Theorem 6.2.1. For example, if we take the Haar wavelet and convolute it with the following function

$$\phi(t) = \begin{Bmatrix} 0, & t < 0 \\ 1, & 0 \leq t \leq 1 \\ 0, & t \geq 1 \end{Bmatrix}, \tag{6.2.9}$$

we obtain a simple wavelet, as shown in Figure 6.3.

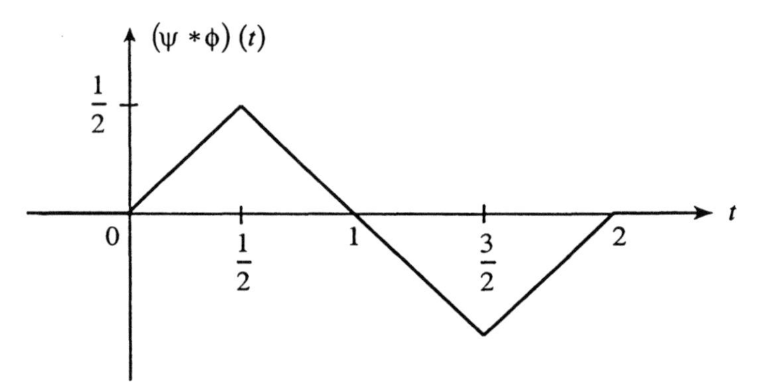

Figure 6.3. The wavelet $(\psi * \phi)(t)$.

Example 6.2.3 The convolution of the Haar wavelet with $\phi(t) = \exp\left(-t^2\right)$ generates a smooth wavelet, as shown in Figure 6.4.

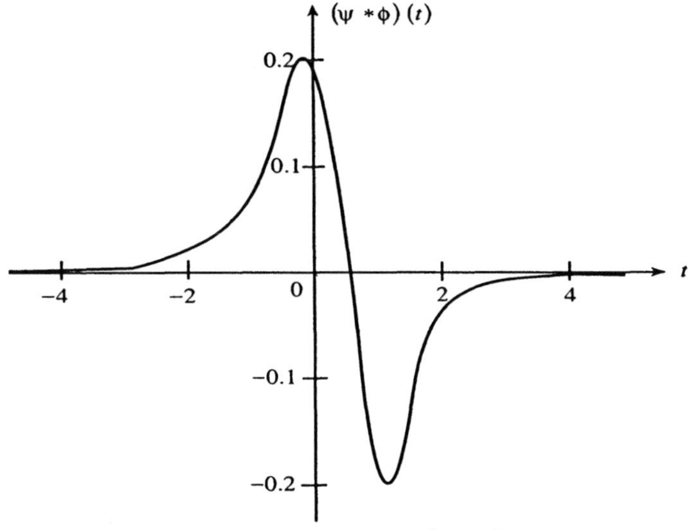

Figure 6.4. The wavelet $(\psi * \phi)(t)$.

In order for the wavelets to be useful analyzing functions, the mother wavelet must have certain properties. One such property is defined by the condition (6.1.4) which guarantees the existence of the inversion formula for the continuous wavelet transform. Condition (6.1.4) is usually referred to as the *admissibility condition* for the mother wavelet. If $\psi \in L^1(\mathbb{R})$, then its Fourier transform $\hat{\psi}$ is continuous. Since $\hat{\psi}$ is continuous, C_ψ can be finite only if $\hat{\psi}(0) = 0$ or, equivalently, $\int_{-\infty}^{\infty} \psi(t)\, dt = 0$. This means that ψ must be an oscillatory function with zero mean. Condition (6.2.1) also imposes a restriction on the rate of decay of $|\hat{\psi}(\omega)|^2$ and is required in finding the inverse of the continuous wavelet transform.

In addition to the admissibility condition, there are other properties that may be useful in particular applications. For example, we may want to require that ψ be n times continuously differentiable or infinitely differentiable. If the Haar wavelet is convoluted $(n+1)$ times with the function ϕ given in Example 6.2.2, then the resulting function $\psi * \phi * \cdots * \phi$ is an n times differentiable wavelet. The function in Figure 6.4 is an infinitely differentiable wavelet. The so-called "Mexican hat wavelet" is another example of an infinitely differentiable (or smooth) wavelet.

Example 6.2.4 (The Mexican Hat Wavelet). The Mexican hat wavelet is defined by the second derivative of a Gaussian function as

$$\psi(t) = \left(1 - t^2\right) \exp\left(-\frac{t^2}{2}\right) = -\frac{d^2}{dt^2} \exp\left(\frac{-t^2}{2}\right) = \psi_{1,0}(t), \qquad (6.2.10)$$

$$\hat{\psi}(\omega) = \hat{\psi}_{1,0}(\omega) = \sqrt{2\pi}\, \omega^2 \exp\left(-\frac{\omega^2}{2}\right). \qquad (6.2.11)$$

In contrast to the Haar wavelet, the Mexican hat wavelet is a C^{∞}-function. It has two vanishing moments. The Mexican hat wavelet $\psi_{1,0}(t)$ and its Fourier transform are shown in Figures 6.5(a) and 6.5(b). This wavelet has excellent localization in time and frequency domains and clearly satisfies the admissibility condition.

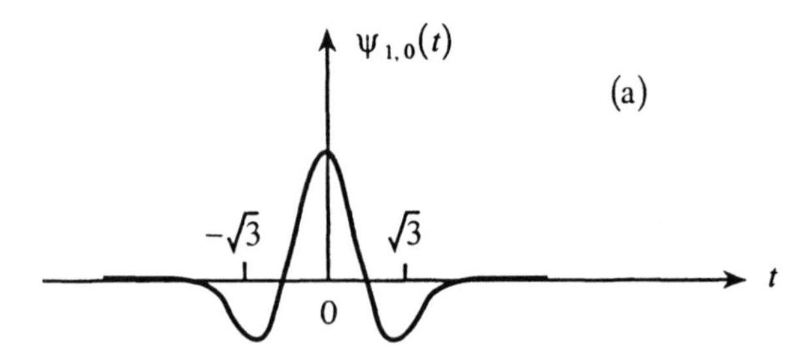

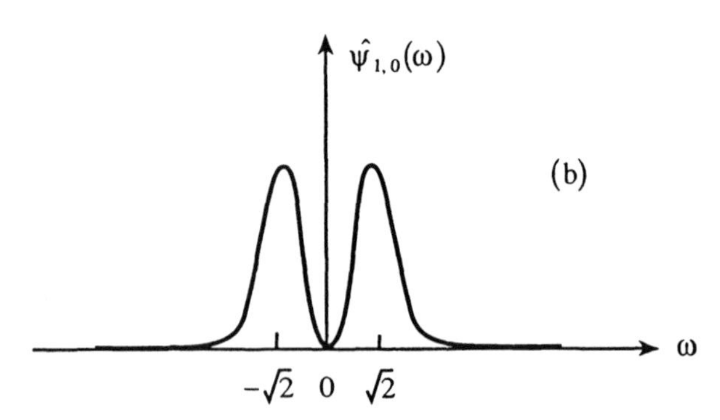

Figure 6.5(a) The Mexican hat wavelet $\psi_{1,0}(t)$ and (b) its Fourier transform $\hat{\psi}_{1,0}(\omega)$.

Two other wavelets, $\psi_{\frac{3}{2},-2}(t)$ and $\psi_{\frac{1}{4},\sqrt{2}}(t)$, from the mother wavelet (6.2.10) can be obtained. These three wavelets, $\psi_{1,0}(t)$, $\psi_{\frac{3}{2},-2}(t)$ and $\psi_{\frac{1}{4},\sqrt{2}}(t)$, are shown in Figure 6.6 (i), (ii), and (iii), respectively.

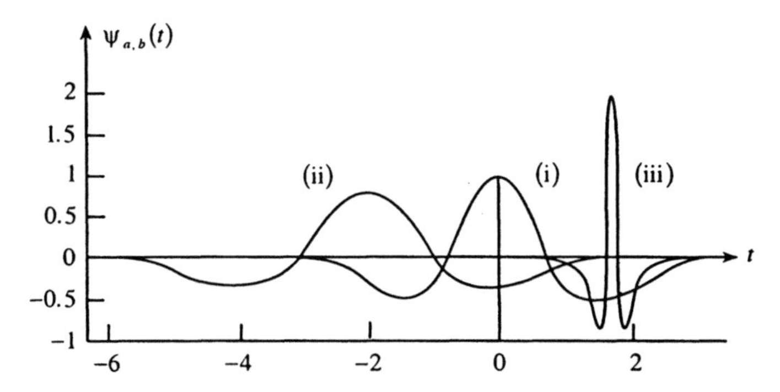

Figure 6.6. Three wavelets $\psi_{1,0}(t)$, $\psi_{\frac{3}{2},-2}(t)$, and $\psi_{\frac{1}{4},\sqrt{2}}(t)$.

Example 6.2.5 (The Morlet Wavelet). The Morlet wavelet is defined by

$$\psi(t) = \exp\left(i\omega_0 t - \frac{t^2}{2}\right), \tag{6.2.12}$$

$$\hat{\psi}(\omega) = \sqrt{2\pi}\,\exp\left[-\frac{1}{2}(\omega - \omega_0)^2\right]. \tag{6.2.13}$$

The Morlet wavelet and its Fourier transform are plotted in Figure 6.7.

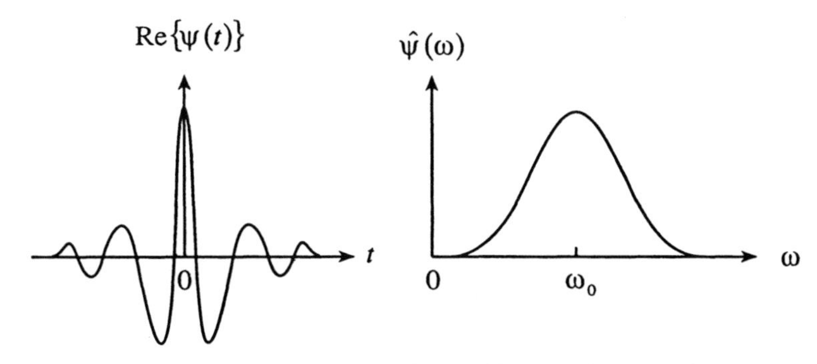

Figure 6.7. The Morlet wavelet and its Fourier transform.

Another desirable property of wavelets is the so-called "localization property." We want ψ to be well-localized in both time and frequency. In other words, ψ and its derivatives must decay very rapidly. For frequency localization, $\hat{\psi}(\omega)$ must decay sufficiently rapidly as $\omega \to \infty$ and $\hat{\psi}(\omega)$ should be flat in the neighborhood of $\omega = 0$. The flatness at $\omega = 0$ is associated with the number of vanishing moments of ψ. The kth moment of ψ is defined by

$$m_k = \int_{-\infty}^{\infty} t^k \psi(t)\, dt. \tag{6.2.14}$$

A wavelet is said to have n vanishing moments if

$$\int_{-\infty}^{\infty} t^k \psi(t)\, dt = 0 \quad \text{for} \quad k = 0, 1, \ldots, n. \tag{6.2.15}$$

Or, equivalently,

$$\left[\frac{d^k \hat{\psi}(\omega)}{d\omega^k}\right]_{\omega=0} = 0 \quad \text{for} \quad k = 0, 1, \ldots, n. \tag{6.2.16}$$

Wavelets with a larger number of vanishing moments result in more flatness when frequency ω is small.

The smoothness and localization properties of wavelet ψ combined with the admissibility condition (6.2.1) suggest that

(i) wavelets are bandpass filters; that is, the frequency response decays sufficiently rapidly as $\omega \to \infty$ and is zero as $\omega \to 0$.

(ii) $\psi(t)$ is the impulse response of the filter which again decays very rapidly as t increases, and it is an oscillatory function with mean zero. Usually, wavelets are assumed to be absolutely square integrable functions, that is, $\psi \in L^2(\mathbb{R})$.

In quantum mechanics, quantities such as $|\psi(t)|^2$ and $|\hat{\psi}(\omega)|^2$ are interpreted as the probability density functions in the time and frequency domains respectively, with mean values defined by

$$\langle t \rangle = \int_{-\infty}^{\infty} t\,|\psi(t)|^2\, dt \quad \text{and} \quad \langle \omega \rangle = \frac{1}{2\pi} \int_{0}^{\infty} \omega\,|\hat{\psi}(\omega)|^2\, d\omega. \tag{6.2.17a,b}$$

The *time resolution* (or the *time spread*) and the *frequency resolution* (or the *frequency spread*) associated with a mother wavelet ψ around the mean values are defined by

$$\sigma_t^2 = \int_{-\infty}^{\infty} (t - \langle t \rangle)^2 |\psi(t)|^2 \, dt, \tag{6.2.18}$$

$$\sigma_\omega^2 = \frac{1}{2\pi} \int_0^{\infty} (\omega - \langle \omega \rangle)^2 |\hat{\psi}(\omega)|^2 \, d\omega. \tag{6.2.19}$$

Thus, for any $\psi \in L^2(\mathbb{R})$, the time and frequency resolutions of the mother wavelet are governed by the Heisenberg uncertainty principle, that is, $\sigma_t \sigma_\omega \geq \frac{1}{2}$.

It is easy to verify that the time-frequency resolution of a wavelet $\psi_{a,b}$ depends on the time-frequency spread of the mother wavelet. We define the energy spread of $\psi_{a,b}$ around b by

$$\sigma_{t,a,b}^2 = \int_{-\infty}^{\infty} (t-b)^2 |\psi_{a,b}(t)|^2 \, dt, \quad (t - b = x)$$

$$= a^2 \int_{-\infty}^{\infty} x^2 |\psi(x)|^2 \, dx = a^2 \sigma_t^2, \tag{6.2.20}$$

where σ_t^2 is defined by (6.2.18) around the zero mean. Clearly, the wavelets have good time resolution for small values of a which correspond to high frequencies or small scales. Scale can be defined as the inverse of frequency.

On the other hand, the Fourier transform $\hat{\psi}_{a,b}(\omega)$ of $\psi_{a,b}(t)$ is given by (6.2.3), so its mean value is $\frac{1}{a}\langle \omega \rangle$. The energy spread of $\hat{\psi}_{a,b}(\omega)$ around $\frac{1}{a}\langle \omega \rangle$ is defined by

$$\sigma_{\omega,a,b}^2 = \frac{1}{2\pi} \int_0^{\infty} \left(\omega - \frac{1}{a}\langle \omega \rangle \right)^2 |\hat{\psi}_{a,b}(\omega)|^2 \, d\omega, \quad (a\omega = x)$$

$$= \frac{1}{2\pi} \int_0^{\infty} \frac{1}{a^2} (x - \langle \omega \rangle)^2 |\hat{\psi}(x)|^2 \, dx = \frac{1}{a^2} \sigma_\omega^2. \tag{6.2.21}$$

This reveals that wavelets have good frequency resolution for large values of the scale a.

Thus, the time-frequency resolution of wavelets $\psi_{a,b}$ is independent of the time location but depends only on the scale a. The energy spread of the wavelet $\psi_{a,b}$ corresponds to a Heisenberg time-frequency rectangle at $\left(b, \frac{1}{a}\langle\omega\rangle\right)$ of sides $a\,\sigma_t$ along the time axis and $\frac{1}{a}\,\sigma_\omega$ along the frequency axis. The area of the rectangle is equal to $\sigma_t\sigma_\omega$ for all scales and is governed by the Heisenberg uncertainty principle, that is, $\sigma_{t,a,b}\,\sigma_{\omega,a,b} = \left(a\,\sigma_t\right)\left(a^{-1}\,\sigma_\omega\right) = \sigma_t\sigma_\omega \geq \frac{1}{2}$.

We close this section by introducing a scaled version of a mother wavelet in the form

$$\psi_a(t) = |a|^{-p}\,\psi\left(\frac{t}{a}\right), \tag{6.2.22}$$

where p is a fixed but arbitrary nonnegative parameter. In particular, when $p = \frac{1}{2}$, the translated version of $\psi_a(t)$ defined by (6.2.22) reduces to wavelets (6.1.4).

Clearly, if $\hat{\psi}(\omega)$ is the Fourier transform of $\psi(t)$, then the Fourier transform of the dilated version of $\psi(t)$ is given by

$$\mathscr{F}\{D_a\,\psi(t)\} = \mathscr{F}\left\{\frac{1}{\sqrt{a}}\,\psi\left(\frac{t}{a}\right)\right\} = D_{\frac{1}{a}}\,\hat{f}(\omega) = \sqrt{a}\,\hat{f}(a\omega), \tag{6.2.23}$$

where $a > 0$. Thus, a contraction in one domain is accompanied by a magnification in the other but in a non-uniform manner over the time-frequency plane. A typical wavelet and its dilations are sketched in Figures 6.8 (a), (b), and (c) together with the corresponding Fourier transforms.

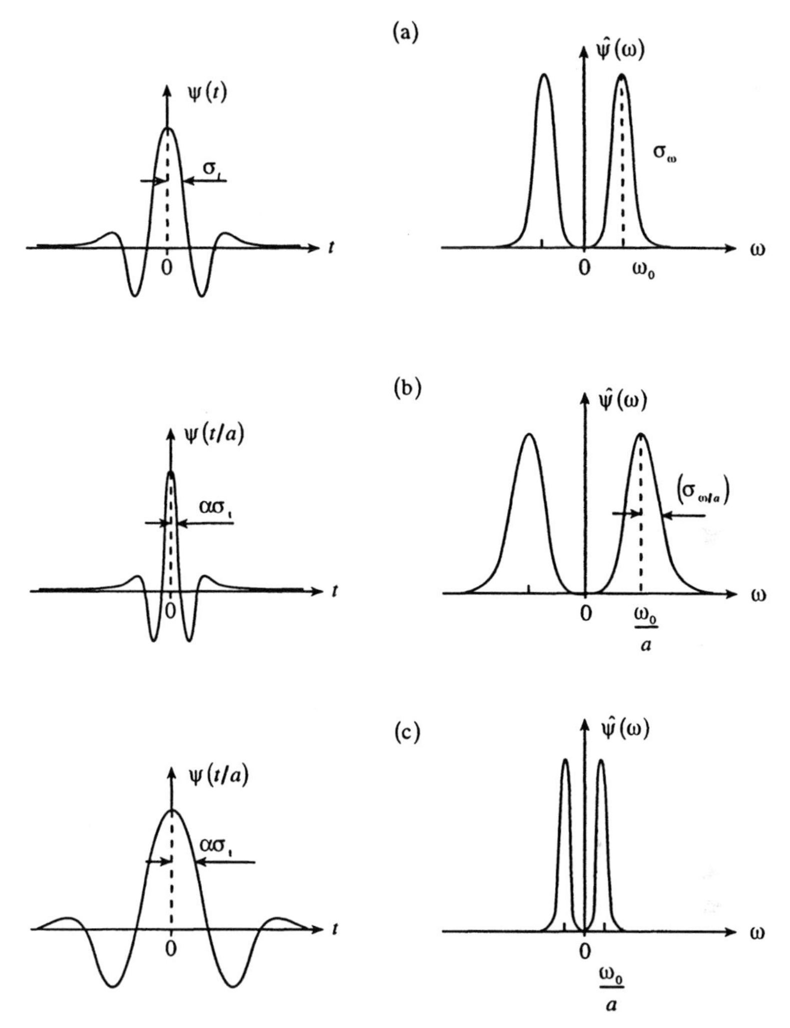

Figure 6.8. Typical wavelet and its dilations with the corresponding Fourier transforms for (a) $a=1$, (b) $0<a<<1$, and (c) $a>>1$ in the time-frequency domains.

If $p=1$ in (6.2.22), the integral

$$\int_{-\infty}^{\infty} \psi_a(t)\,dt = \int_{-\infty}^{\infty} \psi(x)\,dx$$

does not depend on the scaling parameter a. On the other hand, the choice of $p=0$ is found to be convenient for the study of orthonormal bases of wavelets.

However, the specific value of p is completely irrelevant to the general theory of wavelets, so appropriate choices are used in the literature.

For an arbitrary $p \geq 0$, the time localization of signals is obtained by the translated versions of $\psi_a(t)$. If $\psi(t)$ is supported on an interval of length ℓ near $t = 0$, then wavelets can be defined by the translated and scaled versions of the mother wavelet ψ as

$$\psi_{a,b}(t) = \psi_a(t - b) = |a|^{-p} \psi\left(\frac{t - b}{a}\right). \tag{6.2.24}$$

Obviously, this is supported on an interval of length $|a|\ell$ near $t = b$.

If we assume that $\psi \in L^2(\mathbb{R})$, then the square of the norm of $\psi_{a,b}$ is

$$\|\psi_{a,b}\|^2 = |a|^{-2p} \int_{-\infty}^{\infty} \left|\psi\left(\frac{t - b}{a}\right)\right|^2 dt = |a|^{1-2p} \|\psi\|^2. \tag{6.2.25}$$

6.3 Basic Properties of Wavelet Transforms

The following theorem gives several properties of continuous wavelet transforms.

Theorem 6.3.1 If ψ and ϕ are wavelets and f, g are functions which belong to $L^2(\mathbb{R})$, then

(i) (*Linearity*)

$$\mathcal{W}_\psi(\alpha f + \beta g)(a, b) = \alpha\left(\mathcal{W}_\psi f\right)(a, b) + \beta\left(\mathcal{W}_\psi g\right)(a, b), \tag{6.3.1}$$

where α and β are any two scalars.

(ii) (*Translation*)

$$\left(\mathcal{W}_\psi(T_c f)\right)(a, b) = \left(\mathcal{W}_\psi f\right)(a, b - c), \tag{6.3.2}$$

where T_c is the *translation operator* defined by $T_c f(t) = f(t - c)$.

(iii) (*Dilation*)

$$\left(\mathcal{W}_\psi(D_c f)\right)(a, b) = \frac{1}{\sqrt{c}}\left(\mathcal{W}_\psi f\right)\left(\frac{a}{c}, \frac{b}{c}\right), \quad c > 0, \tag{6.3.3}$$

where D_c is a *dilation operator* defined by $D_c f(t) = \frac{1}{c} f\left(\frac{t}{c}\right)$, $c > 0$.

(iv) (*Symmetry*)

$$\left(\mathcal{W}_{\psi} f\right)(a,b) = \overline{\left(\mathcal{W}_{f} \psi\right)\left(\frac{1}{a}, -\frac{b}{a}\right)}, \qquad a \neq 0. \tag{6.3.4}$$

(v) (*Parity*)

$$\left(\mathcal{W}_{P\psi} P f\right)(a,b) = \left(\mathcal{W}_{\psi} f\right)(a, -b), \tag{6.3.5}$$

where P is the *parity operator* defined by $P f(t) = f(-t)$.

(vi) (*Antilinearity*)

$$\left(\mathcal{W}_{\alpha\psi+\beta\phi} f\right)(a,b) = \overline{\alpha}\left(\mathcal{W}_{\psi} f\right)(a,b) + \overline{\beta}\left(\mathcal{W}_{\phi} f\right)(a,b), \tag{6.3.6}$$

for any scalars α, β.

(vii) $$\left(\mathcal{W}_{T_c\psi} f\right)(a,b) = \left(\mathcal{W}_{\psi} f\right)(a, b + ca). \tag{6.3.7}$$

(viii) $$\left(\mathcal{W}_{D_c\psi} f\right)(a,b) = \frac{1}{\sqrt{c}}\left(\mathcal{W}_{\psi} f\right)(ac, b), \qquad c > 0. \tag{6.3.8}$$

Proofs of the above properties are straightforward and are left as exercises.

Theorem 6.3.2 (Parseval's Formula for Wavelet Transforms). If $\psi \in L^2(\mathbb{R})$ and $\left(\mathcal{W}_{\psi} f\right)(a,b)$ is the wavelet transform of f defined by (6.2.4), then, for any functions $f, g \in L^2(\mathbb{R})$, we obtain

$$\int_{-\infty}^{\infty}\int_{-\infty}^{\infty} \left(\mathcal{W}_{\psi} f\right)(a,b)\overline{\left(\mathcal{W}_{\psi} g\right)(a,b)}\, \frac{db\,da}{a^2} = C_{\psi}\left(f, g\right), \tag{6.3.9}$$

where

$$C_{\psi} = \int_{-\infty}^{\infty} \frac{\left|\hat{\psi}(\omega)\right|^2}{|\omega|}\, d\omega < \infty. \tag{6.3.10}$$

Proof. By Parseval's relation (3.4.37) for the Fourier transforms, we have

$$\left(\mathcal{W}_\psi f\right)(a,b) = \int_{-\infty}^{\infty} f(t)\, |a|^{-\frac{1}{2}}\, \overline{\psi\left(\frac{t-b}{a}\right)}\, dt$$

$$= \left(f, \psi_{a,b}\right)$$

$$= \frac{1}{2\pi}\left(\hat{f}, \hat{\psi}_{a,b}\right)$$

$$= \frac{1}{2\pi}\int_{-\infty}^{\infty} \hat{f}(\omega)|a|^{\frac{1}{2}} e^{ib\omega}\, \overline{\hat{\psi}(a\omega)}\, d\omega \qquad \text{by (6.2.3).} \tag{6.3.11}$$

Similarly,

$$\overline{\left(\mathcal{W}_\psi g\right)(a,b)} = \int_{-\infty}^{\infty} \overline{g(t)}\, |a|^{-\frac{1}{2}}\, \psi\left(\frac{t-b}{a}\right) dt$$

$$= \frac{1}{2\pi}\int_{-\infty}^{\infty} \overline{\hat{g}(\sigma)}\, |a|^{\frac{1}{2}}\, e^{-ib\sigma}\, \hat{\psi}(a\sigma)\, d\sigma. \tag{6.3.12}$$

Substituting (6.3.11) and (6.3.12) in the left-hand side of (6.3.9) gives

$$\int_{-\infty}^{\infty}\int_{-\infty}^{\infty} \left(\mathcal{W}_\psi f\right)(a,b)\overline{\left(\mathcal{W}_\psi g\right)(a,b)}\, \frac{db\,da}{a^2}$$

$$= \frac{1}{(2\pi)^2}\int_{-\infty}^{\infty}\int_{-\infty}^{\infty} \frac{db\,da}{a^2} \int_{-\infty}^{\infty}\int_{-\infty}^{\infty} |a|\hat{f}(\omega)\, \overline{\hat{g}}(\sigma)\, \overline{\hat{\psi}}(a\omega)\, \hat{\psi}(a\sigma)$$

$$\times \exp\{ib(\omega-\sigma)\}\, d\omega\, d\sigma,$$

which is, by interchanging the order of integration,

$$= \frac{1}{2\pi}\int_{-\infty}^{\infty} \frac{da}{|a|}\int_{-\infty}^{\infty}\int_{-\infty}^{\infty} \hat{f}(\omega)\, \overline{\hat{g}}(\sigma)\, \overline{\hat{\psi}}(a\omega)\, \hat{\psi}(a\sigma)\, d\omega\, d\sigma$$

$$\times \frac{1}{2\pi}\int_{-\infty}^{\infty} \exp\{ib(\omega-\sigma)\}\, db$$

$$= \frac{1}{2\pi}\int_{-\infty}^{\infty} \frac{da}{|a|}\int_{-\infty}^{\infty}\int_{-\infty}^{\infty} \hat{f}(\omega)\, \overline{\hat{g}}(\sigma)\, \overline{\hat{\psi}}(a\omega)\, \hat{\psi}(a\sigma)\, \delta(\sigma-\omega)\, d\omega\, d\sigma$$

$$= \frac{1}{2\pi}\int_{-\infty}^{\infty} \frac{da}{|a|}\int_{-\infty}^{\infty} \hat{f}(\omega)\, \overline{\hat{g}}(\omega)\, |\hat{\psi}(a\omega)|^2\, d\omega$$

which is, again interchanging the order of integration and putting $a\omega = x$,

$$= \frac{1}{2\pi} \int_{-\infty}^{\infty} \hat{f}(\omega)\, \overline{\hat{g}}(\omega)\, d\omega \cdot \int_{-\infty}^{\infty} \frac{|\hat{\psi}(x)|^2}{|x|}\, dx$$

$$= C_\psi \cdot \frac{1}{2\pi} \left(\hat{f}(\omega), \hat{g}(\omega) \right).$$

Theorem 6.3.3 (Inversion Formula). If $f \in L^2(\mathbb{R})$, then f can be reconstructed by the formula

$$f(t) = \frac{1}{C_\psi} \int_{-\infty}^{\infty}\int_{-\infty}^{\infty} (\mathcal{W}_\psi f)(a,b)\, \psi_{a,b}(t)\, \frac{db\,da}{a^2}, \tag{6.3.13}$$

where the equality holds almost everywhere.

Proof. For any $g \in L^2(\mathbb{R})$, we have, from Theorem 6.3.2,

$$\begin{aligned}
C_\psi (f,g) &= \left(\mathcal{W}_\psi f, \mathcal{W}_\psi g \right) \\
&= \int_{-\infty}^{\infty}\int_{-\infty}^{\infty} (\mathcal{W}_\psi f)(a,b)\, \overline{(\mathcal{W}_\psi g)(a,b)}\, \frac{db\,da}{a^2} \\
&= \int_{-\infty}^{\infty}\int_{-\infty}^{\infty} (\mathcal{W}_\psi f)(a,b)\, \overline{\int_{-\infty}^{\infty} g(t)\, \overline{\psi_{a,b}(t)}\, dt}\, \frac{db\,da}{a^2} \\
&= \int_{-\infty}^{\infty}\int_{-\infty}^{\infty}\int_{-\infty}^{\infty} (\mathcal{W}_\psi f)(a,b)\, \psi_{a,b}(t)\, \frac{db\,da}{a^2}\, \overline{g(t)}\, dt \\
&= \left(\int_{-\infty}^{\infty}\int_{-\infty}^{\infty} (\mathcal{W}_\psi f)(a,b)\, \psi_{a,b}(t)\, \frac{db\,da}{a^2}, g \right).
\end{aligned} \tag{6.3.14}$$

Since g is an arbitrary element of $L^2(\mathbb{R})$, the inversion formula (6.3.13) follows.

If $f = g$ in (6.3.13), then

$$\int_{-\infty}^{\infty}\int_{-\infty}^{\infty} \left| (\mathcal{W}_\psi f)(a,b) \right|^2 \frac{da\,db}{a^2} = C_\psi \|f\|^2 = C_\psi \int_{-\infty}^{\infty} |f(t)|^2\, dt. \tag{6.3.15}$$

This shows that, except for the factor C_ψ, the wavelet transform is an isometry from $L^2(\mathbb{R})$ to $L^2(\mathbb{R}^2)$.

6.4 The Discrete Wavelet Transforms

It has been stated in the last section that the continuous wavelet transform (6.2.4) is a two-parameter representation of a function. In many applications, especially in signal processing, data are represented by a finite number of values, so it is important and often useful to consider discrete versions of the continuous wavelet transform (6.2.4). From a mathematical point of view, a continuous representation of a function of two continuous parameters a, b in (6.2.4) can be converted into a discrete one by assuming that a and b take only integral values. It turns out that it is better to discretize it in a different way. First, we fix two positive constants a_0 and b_0 and define

$$\psi_{m,n}(x) = a_0^{-m/2}\, \psi\left(a_0^{-m}\, x - nb_0\right), \tag{6.4.1}$$

where both m and $n \in \mathbb{Z}$. Then, for $f \in L^2(\mathbb{R})$, we calculate the discrete wavelet coefficients $\left(f, \psi_{m,n}\right)$. The fundamental question is whether it is possible to determine f completely by its wavelet coefficients or discrete wavelet transform which is defined by

$$\left(\mathcal{W}_\psi f\right)(m,n) = \left(f, \psi_{m,n}\right) = \int_{-\infty}^{\infty} f(t)\, \overline{\psi}_{m,n}(t)\, dt$$

$$= a_0^{-\frac{m}{2}} \int_{-\infty}^{\infty} f(t)\, \overline{\psi}\left(a_0^{-m}\, t - n\, b_0\right) dt, \tag{6.4.2}$$

where both f and ψ are continuous, $\psi_{00}(t) = \psi(t)$. It is noted that the discrete wavelet transform (6.4.2) can also be obtained directly from the corresponding continuous version by discretizing the parameters $a = a_0^m$ and $b = n\, b_0\, a_0^m$ (m,n are integers). The discrete wavelet transform represents a function by a countable set of wavelet coefficients, which correspond to points on a two-dimensional grid or lattice of discrete points in the scale-time domain indexed by m and n. If the set $\left\{\psi_{m,n}(t)\right\}$ defined by (6.4.1) is complete in $L^2(\mathbb{R})$ for some choice of ψ, a, and b, then the set is called an *affine wavelet*. Then, we can express any $f(t) \in L^2(\mathbb{R})$ as the superposition

$$f(t) = \sum_{m,n=-\infty}^{\infty} \left(f, \psi_{m,n}\right) \psi_{m,n}(t). \tag{6.4.3}$$

Such complete sets are called *frames*. They are not yet a basis. Frames do not satisfy the Parseval theorem for the Fourier series, and the expansion in terms of frames is not unique. In fact, it can be shown that

$$A\,\|f\|^2 \le \sum_{m,n=-\infty}^{\infty} \left|\left(f, \psi_{m,n}\right)\right|^2 \le B\,\|f\|^2, \tag{6.4.4}$$

where A and B are constants. The set $\{\psi_{m,n}(t)\}$ constitutes a frame if $\psi(t)$ satisfies the admissibility condition and $0 < A < B < \infty$.

For computational efficiency, $a_0 = 2$ and $b_0 = 1$ are commonly used so that results lead to a binary dilation of 2^{-m} and a dyadic translation of $n\,2^m$.

Therefore, a practical sampling lattice is $a = 2^m$ and $b = n\,2^m$ in (6.4.1) so that

$$\psi_{m,n}(t) = 2^{-\frac{m}{2}}\,\psi\left(2^{-m}\,t - n\right). \tag{6.4.5}$$

With this octave time scale and dyadic translation, the sampled values of $(a, b) = (2^m, n\,2^m)$ are shown in Figure 6.9, which represents the dyadic sampling grid diagram for the discrete wavelet transform. Each node corresponds to a wavelet basis function $\psi_{m,n}(t)$ with scale 2^{-m} and time shift $n\,2^{-m}$.

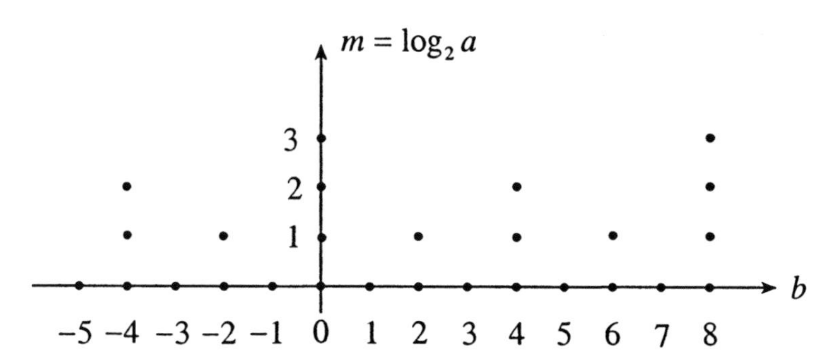

Figure 6.9. Dyadic sampling grid for the discrete wavelet transform.

The answer to the preceding question is positive if the wavelets form a complete system in $L^2(\mathbb{R})$. The problem is whether there exists another function $g \in L^2(\mathbb{R})$ such that

$$\left(f,\psi_{m,n}\right)=\left(g,\psi_{m,n}\right) \qquad \text{for all} \qquad m,n\in\mathbb{Z}$$

implies $f=g$.

In practice, we expect much more than that: we want $\left(f,\psi_{m,n}\right)$ and $\left(g,\psi_{m,n}\right)$ to be "close" if f and g are "close." This will be guaranteed if there exists a $B>0$ independent of f such that

$$\sum_{m,n=-\infty}^{\infty}\left|\left(f,\psi_{m,n}\right)\right|^2 \le B\,\|f\|^2. \tag{6.4.6}$$

Similarly, we want f and g to be "close" if $\left(f,\psi_{m,n}\right)$ and $\left(g,\psi_{m,n}\right)$ are "close." This is important because we want to be sure that when we neglect some small terms in the representation of f in terms of $\left(f,\psi_{m,n}\right)$, the reconstructed function will not differ much from f. The representation will have this property if there exists an $A>0$ independent of f, such that

$$A\,\|f\|^2 \le \sum_{m,n=-\infty}^{\infty}\left|\left(f,\psi_{m,n}\right)\right|^2. \tag{6.4.7}$$

These two requirements are best studied in terms of the so-called frames.

Definition 6.4.1 (Frames). A sequence $\{\phi_1,\phi_2,\ldots\}$ in a Hilbert space H is called a *frame* if these exist $A,B>0$ such that

$$A\,\|f\|^2 \le \sum_{n=1}^{\infty}\left|\left(f,\phi_n\right)\right|^2 \le B\,\|f\|^2 \tag{6.4.8}$$

for all $f\in H$. The constants A and B are called *frame bounds*. If $A=B$, then the frame is called *tight*.

If $\{\phi_n\}$ is an orthonormal basis, then it is a tight frame since $\sum_{n=1}^{\infty}\left|\left(f,\phi_n\right)\right|^2=\|f\|^2$ for all $f\in H$. The vectors $(1,0)$, $\left(-\frac{1}{2},\frac{\sqrt{3}}{2}\right)$, $\left(-\frac{1}{2},-\frac{\sqrt{3}}{2}\right)$ form a tight frame in $\mathbb{C}^2$ which is not a basis.

As pointed out above, we want the family of functions $\psi_{m,n}$ to form a frame in $L^2(\mathbb{R})$. Obviously, the double indexing of the functions is irrelevant. The following theorem gives fairly general sufficient conditions for a sequence $\left(\psi_{m,n}\right)$ to constitute a frame in $L^2(\mathbb{R})$.

Theorem 6.4.1 If ψ and a_0 are such that

(i)
$$\inf_{1\le|\omega|\le a_0} \sum_{m=-\infty}^{\infty} \left|\hat{\psi}\left(a_0^m \omega\right)\right|^2 > 0,$$

(ii)
$$\sup_{1\le|\omega|\le a_0} \sum_{m=-\infty}^{\infty} \left|\hat{\psi}\left(a_0^m \omega\right)\right|^2 \ge 0,$$

and

(iii)
$$\sup_{\omega\in\mathbf{R}} \sum_{m=-\infty}^{\infty} \left|\hat{\psi}\left(a_0^m \omega\right)\right| \left|\hat{\psi}\left(a_0^m \omega + x\right)\right| \le C\left(1+|x|\right)^{-(1+\varepsilon)}$$

for some $\varepsilon > 0$ and some constant C, then there exists $\tilde{b}$ such that $\psi_{m,n}$ form a frame in $L^2(\mathbb{R})$ for any $b_0 \in \left(0, \tilde{b}\right)$.

Proof. Suppose $f \in L^2(\mathbb{R})$. Then,

$$\sum_{m,n=-\infty}^{\infty} \left|\langle f, \psi_{m,n}\rangle\right|^2 = \sum_{m,n=-\infty}^{\infty} \left|\int_{-\infty}^{\infty} f(x)\, a_0^{-m/2}\, \overline{\psi\left(a_0^m x - nb_0\right)}\, dx\right|^2$$

$$\sum_{m,n=-\infty}^{\infty} \left|\int_{-\infty}^{\infty} \hat{f}(\omega)\, a_0^{m/2}\, \overline{\hat{\psi}\left(a_0^m \omega\right)}\, e^{ib_0 a_0^m n\omega}\, d\omega\right|^2 = P$$

by the general Parseval relation (see Theorem 4.11.13 of Debnath and Mikusinski, 1999), basic properties of the Fourier transform (see Theorem 4.11.5 of Debnath and Mikusinski, 1999), and the fact that we sum over all integers. Since, for any $s > 0$, the integral $\int_{-\infty}^{\infty} g(t)\, dt$ can be written as

$$\sum_{l=-\infty}^{\infty} \int_0^s g(t + ls)\, dt,$$

by taking $s = \dfrac{2\pi}{b_0 a_0^m}$, we obtain

$$P=\sum_{m,n=-\infty}^{\infty} a_0^m \left| \sum_{l=-\infty}^{\infty} \int_0^s e^{2\pi i n\omega/s} \hat{f}(\omega+ls) \overline{\hat{\psi}\left(a_0^m(\omega+ls)\right)} \, d\omega \right|^2$$

$$=\sum_{m,n=-\infty}^{\infty} a_0^m \left| \int_0^s e^{2\pi i n\omega/s} \left(\sum_{l=-\infty}^{\infty} \hat{f}(\omega+ls) \overline{\hat{\psi}\left(a_0^m(\omega+ls)\right)} \right) d\omega \right|^2$$

$$=\sum_{m=-\infty}^{\infty} a_0^m \, s \int_0^s \left| \sum_{l=-\infty}^{\infty} \hat{f}(\omega+ls) \overline{\hat{\psi}\left(a_0^m(\omega+ls)\right)} \right|^2 d\omega = Q$$

by Parseval's formula for trigonometric Fourier series.

Since

$$\left| \sum_{l=-\infty}^{\infty} \hat{f}(\omega+ls) \overline{\hat{\psi}\left(a_0^m(\omega+ls)\right)} \right|^2$$

$$=\left(\sum_{l=-\infty}^{\infty} \hat{f}(\omega+ls) \overline{\hat{\psi}\left(a_0^m(\omega+ls)\right)} \right) \left(\sum_{k=-\infty}^{\infty} \overline{\hat{f}(\omega+ks)} \hat{\psi}\left(a_0^m(\omega+ks)\right) \right)$$

and

$$F(\omega)=\sum_{k=-\infty}^{\infty} \overline{\hat{f}(\omega+ks)} \hat{\psi}\left(a_0^m(\omega+ks)\right)$$

is a periodic function with a period of s, we have

$$\int_0^s \left(\sum_{l=-\infty}^{\infty} \hat{f}(\omega+ls) \overline{\hat{\psi}\left(a_0^m(\omega+ls)\right)} \right) F(\omega) \, d\omega$$

$$=\int_{-\infty}^{\infty} \hat{f}(\omega) \overline{\hat{\psi}\left(a_0^m \omega\right)} F(\omega) \, d\omega$$

$$=\sum_{k=-\infty}^{\infty} \int_{-\infty}^{\infty} \hat{f}(\omega) \overline{\hat{\psi}\left(a_0^m \omega\right)} \overline{\hat{f}(\omega+ks)} \hat{\psi}\left(a_0^m(\omega+ks)\right) d\omega.$$

Consequently,

$$Q=\frac{2\pi}{b_0} \sum_{m,k=-\infty}^{\infty} \int_{-\infty}^{\infty} \hat{f}(\omega) \overline{\hat{f}(\omega+ks)} \overline{\hat{\psi}\left(a_0^m s\right)} \hat{\psi}\left(a_0^m(\omega+ks)\right) d\omega$$

$$=\frac{2\pi}{b_0} \int_{-\infty}^{\infty} \left|\hat{f}(\omega)\right|^2 \sum_{m=-\infty}^{\infty} \left|\hat{\psi}\left(a_0^m \omega\right)\right|^2 d\omega$$

$$+\frac{2\pi}{b_0} \sum_{\substack{m,k=-\infty \\ k\neq 0}}^{\infty} \int_{-\infty}^{\infty} \hat{f}(\omega) \overline{\hat{f}(\omega+ks)} \overline{\hat{\psi}\left(a_0^m s\right)} \hat{\psi}\left(a_0^m(\omega+ks)\right) d\omega.$$

To find a bound on the second summation, we apply the Schwarz inequality:

$$\left|\left(\frac{2\pi}{b_0}\right)\sum_{\substack{m,k=-\infty\\k\neq 0}}^{\infty}\int_{-\infty}^{\infty}\hat{f}(\omega)\,\overline{\hat{f}(\omega+ks)\,\hat{\psi}\left(a_0^m\,\omega\right)}\,\hat{\psi}\left(a_0^m(\omega+ks)\right)d\omega\right|$$

$$\leq\left(\frac{2\pi}{b_0}\right)\sum_{\substack{m,k=-\infty\\k\neq 0}}^{\infty}\left(\int_{-\infty}^{\infty}\left|\hat{f}(\omega)\right|^2\left|\hat{\psi}\left(a_0^m\,\omega\right)\right|\left|\hat{\psi}\left(a_0^m(\omega+ks)\right)\right|d\omega\right)^{1/2}$$

$$\times\left(\int_{-\infty}^{\infty}\left|\hat{f}(\omega+ks)\right|^2\left|\hat{\psi}\left(a_0^m\,\omega\right)\right|\left|\hat{\psi}\left(a_0^m(\omega+ks)\right)\right|d\omega\right)^{1/2}=R.$$

Then, by first changing the variables in the second factor and using Hölder's inequality (see Theorem 1.2.1 of Debnath and Mikusinski, 1999), we have

$$R=\left(\frac{2\pi}{b_0}\right)\sum_{\substack{m,k=-\infty\\k\neq 0}}^{\infty}\left(\int_{-\infty}^{\infty}\left|\hat{f}(\omega)\right|^2\left|\hat{\psi}\left(a_0^m\,\omega\right)\right|\left|\hat{\psi}\left(a_0^m(\omega+ks)\right)\right|d\omega\right)^{1/2}$$

$$\times\left(\int_{-\infty}^{\infty}\left|\hat{f}(\omega)\right|^2\left|\hat{\psi}\left(a_0^m(\omega-ks)\right)\right|\left|\hat{\psi}\left(a_0^m\,\omega\right)\right|d\omega\right)^{1/2}$$

$$\leq\left(\frac{2\pi}{b_0}\right)\sum_{\substack{k=-\infty\\k\neq 0}}^{\infty}\left(\int_{-\infty}^{\infty}\left|\hat{f}(\omega)\right|^2\sum_{m=-\infty}^{\infty}\left|\hat{\psi}\left(a_0^m\,\omega\right)\right|\left|\hat{\psi}\left(a_0^m(\omega+ks)\right)\right|d\omega\right)^{1/2}$$

$$\times\left(\int_{-\infty}^{\infty}\left|\hat{f}(\omega)\right|^2\sum_{m=-\infty}^{\infty}\left|\hat{\psi}\left(a_0^m(\omega-ks)\right)\right|\left|\hat{\psi}\left(a_0^m\,\omega\right)\right|d\omega\right)^{1/2}=S.$$

If we denote

$$\beta(\xi)=\sup_{\omega\in\mathbb{R}}\sum_{m=-\infty}^{\infty}\left|\hat{\psi}\left(a_0^m\,\omega\right)\right|\left|\hat{\psi}\left(a_0^m\,\omega+\xi\right)\right|,$$

then

$$S=\left(\frac{2\pi}{b_0}\right)\|f\|^2\sum_{\substack{k=-\infty\\k\neq 0}}^{\infty}\left[\beta\left(a_0^m ks\right)\beta\left(-a_0^m ks\right)\right]^{1/2}$$

$$=\left(\frac{2\pi}{b_0}\right)\|f\|^2\sum_{\substack{k=-\infty\\k\neq 0}}^{\infty}\left[\beta\left(\frac{2\pi k}{b_0}\right)\beta\left(-\frac{2\pi k}{b_0}\right)\right]^{1/2}.$$

Consequently, if we denote

$$A=\left(\frac{2\pi}{b_0}\right)\left\{\sup_{\omega\in R}\sum_{m=-\infty}^{\infty}\left|\hat{\psi}\left(a_0^m\,\omega\right)\right|^2-\sum_{\substack{k=-\infty\\k\neq 0}}^{\infty}\left[\beta\left(\frac{2\pi k}{b_0}\right)\beta\left(-\frac{2\pi k}{b_0}\right)\right]^{1/2}\right\}$$

and

$$B=\left(\frac{2\pi}{b_0}\right)\left\{\inf_{\omega\in R}\sum_{m=-\infty}^{\infty}\left|\hat{\psi}\left(a_0^m\,\omega\right)\right|^2+\sum_{\substack{k=-\infty\\k\neq 0}}^{\infty}\left[\beta\left(\frac{2\pi k}{b_0}\right)\beta\left(-\frac{2\pi k}{b_0}\right)\right]^{1/2}\right\},$$

we conclude

$$A\,\|f\|^2\leq\sum_{m,n=-\infty}^{\infty}\left|\left\langle f,\psi_{m,n}\right\rangle\right|^2\leq B\,\|f\|^2.$$

Since $\beta(\xi)\leq C\left(1+|\xi|\right)^{-(1+\varepsilon)}$, we find

$$\begin{aligned}\sum_{\substack{k=-\infty\\k\neq 0}}^{\infty}\left[\beta\left(\frac{2\pi k}{b_0}\right)\beta\left(-\frac{2\pi k}{b_0}\right)\right]^{1/2}&=2\sum_{k=1}^{\infty}\left[\beta\left(\frac{2\pi k}{b_0}\right)\beta\left(-\frac{2\pi k}{b_0}\right)\right]^{1/2}\\&\leq 2C\sum_{k=1}^{\infty}\left(1+\frac{2\pi k}{b_0}\right)^{-(1+\varepsilon)}\\&\leq 2C\int_0^{\infty}\left(1+\frac{2\pi k}{b_0}\right)^{-(1+\varepsilon)}dt\\&=\frac{Cb_0}{\pi\varepsilon}.\end{aligned}$$

Since $\left(\dfrac{Cb_0}{\pi\varepsilon}\right)\to 0$ as $b_0\to 0$ and $\inf\limits_{1\leq|\omega|\leq a_0}\sum\limits_{m=-\infty}^{\infty}\left|\hat{\psi}\left(a_0^m\,\omega\right)\right|^2>0$, there exists $\tilde{b}$ such that $A>0$ for any $b_0\in\left(0,\tilde{b}\right)$. Moreover, since $\sup\limits_{1\leq|\omega|\leq a_0}\sum\limits_{m=-\infty}^{\infty}\left|\hat{\psi}\left(a_0^m\,\omega\right)\right|^2<\infty$, we also have $B<\infty$ for all $b_0\in\left(0,\tilde{b}\right)$. Thus, $\psi_{m,n}$ constitute a frame for all such b_0. This completes the proof.

The major problem of this section is reconstruction of f from $\left(f,\psi_{m,n}\right)$ and representation of f in terms of $\psi_{m,n}$. For a complete orthonormal system $\{\phi_n\}$, both questions are answered by the equality

$$f=\sum_{n=1}^{\infty}\left(f,\phi_{m,n}\right)\phi_n. \tag{6.4.9}$$

However, since we do not have orthogonality, the problem is more complete for frames.

Definition 6.4.2 (Frame Operator). Let $\{\phi_1, \phi_2, \ldots\}$ be a frame in a Hilbert space H. The operator F from H into l^2 defined by

$$F\{f\} = \{(f, \phi_n)\}$$

is called a *frame operator*.

Lemma 6.4.1 Let F be a frame operator. Then, F is a linear, invertible, and bounded operator. Its inverse F^{-1} is also a bounded operator.

The proof is easy and left as an exercise.

Consider the adjoint operator F^* of a frame operator F associated with frame $\{\phi_n\}$. For any $\{c_n\} \in l^2$, we have

$$\left(F^*(c_n), f\right) = \left((c_n), F f\right) = \sum_{n=1}^{\infty} c_n (\phi_n, f) = \left(\sum_{n=1}^{\infty} c_n \phi_n, f\right).$$

Thus, the adjoint operator of a frame operator has the form

$$F^*(c_n) = \sum_{n=1}^{\infty} c_n \phi_n. \tag{6.4.10}$$

Since

$$\sum_{n=1}^{\infty} |(f, \phi_n)|^2 = \|F f\|^2 = \left(F^* F f, f\right),$$

we note that the condition (6.4.4) can be expressed as

$$A \mathcal{I} \le F^* F \le B \mathcal{I},$$

where the inequality $\le$ is to be understood in the sense defined in Section 4.6 (see Debnath and Mikusinski, 1999).

Theorem 6.4.2 Let $\{\phi_1, \phi_2, \phi_3, \ldots\}$ be frame bounds A and B and let F be the associated frame operator. Define

$$\tilde{\phi}_n = \left(F^* F\right)^{-1} \phi_n.$$

Then, $\{\tilde{\phi}_n\}$ is a frame with frame bounds $\frac{1}{B}$ and $\frac{1}{A}$.

Proof. By Corollary 4.5.1 as stated by Debnath and Mikusinski (1999), we have $(F^* F)^{-1} = ((F^* F)^{-1})^*$. Consequently,

$$(f, \tilde{\phi}_n) = (f, (F^* F)^{-1} \phi_n) = ((F^* F)^{-1} f, \phi_n)$$

and then

$$\begin{aligned}\sum_{n=1}^{\infty} \left|\left(f, \{\tilde{\phi}_n\}\right)\right|^2 &= \sum_{n=1}^{\infty} \left|\left((F^* F)^{-1} f, \phi_n\right)\right|^2 \\ &= \left\|F(F^* F)^{-1} f\right\|^2 \\ &= \left(F(F^* F)^{-1} f, F(F^* F)^{-1} f\right) \\ &= \left((F^* F)^{-1} f, f\right).\end{aligned}$$

Now, since $A\mathcal{I} \le F^* F \le B\mathcal{I}$, Theorem 4.6.5 proved by Debnath and Mikusinski (1999) implies

$$\frac{1}{B}\mathcal{I} \le (F^* F)^{-1} \le \frac{1}{A}\mathcal{I},$$

which leads to the inequality

$$\frac{1}{B}\|f\|^2 \le \sum_{n=1}^{\infty} \left|\left(f, \{\tilde{\phi}_n\}\right)\right|^2 \le \frac{1}{A}\|f\|^2.$$

This proves the theorem. The sequence $(\tilde{\phi}_n)$ is called the *dual frame*.

Lemma 6.4.2 Let F be the frame operator associated with the frame $\{\phi_1, \phi_2, \phi_3, \ldots\}$ and $\tilde{F}$ be the frame operator associated with the dual frame $\{\tilde{\phi}_1, \tilde{\phi}_2, \tilde{\phi}_3, \ldots\}$. Then,

$$\tilde{F}^* F = \mathcal{I} = F^* \tilde{F}.$$

Proof. Since

$$F\left(F^* F\right)^{-1} f = \left\{\left(\left(F^* F\right)^{-1} f, \phi_n\right)\right\} = \left\{\left(f, \tilde{\phi}_n\right)\right\} = \tilde{F} f, \tag{6.4.11}$$

we have

$$\tilde{F}^* F = \left(F\left(F^* F\right)^{-1}\right)^* F = \left(F^* F\right)^{-1} F^* F = \mathcal{I}$$

and

$$F^* \tilde{F} = F^* F\left(F^* F\right)^{-1} = \mathcal{I}.$$

Now, we are ready to state and prove the main theorem, which answers the question of reconstructability of f from the sequence $\left\{\left(f, \phi_n\right)\right\}$.

Theorem 6.4.3 Let $\{\phi_1, \phi_2, \phi_3, \ldots\}$ constitute a frame in a Hilbert space H, and let $\left\{\tilde{\phi}_1, \tilde{\phi}_2, \tilde{\phi}_3, \ldots\right\}$ be the dual frame. Then, for any $f \in H$,

$$f = \sum_{n=1}^{\infty} \left(f, \phi_n\right) \tilde{\phi}_n \tag{6.4.12}$$

and

$$f = \sum_{n=1}^{\infty} \left(f, \tilde{\phi}_n\right) \phi_n. \tag{6.4.13}$$

Proof. Let f be the frame operator associated with $\{\phi_n\}$, and let $\tilde{F}$ be the frame operator associated with the dual frame $\left\{\tilde{\phi}_n\right\}$. Since $I = \tilde{F}^* F$, for any $f \in H$, we have

$$f = \tilde{F}^* F f = \tilde{F}^* \left\{\left(f, \phi_n\right)\right\} = \sum_{n=1}^{\infty} \left(f, \phi_n\right) \tilde{\phi}_n$$

by (6.4.10). The proof of the other equality is similar.

Using the definition of mother wavelet (6.1.4), we can introduce a family Ψ of vectors $\psi_{a,b} \in L^2$ by

$$\Psi = \left\{\psi_{a,b} \middle| \left(a, b\right) \in \mathbb{R}^2 \right\}. \tag{6.4.14}$$

We can then define a frame operator T which transforms a time signal $f \in L^2$ into a function Tf so that

$$Tf(a,b) = (f, \psi_{a,b}) = \mathcal{W}[f](a,b). \tag{6.4.15}$$

Thus, the wavelet transform can be interpreted as the frame operator T corresponding to the family Ψ. In view of the measure $d\mu$ defined in the (a,b) plane by

$$d\mu = d\mu(a,b) = \frac{1}{|a|^2}\, da\, db, \tag{6.4.16}$$

we interpret the integral in (6.3.9) as the inner product in a Hilbert space $H = L^2(\mathbb{R}^2, d\mu)$ so that (6.3.9) can be expressed in terms of the norm as

$$\|\mathcal{W}f\|^2 = C_\psi \|f\|^2 \tag{6.4.17}$$

for all $f \in L^2$ and C_ψ is defined by (6.3.10). Thus, (6.4.17) can be interpreted in terms of frame. The family Ψ represents a tight frame for any mother wavelet with frame constant C_ψ.

6.5 Orthonormal Wavelets

Since the discovery of wavelets, orthonormal wavelets with good time-frequency localization are found to play an important role in wavelet theory and have a great variety of applications. In general, the theory of wavelets begins with a single function $\psi \in L^2(\mathbb{R})$, and a family of functions $\psi_{m,n}$ is generated from this single function ψ by the operation of binary dilations (that is, dilation by 2^m) and dyadic translation of $n2^{-m}$ so that

$$\begin{aligned}\psi_{m,n}(x) &= 2^{m/2}\,\psi\left(2^m\left(x - \frac{n}{2^m}\right)\right), \quad m,n \in \mathbb{Z} \\ &= 2^{m/2}\,\psi\left(2^m x - n\right),\end{aligned} \tag{6.5.1}$$

where the factor $2^{m/2}$ is introduced to ensure orthonormality.

A situation of interest in applications is to deal with an orthonormal family $\{\psi_{m,n}\}$, that is,

$$(\psi_{m,n}, \psi_{k,\ell}) = \int_{-\infty}^{\infty} \psi_{m,n}(x)\, \psi_{k,\ell}(x)\, \psi_{k,\ell}(x)\, dx = \delta_{m,k}\,\delta_{n,\ell}, \tag{6.5.2}$$

where $m, n, k, \ell \in \mathbb{Z}$.

To show how the inner products behave in this formalism, we prove the following lemma.

Lemma 6.5.1 If ψ and $\phi \in L^2(\mathbb{R})$, then

$$\left(\psi_{m,k},\ \phi_{m,\ell}\right) = \left(\psi_{n,k},\ \phi_{n,\ell}\right), \tag{6.5.3}$$

for all $m,\ n,\ k,\ \ell \in \mathbb{Z}$.

Proof. We have

$$\left(\psi_{m,k},\ \psi_{m,\ell}\right) = \int_{-\infty}^{\infty} 2^m\, \psi\left(2^m x - k\right) \phi\left(2^m x - \ell\right) dx$$

which is, by letting $2^m x = 2^n t$,

$$= \int_{-\infty}^{\infty} 2^n\, \psi\left(2^n t - k\right) \phi\left(2^n t - \ell\right) dt$$

$$= \left(\psi_{n,k},\ \phi_{n,\ell}\right).$$

Moreover,

$$\left\|\psi_{m,n}\right\| = \left\|\psi\right\|.$$

Definition 6.5.1 (Orthonormal Wavelet). A wavelet $\psi \in L^2(\mathbb{R})$ is called *orthonormal* if the family of functions $\psi_{m,n}$ generated from ψ by (6.5.1) is orthonormal.

As in the classical Fourier series, the wavelet series for a function $f \in L^2(\mathbb{R})$ based on a given orthonormal wavelet ψ is given by

$$f(x) = \sum_{m,n=-\infty}^{\infty} c_{m,n}\, \psi_{m,n}(x), \tag{6.5.4}$$

where the wavelet coefficients $c_{m,n}$ are given by

$$c_{m,n} = \left(f,\ \psi_{m,n}\right) \tag{6.5.5}$$

and the double wavelet series (6.5.4) converges to the function f in the L^2-norm.

The simplest example of an orthonormal wavelet is the classic Haar wavelet (6.2.6). To prove this fact, we note that the norm of ψ defined by (6.2.6) is one and the same for $\psi_{m,n}$ defined by (6.5.1). We have

$$\left(\psi_{m,n}, \psi_{k,\ell}\right) = \int_{-\infty}^{\infty} 2^{m/2}\, \psi\left(2^m x - n\right) 2^{k/2}\, \psi\left(2^k x - \ell\right) dx$$

which is, by the change of variables $2^m x - n = t$,

$$= 2^{k/2}\, 2^{-m/2} \int_{-\infty}^{\infty} \psi(t)\, \psi\left(2^{k-m}(t+n) - \ell\right) dt. \tag{6.5.6}$$

For $m = k$, this result gives

$$\left(\psi_{m,n}, \psi_{m,\ell}\right) = \int_{-\infty}^{\infty} \psi(t)\, \psi(t + n - \ell)\, dt = \delta_{0,n-\ell} = \delta_{n,\ell}, \tag{6.5.7}$$

where $\psi(t) \neq 0$ in $0 \leq t < 1$ and $\psi\left(t - \overline{\ell - n}\right) \neq 0$ in $\ell - n \leq t < 1 + \ell - n$, and these intervals are disjoint from each other unless $n = \ell$.

We now consider the case $m \neq k$. In view of symmetry, it suffices to consider the case $m > k$. Putting $r = m - k > 0$ in (6.5.6), we can complete the proof by showing that, for $k \neq m$,

$$\left(\psi_{m,n}, \psi_{k,\ell}\right) = 2^{r/2} \int_{-\infty}^{\infty} \psi(t)\, \psi\left(2^r t + s\right) dt = 0, \tag{6.5.8}$$

where $s = 2^r n - \ell \in \mathbb{Z}$.

In view of the definition of the Haar wavelet ψ, we must prove that the integral in (6.5.8) vanishes for $k \neq m$. In other words, it suffices to show

$$\int_0^{\frac{1}{2}} \psi\left(2^r t + s\right) dt - \int_{\frac{1}{2}}^{1} \psi\left(2^r t + s\right) dt = 0.$$

Invoking a simple change of variables, $2^r t + s = x$, we find

$$\int_s^a \psi(x)\, dx - \int_a^b \psi(x)\, dx = 0, \tag{6.5.9}$$

where $a = s + 2^{r-1}$ and $b = s + 2^r$.

A simple argument reveals that $[s, a]$ contains the support $[0,1]$ of ψ so that the first integral in (6.5.9) is identically zero. Similarly, the second integral is also zero. This completes the proof that the Haar wavelet ψ is orthonormal.

Example 6.5.1 (Discrete Haar Wavelet). The discrete Haar wavelet is defined by

$$\psi_{m,n}(t) = 2^{-m/2}\,\psi\left(2^{-m}t - n\right)$$
$$= \begin{Bmatrix} 1, & 2^m n \le t < 2^m n + 2^{m-1} \\ -1, & 2^m n + 2^{m-1} \le t < 2^m n + 2^m \\ 0, & \text{otherwise} \end{Bmatrix}, \tag{6.5.10}$$

where ψ is the Haar wavelet defined by (6.2.6).

Since $\{\psi_{m,n}(t)\}$ is an orthonormal set, any function $f \in L^2(\mathbb{R})$ can be expanded in the wavelet series in the form

$$f(t) = \sum_{m,n=-\infty}^{\infty} \left(f,\, \psi_{m,n}\right)\psi_{m,n}, \tag{6.5.11}$$

where the coefficients $\left(f, \psi_{m,n}\right)$ satisfy (6.4.4) with $A = B = 1$. To prove this, we assume

$$f(t) = \begin{Bmatrix} a, & 0 \le t < \dfrac{1}{2} \\ b, & \dfrac{1}{2} \le t < 1 \\ 0, & \text{otherwise} \end{Bmatrix}. \tag{6.5.12}$$

Evidently, it follows that

$$\left(f,\, \psi_{m,n}\right) = 0, \quad \text{for} \quad m < 0 \text{ or } n \neq 0,$$

and

$$\left(f,\, \psi_{0,0}\right) = \frac{1}{2}(a - b), \tag{6.5.13}$$

$$\left(f,\, \psi_{1,0}\right) = \frac{1}{\sqrt{2}}\left(\frac{a}{2} + \frac{b}{2}\right), \tag{6.5.14}$$

$$\left(f,\, \psi_{2,0}\right) = \frac{1}{2}\left(\frac{a}{2} + \frac{b}{2}\right), \tag{6.5.15}$$

$$\cdots$$

$$\left(f,\, \psi_{m,0}\right) = 2^{-\frac{m}{2}}\left(\frac{a}{2} + \frac{b}{2}\right). \tag{6.5.16}$$

Consequently,

$$\left(f, \psi_{m,0}\right) \psi_{0,0}(t) = \begin{Bmatrix} \frac{1}{2}(a-b), & 0 \le t < \frac{1}{2} \\ -\frac{1}{2}(a-b), & \frac{1}{2} \le t < 1 \end{Bmatrix}, \tag{6.5.17}$$

and for $m \ge 1$,

$$\left(f, \psi_{m,0}\right) \psi_{m,0}(t) = 2^{-m}\left(\frac{a}{2} + \frac{b}{2}\right), \qquad 0 \le t \le 1. \tag{6.5.18}$$

Finally, it turns out that

$$\sum_{m=0}^{\infty} \left(f, \psi_{m,0}\right) \psi_{m,0}(t) = \begin{Bmatrix} \frac{1}{2}(a-b) + \frac{1}{2}(a+b) \sum_{m=1}^{\infty} 2^{-m}, \ 0 \le t < \frac{1}{2} \\ -\frac{1}{2}(a-b) + \frac{1}{2}(a+b) \sum_{m=1}^{\infty} 2^{-m}, \ \frac{1}{2} \le t < 1 \end{Bmatrix}. \tag{6.5.19}$$

Since $\sum_{m=1}^{\infty} 2^{-m} = 1$, result (6.5.19) reduces to

$$\sum_{m=0}^{\infty} \left(f, \psi_{m,0}\right) \psi_{m,0}(t) = \begin{Bmatrix} a, & 0 \le t < \frac{1}{2} \\ b, & \frac{1}{2} \le t < 1 \end{Bmatrix} \tag{6.5.20}$$

which confirms (6.5.11).

Moreover, it follows from (6.5.13) and (6.5.16) that

$$\begin{aligned} \sum_{m=0}^{\infty} \left|\left(f, \psi_{m,0}\right)\right|^2 &= \left|\left(f, \psi_{0,0}\right)\right|^2 + \sum_{m=1}^{\infty} \left|\left(f, \psi_{m,0}\right)\right|^2 \\ &= \left(\frac{a}{2} - \frac{b}{2}\right)^2 + \left(\frac{a}{2} + \frac{b}{2}\right)^2 \\ &= \frac{1}{2}\left(a^2 + b^2\right) = \int_0^1 f^2(t)\, dt. \end{aligned} \tag{6.5.21}$$

This verifies (6.4.4).

Example 6.5.2 (The Discrete Shannon Wavelet). The Shannon function ψ whose Fourier transform satisfies

$$\hat{\psi}(\omega) = \hat{\chi}_I(\omega), \tag{6.5.22}$$

where $I=[-2\pi,-\pi]\cup[\pi,2\pi]$, is called the *Shannon wavelet.* Thus, this wavelet $\psi(t)$ can directly be obtained from the inverse Fourier transform of $\hat{\psi}(\omega)$ so that

$$\psi(t)=\frac{1}{2\pi}\int_{-\infty}^{\infty}e^{i\omega t}\,\hat{\psi}(\omega)\,d\omega$$

$$=\frac{1}{2\pi}\left[\int_{-2\pi}^{-\pi}e^{i\omega t}\,d\omega+\int_{\pi}^{2\pi}e^{i\omega t}\,d\omega\right]$$

$$=\frac{1}{\pi t}\left(\sin 2\pi t-\sin\pi t\right)=\frac{\sin\left(\frac{\pi t}{2}\right)}{\left(\frac{\pi t}{2}\right)}\cos\left(\frac{3\pi t}{2}\right). \tag{6.5.23}$$

This function ψ is orthonormal to its translates by integers. This follows from Parseval's relation

$$\left(\psi(t),\psi(t-n)\right)=\frac{1}{2\pi}\left(\hat{\psi},\,e^{-in\omega}\,\hat{\psi}\right)$$

$$=\frac{1}{2\pi}\int_{-\infty}^{\infty}\hat{\psi}(\omega)\,e^{in\omega}\,\overline{\hat{\psi}}(\omega)\,d\omega$$

$$=\frac{1}{2\pi}\int_{-2\pi}^{2\pi}e^{in\omega}\,d\omega=\delta_{0,n}.$$

It can easily be verified that the wavelet basis is now given by

$$\psi_{m,n}(t)=2^{-m/2}\,\psi\left(2^{-m}t-n-\frac{1}{2}\right),\qquad m,n\in\mathbb{Z},$$

where $\psi\left(t-n-\frac{1}{2}\right)$, $n\in\mathbb{Z}$ is an orthonormal basis for ω_0 and $\psi_{m,n}(t)$, $n\in\mathbb{Z}$ is a basis for functions supported on the interval

$$\left[-2^{-n+1}\pi,\,-2^{-m}\pi\right]\cup\left[2^{-m}\pi,2^{-m+1}\pi\right].$$

Since m may be an arbitrarily large integer, we have a basis for $L^2(\mathbb{R})$ functions. Both $\psi(t)$ and $\hat{\psi}(\omega)$ are shown in Figure 6.10.

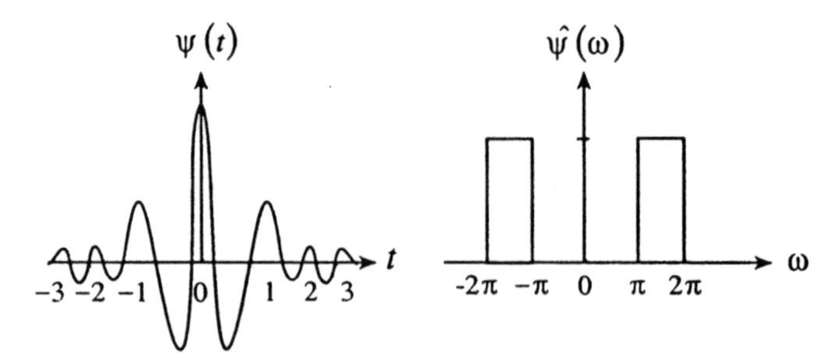

Figure 6.10. The Shannon wavelet and its Fourier transform.

It may be observed that the Shannon wavelet is not well-localized (noncompact) in the time domain and decays as fast as t^{-1}, and hence, it has poor time localization. However, its Fourier transform is band-limited (compact support) and hence has good frequency localization. These features exhibit a striking contrast with the Haar wavelet.

With the dyadic sampling lattice $a = 2^m$ and $b = 2^m n$, the discrete Shannon wavelet is given by

$$\psi_{m,n}(t) = 2^{-\frac{m}{2}} \frac{\sin\left\{\frac{\pi}{2}\left(2^{-m}t - n\right)\right\}}{\frac{\pi}{2}\left(2^{-m}t - n\right)} \cos\left\{\frac{3\pi}{2}\left(2^{-m}t - n\right)\right\}. \tag{6.5.24}$$

Its Fourier transform is

$$\hat{\psi}_{m,n}(\omega) = \begin{cases} 2^{m/2} \exp\left(-i\omega n 2^m\right), & 2^{-m}\pi < |\omega| < 2^{-m+1}\cdot\pi \\ 0, & \text{otherwise} \end{cases}. \tag{6.5.25}$$

Evidently, $\hat{\psi}_{m,n}(\omega)$ and $\hat{\psi}_{k,\ell}(\omega)$ do not overlap for $m \neq k$. Hence, by the Parseval relation (3.4.37), it turns out that, for $m \neq k$,

$$\left(\psi_{m,n}, \psi_{k,\ell}\right) = \frac{1}{2\pi}\left(\hat{\psi}_{m,n}, \hat{\psi}_{k,\ell}\right) = 0. \tag{6.5.26}$$

For $m = k$, we have

$$\begin{aligned}\left(\psi_{m,n},\psi_{k,\ell}\right) &= \frac{1}{2\pi}\left(\hat{\psi}_{m,n},\hat{\psi}_{m,\ell}\right)\\ &= \frac{1}{2\pi}\,2^{-m}\int_{-\infty}^{\infty}\exp\left\{-i\omega 2^{-m}(n-\ell)\right\}\left|\hat{\psi}\left(2^{-m}\omega\right)\right|^2 d\omega\\ &= \frac{1}{2\pi}\int_{-\infty}^{\infty}\exp\left\{-i\sigma(n-\ell)\right\}d\sigma = \delta_{n,\ell}.\end{aligned} \tag{6.5.27}$$

This shows that $\left\{\psi_{m,n}(t)\right\}$ is an orthonormal set.

6.6 Exercises

1. Discuss the scaled and translated versions of the mother wavelet $\psi(t) = t\exp\left(-t^2\right)$.

2. Show that the Fourier transform of the normalized Mexican hat wavelet

$$\psi(t) = \frac{2}{\pi^{\frac{1}{4}}\sqrt{3a}}\left(1-\frac{t^2}{a^2}\right)\exp\left(-\frac{t^2}{2a^2}\right)$$

is

$$\hat{\psi}(\omega) = \sqrt{\frac{8}{3}}\,a^{5/2}\,\pi^{1/4}\,\omega^2\exp\left(-\frac{a^2\omega^2}{2}\right).$$

3. Show that the continuous wavelet transform can be expressed as a convolution, that is,

$$\mathcal{W}_{\psi}[f](a,b) = \left(f*\psi_a\right)(b),$$

where

$$\psi_a(t) = \frac{1}{\sqrt{a}}\,\overline{\psi}\left(-\frac{t}{a}\right).$$

What is the physical significance of the convolution?

4. If f is a homogeneous function of degree n, show that

$$\left(\mathcal{W}_{\psi}f\right)(\lambda a,\lambda b) = \lambda^{n+\frac{1}{2}}\left(\mathcal{W}_{\psi}f\right)(a,b).$$

5. Prove that the vectors $(1,0), \left(-\frac{1}{2}, \frac{\sqrt{3}}{2}\right), \left(-\frac{1}{2}, -\frac{\sqrt{3}}{2}\right)$ form a tight frame in $\mathbb{C}$.

6. If $\{\phi_n\}$ is a tight frame in a Hilbert space H with frame bound A, show that

$$A(f,g) = \sum_{n=1}^{\infty} (f,\phi_n)(\phi_n,g)$$

for all $f, g \in H$.

7. If $\{\phi_n\}$ is a tight frame in a Hilbert space H with frame bound 1, show that $\{\phi_n\}$ is an orthonormal basis in H.

8. Show that

$$\int_{-\infty}^{\infty} \frac{\sin \pi x}{\pi x} \cdot \frac{\sin \pi (2x-n)}{\pi (2x-n)} \, dx = \frac{1}{2\pi n} \sin\left(\frac{n\pi}{2}\right).$$

9. Show that the Fourier transform of one-cycle of the sine function

$$f(t) = \sin t, \qquad |t| < \pi;$$

is

$$\hat{f}(\omega) = \frac{2i}{(\omega^2 - 1)} \sin \pi \omega.$$

10. For the Shannon wavelet

$$\psi(t) = \frac{\sin\left(\frac{\pi t}{2}\right)}{\left(\frac{\pi t}{2}\right)} \cos\left(\frac{3\pi t}{2}\right),$$

show that its Fourier transform is

$$\hat{\psi}(\omega) = \begin{cases} 1, & \pi < |\omega| < 2\pi \\ 0, & \text{otherwise} \end{cases}.$$

11. Show that the Fourier transform of the wavetrain

$$f(t) = \frac{1}{\sqrt{2\pi}} \frac{1}{\sigma} \exp\left(-\frac{t^2}{2\sigma^2}\right) \cos \omega_0 t$$

is

$$\hat{f}(\omega) = \frac{1}{2}\left[\exp\left\{-\frac{\sigma^2}{2}(\omega - \omega_0)^2\right\} + \exp\left\{-\frac{\sigma^2}{2}(\omega + \omega_0)^2\right\}\right].$$

Explain the physical features of $\hat{f}(\omega)$.

12. Show that the Fourier transform of

$$f(t) = \frac{1}{\sqrt{2\pi}}\,\chi_a(t)\,e^{i\omega_0 t}$$

is

$$\hat{f}(\omega) = \sqrt{\frac{2}{\pi}}\,\frac{\sin\left[a(\omega - \omega_0)\right]}{(\omega - \omega_0)}.$$

Explain the features of $\hat{f}(\omega)$.

13. If

$$\psi\left(\frac{t-b}{a}\right) = \begin{Bmatrix} 1, & b \le t < b + \frac{a}{2} \\ -1, & b + \frac{a}{2} \le t < b + a \\ 0, & \text{otherwise} \end{Bmatrix},$$

where $a > 0$, show that

$$\mathcal{W}_\psi[f](a,b) = \frac{1}{\sqrt{a}}\int_b^{b+\frac{a}{2}}\left[f(t) - f\left(t + \frac{a}{2}\right)\right]dt.$$

14. Suppose ψ_1 and ψ_2 are two wavelets and the integral

$$\int_{-\infty}^{\infty}\frac{\overline{\hat{\psi}_1(\omega)}\,\hat{\psi}_2(\omega)}{|\omega|}\,d\omega = C_{\psi_1\psi_2} < \infty.$$

If $\mathcal{W}_{\psi_1}[f](a,b)$ and $\mathcal{W}_{\psi_2}[g](a,b)$ denote wavelet transforms, show that

$$\left(\mathcal{W}_{\psi_1}f,\,\mathcal{W}_{\psi_2}g\right) = C_{\psi_2\psi_2}(f,g),$$

where $f,\, g \in L^2(\mathbb{R})$.

15. The Meyer wavelet ψ is defined by its Fourier transform

$$\hat{\psi}(\omega) = \begin{cases} \dfrac{1}{\sqrt{2\pi}} \exp\left(\dfrac{i\omega}{2}\right) \sin\left\{\dfrac{\pi}{2} \nu\left(\dfrac{3}{2\pi}|\omega| - 1\right)\right\}, & \dfrac{2\pi}{3} \le |\omega| \le \dfrac{4\pi}{3} \\ \dfrac{1}{\sqrt{2\pi}} \exp\left(\dfrac{i\omega}{2}\right) \cos\left\{\dfrac{\pi}{2} \nu\left(\dfrac{3}{4\pi}|\omega| - 1\right)\right\}, & \dfrac{4\pi}{3} \le |\omega| \le \dfrac{8\pi}{3} \end{cases},$$

where ν is a C^k or C^∞ function satisfying

$$\nu(x) = \begin{cases} 0 & \text{if } x \le 0 \\ 1 & \text{if } x \ge 1 \end{cases}$$

and the property

$$\nu(x) + \nu(1-x) = 1.$$

Show that $\psi_{m,n}(t) = 2^{-m/2}\,\psi\left(2^{-m}t - n\right)$ constitutes an orthonormal basis.

Chapter 7

Multiresolution Analysis and Construction of Wavelets

"Multiresolution analysis provides a natural framework for the understanding of wavelet bases, and for the construction of new examples. The history of the formulation of multiresolution analysis is a beautiful example of applications stimulating theoretical development."

Ingrid Daubechies

7.1 Introduction

The concept of multiresolution is intuitively related to the study of signals or images at different levels of resolution – almost like a pyramid. The resolution of a signal is a qualitative description associated with its frequency content. For a low-pass signal, the lower its frequency content (the narrower the bandwidth), the coarser is its resolution. In signal processing, a low-pass and subsampled version of a signal is usually a good coarse approximation for many real world signals. Multiresolution is especially evident in image processing and computer vision, where coarse versions of an image are often used as a first approximation in computational algorithms.

In 1986, Stéphane Mallat and Yves Meyer first formulated the idea of multiresolution analysis (MRA) in the context of wavelet analysis. This is a new and remarkable idea which deals with a general formalism for construction of an orthogonal basis of wavelets. Indeed, multiresolution analysis is central to all constructions of wavelet bases. Mallat's brilliant work (1989a,b,c) has been the major source of many new developments in wavelet analysis and its wide variety of applications.

Mathematically, the fundamental idea of multiresolution analysis is to represent a function (or signal) f as a limit of successive approximations, each of which is a finer version of the function f. These successive approximations correspond to different levels of resolutions. Thus, multiresolution analysis is a formal approach to constructing orthogonal wavelet bases using a definite set of rules and procedures. The key feature of this analysis is to describe mathematically the process of studying signals or images at different scales. The basic principle of the MRA deals with the decomposition of the whole function space into individual subspaces $V_n \subset V_{n+1}$ so that the space V_{n+1} consists of all rescaled functions in V_n. This essentially means a decomposition of each function (or signal) into components of different scales (or frequencies) so that an individual component of the original function f occurs in each subspace. These components can describe finer and finer versions of the original function f. For example, a function is resolved at scales $\Delta t = 2^0, 2^{-1}, \cdots, 2^{-n}$. In audio signals, these scales are basically *octaves* which represent higher and higher frequency components. For images and, indeed, for all signals, the simultaneous existence of a multiscale may also be referred to as *multiresolution*. From the point of view of practical application, MRA is really an effective mathematical framework for hierarchical decomposition of an image (or signal) into components of different scales (or frequencies).

In general, frames have many of the properties of bases, but they lack a very important property of orthogonality. If the condition of orthogonality

$$\left(\phi_{k,\ell},\, \phi_{m,n}\right) = 0 \quad \text{for all} \quad (k,\ell) \neq (m,n) \tag{7.1.1}$$

is satisfied, the reconstruction of the function f from $\left(f,\, \phi_{m,n}\right)$ is much simpler and, for any $f \in L^2(\mathbb{R})$, we have the following representation

$$f = \sum_{m,\, n=-\infty}^{\infty} \left(f,\, \phi_{m,n}\right) \phi_{m,n}, \tag{7.1.2}$$

where

$$\phi_{m,n}(x) = 2^{-m/2}\, \phi\left(2^{-m} x - n\right) \tag{7.1.3}$$

is an orthonormal basis of V_m.

This chapter deals with the idea of multiresolution analysis with examples. Special attention is given to properties of scaling functions and orthonormal wavelet bases. This is followed by a method of constructing orthonormal bases of wavelets from MRA. Special attention is also given to the Daubechies wavelets with compact support and the Daubechies algorithm. Included are discrete wavelet transforms and Mallat's pyramid algorithm.

7.2 Definition of Multiresolution Analysis and Examples

Definition 7.2.1. (Multiresolution Analysis). *A multiresolution analysis* (MRA) consists of a sequence $\{V_m : m \in \mathbb{Z}\}$ of embedded closed subspaces of $L^2(\mathbb{R})$ that satisfy the following conditions:

(i) $\cdots \subset V_{-2} \subset V_{-1} \subset V_{-1} \subset V_0 \subset V_1 \subset V_2 \subset \cdots V_m \subset V_{m+1} \cdots,$

(ii) $\bigcup_{m=-\infty}^{\infty} V_m \mathbb{R}$ is dense in $L^2(\mathbb{R})$, that is, $\overline{\bigcup_{m=-\infty}^{\infty} V_m} = L^2(\mathbb{R})$,

(iii) $\bigcap_{m=-\infty}^{\infty} V_m = \{0\}$,

(iv) $f(x) \in V_m$ if and only if $f(2x) \in V_{m+1}$ for all $m \in \mathbb{Z}$,

(v) there exists a function $\phi \in V_0$ such that $\{\phi_{0,n} = \phi(x-n),\ n \in \mathbb{Z}\}$ is an orthonormal basis for V_0, that is,

$$\|f\|^2 = \int_{-\infty}^{\infty} |f(x)|^2 dx = \sum_{n=-\infty}^{\infty} |\langle f, \phi_{0,n}\rangle|^2 \quad \text{for all } f \in V_0.$$

The function ϕ is called the *scaling function* or *father wavelet.* If $\{V_m\}$ is a multiresolution of $L^2(\mathbb{R})$ and if V_0 is the closed subspace generated by the integer translates of a single function ϕ, then we say that ϕ generates the multiresolution analysis.

Sometimes, condition (v) is relaxed by assuming that $\{\phi(x-n),\ n \in \mathbb{Z}\}$ is a *Riesz basis* for V_0, that is, for every $f \in V_0$, there exists a unique sequence $\{c_n\}_{n=-\infty}^{\infty} \in \ell^2(\mathbb{Z})$ such that

$$f(x) = \sum_{n=-\infty}^{\infty} c_n \phi(x-n)$$

with convergence in $L^2(\mathbb{R})$ and there exist two positive constants A and B independent of $f \in V_0$ such that

$$A \sum_{n=-\infty}^{\infty} |c_n|^2 \le \|f\|^2 \le B \sum_{n=-\infty}^{\infty} |c_n|^2,$$

where $0 < A < B < \infty$. In this case, we have a multiresolution analysis with a Riesz basis.

Note that condition (v) implies that $\{\phi(x-n),\ n \in \mathbb{Z}\}$ is a Riesz basis for V_0 with $A = B = 1$.

Since $\phi_{0,n}(x) \in V_0$ for all $n \in \mathbb{Z}$. Further, if $n \in \mathbb{Z}$, it follows from (iv) that

$$\phi_{m,n}(x) = 2^{m/2} \phi\left(2^m x - n\right),\quad m \in \mathbb{Z} \tag{7.2.1}$$

is an orthonormal basis for V_m.

Consequences of Definition 7.2.1.

1. A repeated application of condition (iv) implies that $f \in V_m$ if and only if $f\left(2^k x\right) \in V_{m+k}$ for all $m, k \in \mathbb{Z}$. In other words, $f \in V_m$ if and only if $f\left(2^{-m} x\right) \in V_0$ for all $m \in \mathbb{Z}$.

   This shows that functions in V_m are obtained from those in V_0 through a scaling 2^{-m}. If the scale $m = 0$ is associated with V_0, then the scale 2^{-m} is associated with V_m. Thus, subspaces V_m are just scaled versions of the central space V_0, which is invariant under translation by integers, that is, $T_n V_0 = V_0$ for all $n \in \mathbb{Z}$.

2. It follows from Definition 7.2.1 that a multiresolution analysis is completely determined by the scaling function ϕ but not conversely. For a given $\phi \in V_0$, we first define

$$V_0 = \left\{ f(x) = \sum_{n=-\infty}^{\infty} c_n \phi_{0,n} = \sum_{n=-\infty}^{\infty} c_n \phi(x-n) \ :\ \{c_n\} \in \ell^2(\mathbb{Z}) \right\}.$$

Condition (iv) implies that V_0 has an orthonormal basis $\{\phi_{0,n}\} = \{\phi(x-n)\}$. Then, V_0 consists of all functions $f(x) = \sum_{n=-\infty}^{\infty} c_n \phi(x-n)$ with finite energy $\|f\|^2 = \sum_{n=-\infty}^{\infty} |c_n|^2 < \infty$. Similarly, the space V_m has the orthonormal basis $\phi_{m,n}$ given by (7.2.1) so that $f_m(x)$ is given by

$$f_m(x) = \sum_{n=-\infty}^{\infty} c_{mn}\ \phi_{m,n}(x) \tag{7.2.2}$$

with the finite energy

$$\|f_m\|^2 = \sum_{n=-\infty}^{\infty} |c_{mn}|^2 < \infty.$$

Thus, f_m represents a typical function in the space V_m. It builds in self-invariance and scale invariance through the basis $\{\phi_{m,n}\}$.

3. Conditions (ii) and (iii) can be expressed in terms of the orthogonal projections P_m onto V_m, that is, for all $f \in L^2(\mathbb{R})$,

$$\lim_{m\to-\infty} P_m f = 0 \quad \text{and} \quad \lim_{m\to+\infty} P_m f = f. \tag{7.2.3a,b}$$

The projection $P_m f$ can be considered as an approximation of f at the scale 2^{-m}. Therefore, the successive approximations of a given function f are defined as the orthogonal projections P_m onto the space V_m:

$$P_m f = \sum_{n=-\infty}^{\infty} (f, \phi_{m,n})\ \phi_{m,n}, \tag{7.2.4}$$

where $\phi_{m,n}(x)$ given by (7.2.1) is an orthonormal basis for V_m.

4. Since $V_0 \subset V_1$, the scaling function ϕ that leads to a basis for V_0 is also V_1. Since $\phi \in V_1$ and $\phi_{1,n}(x) = \sqrt{2}\ \phi(2x-n)$ is an orthonormal basis for V_1, ϕ can be expressed in the form

$$\phi(x) = \sum_{n=-\infty}^{\infty} c_n \phi_{1,n}(x) = \sqrt{2} \sum_{n=-\infty}^{\infty} c_n\, \phi(2x-n), \tag{7.2.5}$$

where

$$c_n = (\phi, \phi_{1,n}) \text{ and } \sum_{n=-\infty}^{\infty} |c_n|^2 = 1.$$

Equation (7.2.5) is called the *dilation equation*. It involves both x and $2x$ and is often referred to as the *two-scale equation* or *refinement equation* because it displays $\phi(x)$ in the refined space V_1. The space V_1 has the finer scale 2^{-1} and it contains $\phi(x)$ which has scale 1.

All of the preceding facts reveal that multiresolution analysis can be described at least three ways so that we can specify

(a) the subspaces V_m,

(b) the scaling function ϕ,

(c) the coefficient c_n in the dilation equation (7.2.5).

The real importance of a multiresolution analysis lies in the simple fact that it enables us to construct an orthonormal basis for $L^2(\mathbb{R})$. In order to prove this statement, we first assume that $\{V_m\}$ is a multiresolution analysis. Since $V_m \subset V_{m+1}$, we define W_m as the orthogonal complement of V_m in V_{m+1} for every $m \in \mathbb{Z}$ so that we have

$$\begin{aligned} V_{m+1} &= V_m \oplus W_m \\ &= (V_{m-1} \oplus W_{m-1}) \oplus W_m \\ &= \cdots \\ &= V_0 \oplus W_0 \oplus W_1 \oplus \cdots \oplus W_m \\ &= V_0 \oplus \left(\bigoplus_{m=0}^{m} W_m \right) \end{aligned} \tag{7.2.6}$$

and $V_n \perp W_m$ for $n \neq m$.

Since $\bigcup_{m=-\infty}^{\infty} V_m$ is dense in $L^2(\mathbb{R})$, we may take the limit as $m \to \infty$ to obtain

$$V_0 \oplus \left(\bigoplus_{m=0}^{\infty} W_m \right) = L^2(\mathbb{R}).$$

Similarly, we may go in the other direction to write

$$\begin{aligned} V_0 &= V_{-1} \oplus W_{-1} \\ &= (V_{-2} \oplus W_{-2}) \oplus W_{-1} \\ &= \cdots \\ &= V_{-m} \oplus W_{-m} \oplus \cdots \oplus W_{-1}. \end{aligned}$$

We may again take the limit as $m \to \infty$. Since $\bigcap_{m\in\mathbb{Z}} V_m = \{0\}$, it follows that $V_{-m} = \{0\}$. Consequently, it turns out that

$$\bigoplus_{m=-\infty}^{\infty} W_m = L^2(\mathbb{R}). \tag{7.2.7}$$

We include here a pictorial representation of $V_1 = V_0 \oplus W_0$ in Figure 7.1.

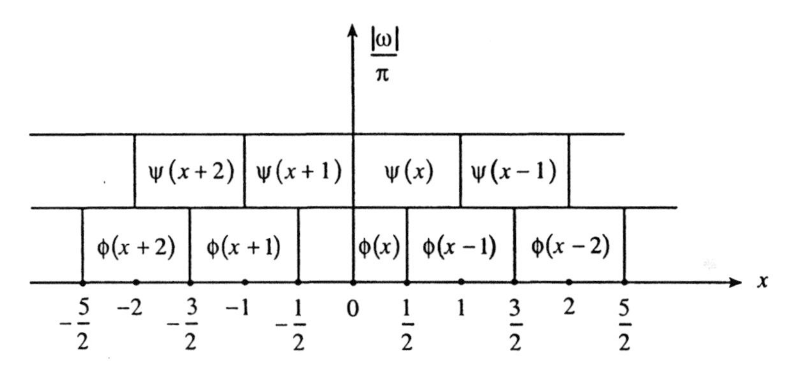

Figure 7.1. Pictorial representation of $V_1 = V_0 \oplus W_0$.

Finally, the difference between the two successive approximations $P_m f$ and $P_{m+1} f$ is given by the orthogonal projection $Q_m f$ of f onto the orthogonal complement W_m of V_m in V_{m+1} so that

$$Q_m f = P_{m+1} f - P_m f.$$

It follows from conditions (i)–(v) in Definition 7.2.1 that the spaces W_m are also scaled versions of W_0 and, for $f \in L^2(\mathbb{R})$,

$$f \in W_m \text{ if and only if } f\left(2^{-m} x\right) \in W_0 \text{ for all } m \in \mathbb{Z}, \tag{7.2.8}$$

and they are translation-invariant for the discrete translations $n \in \mathbb{Z}$, that is,

$$f \in W_0 \text{ if and only if } f(x-n) \in W_0,$$

and they are mutually orthogonal spaces generating all of $L^2(\mathbb{R})$,

$$\left.\begin{aligned} W_m &\perp W_k \quad \text{for } m \neq k, \\ \bigoplus_{m\in\mathbb{Z}} W_m &= L^2(\mathbb{R}). \end{aligned}\right\} \tag{7.2.9a,b}$$

Moreover, there exists a function $\psi \in W_0$ such that $\psi_{0,n}(x) = \psi(x-n)$ constitutes an orthonormal basis for W_0. It follows from (7.2.8) that

$$\psi_{m,n}(x) = 2^{m/2}\,\psi\left(2^m x - n\right), \quad \text{for } n \in \mathbb{Z} \tag{7.2.10}$$

constitute an orthonormal basis for W_m. Thus, the family $\psi_{m,n}(x)$ represents an orthonormal basis of wavelets for $L^2(\mathbb{R})$. Each $\psi_{m,n}(x)$ is represented by the point (p,s), where $p = \left(n + \frac{1}{2}\right)2^m$ and $s = 2^m$, $(m,\ n \in \mathbb{Z})$ in the position-scale plane, as shown in Figure 7.2. Since scale is the inverse of the frequency, small scales 2^m (or high frequencies 2^{-m}) are near the position axis.

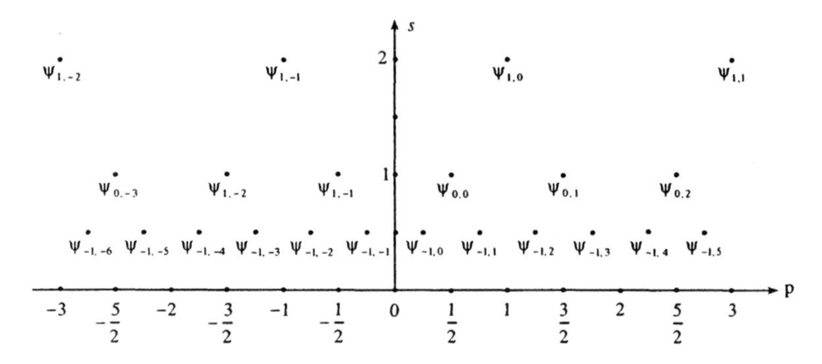

Figure 7.2. Dyadic grid representation.

Example 7.2.1 (Characteristic Function). We assume that $\phi = \chi_{[0,1]}$ is the characteristic function of the interval $[0,1]$. Define spaces V_m by

$$V_m = \left\{ \sum_{k=-\infty}^{\infty} c_k\, \phi_{m,k} \,:\, \{c_k\} \in \ell^2(\mathbb{Z}) \right\},$$

where

$$\phi_{m,n}(x) = 2^{-m/2}\phi\left(2^{-m}x - n\right).$$

The spaces V_m satisfy all the conditions of Definition 7.2.1, and so, $\{V_m\}$ is a multiresolution analysis.

Example 7.2.2 (Piecewise Constant Function). Consider the space V_m of all functions in $L^2(\mathbb{R})$ which are constant on intervals $[2^{-m}n,\ 2^{-m}(n+1)]$, where $n \in \mathbb{Z}$. Obviously, $V_m \subset V_{m+1}$ because any function that is constant on intervals of length 2^{-m} is automatically constant on intervals of half that length. The space V_0 contains all functions $f(x)$ in $L^2(\mathbb{R})$ that are constant on $n \le x < n+1$. The function $f(2x)$ in V_1 is then constant on $\frac{n}{2} \le x < \frac{n+1}{2}$. Intervals of length 2^{-m} are usually referred to as *dyadic intervals.* A sample function in spaces V_m is shown in Figure 7.3.

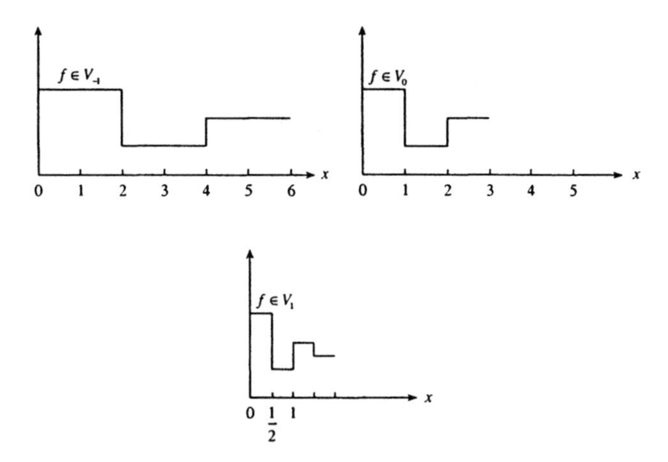

Figure 7.3. Piecewise constant functions in $V_{-1}, V_0,$ and V_1.

Clearly, the piecewise constant function space V_m satisfies the conditions (i)–(iv) of a multiresolution analysis. It is easy to guess a scaling function ϕ in V_0 which is orthogonal to its translates. The simplest choice for ϕ is the characteristic function so that $\phi(x) = \chi_{[0,1]}(x)$. Therefore, any function $f \in V_0$ can be expressed in terms of the scaling function ϕ as

$$f(x) = \sum_{n=-\infty}^{\infty} c_n\, \phi(x-n).$$

Thus, the condition (v) is satisfied by the characteristic function $\chi_{[0,1]}$ as the scaling function. As we shall see later, this MRA is related to the classic Haar wavelet.

7.3 Properties of Scaling Functions and Orthonormal Wavelet Bases

Theorem 7.3.1. For any function $\phi \in L^2(\mathbb{R})$, the following conditions are equivalent.

(a) The system $\{\phi_{0,n} \equiv \phi(x-n), \quad n \in \mathbb{Z}\}$ is orthonormal.

(b) $\sum_{k=-\infty}^{\infty} \left|\hat{\phi}(\omega + 2k\pi)\right|^2 = 1$ almost everywhere (a.e.).

Proof. Obviously, the Fourier transform of $\phi_{0,n}(x) = \phi(x-n)$ is

$$\hat{\phi}_{0,n}(\omega) = \exp(-in\omega)\hat{\phi}(\omega).$$

In view of the general Parseval relation (3.4.37) for the Fourier transform, we have

$$\begin{aligned}
\left(\phi_{0,n}, \phi_{0,m}\right) &= \left(\phi_{0,0}, \phi_{0,m-n}\right) \\
&= \frac{1}{2\pi}\left(\hat{\phi}_{0,0}, \hat{\phi}_{0,m-n}\right) \\
&= \frac{1}{2\pi}\int_{-\infty}^{\infty} \exp\{-i(m-n)\omega\} \cdot \left|\hat{\phi}(\omega)\right|^2 d\omega \\
&= \frac{1}{2\pi}\sum_{k=-\infty}^{\infty} \int_{2\pi k}^{2\pi(k+1)} \exp\{-i(m-n)\omega\} \left|\hat{\phi}(\omega)\right|^2 d\omega \\
&= \frac{1}{2\pi}\int_{0}^{2\pi} \exp\{-i(m-n)\omega\} \sum_{k=-\infty}^{\infty} \left|\hat{\phi}(\omega + 2\pi k)\right|^2 d\omega.
\end{aligned}$$

Thus, it follows from the completeness of $\{\exp(-in\omega), n \in \mathbb{Z}\}$ in $L^2(0, 2\pi)$ that

$$\left(\phi_{0,n}, \phi_{0,m}\right) = \delta_{n,m}$$

if and only if

$$\sum_{k=-\infty}^{\infty} \left|\hat{\phi}(\omega + 2\pi k)\right|^2 = 1 \qquad \text{almost everywhere.}$$

Theorem 7.3.2. For any two functions $\phi,\ \psi \in L^2(\mathbb{R})$, the sets of functions $\{\phi_{0,n} \equiv \phi_1(x-n),\ n \in \mathbb{Z}\}$ and $\{\psi_{0,m} \equiv \psi(x-m),\ m \in \mathbb{Z}\}$ are biorthogonal, that is,

$$\left(\phi_{0,n},\ \psi_{0,m}\right) = 0 \qquad \text{for all } n,\ m \in \mathbb{Z},$$

if and only if

$$\sum_{k=-\infty}^{\infty} \hat{\phi}(\omega + 2\pi k)\ \overline{\hat{\psi}}(\omega + 2\pi k) = 0 \qquad \text{almost everywhere.}$$

Proof. We apply arguments similar to those stated in the proof of Theorem 7.3.1 to obtain

$$\begin{aligned}
\left(\phi_{0,n},\ \psi_{0,m}\right) &= \left(\phi_{0,0},\ \psi_{0,m-n}\right) \\
&= \frac{1}{2\pi}\left(\hat{\phi}_{0,0},\ \hat{\psi}_{0,m-n}\right) \\
&= \frac{1}{2\pi}\int_{-\infty}^{\infty} \exp\{i(n-m)\omega\}\ \hat{\phi}(\omega)\ \overline{\hat{\psi}}(\omega)\ d\omega \\
&= \frac{1}{2\pi}\sum_{k=-\infty}^{\infty}\ \int_{2\pi k}^{2\pi(k+1)} \exp\{i(n-m)\omega\}\ \hat{\phi}(\omega)\ \overline{\hat{\psi}}(\omega)\ d\omega \\
&= \frac{1}{2\pi}\int_{0}^{2\pi} \exp\{i(n-m)\omega\}\left[\sum_{k=-\infty}^{\infty} \phi(\omega + 2\pi k)\ \overline{\hat{\psi}}(\omega + 2\pi k)\right] d\omega.
\end{aligned}$$

Thus,

$$\left(\phi_{0,n},\ \psi_{0,m}\right) = 0 \ \text{ for all } n \text{ and } m$$

if and only if

$$\sum_{k=-\infty}^{\infty} \hat{\phi}(\omega + 2\pi k)\ \overline{\hat{\psi}}(\omega + 2\pi k) = 0 \qquad \text{almost everywhere.}$$

We next proceed to the construction of a mother wavelet by introducing an important generating function $\hat{m}(\omega) \in L^2[0,\ 2\pi]$ in the following lemma.

Lemma 7.3.1. The Fourier transform of the scaling function ϕ satisfies the following conditions:

$$\sum_{k=-\infty}^{\infty} \left|\hat{\phi}(\omega + 2\pi k)\right|^2 = 1 \quad \text{almost everywhere,} \tag{7.3.1}$$

$$\hat{\phi}(\omega) = \hat{m}\left(\frac{\omega}{2}\right) \hat{\phi}\left(\frac{\omega}{2}\right), \tag{7.3.2}$$

where

$$\hat{m}(\omega) = \frac{1}{\sqrt{2}} \sum_{n=-\infty}^{\infty} c_n \exp(-in\omega) \tag{7.3.3}$$

is a 2π - periodic function and satisfies the so-called *orthogonality condition*

$$|\hat{m}(\omega)|^2 + |\hat{m}(\omega + \pi)|^2 = 1 \quad \text{a.e.} \tag{7.3.4}$$

Remark. The Fourier transform $\hat{\phi}$ of the scaling function ϕ satisfies the functional equation (7.3.2). The function $\hat{m}$ is called the *generating function* of the multiresolution analysis. This function is often called the *discrete Fourier transform* of the sequence $\{c_n\}$. In signal processing, $\hat{m}(\omega)$ is called the transfer function of a *discrete filter* with impulse response $\{c_n\}$ or the *low-pass filter* associated with the scaling function ϕ.

Proof. Condition (7.3.1) follows from Theorem 7.3.1.

To establish (7.3.2), we first note that $\phi \in V_1$ and

$$\phi_{1,n}(x) = \sqrt{2}\, \phi\,(2x - n)$$

is an orthonormal basis for V_1. Thus, the scaling function ϕ has the following representation

$$\phi(x) = \sqrt{2} \sum_{n=-\infty}^{\infty} c_n\, \phi(2x - n), \tag{7.3.5}$$

where $c_n = (\phi, \phi_{1,n})$ and $\sum_{n=-\infty}^{\infty} |c_n|^2 < \infty.$

The Fourier transform of (7.3.5) gives

$$\hat{\phi}(\omega) = \frac{1}{\sqrt{2}} \sum_{n=-\infty}^{\infty} c_n \exp\left(-\frac{i\omega n}{2}\right) \hat{\phi}\left(\frac{\omega}{2}\right) = \hat{m}\left(\frac{\omega}{2}\right) \hat{\phi}\left(\frac{\omega}{2}\right). \tag{7.3.6}$$

This proves the functional equation (7.3.2).

To verify the orthogonality condition (7.3.4), we substitute (7.3.2) in (7.3.1) so that condition (7.3.1) becomes

$$1 = \sum_{k=-\infty}^{\infty} \left|\hat{\phi}(\omega + 2k\pi)\right|^2$$

$$= \sum_{k=-\infty}^{\infty} \left|\hat{m}\left(\frac{\omega}{2} + k\pi\right)\right|^2 \left|\hat{\phi}\left(\frac{\omega}{2} + k\pi\right)\right|^2.$$

This is true for any ω and hence, replacing ω by 2ω gives

$$1 = \sum_{k=-\infty}^{\infty} \left|\hat{m}(\omega + k\pi)\right|^2 \left|\hat{\phi}(\omega + k\pi)\right|^2. \tag{7.3.7}$$

We now split the above infinite sum over k into even and odd integers and use the 2π- periodic property of the function $\hat{m}$ to obtain

$$1 = \sum_{k=-\infty}^{\infty} \left|\hat{m}(\omega + 2\pi k)\right|^2 \left|\hat{\phi}(\omega + 2\pi k)\right|^2$$

$$+ \sum_{k=-\infty}^{\infty} \left|\hat{m}(\omega + (2k+1)\pi)\right|^2 \left|\hat{\phi}(\omega + (2k+1)\pi)\right|^2$$

$$= \sum_{k=-\infty}^{\infty} \left|\hat{m}(\omega)\right|^2 \left|\hat{\phi}(\omega + 2\pi k)\right|^2 + \sum_{k=-\infty}^{\infty} \left|\hat{m}(\omega + \pi)\right|^2 \left|\hat{\phi}(\omega + \pi + 2k\pi)\right|^2$$

$$= \left|\hat{m}(\omega)\right|^2 + \left|\hat{m}(\omega + \pi)\right|^2$$

by (7.3.1) used in its original form and ω replaced by $(\omega + \pi)$. This leads to the desired condition (7.3.4).

Remark. Since $\left|\hat{\phi}(0)\right| = 1 \neq 0$, $\hat{m}(0) = 1$ and $\hat{m}(\pi) = 0$. This implies that $\hat{m}$ can be considered as a low-pass filter because the transfer function passes the frequencies near $\omega = 0$ and cuts off the frequencies near $\omega = \pi$.

Lemma 7.3.2. The function $\hat{\phi}$ can be represented by the infinite product

$$\hat{\phi}(\omega) = \prod_{k=1}^{\infty} \hat{m}\left(\frac{\omega}{2^k}\right). \tag{7.3.8}$$

Proof. A simple iteration of (7.3.2) gives

$$\hat{\phi}(\omega)=\hat{m}\left(\frac{\omega}{2}\right)\hat{\phi}\left(\frac{\omega}{2}\right)=\hat{m}\left(\frac{\omega}{2}\right)\left[\hat{m}\left(\frac{\omega}{4}\right)\hat{\phi}\left(\frac{\omega}{4}\right)\right]$$

which is, by the $(k-1)$th iteration,

$$=\hat{m}\left(\frac{\omega}{2}\right)\hat{m}\left(\frac{\omega}{4}\right)\cdots\hat{m}\left(\frac{\omega}{2^k}\right)\cdot\hat{\phi}\left(\frac{\omega}{2^k}\right)$$

$$=\prod_{k=1}^{k}\hat{m}\left(\frac{\omega}{2^k}\right)\hat{\phi}\left(\frac{\omega}{2^k}\right). \tag{7.3.9}$$

Since $\hat{\phi}(0)=1$ and $\hat{\phi}(\omega)$ is continuous, we obtain

$$\lim_{k\to\infty}\hat{\phi}\left(\frac{\omega}{2^k}\right)=\hat{\phi}(0)=1.$$

The limit of (7.3.9) as $k\to\infty$ gives (7.3.8).

We next prove the following major technical lemma.

Lemma *7.3.3.* The Fourier transform of any function $f\in W_0$ can be expressed in the form

$$\hat{f}(\omega)=\hat{v}(\omega)\exp\left(\frac{i\omega}{2}\right)\overline{\hat{m}\left(\frac{\omega}{2}+\pi\right)}\hat{\phi}\left(\frac{\omega}{2}\right), \tag{7.3.10}$$

where $\hat{v}(\omega)$ is a 2π- periodic function and the factor $\exp\left(\frac{i\omega}{2}\right)\overline{\hat{m}}\left(\frac{\omega}{2}+\pi\right)\hat{\phi}\left(\frac{\omega}{2}\right)$ is independent of *f*.

Proof. Since $f\in W_0$, it follows from $V_1=V_0\oplus W_0$ that $f\in V_1$ and is orthogonal to V_0. Thus, it follows from $V_1=V_0\oplus W_0$ that $f\in V_0$ and is orthogonal to V_0. Thus, the function *f* can be expressed in the form

$$f(x)=\sum_{n=-\infty}^{\infty}c_n\,\phi_{1,n}(x)=\sqrt{2}\sum_{n=-\infty}^{\infty}c_n\,\phi(2x-n), \tag{7.3.11}$$

where $c_n=(f,\phi_{1,n})$.

We use an argument similar to that in Lemma 7.3.1 to obtain the result

$$\hat{f}(\omega)=\frac{1}{\sqrt{2}}\sum_{n=-\infty}^{\infty}c_n\exp\left(-\frac{in\omega}{2}\right)\hat{\phi}\left(\frac{\omega}{2}\right)=\hat{m}_f\left(\frac{\omega}{2}\right)\hat{\phi}\left(\frac{\omega}{2}\right), \tag{7.3.12}$$

where the function $\hat{m}_f$ is given by

$$\hat{m}_f(\omega) = \frac{1}{\sqrt{2}} \sum_{n=-\infty}^{\infty} c_n \exp(-i n \omega). \tag{7.3.13}$$

Evidently, $\hat{m}_f$ is a 2π- periodic function which belongs to $L^2(0,2\pi)$. Since $f \perp V_0$, we have

$$\int_{-\infty}^{\infty} \hat{f}(\omega)\, \bar{\hat{\phi}}(\omega) \exp(i n \omega)\, d\omega = 0$$

and hence,

$$\int_{-\infty}^{\infty} \left\{ \sum_{k=-\infty}^{\infty} \hat{f}(\omega + 2\pi k)\, \bar{\hat{\phi}}(\omega + 2\pi k) \right\} e^{in\omega} d\omega = 0. \tag{7.3.14}$$

Consequently,

$$\sum_{k=-\infty}^{\infty} \hat{f}(\omega + 2\pi k)\, \bar{\hat{\phi}}(\omega + 2\pi k) = 0. \tag{7.3.15}$$

We now substitute (7.3.12) and (7.3.2) into (7.3.15) to obtain

$$0 = \sum_{k=-\infty}^{\infty} \hat{m}_f\left(\frac{\omega}{2} + \pi k\right)\, \bar{\hat{m}}\left(\frac{\omega}{2} + \pi k\right) \left|\hat{\phi}\left(\frac{\omega}{2} + \pi k\right)\right|^2,$$

which is, by splitting the sum into even and odd integers k and then using the 2π- periodic property of the function $\hat{m}$,

$$0 = \sum_{k=-\infty}^{\infty} \hat{m}_f\left(\frac{\omega}{2} + 2\pi k\right)\, \bar{\hat{m}}\left(\frac{\omega}{2} + 2\pi k\right) \left|\hat{\phi}\left(\frac{\omega}{2} + 2\pi k\right)\right|^2$$

$$+ \sum_{k=-\infty}^{\infty} \hat{m}_f\left(\frac{\omega}{2} + \pi + 2\pi k\right)\, \bar{\hat{m}}\left(\frac{\omega}{2} + \pi + 2\pi k\right) \left|\hat{\phi}\left(\frac{\omega}{2} + \pi + 2\pi k\right)\right|^2$$

$$= \hat{m}_f\left(\frac{\omega}{2}\right)\, \bar{\hat{m}}\left(\frac{\omega}{2}\right) \sum_{k=-\infty}^{\infty} \left|\hat{\phi}\left(\frac{\omega}{2} + 2k\pi\right)\right|^2$$

$$+ \hat{m}_f\left(\frac{\omega}{2} + \pi\right)\, \bar{\hat{m}}\left(\frac{\omega}{2} + \pi\right) \sum_{k=-\infty}^{\infty} \left|\hat{\phi}\left(\frac{\omega}{2} + \pi + 2\pi k\right)\right|^2,$$

which is, due to orthonormality of the system $\{\phi_{0,k}(x)\}$ and (7.3.1),

$$= \left\{ \hat{m}_f\left(\frac{\omega}{2}\right) \bar{\hat{m}}\left(\frac{\omega}{2}\right) + \hat{m}_f\left(\frac{\omega}{2} + \pi\right) \bar{\hat{m}}\left(\frac{\omega}{2} + \pi\right) \right\}.1. \tag{7.3.16}$$

Finally, replacing ω by 2ω in (7.3.16) gives

$$\hat{m}_f(\omega)\, \bar{\hat{m}}(\omega) + \hat{m}_f(\omega + \pi)\, \bar{\hat{m}}(\omega + \pi) = 0 \quad \text{a.e.} \tag{7.3.17}$$

Or, equivalently,

$$\begin{vmatrix} \hat{m}_f(\omega) & \overline{\hat{m}}(\omega+\pi) \\ -\hat{m}_f(\omega+\pi) & \overline{\hat{m}}(\omega) \end{vmatrix} = 0.$$

This can be interpreted as the linear dependence of two vectors

$$\left(\hat{m}_f(\omega), \;\; -\hat{m}_f(\omega+\pi)\right) \text{ and } \left(\overline{\hat{m}}(\omega+\pi), \;\; \overline{\hat{m}}(\omega)\right).$$

Hence, there exists a function $\hat{\lambda}$ such that

$$\hat{m}_f(\omega) = \hat{\lambda}(\omega) \; \overline{\hat{m}}(\omega+\pi) \quad \text{a.e.} \tag{7.3.18}$$

Since both $\hat{m}$ and $\hat{m}_f$ are 2π-periodic functions, so is $\hat{\lambda}$. Further, substituting (7.3.18) into (7.3.17) gives

$$\hat{\lambda}(\omega) + \hat{\lambda}(\omega+\pi) = 0 \quad \text{a.e.} \tag{7.3.19}$$

Thus, there exists a 2π- periodic function $\hat{v}$ defined by

$$\hat{\lambda}(\omega) = \exp(i\omega) \; \hat{v}(2\omega). \tag{7.3.20}$$

Finally, a simple combination of (7.3.12), (7.3.18), and (7.3.20) gives the desired representation (7.3.10). This completes the proof of Lemma 7.3.3.

Now, we return to the main problem of constructing a mother wavelet $\psi(x)$. Suppose that there is a function ψ such that $\{\psi_{0,n} : n \in \mathbb{Z}\}$ is a basis for the space W_0. Then, every function $f \in W_0$ has a series representation

$$f(x) = \sum_{n=-\infty}^{\infty} h_n \psi_{0,n} = \sum_{n=-\infty}^{\infty} h_n \psi(x-n), \tag{7.3.21}$$

where

$$\sum_{n=-\infty}^{\infty} |h_n|^2 < \infty.$$

Application of the Fourier transform to (7.3.21) gives

$$\hat{f}(\omega) = \left(\sum_{n=-\infty}^{\infty} h_n \; e^{-in\omega} \right) \hat{\psi}(\omega) = \hat{h}(\omega) \; \hat{\psi}(\omega), \tag{7.3.22}$$

where the function $\hat{h}$ is

$$\hat{h}(\omega) = \sum_{n=-\infty}^{\infty} h_n \; \exp(-in\omega), \tag{7.3.23}$$

and it is a square integrable and 2π-periodic function in $[0,\ 2\pi]$. When (7.3.22) is compared with (7.3.10), we see that $\hat{\psi}(\omega)$ should be

$$\hat{\psi}(\omega)=\exp\left(\frac{i\omega}{2}\right)\ \overline{\hat{m}}\left(\frac{\omega}{2}+\pi\right)\ \hat{\phi}\left(\frac{\omega}{2}\right) \tag{7.3.24}$$

$$=\hat{m}_1\left(\frac{\omega}{2}\right)\ \hat{\phi}\left(\frac{\omega}{2}\right), \tag{7.3.25}$$

where the function $\hat{m}_1$ is given by

$$\hat{m}_1(\omega)=\exp(i\omega)\ \overline{\hat{m}}(\omega+\pi). \tag{7.3.26}$$

Thus, the function $\hat{m}_1(\omega)$ is called the *filter* conjugate to $\hat{m}(\omega)$ and hence, $\hat{m}$ and $\hat{m}_1$ are called *conjugate quadratic filters* (CQF) in signal processing.

Finally, substituting (7.3.3) into (7.3.24) gives

$$\hat{\psi}(\omega)=\exp\left(\frac{i\omega}{2}\right)\cdot\frac{1}{\sqrt{2}}\sum_{n=-\infty}^{\infty}\bar{c}_n\exp\left\{in\left(\frac{\omega}{2}+\pi\right)\right\}\hat{\phi}\left(\frac{\omega}{2}\right)$$

$$=\frac{1}{\sqrt{2}}\sum_{n=-\infty}^{\infty}\bar{c}_n\exp\left[in\pi+i(n+1)\frac{\omega}{2}\right]\hat{\phi}\left(\frac{\omega}{2}\right).$$

which is, by putting $n=-(k+1)$,

$$=\frac{1}{\sqrt{2}}\sum_{k=-\infty}^{\infty}\bar{c}_{-k-1}\,(-1)^k\exp\left(-\frac{ik\omega}{2}\right)\cdot\hat{\phi}\left(\frac{\omega}{2}\right). \tag{7.3.27}$$

Invoking the inverse Fourier transform to (7.3.27) with k replaced by n gives the mother wavelet

$$\psi(x)=\sqrt{2}\sum_{n=-\infty}^{\infty}(-1)^{n-1}\,\bar{c}_{-n-1}\,\phi(2x-n) \tag{7.3.28}$$

$$=\sqrt{2}\sum_{n=-\infty}^{\infty}d_n\,\phi(2x-n), \tag{7.3.29}$$

where the coefficients d_n are given by

$$d_n=(-1)^{n-1}\,\bar{c}_{-n-1}. \tag{7.3.30}$$

Thus, the representation (7.3.29) of a mother wavelet ψ has the same structure as that of the father wavelet ϕ given by (7.3.5).

Remarks.

1. The mother wavelet ψ associated with a given MRA is not unique because

$$d_n = (-1)^{n-1}\, \bar{c}_{2N-1-n} \tag{7.3.31}$$

defines the same mother wavelet (7.3.28) with suitably selected $N \in \mathbb{Z}$. This wavelet with coefficients d_n given by (7.3.31) has the Fourier transform

$$\hat{\psi}(\omega) = \exp\left\{(2N-1)\frac{i\omega}{2}\right\}\, \bar{m}\left(\frac{\omega}{2}+\pi\right)\hat{\phi}\left(\frac{\omega}{2}\right). \tag{7.3.32}$$

The nonuniqueness property of ψ allows us to define another form of ψ, instead of (7.3.28), by

$$\psi(x) = \sqrt{2}\sum_{n=-\infty}^{\infty} d_n\, \phi(2x-n), \tag{7.3.33}$$

where a slightly modified d_n is

$$d_n = (-1)^n\, \bar{c}_{1-n}. \tag{7.3.34}$$

In practice, any one of the preceding formulas for d_n can be used to find a mother wavelet.

2. The orthogonality condition (7.3.4) together with (7.3.2) and (7.3.24) implies

$$\left|\hat{\phi}(\omega)\right|^2 + \left|\hat{\psi}(\omega)\right|^2 = \left|\hat{\phi}\left(\frac{\omega}{2}\right)\right|^2. \tag{7.3.35}$$

Or, equivalently,

$$\left|\hat{\phi}\left(2^m\omega\right)\right|^2 + \left|\hat{\psi}\left(2^m\omega\right)\right|^2 = \left|\hat{\phi}\left(2^{m-1}\omega\right)\right|^2. \tag{7.3.36}$$

Summing both sides of (7.3.36) from $m=1$ to ∞ leads to the result

$$\left|\hat{\phi}(\omega)\right|^2 = \sum_{m=1}^{\infty}\left|\hat{\psi}\left(2^m\omega\right)\right|^2. \tag{7.3.37}$$

3. If ϕ has a compact support, the series (7.3.29) for the mother wavelet ψ terminates and consequently, ψ is represented by a finite linear combination of translated versions of $\phi(2x)$.

Finally, all of the above results lead to the main theorem of this section.

Theorem 7.3.3. If $\{V_n\}$, $n \in \mathbb{Z}$ is a multiresolution analysis with the scaling function ϕ, then there is a mother wavelet ψ given by

$$\psi(x) = \sqrt{2} \sum_{n=-\infty}^{\infty} (-1)^{n-1} \bar{c}_{-n-1} \phi(2x-n), \tag{7.3.38}$$

where the coefficients c_n are given by

$$c_n = (\phi, \phi_{1,n}) = \sqrt{2} \int_{-\infty}^{\infty} \phi(x) \overline{\phi(2x-n)} \, dx. \tag{7.3.39}$$

That is, the system $\{\psi_{m,n}(x) : m, n \in \mathbb{Z}\}$ is an orthonormal basis of $L^2(\mathbb{R})$.

Proof. First, we have to verify that $\{\psi_{m,n}(x) : m, n \in \mathbb{Z}\}$ is an orthonormal set. Indeed, we have

$$\int_{-\infty}^{\infty} \psi(x-k) \overline{\psi(x-\ell)} \, dx = \frac{1}{2\pi} \int_{-\infty}^{\infty} \exp[-i\omega(k-\ell)] |\hat{\psi}(\omega)|^2 d\omega$$

$$= \frac{1}{2\pi} \int_{0}^{2\pi} \exp[-i\omega(k-\ell)] \sum_{k=-\infty}^{\infty} |\hat{\psi}(\omega + 2\pi k)|^2 d\omega$$

$$\sum_{k=-\infty}^{\infty} |\hat{\psi}(\omega + 2\pi k)|^2 = \sum_{k=-\infty}^{\infty} \left| \hat{m}\left(\frac{\omega}{2} + (k+1)\pi\right) \right|^2 \left| \hat{\phi}\left(\frac{\omega}{2} + k\pi\right) \right|^2$$

which is, by splitting the sum into even and odd integers k,

$$= \sum_{k=-\infty}^{\infty} \left| \hat{m}\left(\frac{\omega}{2} + (2k+1)\pi\right) \right|^2 \left| \hat{\phi}\left(\frac{\omega}{2} + k\pi\right) \right|^2$$

$$+ \sum_{k=-\infty}^{\infty} \left| \hat{m}\left(\frac{\omega}{2} + (2k+2)\pi\right) \right|^2 \left| \hat{\phi}\left(\frac{\omega}{2} + (2k+1)\pi\right) \right|^2$$

$$= \left| \hat{m}\left(\frac{\omega}{2} + \pi\right) \right|^2 \sum_{k=-\infty}^{\infty} \left| \hat{\phi}\left(\frac{\omega}{2} + 2k\pi\right) \right|^2$$

$$+ \left| \hat{m}\left(\frac{\omega}{2}\right) \right|^2 \sum_{k=-\infty}^{\infty} \left| \hat{\phi}\left(\frac{\omega}{2} + (2k+1)\pi\right) \right|^2$$

$$= \left| \hat{m}\left(\frac{\omega}{2}\right) \right|^2 + \left| \hat{m}\left(\frac{\omega}{2} + \pi\right) \right|^2 = 1 \qquad \text{by (7.3.4).}$$

Thus, we find

$$\int_{-\infty}^{\infty} \psi(x-k)\, \overline{\psi(x-\ell)}\, dx = \delta_{k,\ell}$$

This shows that $\{\psi_{m,n} : m, n \in \mathbb{Z}\}$ is an orthonormal system. In view of Lemma 7.3.2 and our discussion preceding this theorem, to prove that it is a basis, it suffices to show that function $\hat{v}$ in (7.3.20) is square integrable over $[0, 2\pi]$. In fact,

$$\begin{aligned}
\int_0^{2\pi} |\hat{v}(\omega)|^2\, d\omega &= 2\int_0^{\pi} \left|\hat{\lambda}(\omega)\right|^2 d\omega \\
&= 2\int_0^{\pi} \left|\hat{\lambda}(\omega)\right|^2 \left\{|\hat{m}(\omega+\pi)|^2 + |\hat{m}(\omega)|^2\right\} d\omega, \quad \text{by (7.3.4)} \\
&= 2\int_0^{2\pi} \left|\hat{\lambda}(\omega)\right|^2 |\hat{m}(\omega+\pi)|^2\, d\omega \\
&= 2\int_0^{2\pi} |\hat{m}_f(\omega)|^2\, d\omega, \quad \text{by (7.3.18)} \\
&= 2\pi \sum_{n=-\infty}^{\infty} |c_n|^2, \quad c_n = (f, \phi_{1,n}) \\
&= 2\pi\, \|f\|^2 < \infty.
\end{aligned}$$

This completes the proof.

Example 7.3.1 (The Shannon Wavelet). We consider the Fourier transform $\hat{\phi}$ of a scaling function ϕ defined by

$$\hat{\phi}(\omega) = \chi_{[-\pi,\pi]}(\omega)$$

so that

$$\phi(x) = \frac{1}{2\pi} \int_{-\pi}^{\pi} e^{i\omega x}\, d\omega = \frac{\sin \pi x}{\pi x}.$$

This is also known as the Shannon sampling function. Both $\phi(x)$ and $\hat{\phi}(\omega)$ have been introduced in Chapter 3 (see Figure 3.12 with $\omega_0 = \pi$). Clearly, the Shannon scaling function does not have finite support. However, its Fourier transform has a finite support (band-limited) in the frequency domain and has good frequency localization. Evidently, the system

$$\phi_{0,k}(x) = \phi(x-k) = \frac{\sin \pi(x-k)}{\pi(x-k)}, \quad k \in \mathbb{Z}$$

is orthonormal because

$$\begin{aligned}\left(\phi_{0,k}, \phi_{0,\ell}\right) &= \frac{1}{2\pi}\left(\hat{\phi}_{0,k}, \hat{\phi}_{0,\ell}\right) \\ &= \frac{1}{2\pi}\int_{-\infty}^{\infty} \hat{\phi}_{0,k}(\omega)\,\overline{\hat{\phi}}_{0,\ell}(\omega)\,d\omega \\ &= \frac{1}{2\pi}\int_{-\infty}^{-\infty} \exp\{-i(k-\ell)\,\omega\}\,d\omega = \delta_{k,\ell}.\end{aligned}$$

In general, we define, for $m = 0$,

$$V_0 = \left\{\sum_{k=-\infty}^{\infty} c_k \frac{\sin \pi(x-k)}{\pi(x-k)} : \sum_{k=-\infty}^{\infty} |c_k|^2 < \infty\right\},$$

and, for other $m \neq 0, \quad m \in \mathbb{Z}$,

$$V_m = \left\{\sum_{k=-\infty}^{\infty} c_k \frac{2^{m/2}\sin \pi\left(2^m x - k\right)}{\pi\left(2^m x - k\right)} : \sum_{k=-\infty}^{\infty} |c_k|^2 < \infty\right\}.$$

It is easy to check that all conditions of Definition 7.3.1 are satisfied. We next find out the coefficients c_k defined by

$$\begin{aligned}c_k = \left(\phi, \phi_{1,n}\right) &= \sqrt{2}\int_{-\infty}^{\infty} \frac{\sin \pi x}{\pi x} \cdot \frac{\sin \pi(2x-k)}{\pi(2x-k)}\,dx \\ &= \left\{\begin{matrix} \dfrac{1}{\sqrt{2}}, & k = 0 \\ \dfrac{\sqrt{2}}{\pi k}\sin\left(\dfrac{\pi k}{2}\right), & k \neq 0 \end{matrix}\right\}.\end{aligned}$$

Consequently, we can use the formula (7.3.38) to find the Shannon mother wavelet

$$\begin{aligned}\psi(x) &= \sum_{n=-\infty}^{\infty} (-1)^{n-1} c_{-n-1}\,\phi(2x-n) \\ &= \frac{1}{\sqrt{2}}\frac{\sin \pi(2x+1)}{\pi(2x+1)} + \frac{\sqrt{2}}{\pi}\sum_{n\neq -1} \frac{(-1)^{n-1}}{(n+1)}\cos\left(\frac{n\pi}{2}\right)\frac{\sin \pi(2x-n)}{\pi(2x-n)}.\end{aligned}$$

Obviously, the system $\left\{\psi_{m,n} : m,n \in \mathbb{Z}\right\}$ is an orthonormal basis in $L^2(\mathbb{R})$. It is known as the *Shannon system*.

Theorem 7.3.4. If ϕ is a scaling function for a multiresolution analysis and $\hat{m}(\omega)$ is the associated low-pass filter, then a function $\psi \in W_0$ is an orthonormal wavelet for $L^2(\mathbb{R})$ if and only if

$$\hat{\psi}(\omega) = \exp\left(\frac{i\omega}{2}\right)\ \hat{v}(\omega)\ \overline{\hat{m}}\left(\frac{\omega}{2}+\pi\right)\hat{\phi}\left(\frac{\omega}{2}\right) \tag{7.3.40}$$

for some 2π-periodic function $\hat{v}$ such that $|\hat{v}(\omega)| = 1$.

Proof. It is enough to prove that all orthonormal wavelets $\psi \in W_0$ can be represented by (7.3.40). For any $\psi \in W_0$, by Lemma 7.3.4, there must be a 2π-periodic function $\hat{v}$ such that

$$\hat{\psi}(\omega) = \exp\left(\frac{i\omega}{2}\right)\ \hat{v}(\omega)\ \overline{\hat{m}}\left(\frac{\omega}{2}+\pi\right)\ \hat{\phi}\left(\frac{\omega}{2}\right).$$

If ψ is an orthonormal wavelet, then the orthonormality of $\{\psi(x-k),\ k\in\mathbb{Z}\}$ leads to

$$\begin{aligned} 1 &= \sum_{k=-\infty}^{\infty} |\hat{\psi}(\omega + 2\pi k)|^2 \\ &= |\hat{v}(\omega)|^2 \sum_{k=-\infty}^{\infty} \left|\hat{m}\left(\frac{\omega}{2} + k\pi + \pi\right)\right|^2 \left|\hat{\phi}\left(\frac{\omega}{2} + \pi k\right)\right|^2 \end{aligned}$$

which is, splitting the sum into even and odd integers k,

$$\begin{aligned} = |\hat{v}(\omega)^2| &\left\{ \sum_{k=-\infty}^{\infty} \left|\hat{m}\left(\frac{\omega}{2} + 2k\pi + \pi\right)\right|^2 \left|\hat{\phi}\left(\frac{\omega}{2} + 2k\pi\right)\right|^2 \right. \\ &\left. + \sum_{k=-\infty}^{\infty} \left|\hat{m}\left(\frac{\omega}{2} + (2k+1)\pi + \pi\right)\right| \left|\hat{\phi}\left(\frac{\omega}{2} + (2k+1)\pi\right)\right|^2 \right\} \end{aligned}$$

which is, by (7.3.1) and the 2π-periodic property of $\hat{m}$,

$$\begin{aligned} &= |\hat{v}(\omega)|^2 \left\{ \left|\hat{m}\left(\frac{\omega}{2}\right)\right|^2 + \left|\hat{m}\left(\frac{\omega}{2} + \pi\right)\right|^2 \right\} \\ &= |\hat{v}(\omega)|^2, \qquad \text{by} \quad (7.3.4). \end{aligned}$$

If the scaling function ϕ of an MRA is not an orthonormal basis of V_0 but rather is a Riesz basis, we can use the following orthonormalization process to generate an orthonormal basis.

Theorem 7.3.5. (Orthonormalization Process). If $\phi \in L^2(\mathbb{R})$ and if $\{\phi(x-n),\ n \in \mathbb{Z}\}$ is a Riesz basis, that is, there exists two constants $A,\ B > 0$ such that

$$0 < A \leq \sum_{k=-\infty}^{\infty} \left|\hat{\phi}(\omega + 2\pi k)\right|^2 \leq B < \infty, \tag{7.3.41}$$

then $\left\{\tilde{\phi}(x-n),\ n \in \mathbb{Z}\right\}$ is an orthonormal basis of V_0 with

$$\hat{\tilde{\phi}}(\omega) = \frac{\hat{\phi}(\omega)}{\sqrt{\hat{\Phi}(\omega)}}, \tag{7.3.42}$$

where the function $\hat{\Phi}$ is

$$\hat{\Phi}(\omega) = \sum_{k=-\infty}^{\infty} \left|\phi(\omega + 2\pi k)\right|^2. \tag{7.3.43}$$

Proof. It follows from inequality (7.3.41) that $\tilde{\phi} \in L^2(\mathbb{R})$. It also follows from (7.3.42) that

$$\sum_{k=-\infty}^{\infty} \left|\hat{\tilde{\phi}}(\omega + 2\pi k)\right|^2 = 1.$$

We consider a 2π-periodic function $\hat{g}$ defined by

$$\hat{g}(\omega) = \frac{1}{\sqrt{\hat{\Phi}(\omega)}}$$

so that $\hat{g}$ can be expanded as a Fourier series

$$\hat{g}(\omega) = \sum_{k=-\infty}^{\infty} g_k \exp(-ik\omega).$$

The inverse Fourier transform gives g in terms of the Dirac delta function of the form

$$g(t) = \sum_{k=-\infty}^{\infty} g_k\ \delta(t-k).$$

Applying the convolution theorem to (7.3.42) gives

$$\tilde{\phi}(x) = \phi(x) * g(x) = \int_{-\infty}^{\infty} \phi(x-t)\, g(t)\, dt$$

$$= \int_{-\infty}^{\infty} \phi(x-t) \sum_{k=-\infty}^{\infty} g_k\, \delta(t-k)\, dt$$

$$= \sum_{k=-\infty}^{\infty} g_k\, \phi(x-k).$$

This shows that $\{\phi(x-n),\ n \in \mathbb{Z}\}$ belongs to V_0. Thus, the function $\tilde{\phi}$ satisfies condition (b) of Theorem 7.3.1. Therefore, $\{\tilde{\phi}(x-n),\ n \in \mathbb{Z}\}$ is an orthonormal set.

It is easy to show that the span of $\{\tilde{\phi}(x-n),\ n \in \mathbb{Z}\} = \tilde{V}_0$ is the same as the span of $\{\phi(x-n),\ n \in \mathbb{Z}\} = V_0$. Hence, the multiresolution analysis is preserved under this orthonormalization process.

We describe another approach to constructing a multiresolution analysis, which begins with a function $\phi \in L^2(\mathbb{R})$ that satisfies the following relations

$$\phi(x) = \sum_{n=-\infty}^{\infty} c_n\, \phi(2x-n), \qquad \sum_{n=-\infty}^{\infty} |c_n|^2 < \infty, \tag{7.3.44}$$

and

$$0 < A \le \sum_{k=-\infty}^{\infty} \left|\hat{\phi}(\omega + 2\pi k)\right|^2 \le B < \infty, \tag{7.3.45}$$

where A and B are constants.

We define V_0 as the closed span of $\{\phi(x-n),\ n \in \mathbb{Z}\}$ and V_m as the span of $\{\phi_{m,n}(x),\ n \in \mathbb{Z}\}$. It follows from relation (7.3.45) that $\{V_m\}$ satisfies property (i) of the multiresolution analysis. In order to ensure that properties (ii) and (iii) of the MRA are satisfied, we further assume that $\hat{\phi}(\omega)$ is continuous and bounded with $\hat{\phi}(0) \ne 0$.

If $\left|\hat{\phi}(\omega)\right| \le C(1+|\omega|)^{-2^{-1}-\varepsilon}$, where $\varepsilon > 0$, then

$$\hat{\phi}(\omega) = \sum_{k=-\infty}^{\infty} \left|\hat{\phi}(\omega + 2\pi k)\right|^2$$

is continuous.

This ensures that the orthonormalization process can be used. Therefore, we assume

$$\hat{\tilde{\phi}}(\omega)=\frac{\hat{\phi}(\omega)}{\sqrt{\Phi(\omega)}} \quad \text{and} \quad \hat{\tilde{m}}\left(\frac{\omega}{2}\right)=\frac{\hat{\tilde{\phi}}(\omega)}{\hat{\tilde{\phi}}\left(\frac{\omega}{2}\right)}. \tag{7.3.46a,b}$$

Using (7.3.2) in (7.3.46b) gives

$$\hat{\tilde{m}}\left(\frac{\omega}{2}\right)=\left\{\frac{\hat{\Phi}\left(\frac{\omega}{2}\right)}{\hat{\Phi}(\omega)}\right\}^{\frac{1}{2}} \hat{m}\left(\frac{\omega}{2}\right). \tag{7.3.47}$$

We now recall (7.3.24) to obtain $\hat{\tilde{\psi}}(\omega)$ as

$$\hat{\tilde{\psi}}(\omega)=\exp\left(\frac{i\omega}{2}\right) \overline{\hat{\tilde{m}}\left(\frac{\omega}{2}+\pi\right)}\, \hat{\tilde{\phi}}\left(\frac{\omega}{2}\right), \tag{7.3.48}$$

which is, by (7.3.46a) and (7.3.47),

$$=\exp\left(\frac{i\omega}{2}\right)\left\{\frac{\hat{\Phi}\left(\frac{\omega}{2}+\pi\right)}{\hat{\Phi}(\omega)\,\hat{\Phi}\left(\frac{\omega}{2}\right)}\right\}^{\frac{1}{2}}\left\{\frac{\hat{\phi}(\omega+2\pi)\,\hat{\phi}\left(\frac{\omega}{2}\right)}{\hat{\phi}\left(\frac{\omega}{2}+\pi\right)}\right\}. \tag{7.3.49}$$

We introduce a complex function P defined by

$$P(z)=\frac{1}{2}\sum_{n=-\infty}^{\infty} c_n z^n, \qquad z\in\mathbb{C}, \tag{7.3.50}$$

where $z=\exp(-i\omega)$ and $|z|=1$.

We assume that $\sum_{n=-\infty}^{\infty}|c_n|<\infty$ so that the series defining P converges absolutely and uniformly on the unit circle in $\mathbb{C}$. Thus, P is continuous on the unit circle, $|z|=1$.

Since $P(z)=\frac{1}{2}\sum_{n=-\infty}^{\infty} c_n e^{-in\omega}=\hat{m}(\omega)$, it follows that

$$\hat{m}(\omega+\pi)=\frac{1}{2}\sum_{n=-\infty}^{\infty} c_n e^{-in\omega}\cdot e^{-in\pi}=\frac{1}{2}\sum_{n=-\infty}^{\infty} c_n(-z)^n=P(-z). \tag{7.3.51}$$

Consequently, the orthogonality condition (7.3.4) is equivalent to

$$|P(z)|^2+|P(-z)|^2=1. \tag{7.3.52}$$

Lemma 7.3.4. Suppose ϕ is a function in $L^1(\mathbb{R})$ which satisfies the two-scale relation

$$\phi(x)=\sum_{n=-\infty}^{\infty} c_n\,\phi(2x-n) \quad \text{with} \quad \sum_{n=-\infty}^{\infty} |c_n|<\infty. \tag{7.3.53}$$

(i) If the function P defined by (7.3.50) satisfies (7.3.52) for all z on the unit circle, $|z|=1$, and if $\hat{\phi}(0)\neq 0$, then $P(1)=1$ and $P(-1)=0$.

(ii) If $P(-1)=0$, then $\hat{\phi}(n)=0$ for all nonzero integers n.

Proof. We know that the relation

$$\hat{\phi}(\omega)=\hat{m}\left(\frac{\omega}{2}\right)\hat{\phi}\left(\frac{\omega}{2}\right)=P\left(e^{\frac{-i\omega n}{2}}\right)\hat{\phi}\left(\frac{\omega}{2}\right) \tag{7.3.54}$$

holds for all $\omega\in\mathbb{R}$. Putting $\omega=0$ leads to $P(1)=1$. It follows from equation (7.3.52) with $z=1$ that $P(-1)=0$.

The proof of part (ii) is left to the reader as an exercise.

We close this section by describing some properties of the coefficients of the scaling function. The coefficients c_n determine all the properties of the scaling function ϕ and the wavelet function ψ. In fact, Mallat's multiresolution algorithm uses the c_n to calculate the wavelet transform without explicit knowledge of ψ. Furthermore, both ϕ and ψ can be reconstructed from the c_n and this in fact is central to Daubechies' wavelet analysis.

Lemma 7.3.5. If c_n are coefficients of the scaling function defined by (7.3.5), then

(i) $\displaystyle\sum_{n=-\infty}^{\infty} c_n=\sqrt{2}$, (ii) $\displaystyle\sum_{n=-\infty}^{\infty} (-1)^n c_n=0$,

$$\sum_{n=-\infty}^{\infty} c_{2n}=\frac{1}{\sqrt{2}}=\sum_{n=-\infty}^{\infty} c_{2n+1},$$

$$\sum_{n=-\infty}^{\infty} (-1)^n n^m c_n=0 \quad \text{for} \quad m=0,1,2,\cdots,(p-1).$$

Proof. It follows from (7.3.2) and (7.3.3) that $\hat{\phi}(0)=0$ and $\hat{m}(0)=1$. Putting $\omega=0$ in (7.3.3) gives (i).

Since $\hat{m}(0)=1$, (7.3.4) implies that $\hat{m}(\pi)=0$ which gives (ii).

Then, (iii) is a simple consequence of (i) and (ii).

To prove (iv), we recall (7.3.8) and (7.3.3) so that

$$\hat{\phi}(\omega)=\hat{m}\left(\frac{\omega}{2}\right)\hat{m}\left(\frac{\omega}{2^2}\right)\cdots$$

and

$$\hat{m}\left(\frac{\omega}{2^k}\right)=\frac{1}{\sqrt{2}}\sum_{n=-\infty}^{\infty}c_n\exp\left(-\frac{in\omega}{2^k}\right).$$

Clearly,

$$\hat{\phi}(2\pi)=\hat{m}(\pi)\,\hat{m}\left(\frac{\omega}{2}\right).$$

According to Strang's (1989) accuracy condition, $\hat{\phi}(\omega)$ must have zeros of the highest possible order when $\omega=2\pi,\,4\pi,\,6\pi,\,\ldots$. Thus,

$$\hat{\phi}(2\pi)=\hat{m}(\pi)\,\hat{m}\left(\frac{\pi}{2}\right)\hat{m}\left(\frac{\pi}{2^2}\right)\cdots,$$

and the first factor $\hat{m}(\omega)$ will be zero of order p at $\omega=\pi$ if

$$\frac{d^m\,\hat{m}(\omega)}{d\omega^m}=0\quad\text{for}\quad m=0,1,2,\cdots(p-1),$$

which gives

$$\sum_{n=-\infty}^{\infty}c_n(-in)^m\,e^{-in\pi}=0\quad\text{for}\quad m=0,1,2,\cdots(p-1).$$

Or, equivalently,

$$\sum_{n=-\infty}^{\infty}(-1)^n\,n^m\,c_n=0,\quad\text{for }m=0,1,2,\cdots(p-1).$$

From the fact that the scaling function $\phi(x)$ is orthonormal to itself in any translated position, we can show that

$$\sum_{n=-\infty}^{\infty}c_n^2=1. \tag{7.3.55}$$

This can be seen by using $\phi(x)$ from (7.3.5) to obtain

$$\int_{-\infty}^{\infty} \phi^2(x)\, dx = 2 \sum_m \sum_n c_m c_n \int_{-\infty}^{\infty} \phi(2x-m)\, \phi(2x-n)\, dx$$

where the integral on the right-hand side vanishes due to orthonormality unless $m = n$, giving

$$\int_{-\infty}^{\infty} \phi^2(x)\, dx = 2 \sum_{n=-\infty}^{\infty} c_n^2 \int_{-\infty}^{\infty} \phi^2(2x-n)\, dx$$

$$= 2 \sum_{n=-\infty}^{\infty} c_n^2 \cdot \frac{1}{2} \int_{-\infty}^{\infty} \phi^2(t)\, dt$$

whence follows (7.3.55).

Finally, we prove

$$\sum_k c_k\, c_{k+2n} = \delta_{0n}\,. \tag{7.3.56}$$

We use the scaling function ϕ defined by (7.3.5) and the corresponding wavelet given by (7.3.29) with (7.3.31), that is,

$$\psi(x) = \sqrt{2} \sum_{n=-\infty}^{\infty} (-1)^{n-1}\, c_{2N-1-n}\, \phi(2x-n)$$

which is, by substituting $2N-1-n = k$,

$$= \sqrt{2} \sum_{k=-\infty}^{\infty} (-1)^k c_k\, \phi(2x+k-2N+1). \tag{7.3.57}$$

We use the fact that mother wavelet $\psi(x)$ is orthonormal to its own translate $\psi(x-n)$ so that

$$\int_{-\infty}^{\infty} \psi(x)\, \psi(x-n)\, dx = \delta_{0n}\,. \tag{7.3.58}$$

Substituting (7.3.57) to the left-hand side of (7.3.58) gives

$$\int_{-\infty}^{\infty} \psi(x)\, \psi(x-n)\, dx$$

$$= 2 \sum_k \sum_m (-1)^{k+m} c_k c_m \int_{-\infty}^{\infty} \phi(2x+k-2N+1)\phi(2x+m-2N+1-2n)\, dx,$$

where the integral on the right-hand side is zero unless $k = m-2n$ so that

$$\int_{-\infty}^{\infty} \psi(n)\,\psi(x-n)\,dx = 2\sum_k (-1)^{2(k+n)} c_k\, c_{k+2n} \cdot \frac{1}{2}\int_{-\infty}^{\infty} \phi^2(t)\,dt.$$

This means that

$$\sum_k c_k\, c_{k+2n} = 0 \quad \text{for all} \quad n \neq 0.$$

7.4 Construction of Orthonormal Wavelets

We now use the properties of scaling functions and filters for constructing orthonormal wavelets.

Example 7.4.1 **(*The Haar Wavelet*).** Example 7.2.2 shows that spaces of piecewise constant functions constitute a multiresolution analysis with the scaling function $\phi = \chi_{[0,1)}$. Moreover, ϕ satisfies the dilation equation

$$\phi(x) = \sqrt{2}\sum_{n=-\infty}^{\infty} c_n\,\phi(2x-n), \tag{7.4.1}$$

where the coefficients c_n are given by

$$c_n = \sqrt{2}\int_{-\infty}^{\infty} \phi(x)\,\phi(2x-n)\,dx. \tag{7.4.2}$$

Evaluating this integral with $\phi = \chi_{[0,1)}$ gives c_n as follows:

$$c_0 = c_1 = \frac{1}{\sqrt{2}} \quad \text{and} \quad c_n = 0 \quad \text{for} \quad n \neq 0,1.$$

Consequently, the dilation equation becomes

$$\phi(x) = \phi(2x) + \phi(2x-1). \tag{7.4.3}$$

This means that $\phi(x)$ is a linear combination of the even and odd translates of $\phi(2x)$ and satisfies a very simple *two-scale relation* (7.4.3), as shown in Figure 7.4.

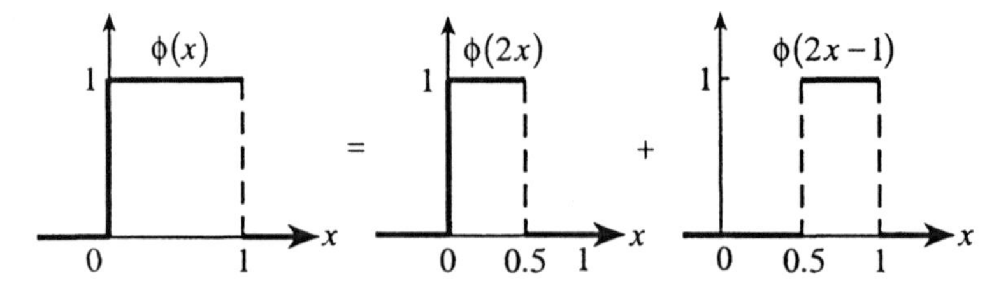

Figure 7.4. Two-scale relation of $\phi(x) = \phi(2x) + \phi(2x-1)$.

In view of (7.3.34), we obtain

$$d_0 = c_1 = \frac{1}{\sqrt{2}} \quad \text{and} \quad d_1 = -c_0 = -\frac{1}{\sqrt{2}}.$$

Thus, the Haar mother wavelet is obtained from (7.3.33) as a simple *two-scale relation*

$$\psi(x) = \phi(2x) - \phi(2x-1) \tag{7.4.4}$$

$$= \chi_{[0,.5]}(x) - \chi_{[.5,1]}(x)$$

$$= \begin{Bmatrix} +1, & 0 \le x < \frac{1}{2} \\ -1, & \frac{1}{2} \le x < 1 \\ 0, & \text{otherwise} \end{Bmatrix}. \tag{7.4.5}$$

This two-scale relation (7.4.4) of ψ is represented in Figure 7.5.

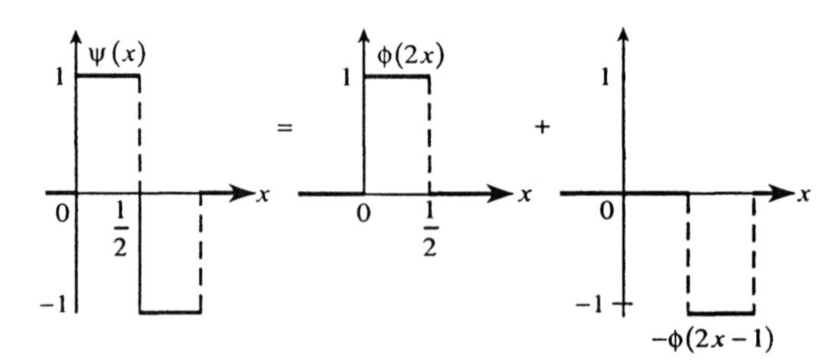

Figure 7.5. Two-scale relation of $\psi(x) = \phi(2x) - \phi(2x-1)$.

Alternatively, the Haar wavelet can be obtained from the Fourier transform of the scaling function $\phi = \chi_{[0,1]}$ so that

$$\hat{\phi}(\omega) = \hat{\chi}_{[0,1]}(\omega) = \exp\left(-\frac{i\omega}{2}\right) \frac{\sin\left(\frac{\omega}{2}\right)}{\left(\frac{\omega}{2}\right)}$$

$$= \exp\left(-\frac{i\omega}{4}\right) \cos\left(\frac{\omega}{4}\right) \cdot \exp\left(-\frac{i\omega}{4}\right) \frac{\sin\left(\frac{\omega}{4}\right)}{\left(\frac{\omega}{4}\right)}$$

$$= \hat{m}\left(\frac{\omega}{2}\right) \cdot \hat{\phi}\left(\frac{\omega}{2}\right), \tag{7.4.6}$$

where the associated filter $\hat{m}(\omega)$ and its complex conjugate are given by

$$\hat{m}(\omega) = \exp\left(-\frac{i\omega}{2}\right) \cos\left(\frac{\omega}{2}\right) = \frac{1}{2}\left(1 + e^{-i\omega}\right), \tag{7.4.7}$$

$$\overline{\hat{m}}(\omega) = \exp\left(\frac{i\omega}{2}\right) \cos\left(\frac{\omega}{2}\right) = \frac{1}{2}\left(1 + e^{i\omega}\right). \tag{7.4.8}$$

Thus, the Haar wavelet can be obtained form (7.3.24) or (7.3.40) and is given by

$$\hat{\psi}(\omega) = \hat{\upsilon}(\omega) \exp\left(\frac{i\omega}{2}\right) \overline{\hat{m}}\left(\frac{\omega}{2} + \pi\right) \hat{\phi}\left(\frac{\omega}{2}\right)$$

$$= \hat{v}(\omega) \cdot \exp\left(\frac{i\omega}{2}\right) \cdot \frac{1}{2}\left(1 - e^{\frac{i\omega}{2}}\right) \cdot \hat{\phi}\left(\frac{\omega}{2}\right)$$

where $\hat{v}(\omega) = -i\exp(-i\omega)$ is chosen to find the exact result (7.4.4). Using this value for $\hat{v}(\omega)$, we obtain

$$\hat{\psi}(\omega) = \frac{1}{2}\hat{\phi}\left(\frac{\omega}{2}\right) - \frac{1}{2}\exp\left(-\frac{i\omega}{2}\right) \hat{\phi}\left(\frac{\omega}{2}\right)$$

so that the inverse Fourier transform gives the exact result (7.4.4) as

$$\psi(x) = \phi(2x) - \phi(2x - 1).$$

On the other hand, using (7.3.24) also gives the Haar wavelet as

$$\hat{\psi}(\omega) = \exp\left(\frac{i\omega}{2}\right) \bar{\hat{m}}\left(\frac{\omega}{2}+\pi\right) \hat{\phi}\left(\frac{\omega}{2}\right)$$

$$= \exp\left(\frac{i\omega}{2}\right) \cdot \frac{1}{2}\left(1 - e^{\frac{i\omega}{2}}\right) \cdot \frac{\left\{1 - \exp\left(-\frac{i\omega}{2}\right)\right\}}{\left(\frac{i\omega}{2}\right)} \tag{7.4.9}$$

$$= \frac{i}{\omega}\left(1 - e^{\frac{i\omega}{2}}\right)^2$$

$$= \frac{i}{\omega}\left(e^{\frac{i\omega}{4}} \cdot e^{-\frac{i\omega}{4}} - e^{\frac{i\omega}{4}} \cdot e^{\frac{i\omega}{4}}\right)^2$$

$$= -i \exp\left(\frac{i\omega}{2}\right) \cdot \left[\frac{\sin^2\left(\frac{\omega}{4}\right)}{\left(\frac{\omega}{4}\right)}\right]$$

$$= \left\{i \exp\left(-\frac{i\omega}{2}\right) \frac{\sin^2\left(\frac{\omega}{4}\right)}{\frac{\omega}{4}}\right\} \{-\exp(-i\omega)\}. \tag{7.4.10}$$

This corresponds to the same Fourier transform (6.2.7) of the Haar wavelet (7.4.5) except for the factor $-\exp(-i\omega)$. This means that this factor induces a translation of the Haar wavelet to the left by one unit. Thus, we have chosen $\hat{v}(\omega) = -\exp(-i\omega)$ in (7.3.40) to find the same value (7.4.5) for the classic Haar wavelet.

Example 7.4.2 (Cardinal B-splines and Spline Wavelets). The cardinal B-splines (basis splines) consist of functions in $C^{n-1}(\mathbb{R})$ with equally spaced integer knots that coincide with polynomials of degree n on the intervals $\left[2^{-m}k,\ w^{-m}(k+1)\right]$. These B-splines of order n with compact support generate a linear space V_0 in $L^2(\mathbb{R})$. This leads to a multiresolution analysis $\{V_m,\ m \in \mathbb{Z}\}$ by defining $f(x) \in V_m$ if and only if $f(2x) \in V_{m+1}$.

The cardinal B-splines $B_n(x)$ of order n are defined by the following convolution product

$$B_1(x) = \chi_{[0,1]}(x), \tag{7.4.11}$$

$$B_n(x) = B_1(x) * B_1(x) * \cdots * B_1(x) = B_1(x) * B_{n-1}(x), \quad (n \geq 2), \tag{7.4.12}$$

where n factors are involved in the convolution product. Obviously,

$$B_n(x) = \int_{-\infty}^{\infty} B_{n-1}(x-t)\, B_1(t)\, dt = \int_0^1 B_{n-1}(x-t)\, dt = \int_{x-1}^{x} B_{n-1}(t)\, dt. \tag{7.4.13}$$

Using the formula (7.4.13), we can obtain the explicit representation of splines $B_2(x)$, $B_3(x)$, and $B_4(x)$ as follows:

$$B_2(x) = \int_{x-1}^{x} B_1(t)\, dt = \int_{x-1}^{x} \chi_{[0,1]}(t)\, dt.$$

Evidently, it turns out that

$$B_2(x) = 0 \quad \text{for} \quad x \leq 0.$$

$$B_2(x) = \int_0^x dt = x \quad \text{for} \quad 0 \leq n \leq 1, \quad (x-1 \leq 0).$$

$$B_2(x) = \int_{x-1}^{1} dt = 2 - x \quad \text{for} \quad 1 \leq n \leq 2, \quad (- \leq x-1 \leq 1 \leq x).$$

$$B_2(x) = 0 \quad \text{for} \quad 2 \leq x, \quad (1 \leq x-1).$$

Or, equivalently,

$$B_2(x) = x\, \chi_{[0,1]}(x) + (2-x)\, \chi_{[1,2]}(x). \tag{7.4.14}$$

Similarly, we find

$$B_3(x) = \int_{x-1}^{x} B_2(x)\, dx.$$

More explicitly,

$$B_3(x) = 0 \quad \text{for} \quad x \leq 0.$$

$$B_3(x) = \int_0^x t\, dt = \frac{x^2}{2} \quad \text{for} \quad 0 \leq x \leq 1, \quad (x-1 \leq 0 \leq x \leq 1).$$

$$B_3(x) = \int_{x-1}^{1} t\,dt + \int_{1}^{x} (2-t)\,dt \quad \text{for} \quad 1 \le x \le 2, \ (0 \le x-1 \le 1 \le x \le 2)$$

$$= \frac{1}{2}\left(6x - 2x^2 - 3\right) \quad \text{for} \quad 1 \le x \le 2.$$

$$B_3(x) = \int_{x-1}^{2} (2-t)\,dt = \frac{1}{2}(x-3)^2 \quad \text{for} \quad 2 \le x \le 3, (1 \le x-1 \le 2 \le x \le 3).$$

$$B_3(x) = 0 \quad \text{for} \quad x \ge 3, \ (2 \le x-1).$$

Or, equivalently,

$$B_3(x) = \frac{x^2}{2}\,\chi_{[0,1]} + \frac{1}{2}\left(6x - 2x^2 - 3\right)\chi_{[1,2]} + \frac{1}{2}(x-3)^2\,\chi_{[2,3]}. \tag{7.4.15}$$

Finally, we have

$$B_4(x) = \int_{x-1}^{x} B_3(t)\,dt.$$

$$B_4(x) = 0 \quad \text{for} \quad x-1 \le -1 \le x \le 0.$$

$$B_4(x) = \int_{0}^{x}\left(\frac{1}{2}t^2\right) dt = \frac{1}{6}x^3 \quad \text{for} \quad -1 \le x-1 \le 0 \le x < 1.$$

$$B_4(x) = \int_{x-1}^{1}\left(\frac{1}{2}t^2\right) dt + \int_{1}^{x}\left(-\frac{3}{2} + 3t - t^2\right) dt \quad \text{for} \quad 1 \le x \le 2,$$

$$(0 \le x-1 \le 1 \le x \le 2)$$

$$= \frac{2}{3} - 2x + 2n^2 - \frac{1}{3}x^3 \quad \text{for} \quad 1 \le x \le 2.$$

$$\mathrm{B}_4(x) = \int_{x-1}^{2}\left(-\frac{3}{2} + 3t - t^2\right) dt + \frac{1}{2}\int_{2}^{x}(3-t)^2\,dt \quad \text{for} \quad 1 \le x-1 \le 2 \le x \le 3$$

$$= \frac{1}{2}\left(x^3 - 2x^2 + 20x - 13\right) \quad \text{for} \quad 2 \le x \le 3.$$

Or, equivalently,

$$B_4(x) = \frac{1}{6}x^3\,\chi_{[0,1]} + \frac{1}{3}\left(2 - 6x + 6x^2 - x^3\right)\chi_{[1,2]}$$

$$+ \frac{1}{2}\left(x^3 - 2x^2 + 20x - 13\right)\chi_{[2,3]}. \tag{7.4.16}$$

In order to obtain the two-scale relation for the B-splines of order n, we apply the Fourier transform of (7.4.11) so that

$$\hat{B}_1(\omega) = \exp\left(-\frac{i\omega}{2}\right) \frac{\sin\left(\frac{\omega}{2}\right)}{\left(\frac{\omega}{2}\right)} = \exp\left(-\frac{i\omega}{2}\right) \operatorname{sinc}\left(\frac{\omega}{2}\right), \tag{7.4.17a}$$

$$= \frac{1}{i\omega}\left(1 - e^{-i\omega}\right) = \int_0^1 e^{-i\omega t}\, dt, \tag{7.4.17b}$$

where the sinc function, sinc (x) is defined by

$$\operatorname{sinc}(x) = \begin{Bmatrix} \dfrac{\sin x}{x}, & x \neq 0 \\ 1 & x = 0 \end{Bmatrix}. \tag{7.4.18}$$

We can also express (7.4.17a) in terms of $z = \exp\left(-\frac{i\omega}{2}\right)$ as

$$\hat{B}_1(\omega) = \frac{1}{2}(1+z)\, \hat{B}_1\left(\frac{\omega}{2}\right). \tag{7.14.19}$$

Application of the convolution theorem of the Fourier transform to (7.4.12) gives

$$\hat{B}_n(\omega) = \left\{\hat{B}_1(\omega)\right\}^n = \hat{B}_1(\omega)\, \hat{B}_{n-1}(\omega), \tag{7.14.20}$$

$$= \left(\frac{1+z}{2}\right)^n \left\{\hat{B}_1\left(\frac{\omega}{2}\right)\right\}^n = \left(\frac{1+z}{2}\right)^n \hat{B}_n\left(\frac{\omega}{2}\right), \tag{7.14.21}$$

$$= \hat{M}_n\left(\frac{\omega}{2}\right) \hat{B}_n\left(\frac{\omega}{2}\right), \tag{7.4.22}$$

where the associated filter $\hat{M}_n$ is given by

$$\hat{M}_n\left(\frac{\omega}{2}\right) = \left(\frac{1+z}{2}\right)^n = \frac{1}{2^n}\left(1 + e^{-\frac{i\omega}{2}}\right)^n = e^{-\frac{i\omega n}{2}}\left(\cos\frac{\omega}{2}\right)^n$$

$$= \frac{1}{2^n}\sum_{k=0}^{n}\binom{n}{k} \exp\left(-\frac{in\omega}{2}\right), \tag{7.4.23}$$

which is, by definition of $\hat{M}_n\left(\frac{\omega}{2}\right)$,

$$= \frac{1}{\sqrt{2}}\sum_{k=-\infty}^{\infty} c_{n,k} \exp\left(-\frac{ik\omega}{2}\right). \tag{7.4.24}$$

Obviously, the coefficients $c_{n,k}$ are given by

$$c_{n,k} = \begin{Bmatrix} \frac{\sqrt{2}}{2^n}\binom{n}{k}, & 0 \le k \le n \\ 0, & \text{otherwise} \end{Bmatrix}. \tag{7.4.25}$$

Therefore, the spline function in the time domain is

$$B_n(x) = \sqrt{2}\sum_{k=0}^{\infty} c_{n,k}\ \phi(2x-k) = \sum_{k=0}^{n} 2^{1-n}\binom{n}{k} B_n(2x-k). \tag{7.4.26}$$

This may be referred to as the two-scale relation for the B-splines of order n.

In view of (7.4.17a), it follows that

$$\left|\hat{B}_n(\omega)\right| = \left|\text{sinc}\left(\frac{\omega}{2}\right)\right|^n, \tag{7.4.27}$$

where $\text{sinc}(x)$ is defined by (7.14.18). Thus, for each $n \ge 1$, $\hat{B}_n(\omega)$ is a first-order *Butterworth filter* which satisfies the following conditions

$$\left|\hat{B}_n(0)\right| = 1, \left[\frac{d}{d\omega}\left|\hat{B}_n(\omega)\right|\right]_{\omega=0} = 0, \text{ and } \left[\frac{d^2}{d\omega^2}\left|\hat{B}_n^2(\omega)\right|\right]_{\omega=0} \ne 0. \tag{7.4.28}$$

The graphical representation of $B_n(x)$ and their filter characteristics $\left|\hat{B}_n(\omega)\right|$ are shown in Figures 7.6 and 7.7.

It is evident from (7.4.27) that

$$\left|\hat{B}_n(\omega+2\pi k)\right|^2 = \frac{\sin^{2n}\left(\frac{\omega}{2}+\pi k\right)}{\left(\frac{\omega}{2}+\pi k\right)^{2n}} = \frac{\sin^{2n}\left(\frac{\omega}{2}\right)}{\left(\frac{\omega}{2}+\pi k\right)^{2n}}. \tag{7.4.29}$$

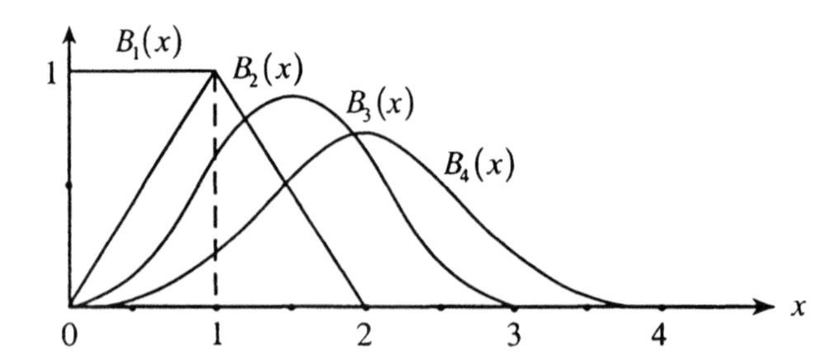

Figure 7.6. Cardinal B-spline functions.

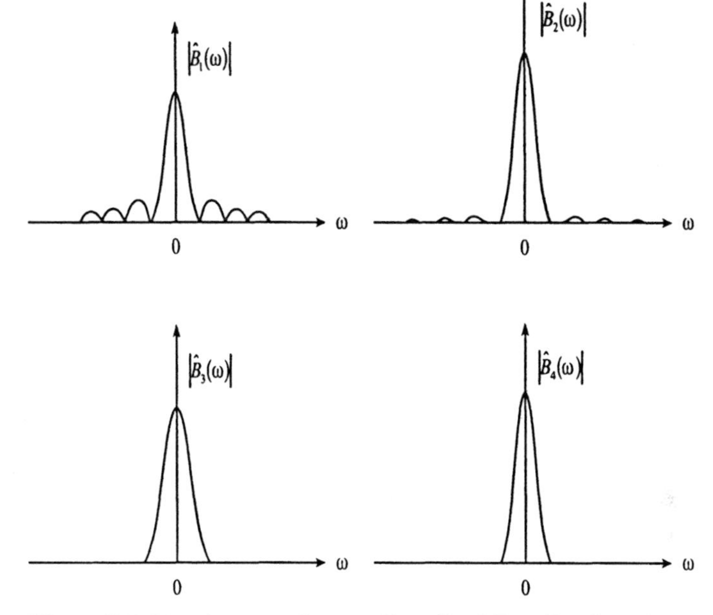

Figure 7.7. Fourier transforms of cardinal B-spline functions.

We replace ω by 2ω in (7.4.29) and then sum the result over all integers k to obtain

$$\sum_{k=-\infty}^{\infty} \left|\hat{B}_n\left(2\omega + 2\pi k\right)\right|^2 = \sin^{2n}\omega \sum_{k=-\infty}^{\infty} \frac{1}{\left(\omega + \pi k\right)^{2n}}. \tag{7.4.30}$$

It is well known in complex analysis that

$$\sum_{k=-\infty}^{\infty} \frac{1}{\left(\omega + \pi k\right)} = \cot\omega, \tag{7.4.31}$$

which leads to the following result after differentiating $(2n-1)$ times

$$\sum_{k=-\infty}^{\infty} \frac{1}{\left(\omega + \pi k\right)^{2n}} = -\frac{1}{(2n-1)!} \frac{d^{2n-1}}{d\omega^{2n-1}} \left(\cot\omega\right). \tag{7.4.32}$$

Substituting this result in (7.4.30) yields

$$\sum_{k=-\infty}^{\infty} \left|\hat{B}_n\left(2\omega + 2\pi k\right)\right|^2 = -\frac{\sin^{2n}(\omega)}{(2n-1)!} \frac{d^{2n-1}}{d\omega^{2n-1}} \left(\cot\omega\right). \tag{7.4.33}$$

These results are used to find the Franklin wavelets and the Battle-Lemarié wavelets.

When $n = 1$, (7.4.32) gives another useful identity

$$\sum_{k=-\infty}^{\infty} \frac{1}{(\omega + 2\pi k)^2} = \frac{1}{4} \operatorname{cosec}^2\left(\frac{\omega}{2}\right). \tag{7.4.34}$$

Summing (7.4.29) over all integers k and using (7.4.34) leads to

$$\sum_{k=-\infty}^{\infty} \left|\hat{B}_1(\omega + 2\pi k)\right|^2 = 4\sin^2\left(\frac{\omega}{2}\right) \sum_{k=-\infty}^{\infty} \frac{1}{(\omega + 2\pi k)^2} = 1. \tag{7.4.35}$$

This shows that the first order B-spline $B_1(x)$ defined by (7.4.11) is a scaling function that generates the classic Haar wavelet.

Example 7.4.3 (The Franklin Wavelet). The Franklin wavelet is generated by the second order $(n = 2)$ splines.

Differentiating (7.4.34) $(2n - 2)$ times gives the result

$$\sum_{k=-\infty}^{\infty} \frac{1}{(\omega + 2\pi k)^{2n}} = \frac{1}{4(2n-1)!} \frac{d^{2n-2}}{d\omega^{2n-2}} \left[\operatorname{cosec}^2\left(\frac{\omega}{2}\right)\right]. \tag{7.4.36}$$

When $n = 2$, (7.4.36) yields the identity

$$\sum_{k=-\infty}^{\infty} \frac{1}{(\omega + 2\pi k)^4} = \frac{1}{\left(2\sin\frac{\omega}{2}\right)^4} \cdot \left\{1 - \frac{2}{3}\sin^2\left(\frac{\omega}{2}\right)\right\}. \tag{7.4.37}$$

For $n = 2$, we sum (7.4.29) over all integers k so that

$$\begin{aligned} \sum_{k=-\infty}^{\infty} \left|\hat{B}_2(\omega + 2\pi k)\right|^2 &= 16\sin^4\left(\frac{\omega}{2}\right) \sum_{k=-\infty}^{\infty} \frac{1}{(\omega + 2\pi k)^4} \\ &= \left\{1 - \frac{2}{3}\sin^2\frac{\omega}{2}\right\}. \end{aligned} \tag{7.4.38}$$

Or, equivalently,

$$\left[\left\{1 - \frac{2}{3}\sin^2\frac{\omega}{2}\right\}^{-\frac{1}{2}}\right]^2 \sum_{k=-\infty}^{\infty} \left|\hat{B}_2(\omega + 2\pi k)\right|^2 = 1. \tag{7.4.39}$$

Thus, the condition of orthonormality (7.4.33) ensures that the scaling function ϕ has the Fourier transform

$$\hat{\phi}(\omega) = \left(\frac{\sin \frac{\omega}{2}}{\frac{\omega}{2}} \right)^2 \left(1 - \frac{2}{3} \sin^2 \frac{\omega}{2} \right)^{-\frac{1}{2}}. \tag{7.4.40}$$

Thus, the filter associated with this scaling function ϕ is obtained from (7.3.2) so that

$$\hat{m}(\omega) = \frac{\phi(2\omega)}{\phi(\omega)} = \left(\frac{\sin \omega}{2 \sin \frac{\omega}{2}} \right)^2 \left[\frac{1 - \frac{2}{3} \sin^2 \frac{\omega}{2}}{1 - \frac{2}{3} \sin^2 \omega} \right]^{\frac{1}{2}}$$

$$= \cos^2 \left(\frac{\omega}{2} \right) \left[\frac{1 - \frac{2}{3} \sin^2 \frac{\omega}{2}}{1 - \frac{2}{3} \sin^2 \omega} \right]^{\frac{1}{2}}. \tag{7.4.41}$$

Finally, the Fourier transform of the orthonormal wavelet ψ is obtained from (7.3.24) so that

$$\hat{\psi}(2\omega) = \hat{m}_1(\omega)\, \hat{\phi}(\omega) = e^{i\omega}\, \overline{\hat{m}}\, (\omega + \pi)\, \hat{\phi}(\omega) \tag{7.4.42}$$

$$= e^{i\omega} \left[\frac{1 - \frac{2}{3} \sin^2 \frac{\omega}{2}}{1 - \frac{2}{3} \sin^2 \omega} \right]^{\frac{1}{2}} \left(\frac{\sin^2 \frac{\omega}{2}}{\frac{\omega}{2}} \right)^2 \left\{ 1 - \frac{2}{3} \sin^2 \frac{\omega}{2} \right\}^{-\frac{1}{2}}$$

$$= e^{i\omega} \frac{\sin^4 \left(\frac{\omega}{2} \right)}{\left(\frac{\omega}{2} \right)^2} \left[\frac{1 - \frac{2}{3} \cos^2 \frac{\omega}{4}}{\left(1 - \frac{2}{3} \sin^2 \frac{\omega}{2} \right) \left(1 - \frac{2}{3} \sin^2 \frac{\omega}{4} \right)} \right]^{\frac{1}{2}}. \tag{7.4.43}$$

This is known as the *Franklin wavelet* generated by the second order spline function $B_2(x)$. The scaling function ϕ for the Franklin wavelet, the magnitude of its Fourier transform, $|\hat{\phi}(\omega)|$, the Franklin wavelet ψ, and the magnitude of its Fourier transform $|\hat{\psi}(\omega)|$ are shown in Figures 7.8 and 7.9.

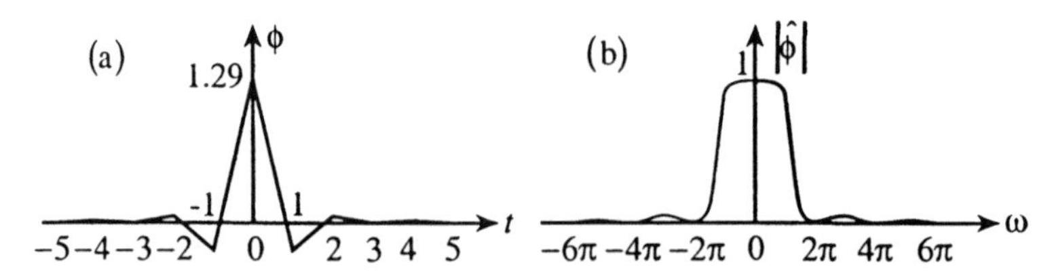

Figure 7.8. (a) Scaling function of the Franklin wavelet ϕ,

(b) The Fourier transform $|\hat{\phi}|$.

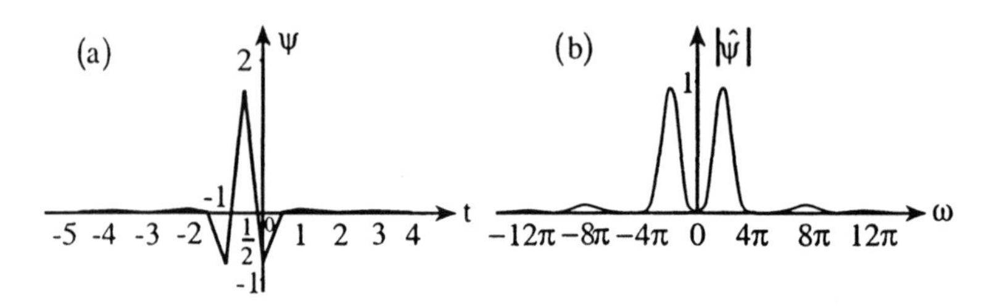

Figure 7.9. (a) The Franklin wavelet ψ,

(b) The Fourier transform $|\hat{\psi}|$.

Example 7.4.3 (The Battle-Lemarié Wavelet). The Fourier transform $\hat{\phi}(\omega)$ associated with the nth order spline function $B_n(x)$ is

$$\hat{\phi}(\omega) = \frac{\hat{B}_n(\omega)}{\left\{\sum_{k=-\infty}^{\infty} \left|\hat{B}_n(\omega + 2k\pi)\right|^2\right\}^{\frac{1}{2}}}, \tag{7.4.44}$$

where $\hat{B}_n(\omega)$ is given by (7.14.20) and

$$\left|\hat{B}_n(\omega + 2k\pi)\right|^2 = \left\{\frac{\sin\left(\frac{\omega}{2} + k\pi\right)}{\left(\frac{\omega + 2k\pi}{2}\right)}\right\}^{2n}, \tag{7.4.45}$$

and

$$\left\{\sum \left|\hat{B}_n(\omega + 2k\pi)\right|^2\right\}^{\frac{1}{2}} = \frac{2^n \sin^n\left(\frac{\omega}{2}\right)}{\sqrt{\hat{S}_{2n}(\omega)}}, \tag{7.4.46}$$

with

$$\hat{S}_{2n}(\omega) = \sum_{k=-\infty}^{\infty} \frac{1}{(\omega + 2k\pi)^{2n}}. \tag{7.4.47}$$

Consequently, (7.4.44) can be expressed in the form

$$\hat{\phi}(\omega) = \frac{\exp\left(-\frac{i\varepsilon\omega}{2}\right)}{\omega^n \sqrt{\hat{S}_{2n}(\omega)}}, \tag{7.4.48}$$

where $\varepsilon = 1$ when n is odd or $\varepsilon = 0$ when n is even, and $\hat{S}_{2n}(\omega)$ can be computed by using the formula (7.4.36).

In particular, when $n = 4$, corresponding to the cubic spline of order four, $\hat{\phi}(\omega)$ is calculated from (7.4.48) by inserting

$$\hat{S}_8(\omega) = \sum_{k=-\infty}^{\infty} \frac{1}{(\omega + 2k\pi)^8} = \frac{\hat{N}_1(\omega) + \hat{N}_2(\omega)}{(105)\left(2\sin\frac{\omega}{2}\right)^8}, \tag{7.4.49}$$

where

$$\hat{N}_1(\omega) = 5 + 30\cos^2\left(\frac{\omega}{2}\right) + 30\left(\sin\frac{\omega}{2}\cos\frac{\omega}{2}\right)^2, \tag{7.4.50}$$

and

$$\hat{N}_2(\omega) = 70\cos^4\left(\frac{\omega}{2}\right) + 2\sin^4\left(\frac{\omega}{2}\right)\cos^2\left(\frac{\omega}{4}\right) + \frac{2}{3}\sin^6\left(\frac{\omega}{2}\right). \tag{7.4.51}$$

Finally, the Fourier transform of the Battle-Lemarié wavelet ψ can be found by using the same formulas stated in Example 7.4.2. The Battle-Lemarié scaling function ϕ and the Battle-Lemarié wavelet ψ are displayed in Figures 7.10(a) and 7.10(b).

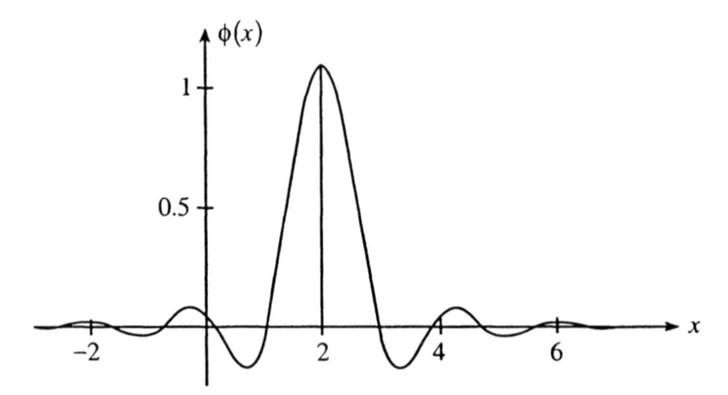

Figure 7.10. (a) The Battle-Lemarié scaling function.

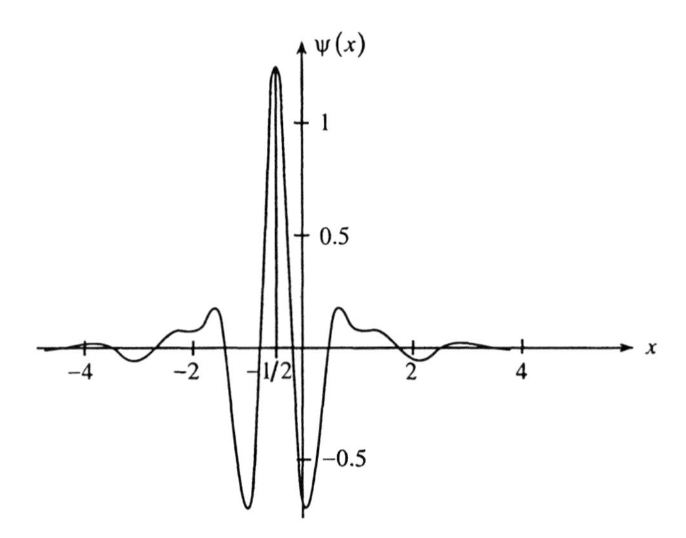

Figure 7.10. (b) The Battle-Lemarié wavelet.

The rest of this section is devoted to the construction of one of the compactly supported orthonormal wavelets first discovered by Daubechies (1988). We suppose that the scaling function ϕ satisfies the two-scale relation

$$\phi(x) = \sum_{n=0}^{1} c_n \phi(2x - n) = c_0 \phi(2x) + c_1 \phi(2x - n) \tag{7.4.52}$$

for almost all $n \in \mathbb{R}$. We want $\{\phi(x-n) : n \in \mathbb{Z}\}$ to be an orthonormal set, and thus, we impose the necessary condition on the function P

$$|P(z)|^2 + |P(-z)|^2 = 1, \quad (z \in \mathbb{C},\ |z| = 1).$$

We also assume $\hat{\phi}(0) = 1$. Then, $P(1) = 1$ and $P(-1) = 0$ by Lemma 7.3.4. Thus, P contains $(1+z)$ as a factor. Since P is a linear polynomial, we construct P with the form

$$P(z) = \frac{(1+z)}{2}\, S(z). \tag{7.4.53}$$

This form ensures that $P(-1) = 0$. The relation $P(1) = 1$ holds if and only if $S(1) = 1$. Indeed, the assumption on P is a particular case of a general procedure where we assume the form

$$P(z) = \left(\frac{1+z}{2}\right)^N S(z), \tag{7.4.54}$$

where N is a positive integer to be selected appropriately.

Writing

$$P(z) = \frac{1}{2}\,(1+z)(p_0 + p_1 z)$$

and using $P(1) = 1$ gives

$$p_0 + p_1 = 1. \tag{7.4.55}$$

The result

$$|P(i)|^2 + |P(-i)|^2 = 1 \tag{7.4.56}$$

leads to another equation for p_0 and p_1

$$\begin{aligned} 1 &= \frac{1}{4}\, |(p_0 - p_1) + i\,(p_0 + p_1)|^2 + |(p_0 - p_1) - i\,(p_0 + p_1)|^2 \\ &= p_0^2 + p_1^2. \end{aligned} \tag{7.4.57}$$

Solving (7.4.55) and (7.4.57) gives either $p_0 = 1$ $p_1 = 0$ or vice versa. However, the values $p_0 = 1$ and $p_1 = 1$ yield

$$P(z) = \frac{1}{2}\,(1+z).$$

Equating this value of P with its definition (7.3.51) leads to $c_0 = 1$ and $c_1 = 1$. Thus, the scaling function (7.4.52) becomes

$$\phi(x) = \phi(2x) + \phi(2x-1).$$

This corresponds to the Haar wavelet.

With $N = 2$, we obtain, from (7.4.54),

$$P(z) = \left(\frac{1+z}{2}\right)^2 S(z) = \left(\frac{1+z}{2}\right)^2 (p_0 + p_1 z), \tag{7.4.58}$$

where p_0 and p_1 are determined from $P(1) = 1$ and (7.4.56). It turns out that

$$p_0 + p_1 = 1, \tag{7.4.59}$$

$$p_0^2 + p_1^2 = 2. \tag{7.4.60}$$

Solving these two equations yields either

$$p_0 = \frac{1}{2}\left(1+\sqrt{3}\right) \text{ and } p_1 = \frac{1}{2}\left(1-\sqrt{3}\right)$$

or vice versa. Consequently, it turns out that

$$P(z) = \frac{1}{4}\left[p_0 + (2p_0 + p_1)z + (p_0 + 2p_1)z^2 + p_1 z^3\right]. \tag{7.4.61}$$

Equating result (7.4.61) with

$$P(z) = \frac{1}{2}\sum_{n=0}^{3} c_n z^n = \frac{1}{2}\left(c_0 + c_1 z + c_2 z^2 + c_3 z^3\right)$$

gives the values for the coefficients

$$c_0 = \frac{1}{4}\left(1+\sqrt{3}\right),\ c_1 = \frac{1}{4}\left(3+\sqrt{3}\right),\ c_2 = \frac{1}{4}\left(3-\sqrt{3}\right),\ c_3 = \frac{1}{4}\left(1-\sqrt{3}\right). \tag{7.4.62}$$

Consequently, the scaling function becomes

$$\phi(x) = \frac{1}{4}\left(1+\sqrt{3}\right)\phi(2x) + \frac{1}{4}\left(3+\sqrt{3}\right)\phi(2x-1) + \frac{1}{4}\left(3-\sqrt{3}\right)\phi(2x-2) + \frac{1}{4}\left(1-\sqrt{3}\right)\phi(2x-3). \tag{7.4.63}$$

Or equivalently,

$$\phi(x) = c_0\,\phi(2x) + (1-c_3)\,\phi(2x-1) + (1-c_0)\,\phi(2x-2) + c_3\,\phi(2x-3). \tag{7.4.64}$$

In the preceding calculation, the factor $\sqrt{2}$ is dropped in the formula (7.3.53) for the scaling function ϕ and hence, we have to drop the factor $\sqrt{2}$ in the wavelet formula (7.3.33) so that $\psi(x)$ takes the form

$$\psi(x) = d_0\,\phi(2x) + d_1\,\phi(2x-1) + d_{-1}\,\phi(2x+1) + d_{-2}\,\phi(2x+2), \quad (7.4.65)$$

where $d_n = (-1)^n\, c_{1-n}$ is used to find $d_0 = c_1$, $d_1 = -c_0$, $d_{-1} = -c_2$, $d_{-2} = c_3$. Consequently, the final form of $\psi(x)$ becomes

$$\psi(x) = \frac{1}{4}\left(1-\sqrt{3}\right)\phi(2x+2) - \frac{1}{4}\left(3-\sqrt{3}\right)\phi(2x+1) + \frac{1}{4}\left(3+\sqrt{3}\right)\phi(2x)$$
$$-\frac{1}{4}\left(1+\sqrt{3}\right)\phi(2x-1). \quad (7.4.66)$$

This is called the *Daubechies wavelet*. Daubechies (1992) has shown that in this family of examples the size of the support of ϕ, ψ is determined by the desired regularity. It turns out that this is a general feature and that a linear relationship between these two quantities support width and regularity, is the best. Daubechies (1992) also proved the following theorem.

Theorem 7.4.1. If $\phi \in C^m$, support $\phi \subset [0, N]$, and $\phi(x) = \sum_{n=0}^{N} c_n\, \phi(2x-n)$, then $N \geq m+2$.

For proof of this theorem, the reader is referred to Daubechies (1992).

7.5 Daubechies' Wavelets and Algorithms

Daubechies (1988, 1992) first developed the theory and construction of orthonormal wavelets with compact support. Wavelets with compact support have many interesting properties. They can be constructed to have a given number of derivatives and to have a given number of vanishing moments.

We assume that the scaling function ϕ satisfies the dilation equation

$$\phi(x) = \sqrt{2} \sum_{n=-\infty}^{\infty} c_n\, \phi(2x-n), \quad (7.5.1)$$

where $c_n = (\phi, \phi_{1,n})$ and $\sum_{n=-\infty}^{\infty} |c_n|^2 \leq \infty$.

If the scaling function ϕ has compact support, then only a finite number of c_n have nonzero values. The associated generating function $\hat{m}$,

$$\hat{m}(\omega) = \frac{1}{\sqrt{2}} \sum_{n=-\infty}^{\infty} c_n \exp(-i\omega n) \tag{7.5.2}$$

is a trigonometric polynomial and it satisfies the identity (7.3.4) with special values $\hat{m}(0) = 1$ and $\hat{m}(\pi) = 0$. If coefficients c_n are real, then the corresponding scaling function as well as the mother wavelet ψ will also be real-valued. The mother wavelet ψ corresponding to ϕ is given by the formula (7.3.24) with $|\hat{\phi}(0)| = 1$. The Fourier transform $\hat{\psi}(\omega)$ of order N is N-times continuously differentiable and it satisfies the moment condition (6.2.16), that is,

$$\hat{\psi}^{(k)}(0) = 0 \text{ for } k = 0, 1, \ldots, m. \tag{7.5.3}$$

It follows that $\psi \in C^m$ implies that $\hat{m}_0$ has a zero at $\omega = \pi$ of order $(m+1)$. In other words,

$$\hat{m}_0(\omega) = \left(\frac{1+e^{-i\omega}}{2}\right)^{m+1} \hat{L}(\omega), \tag{7.5.4}$$

where $\hat{L}$ is a trigonometric polynomial.

In addition to the orthogonality condition (7.3.4), we assume

$$\hat{m}_0(\omega) = \left(\frac{1+e^{-i\omega}}{2}\right)^{N} \hat{L}(\omega), \tag{7.5.5}$$

where $\hat{L}(\omega)$ is 2π-periodic and $\hat{L} \in C^{N-1}$. Evidently,

$$|\hat{m}_0(\omega)|^2 = \hat{m}_0(\omega)\,\hat{m}_0(-\omega) = \left(\frac{1+e^{-i\omega}}{2}\right)^{N} \left(\frac{1+e^{i\omega}}{2}\right)^{N} \hat{L}(\omega)\,\hat{L}(-\omega)$$

$$= \left(\cos^2 \frac{\omega}{2}\right)^{N} |\hat{L}(\omega)|^2, \tag{7.5.6}$$

where $|\hat{L}(\omega)|^2$ is a polynomial in $\cos\omega$, that is,

$$|\hat{L}(\omega)|^2 = Q(\cos\omega).$$

Since $\cos\omega = 1 - 2\sin^2\left(\frac{\omega}{2}\right)$, it is convenient to introduce $x = \sin^2\left(\frac{\omega}{2}\right)$ so that (7.5.6) reduces to the form

$$\left|\hat{m}_0(\omega)\right|^2 = \left(\cos^2\frac{\omega}{2}\right)^N Q(1-2x) = (1-x)^N P(x), \tag{7.5.7}$$

where $P(x)$ is a polynomial in x.

We next use the fact that

$$\cos^2\left(\frac{\omega+\pi}{2}\right) = \sin^2\left(\frac{\omega}{2}\right) = x$$

and

$$\begin{aligned}\left|\hat{L}(\omega+\pi)\right|^2 &= Q(-\cos\omega) = Q(2x-1) \\ &= Q(1-2(1-x)) = P(1-x)\end{aligned} \tag{7.5.8}$$

to express the identity (7.3.4) in terms of x so that (7.3.4) becomes

$$(1-x)^N P(x) + x^N P(1-x) = 1. \tag{7.5.9}$$

Since $(1-x)^N$ and x^N are two polynomials of degree N which are relatively prime, then, by Bezout's theorem (see Daubechies, 1992), there exists a unique polynomial P_N of degree $\le N-1$ such that (7.5.9) holds. An explicit solution for $P_N(x)$ is given by

$$P_N(x) = \sum_{k=0}^{N-1} \binom{N+k-1}{k} x^k, \tag{7.5.10}$$

which is positive for $0 < x < 1$ so that $P_N(x)$ is at least a possible candidate for $\left|\hat{L}(\omega)\right|^2$. There also exist higher degree polynomial solutions $P_N(x)$ of (7.5.9) which can be written as

$$P_N(x) = \sum_{k=0}^{N-1} \binom{N+k-1}{k} x^k + x^N R\left(x - \frac{1}{2}\right), \tag{7.5.11}$$

where R is an odd polynomial.

Since $P_N(x)$ is a possible candidate for $\left|\hat{L}(\omega)\right|^2$ and

$$\hat{L}(\omega)\,\hat{L}(-\omega) = \left|\hat{L}^2(\omega)\right|^2 = Q(\cos\omega) = Q(1-2x) = P_N(x), \tag{7.5.12}$$

the next problem is how to find out $\hat{L}(\omega)$. This can be done by the following lemma:

Lemma 7.5.1 (Riesz Spectral Factorization). If

$$\hat{A}(\omega) = \sum_{k=0}^{n} a_k \cos^k \omega, \tag{7.5.13}$$

where $a_k \in \mathbb{R}$ and $a_n \neq 0$, and if $\hat{A}(\omega) \geq 0$ for real ω with $\hat{A}(0) = 1$, then there exists a trigonometric polynomial

$$\hat{L}(\omega) = \sum_{k=0}^{n} b_k e^{-ik\omega} \tag{7.5.14}$$

with real coefficients b_k with $\hat{L}(0) = 1$ such that

$$\hat{A}(\omega) = \hat{L}(\omega)\, \hat{L}(-\omega) = \left|\hat{L}(\omega)\right|^2 \tag{7.5.15}$$

is identically satisfied for ω.

We refer to Daubechies (1992) for a proof of the Riesz lemma 7.5.1. We also point out that the factorization of $\hat{A}(\omega)$ given in (7.5.15) is not unique.

For a given N, if we select $P = P_N$, then $\hat{A}(\omega)$ becomes a polynomial of degree $N-1$ in $\cos\omega$ and $\hat{L}(\omega)$ is a polynomial of degree $(N-1)$ in $\exp(-i\omega)$. Therefore, the generating function $\hat{m}_0(\omega)$ given by (7.5.5) is of degree $(2N-1)$ in $\exp(-i\omega)$. The interval $[0, 2N-1]$ becomes the support of the corresponding scaling function ${}_N\phi$. The mother wavelet ${}_N\psi$ obtained from ${}_N\phi$ is called the *Daubechies wavelet.*

Example 7.5.1 (The Haar Wavelet). For $N = 1$, it follows from (7.5.10) that $P_1(x) \equiv 1$, and this in turn leads to the fact that $Q(\cos\omega) = 1$, $\hat{L}(\omega) = 1$ so that the generating function is

$$\hat{m}_0(\omega) = \frac{1}{2}\left(1 + e^{-i\omega}\right). \tag{7.5.16}$$

This corresponds to the generating function (7.4.7) for the Haar wavelet.

Example 7.5.2 **(*The Daubechies Wavelet*).** For $N=2$, it follows from (7.5.10) that

$$P_2(x) = \sum_{k=0}^{1} \binom{k+1}{k} x^k = 1 + 2x$$

and hence (7.5.12) gives

$$\left|\hat{L}^2(\omega)\right|^2 = P_2(x) = P_2\left(\sin^2\frac{\omega}{2}\right) = 1 + 2\sin^2\frac{\omega}{2} = (2 - \cos\omega).$$

Using (7.5.14) in Lemma 7.5.1, we obtain that $\hat{L}(\omega)$ is a polynomial of degree $N-1=1$ and

$$\hat{L}(\omega)\,\hat{L}(-\omega) = 2 - \frac{1}{2}\left(e^{i\omega} + e^{-i\omega}\right).$$

It follows from (7.5.14) that

$$\left(b_0 + b_1 e^{-i\omega}\right)\left(b_0 + b_1 e^{i\omega}\right) = 2 - \frac{1}{2}\left(e^{i\omega} + e^{-i\omega}\right). \tag{7.5.17}$$

Equating the coefficients in this identity gives

$$b_0^2 + b_1^2 = 1 \quad \text{and} \quad 2\,b_0 b_1 = -1. \tag{7.5.18}$$

These equations admit solutions as

$$b_0 = \frac{1}{2}\left(1+\sqrt{3}\right) \quad \text{and} \quad b_1 = \frac{1}{2}\left(1-\sqrt{3}\right). \tag{7.5.19}$$

Consequently, the generating function (7.3.5) takes the form

$$\begin{aligned}\hat{m}_0(\omega) &= \left(\frac{1+e^{-i\omega}}{2}\right)^2 \left(b_0 + b_1 e^{-i\omega}\right)\\ &= \frac{1}{8}\left[\left(1+\sqrt{3}\right) + \left(3+\sqrt{3}\right)e^{-i\omega} + \left(3-\sqrt{3}\right)e^{-2i\omega} + \left(1-\sqrt{3}\right)e^{-3i\omega}\right]\end{aligned} \tag{7.5.20}$$

with $\hat{m}_0(0) = 1$.

Comparing coefficients of (7.5.20) with (7.3.3) gives c_n as

$$\left.\begin{aligned} c_0 &= \frac{1}{4\sqrt{2}}\left(1+\sqrt{3}\right),\ c_1 = \frac{1}{4\sqrt{2}}\left(3+\sqrt{3}\right)\\ c_2 &= \frac{1}{4\sqrt{2}}\left(3-\sqrt{3}\right),\ c_3 = \frac{1}{4\sqrt{2}}\left(1-\sqrt{3}\right)\end{aligned}\right\}. \tag{7.5.21}$$

Consequently, the Daubechies scaling function ${}_2\phi(x)$ takes the form, dropping the subscript,

$$\phi(x) = \sqrt{2}\,\left[c_0\,\phi(2x) + c_1\,\phi(2x-1) + c_2\,\phi(2x-2) + c_3\,\phi(2x-3)\right]. \tag{7.5.22}$$

Using (7.3.31) with $N = 2$, we obtain the Daubechies wavelet ${}_2\psi(x)$, dropping the subscript,

$$\begin{aligned}\psi(x) &= \sqrt{2}\,\left[d_0\,\phi(2x) + d_1\,\phi(2x-1) + d_2\,\phi(2x-2) + d_3\,\phi(2x-3)\right] \\ &= \sqrt{2}\,\left[-c_3\,\phi(2x) + c_2\,\phi(2x-1) - c_1\,\phi(2x-2) + c_0\,\phi(2x-3)\right],\end{aligned} \tag{7.5.23}$$

where the coefficients in (7.5.23) are the same as for the scaling function $\phi(x)$, but in reverse order and with alternate terms having their signs changed from plus to minus.

On the other hand, the use of (7.3.29) with (7.3.34) also gives the Daubechies wavelet ${}_2\psi(x)$ in the form

$${}_2\psi(x) = \sqrt{2}\,\left[-c_0\,\phi(2x-1) + c_1\,\phi(2x) - c_2\,\phi(2x+1) + c_3\,\phi(2x+2)\right]. \tag{7.5.24}$$

The wavelet has the same coefficients as ψ given in (7.5.23) except that the wavelet is reversed in sign and runs from $x = -1$ to 2 instead of starting from $x = 0$. It is often referred to as the Daubechies *D4 wavelet* since it is generated by four coefficients.

However, in general, c's (some positive and some negative) in (7.5.22) are numerical constants. Except for a very simple case, it is not easy to solve (7.5.22) directly to find the scaling function $\phi(x)$. The simplest approach is to set up an iterative algorithm in which each new approximation $\phi_m(x)$ is computed from the previous approximation $\phi_{m-1}(x)$ by the scheme

$$\phi_m(x) = \sqrt{2}\,\left[c_0\,\phi_{m-1}(2x) + c_1\,\phi_{m-1}(2x-1) + c_2\,\phi_{m-1}(2x-2) + c_3\,\phi_{m-1}(2x-3)\right]. \tag{7.5.25}$$

This iteration process can be continued until $\phi_m(x)$ becomes indistinguishable from $\phi_{m-1}(x)$. This iterative algorithm is briefly described below starting from the characteristic function

$$\chi_{[0,1]}(x) = \begin{Bmatrix} 1, & 0 \le x < 1 \\ 0, & \text{otherwise} \end{Bmatrix}. \tag{7.5.26}$$

After one iteration the characteristic function over $0 \le x < 1$ assumes the shape of a staircase function over the interval $0 \le x < 2$. In order to describe the algorithm, we select the set of four coefficients c_0, c_1, c_2, c_3 given in (7.5.21), deleting the factor $\frac{1}{\sqrt{2}}$ in each coefficient so that it produces the Daubechies scaling function $\phi(x)$ given by (7.5.22) and the orthonormal Daubechies wavelet $\psi(x)$ (or D4 wavelet) given by (7.5.23) without the factor $\sqrt{2}$.

We represent the characteristic function by the ordinate 1 at $x = 0$. The first iteration generates a new set of four ordinates c_0, c_1, c_2, c_3 at $x = 0.0, 0.5, 1.0, 1.5$. The second iteration with ordinate c_0 at $x = 0$ produces a new set of another four ordinates c_0^2, $c_0 c_1$, $c_1 c_2$, $c_1 c_3$ at $x = 0.00, 0.25, 0.75$, and so on. After completing the second iteration process, there are ten new ordinates c_0^2, $c_0 c_1$, $c_0 c_1 + c_1 c_0$, $c_0 c_3 + c_1^2$, $c_1 c_2 + c_2 c_0$, $c_1 c_3 + c_2 c_1$, $c_2^2 + c_3 c_0$, $c_2 c_3 + c_3 c_1$ $c_3 c_2$, c_3^2 at $x = 0.25, 0.50, 0.75, 1.00, \cdots, 2.25$. This iteration process can be described by the matrix scheme

$$
\left[{}_2\phi\right] = \begin{bmatrix} c_0 & & & \\ c_1 & & & \\ c_2 & c_0 & & \\ c_3 & c_1 & & \\ & c_2 & c_0 & \\ & c_3 & c_1 & \\ & & c_2 & c_0 \\ & & c_3 & c_1 \\ & & & c_2 \\ & & & c_3 \end{bmatrix} \begin{bmatrix} c_0 \\ c_1 \\ c_2 \\ c_3 \end{bmatrix} [1] = M_2 \, M_1 \, [1], \tag{7.5.27}
$$

where M_n represents the matrix of the order $\left(2^{n+1} + 2^n - 2\right) \times \left(2^n + 2^{n-1} - 2\right)$ in which each column has a submatrix of the coefficients c_0, c_1, c_2, c_3 located two places below the submatrix to its left.

We also use the same matrix scheme for developing the Daubechies wavelet ${}_2\psi(x)$ from ${}_2\phi(x)$ which is given by (7.5.22) without the factor $\sqrt{2}$. For simplicity, we assume that only one iteration process gives the final ${}_2\phi(x)$, so this can be described by four ordinates c_0, c_1, c_2, c_3 at $x = 0.0, 0.50, 1.0, 1.50$. In view of (7.5.23), these four ordinates produce ten new ordinates spaced 0.25

apart. The term $-c_3\,\phi(2x)$ in (7.5.23) gives $-c_3\,c_0,\ -c_3\,c_1,\ -c_3\,c_2.\ -c_3^2$; the term $c_2\,\phi(2x-1)$ gives $c_2\,c_0,\ c_2\,c_1,\ c_2^2,\ c_2\,c_3$ shifted two places to the right and so on for the other terms, so that the new ten ordinates for the wavelet are given by $-c_3\,c_0,\ -c_3\,c_1,\ -c_3\,c_2+c_2\,c_0,\ -c_3^2+c_2c_1,\ c_2^2-c_1\,c_0,\ c_2\,c_3-c_1^2,\ -c_1\,c_2+c_0^2,$ $-c_1\,c_3+c_0\,c_1,\ c_0\,c_2,\ c_0\,c_3$. These ordinates are generated by the matrix scheme

$$\left[{}_2\psi\right]=\begin{bmatrix} -c_3 & & & \\ 0 & -c_3 & & \\ c_2 & 0 & -c_3 & \\ 0 & c_2 & 0 & -c_3 \\ -c_1 & 0 & c_2 & 0 \\ 0 & -c_1 & 0 & c_2 \\ c_0 & 0 & -c_1 & 0 \\ & c_0 & 0 & -c_1 \\ & & c_0 & 0 \\ & & & c_0 \end{bmatrix}\begin{bmatrix} c_0 \\ c_1 \\ c_2 \\ c_3 \end{bmatrix}[1]\,. \tag{7.5.28}$$

Or, alternatively, by the matrix scheme

$$\left[{}_2\psi\right]=\begin{bmatrix} c_0 & & & \\ c_1 & & & \\ c_2 & c_0 & & \\ c_3 & c_1 & & \\ & c_2 & c_0 & \\ & c_3 & c_1 & \\ & & c_2 & c_0 \\ & & c_3 & c_1 \\ & & & c_2 \\ & & & c_3 \end{bmatrix}\begin{bmatrix} -c_3 \\ c_2 \\ -c_1 \\ c_0 \end{bmatrix}[1]\,. \tag{7.5.29}$$

Making reference to Newland (1993b), it can be verified that ${}_3\psi(x)$ can be described by the matrix scheme

$$\left[{}_3\psi\right]=M_3\,M_2\begin{bmatrix} -c_3 \\ c_2 \\ -c_1 \\ c_0 \end{bmatrix}[1]\,, \tag{7.5.30}$$

where the matrix M_3 is of order 22×10 with ten submatrices $\left[c_0\ c_1\ c_2\ c_3\right]^T$, each organized two places below its left-hand neighboring matrix.

The matrix scheme (7.5.30) is used to generate wavelets in the inverse discrete wavelet transform (IDWT). All subsequent steps of the iteration use the matrices M_r consisting of submatrices $\left[c_0\ c_1\ c_2\ c_3\right]^T$ staggered vertically two places each. After eight steps leading to 766 ordinates as before, the resulting wavelet is very close to that in Figure 7.11(a).

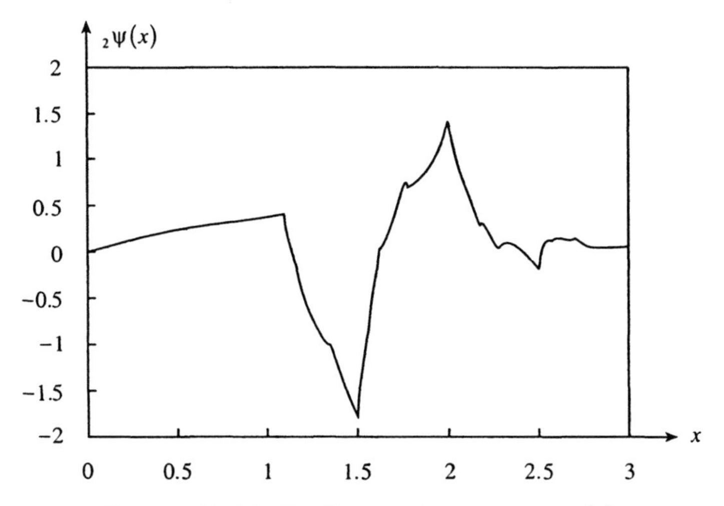

Figure 7.11. (a). The Daubechies wavelet $_2\psi(x)$.

In order to analyze or synthesize a part of a signal by wavelets, Daubechies (1992) considered the scaling function ϕ defined by (7.5.22) as a *building block* so that

$$\phi(x) = 0 \text{ when } x \le 0 \text{ or } x \ge 3. \tag{7.5.31}$$

Daubechies (1992) proved that the scaling function ϕ does not admit any simple algebraic relation in terms of elementary or special functions. She also demonstrated that ϕ satisfies several algebraic relations that play a major role in computational analysis.

Replacing x by $\frac{x}{2}$ in (7.5.22) gives

$$\phi\left(\frac{x}{2}\right) = \sqrt{2} \sum_{k=0}^{3} c_k\, \phi(x-k) \tag{7.5.32}$$

which can be found exactly if $\phi(x)$, $\phi(x-1)$, $\phi(x-2)$, $\phi(x-3)$ are all known. Suppose that we can find $\phi(0)$, $\phi(1)$, $\phi(2)$, $\phi(3)$. It is known that $\phi(-1)$, $\phi(4)$, etc. are all zero. Then, by using (7.5.32), we can calculate

$$\phi\left(\frac{1}{2}\right),\ \phi\left(\frac{3}{2}\right),\ \phi\left(\frac{5}{2}\right).$$

Again, by using (7.5.32) and these new values, we can calculate

$$\phi\left(\frac{1}{4}\right),\ \phi\left(\frac{3}{4}\right),\ \phi\left(\frac{5}{4}\right),\ \phi\left(\frac{7}{4}\right),\ \phi\left(\frac{9}{4}\right),\ \phi\left(\frac{11}{4}\right),$$

and so on.

In order to carry out this recursive process, we set initial values

$$\phi(0)=0,\quad \phi(1)=\frac{1}{2}\left(1+\sqrt{3}\right),\quad \phi(2)=\frac{1}{2}\left(1-\sqrt{3}\right),\quad \phi(3)=0. \tag{7.5.33}$$

For example, for $x=1$, we obtain from (7.5.32) that

$$\phi\left(\frac{1}{2}\right)=\sqrt{2}\,\left[c_0\,\phi(1)+c_1\,\phi(0)+c_2\,\phi(-1)+c_3\,\phi(-2)\right]$$

which is, by (7.5.21) and (7.5.31),

$$=\sqrt{2}\,c_0\,\phi(1)=\frac{\left(1+\sqrt{3}\right)^2}{8}=\frac{1}{4}\left(2+\sqrt{3}\right).$$

Similarly, we can calculate $\phi\left(\frac{3}{2}\right)$, $\phi\left(\frac{5}{2}\right)$ so that

$$x=\frac{1}{2},\qquad \frac{3}{2},\qquad \frac{5}{2},$$

$$\phi(x)=\frac{1}{4}\left(2+\sqrt{3}\right),\quad 0,\quad \frac{1}{4}\left(2-\sqrt{3}\right),$$

and $\phi(x\geq 3)=0$.

A similar calculation gives the values of ϕ at multiples of $\frac{1}{4}$ as given below:

$$x=\frac{1}{4},\qquad \frac{3}{4},\qquad \frac{5}{4},\qquad \frac{7}{4},\qquad \frac{9}{4},$$

$$\phi(x)=\frac{5+3\sqrt{3}}{16},\quad \frac{9+5\sqrt{3}}{16},\quad \frac{2\left(1+\sqrt{3}\right)}{16},\quad \frac{2\left(1-\sqrt{3}\right)}{16},\quad \frac{9-5\sqrt{3}}{16}.$$

The Daubechies wavelet $\psi(x)$ is given by (7.5.24). In view of (7.5.31), it turns out that $\psi(x)=0$ if $2x+2\le 0$ or $2x-1\ge 3$, that is, $\psi(x)=0$ for $x\le -1$ or $x\ge 2$. Hence, ψ can be computed from (7.5.24) with (7.5.21) and (7.5.33). For example,

$$\begin{aligned}\psi(0) &= \sqrt{2}\,\left[c_3\,\phi(2)-c_2\,\phi(1)+c_1\,\phi(0)-c_0\,\phi(-1)\right]\\ &= \sqrt{2}\,\left[c_3\,\phi(2)-c_3\,\phi(1)\right]=\left(\frac{1-\sqrt{3}}{4}\right)\left(\frac{1-\sqrt{3}}{2}\right)-\left(\frac{3-\sqrt{3}}{4}\right)\left(\frac{1+\sqrt{3}}{2}\right)\\ &= \frac{1}{2}\left(1-\sqrt{3}\right).\end{aligned}$$

Consequently, $\psi(x)$ at $x=-1,\ -\frac{1}{2},\ 0,\ 1,\ \frac{3}{2}$ is given as follows:

$$\begin{array}{llllll} x=-1, & -\frac{1}{2}, & 0, & 1, & \frac{3}{2}, \\ \psi(x)=0, & -\frac{1}{4}, & \frac{1}{2}\left(1-\sqrt{3}\right), & -\frac{1}{2}\left(1+\sqrt{3}\right), & -\frac{1}{4}. \end{array}$$

Both Daubechies' scaling function ϕ and Daubechies' wavelet ψ for $N=2$ are shown in Figures 7.11(a) and 7.11(b), respectively.

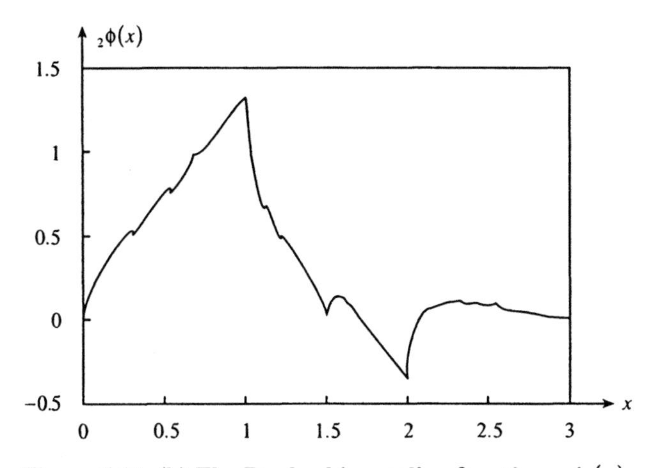

Figure 7.11. (b) The Daubechies scaling function $_2\phi(x)$.

In view of its fractal shape, the Daubechies wavelet ${}_2\psi(x)$ given in Figure 7.11(a) has received tremendous attention so that it can serve as a basis for signal analysis. According to Strang's (1989) analysis, a wavelet expansion based on the D4 wavelet represents a linear function $f(x) = ax$ exactly, where a is a constant. Six wavelet coefficients are needed to represent $f(x) = ax + bx^2$, where a and b are constants. In general, more wavelet coefficients are necessary to represent a polynomial with terms like x^n. Figures 7.12(a) and 7.12(b) exhibit wavelets with $N = 3, 5, 7$, and 10 coefficients. The range of these wavelets is always $(2N - 1)$ unit intervals so that more wavelet coefficients generate longer wavelets. As N increases, wavelets lose their irregular shape and become increasingly smooth with a Gaussian harmonic waveform. For $N = 10$, the frequency of the waveform is not constant and some minor irregularities still persist on the right. Each of the wavelets in Figures 7.12(a) and 7.12(b) represents the basis for a family of wavelets of different levels and different locations along the x-axis. The only difference is that a wavelet with $2N$ coefficients occupies $(2N - 1)$ unit intervals with the exception of the Haar wavelet which occupies one interval. Wavelets at each level overlap one another and the amount of overlap depends on the number of wavelet coefficients involved.

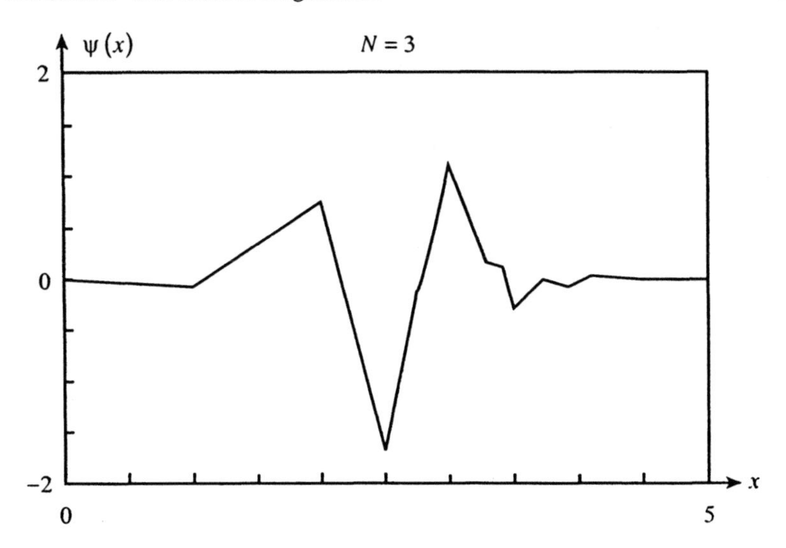

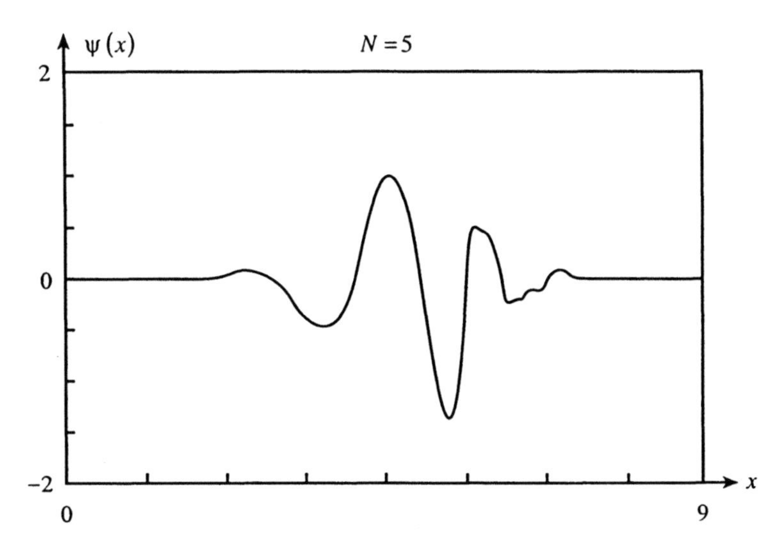

Figure 7.12. (a). Wavelets for $N = 3, 5$ drawn using the Daubechies algorithm.

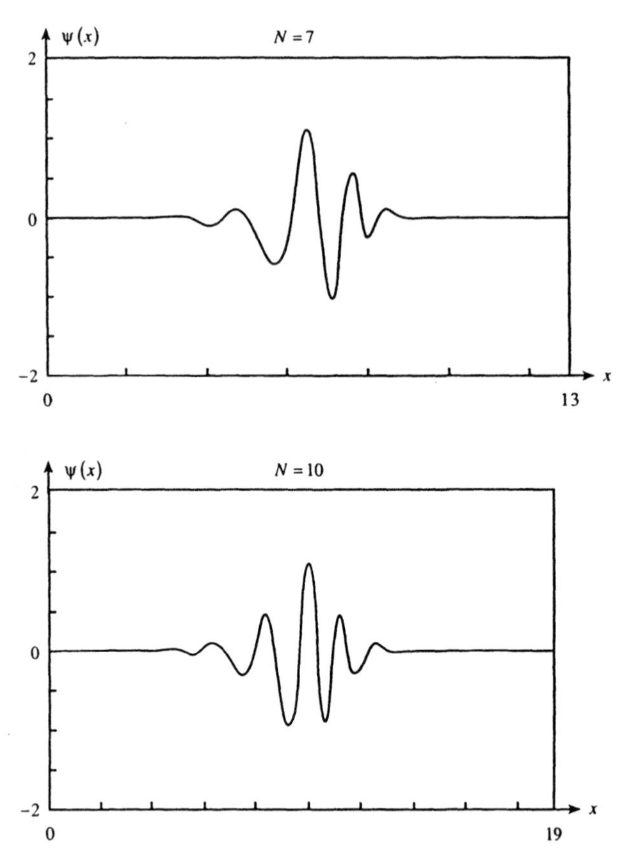

Figure 7.12. (b). Wavelets for $N = 7$, 10 drawn using the Daubechies algorithm.

The recursive method just described above yields the values of the building block $\phi(x)$ and the wavelet $\psi(x)$ only at integral multiples of positive or negative powers of 2. These values are sufficient for equally spaced samples from a signal. Due to the importance of such powers of 2, the idea of a dyadic number and related notation and terminology seem to be useful in wavelet algorithms.

***Definition 7.5.1* (*Dyadic Number*).** A number m is called a *dyadic number* if and only if it is an integral multiple of an integral power of 2.

We denote the set of all dyadic numbers by $\mathbb{D}$ and the set of all integral multiples by $\mathbb{D}_n$ for $n \in \mathbb{N}$. A dyadic number has a finite binary expansion, and a dyadic number in $\mathbb{D}_n$ has a binary expansion with at most n binary digits past the binary point.

Definition 7.5.2. The set of all linear combinations of 1 and $\sqrt{3}$ with dyadic coefficients $p,\, q \in \mathbb{D}$ is denoted by $\mathbb{D}\left[\sqrt{3}\right]$ so that

$$\mathbb{D}\left[\sqrt{3}\right] = \left\{p + q\sqrt{3} \,:\, p,\, q \in \mathbb{D}\right\}.$$

For every integer n, we consider combinations with coefficients in $\mathbb{D}_n$ so that

$$\mathbb{D}_n\left[\sqrt{3}\right] = \left\{p + q\sqrt{3} \,:\, p,\, q \in \mathbb{D}_n\right\}.$$

We define the *conjugate* $\overline{m}$ of m by

$$\overline{\left(p + q\sqrt{3}\right)} = \left(p - q\sqrt{3}\right).$$

The set $\mathbb{D}\left[\sqrt{3}\right]$ is an *integer ring* under ordinary addition and multiplication.

In terms of two quantities

$$a = \frac{1}{4}\left(1 + \sqrt{3}\right) \text{ and } \overline{a} = \frac{1}{4}\left(1 - \sqrt{3}\right), \tag{7.5.34}$$

the scaling function ${}_2\phi$ can be written as

$${}_2\phi(x) = \sqrt{2} \sum_{k=0}^{2N-1} c_k\, \phi(2x - k), \qquad (N = 2) \tag{7.5.35a}$$

$$= a\,\phi(2x) + (1 - \overline{a})\,\phi(2x - 1) + (1 + a)\,\phi(2x - 2) + \overline{a}\,\phi(2x - 3). \tag{7.5.35b}$$

If $0 \le m \le 2N - 1$, (7.5.35) can be rewritten as

$$\phi(m) = \sqrt{2} \sum_{k=0}^{2N-1} c_{2m-k}\, \phi(k). \tag{7.5.36}$$

This system of equations can be written in the matrix form

$$\begin{bmatrix} \phi(0) \\ \phi(1) \\ \phi(2) \\ \phi(3) \end{bmatrix} = \begin{bmatrix} a & 0 & 0 & 0 \\ 1 - a & 1 - \overline{a} & a & 0 \\ 0 & \overline{a} & 1 - a & 1 - \overline{a} \\ 0 & 0 & 0 & \overline{a} \end{bmatrix} \begin{bmatrix} \phi(0) \\ \phi(1) \\ \phi(2) \\ \phi(3) \end{bmatrix}. \tag{7.5.37}$$

This system (7.5.37) has exactly one solution,

$$\phi(0)=0,\quad \phi(1)=2a,\quad \phi(2)=2\bar{a},\quad \phi(3)=0. \tag{7.5.38}$$

We set $\phi(k)=0$ for all remaining values of $k\in\mathbb{Z}$. Then, ϕ can recursively be calculated for all of $\mathbb{D}$ by (7.5.35b).

Finally, we conclude this section by including the Daubechies scaling function ${}_3\phi(x)$ and the Daubechies wavelet ${}_3\psi(x)$ for $N=3$. In this case, (7.5.10) gives

$$P(x)=P_3(x)=1+3x+6x^2, \tag{7.5.39}$$

where

$$x=\sin^2\frac{\omega}{2}=\frac{1}{4}\left(-e^{-i\omega}+2-e^{i\omega}\right) \text{ and}$$

$$x^2=\frac{1}{16}\left(e^{-2i\omega}+4+e^{2i\omega}-4e^{-i\omega}-4e^{i\omega}+2\right).$$

Consequently, (7.5.12) gives the result

$$\left|\hat{L}(\omega)\right|^2=\frac{3}{8}\,e^{-2i\omega}-\frac{9}{4}\,e^{-i\omega}+\frac{19}{4}-\frac{9}{4}\,e^{i\omega}+\frac{3}{8}\,e^{2i\omega}. \tag{7.5.40}$$

In this case,

$$A(\omega)=b_0+b_1\,e^{-i\omega}+b_2\,e^{-2i\omega}, \tag{7.5.41}$$

so that

$$\begin{aligned}\left|\hat{L}(\omega)\right|^2&=A(\omega)\,A(-\omega)=\left(b_0+b_1\,e^{-i\omega}+b_2\,e^{-2i\omega}\right)\left(b_0+b_1\,e^{i\omega}+b_2\,e^{2i\omega}\right)\\&=\left(b_0^2+b_1^2+b_2^2\right)+e^{-i\omega}\left(b_0\,b_1+b_2\,b_1\right)\\&\quad+e^{i\omega}\left(b_0b_1+b_1b_2\right)+b_0\,b_2\,e^{2i\omega}+b_0\,b_2\,e^{-2i\omega}.\end{aligned} \tag{7.5.42}$$

Equating the coefficients in (7.5.40) and (7.5.42) gives

$$b_0^2+b_1^2+b_2^2=\frac{19}{4},\quad b_1\,b_0+b_2\,b_1=-\frac{9}{4},\quad b_2\,b_0=\frac{3}{8}. \tag{7.5.43}$$

In view of the fact that $\left|\hat{L}(0)\right|^2=1$ and $P(0)=1$, the Riesz lemma 7.5.1 ensures that there are real solutions (b_0,b_1,b_2) that satisfy the additional requirement $b_0+b_1+b_2=1$. Eliminating b_1 from this equation and the second equation in (7.5.43) gives

$$b_1^2-b_1-\frac{9}{4}=0$$

so that

$$b_1 = \frac{1}{2}\left(1 \pm \sqrt{10}\right). \tag{7.5.44}$$

Consequently,

$$b_0 + b_2 = \frac{1}{2}\left(1 \mp \sqrt{10}\right). \tag{7.5.45}$$

The plus and the minus signs in these equations result in complex roots for b_0 and b_2. This means that the real root for b_1 corresponds to the minus sign in (7.5.44) so that

$$b_1 = \frac{1}{2}\left(1 - \sqrt{10}\right). \tag{7.5.46}$$

Obviously,

$$b_0 + b_2 = \frac{1}{2}\left(1 + \sqrt{10}\right) \quad \text{and} \quad b_0\, b_2 = \frac{3}{8}$$

lead to the fact that b_0 and b_2 satisfy

$$x^2 - \frac{1}{2}\left(1 + \sqrt{10}\right) x + \frac{3}{8} = 0. \tag{7.5.47}$$

Thus,

$$\left(b_o,\, b_2\right) = \frac{1}{4}\left[\left(1 + \sqrt{10}\right) \pm \sqrt{5 + 2\sqrt{10}}\,\right]. \tag{7.5.48}$$

Consequently, $A(\omega)$ is explicitly known and hence, $\hat{m}(\omega)$ becomes

$$\hat{m}(\omega) = \frac{1}{8}\left[b_0 + \left(3b_0 + b_1\right) e^{-i\omega} + \left(3b_0 + 3b_1 + b_2\right) e^{-2i\omega} \right.$$
$$\left. + \left(b_0 + 3b_1 + 3b_2\right) e^{-3i\omega} + \left(b_1 + 3b_2\right) e^{-4i\omega} + b_2\, e^{-5i\omega}\right], \tag{7.5.49}$$

which is equal to (7.3.3). Equating the coefficients of (7.3.3) and (7.5.49) gives all six c_k's as

$$c_0 = \frac{\sqrt{2}}{8}\, b_0 = \frac{\sqrt{2}}{32}\left[\left(1+\sqrt{10}\right)+\sqrt{5+2\sqrt{10}}\right],$$

$$c_1 = \frac{\sqrt{2}}{8}\left(3b_0 + b_1\right) = \frac{\sqrt{2}}{32}\left[\left(5+\sqrt{10}\right)+3\sqrt{5+2\sqrt{10}}\right],$$

$$c_2 = \frac{\sqrt{2}}{8}\left(3b_0 + 3b_1 + b_2\right) = \frac{\sqrt{2}}{32}\left[\left(5-\sqrt{10}\right)+\sqrt{5+2\sqrt{10}}\right],$$

$$c_3 = \frac{\sqrt{2}}{8}\left(b_0 + 3b_1 + 3b_2\right) = \frac{\sqrt{2}}{32}\left[\left(5-\sqrt{10}\right)-\sqrt{5+2\sqrt{10}}\right],$$

$$c_4 = \frac{\sqrt{2}}{8}\left(b_1 + 3b_2\right) = \frac{\sqrt{2}}{32}\left[\left(5+\sqrt{10}\right)-3\sqrt{5+2\sqrt{10}}\right],$$

$$c_5 = \frac{\sqrt{2}}{8}\, b_2 = \frac{\sqrt{2}}{32}\left[\left(1+\sqrt{10}\right)-\sqrt{5+2\sqrt{10}}\right].$$

Evidently, the Daubechies scaling function ${}_3\phi(x)$ and the Daubechies wavelet ${}_3\psi(x)$ (or simply *D6 wavelet*) can be rewritten as

$$ {}_3\phi(x) = \sqrt{2}\sum_{k=0}^{5} c_k\, \phi(2x-k), \tag{7.5.50}$$

$$ {}_3\psi(x) = \sqrt{2}\sum_{k=0}^{5} d_k\, \phi(2x-k), \tag{7.5.51}$$

where c_k and d_k are explicitly known. Figures 7.13(a) and 7.13(b) exhibit the scaling function ${}_3\phi(x)$ and the wavelet ${}_3\psi(x)$.

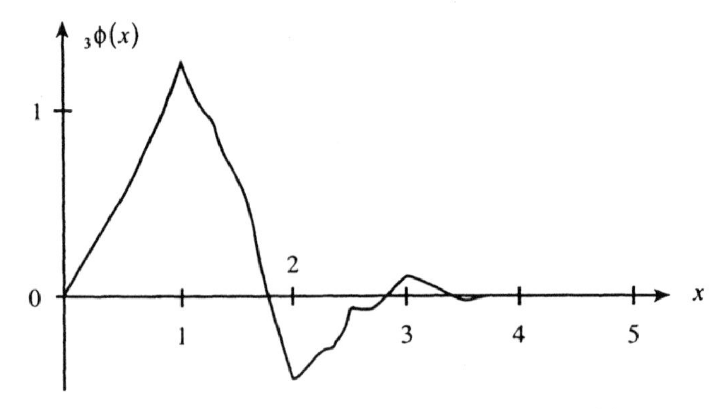

Figure 7.13. (a). The Daubechies scaling function ${}_3\phi(x)$ for $N = 3$.

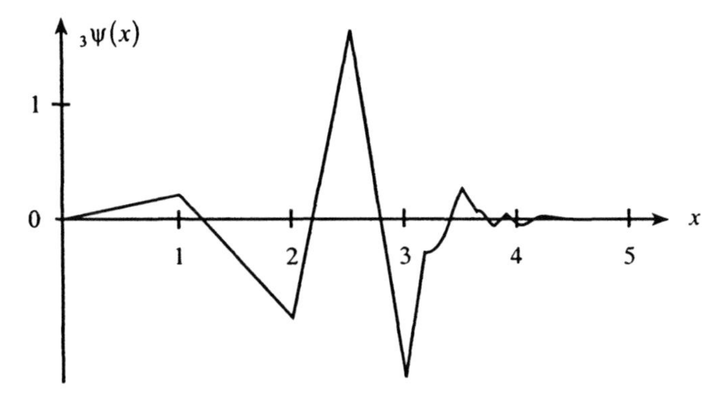

Figure 7.13. (b). The Daubechies wavelet $_3\psi(x)$ for $N=3$.

With a given even number of wavelet coefficients c_k, $k=0,1,\ldots,\ 2N-1$, we can define the scaling function ϕ by

$$\phi(x)=\sqrt{2}\sum_{k=0}^{2N-1} c_k\ \phi(2x-k) \tag{7.5.52}$$

and the corresponding wavelet by

$$\psi(x)=\sqrt{2}\sum_{k=0}^{2N-1}(-1)^k\ c_k\ \phi(2x+k-2N+1), \tag{7.5.53}$$

where the coefficients c_k satisfy the following conditions

$$\sum_{k=0}^{2N-1} c_k=\sqrt{2},\quad \sum_{k=0}^{2N-1}(-1)^k\ k^m\ c_k=0, \tag{7.5.54a,b}$$

where $m=0,1,2,\ldots,N-1$ and

$$\sum_{k=0}^{2N-1} c_k\ c_{k+2m}=0,\quad m\neq 0, \tag{7.5.55}$$

where $m=1,2,3,\ldots,N-1$ and

$$\sum_{k=0}^{2N-1} c_k^2=1. \tag{7.5.56}$$

When $N=1$, two coefficients, c_0 and c_1, satisfy the following equations:

$$c_0+c_1=\sqrt{2},\quad c_0-c_1=0,\quad c_0^2+c_1^2=1$$

which admit solutions

$$c_0 = c_1 = \frac{1}{\sqrt{2}}.$$

They give the classic Haar scaling function and the Haar wavelet.

When $N = 2$, four coefficients c_0, c_1, c_2, c_3 satisfy the following equations:

$$c_0 + c_1 + c_2 + c_3 = \sqrt{2}, \quad c_0 - c_1 + c_2 - c_3 = 0,$$
$$c_0 c_2 + c_1 c_3 = 0, \qquad c_0^2 + c_1^2 + c_2^2 + c_3^2 = 1.$$

These give solutions

$$c_0 = \frac{1}{4\sqrt{2}}\left(1+\sqrt{3}\right), \quad c_1 = \frac{1}{4\sqrt{2}}\left(3+\sqrt{3}\right),$$
$$c_2 = \frac{1}{4\sqrt{2}}\left(3-\sqrt{3}\right), \quad c_3 = \frac{1}{4\sqrt{2}}\left(1-\sqrt{3}\right).$$

These coefficients constitute the Daubechies scaling function (7.5.22) and the Daubechies D4 wavelet (7.5.23) or (7.5.24).

7.6 Discrete Wavelet Transforms and Mallat's Pyramid Algorithm

In harmonic analysis, a signal is decomposed into harmonic functions of different frequencies, whereas in wavelet analysis a signal is decomposed into wavelets of different scales (or levels) along the x-axis. Any arbitrary signal $f(x)$ can be decomposed into wavelet components at different scales as

$$f(x) = \sum_{m=-\infty}^{\infty} \sum_{k=-\infty}^{\infty} c_{m,k}\, \psi\left(2^m x - k\right), \tag{7.6.1}$$

where $c_{m,k}$ are wavelet coefficients.

It is well-known that the Haar wavelet is the simplest orthonormal wavelet defined in Example 6.2.1. This wavelet is a member of a family of similar-shaped wavelets of different horizontal scales, each located at a different position of the x-axis. Obviously, there are two half-length wavelets represented by $\psi(2x)$ and $\psi(2x-1)$ and four quarter-length wavelets represented by $\psi(4x), \psi(4x-1), \psi(4x-2)$, and $\psi(4x-3)$, as shown in Figures 7.14 (a)-(c).

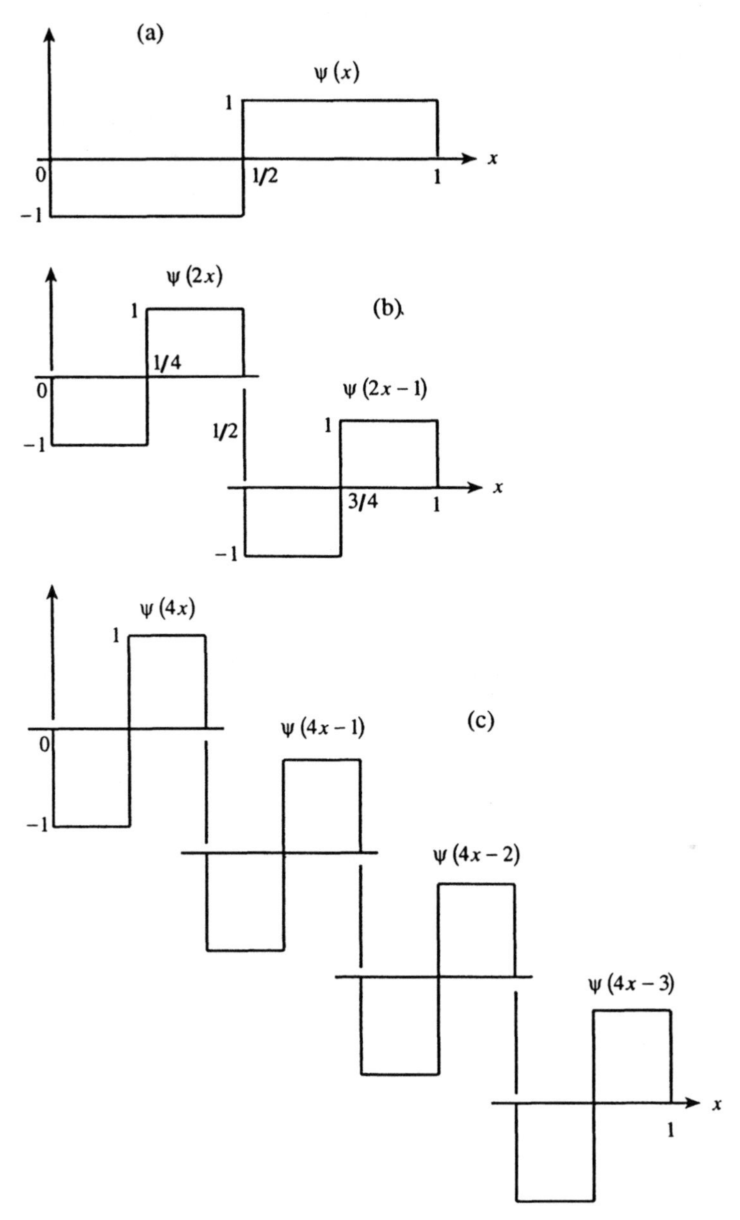

Figure 7.14. (a), (b), and (c). The Haar wavelet at levels $m = 0,1,2$.

The position and scale of each wavelet can be obtained from its argument. For instance, $\psi(2x-1)$ is the same as $\psi(2x)$ except that it is compressed into half the horizontal length and starts at $x = 2^{-1}$ instead of at $x = 0$. The level of the wavelet is determined by how many wavelets fit into the unit interval $0 \le x < 1$. At level 0, there is $2^0 = 1$ wavelet (the Haar wavelet) in each unit interval, as shown in Figure 7.14(a). At level 1, there are $2^1 = 2$ wavelets in the unit interval (see Figure 7.14(b)). At level 2, there are $2^2 = 4$ wavelets in the unit interval, as shown in Figure 7.14(c), and so on. On the other hand, at level -1, there is $2^{-1} = \frac{1}{2}$ a wavelet in the unit interval and at level -2 there is $2^{-2} = \frac{1}{4}$ of a wavelet in the unit interval, and so on.

It is shown in Figure 7.15 that, for all levels less than zero $(m < 0 \text{ or } m \le -1)$, the contribution is constant over each unit interval. Evidently, the sum of the contributions from all of these levels is also constant. It is known that the scaling function $\phi(x)$ for the Haar wavelet is also constant so that $\phi(x) = 1$ for $0 \le x < 1$. Consequently, the representation (7.6.1) becomes

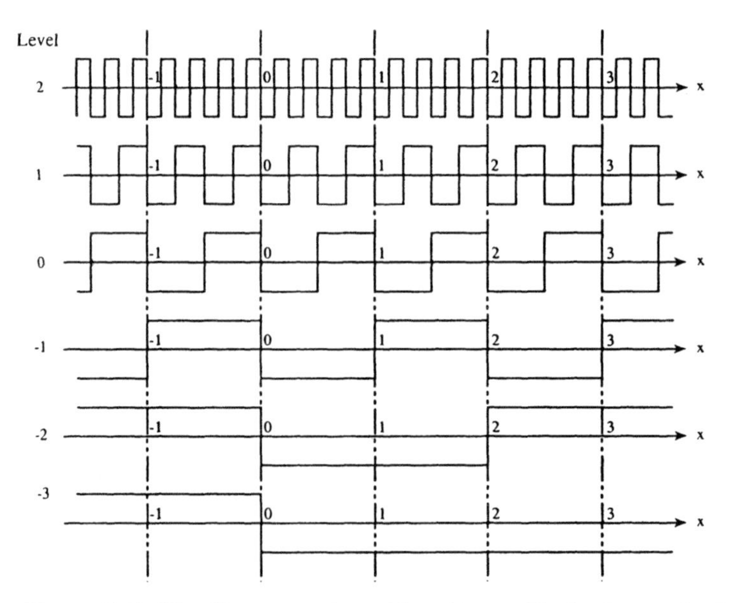

Figure 7.15 The Haar wavelet with unit amplitude drawn for levels $m = -3, -2, -1, 0, 1, 2$. For these levels, there are $2^{-3}, 2^{-2}, 2^{-1}, 2^0, 2^1, 2^2$ wavelets per unit interval.

$$f(x)=\sum_{m=-\infty}^{-1}\sum_{k=-\infty}^{\infty}c_{m,k}\,\psi\left(2^m x-k\right)+\sum_{m=0}^{\infty}\sum_{k=-\infty}^{\infty}c_{m,k}\,\psi\left(2^m x-k\right)$$

$$=\sum_{k=-\infty}^{\infty}c_{\phi,k}\,\phi(x-k)+\sum_{m=0}^{\infty}\sum_{k=-\infty}^{\infty}c_{m,k}\,\psi\left(2^m x-k\right). \tag{7.6.2}$$

Under very general conditions on f and ψ, the wavelet series (7.6.1) and (7.6.2) converge so that they represent a practical basis for signal analysis.

In order to develop a discrete wavelet transform analysis, it is convenient to define $f(x)$ in the unit interval $0\le x<1$. If time t is an independent variable for a signal over duration T, then $x=(t/T)$ and $0\le x<1$ where x is a dimensionless variable. We assume that $f(x), 0\le x<1$, is one period of a periodic signal so that the signal is exactly repeated in adjacent unit intervals to yield

$$F(x)=\sum_{k}f(x-k), \tag{7.6.3}$$

where $f(x)$ is zero outside the interval $0\le x<1$.

We consider the Daubechies wavelet D4, $\psi(x)$, which occupies three unit intervals $0\le x<3$. In the unit interval $0\le x<1$, $f(x)$ will have contributions from the first third of $\psi(x)$, the middle third of $\psi(x+1)$, and the last third of $\psi(x+2)$. When any wavelet that begins in the interval $0\le x<1$ runs off the end $x=1$, it may be assumed to be wrapped around the interval several times if there are many coefficients, so that the wavelet extends over many intervals. With this assumption, the wavelet representation (7.6.2) of $f(x)$ in $0\le x<1$ can be expressed as

$$f(x)=a_0\,\phi(x)+a_1\,\psi(x)+\left[a_2\;a_3\right]\begin{bmatrix}\psi(2x)\\ \psi(2x-1)\end{bmatrix}$$

$$+\left[a_4\;a_5\;a_6\;a_7\right]\begin{bmatrix}\psi(4x)\\ \psi(4x-1)\\ \psi(4x-2)\\ \psi(4x-3)\end{bmatrix}+\cdots$$

$$+\cdots+a_{2^m+k}\psi\left(2^m x-k\right)+\cdots, \tag{7.6.4}$$

where the coefficients $a_1, a_2, a_3\cdots$ represent the amplitudes of each of the wavelets after wrapping to one cycle of the periodic function (7.6.3) in

$0 \le x < 1$. Due to the wrapping process, the scaling function $\phi(x)$ always becomes a constant. The second term $a_1\psi(x)$ is a wavelet of scale zero, the third and fourth terms $a_2\psi(2x)$ and $a_3\psi(2x-1)$ are wavelets of scale one, and the second is translated $\Delta x = 2^{-1}$ with respect to the first. The next four terms represent wavelets of scale two and so on for wavelets of increasingly higher scale. The higher the scale, the finer the detail; so there are more coefficients involved. At scale m, there are 2^m wavelets, each spaced $\Delta x = 2^{-m}$ apart along the x-axis.

In view of orthonormal properties, the coefficients can be obtained from

$$\int \psi(2^m x - k)\, f(x)\, dx = a_{2^m+k} \int \psi^2(2^m x - k)\, dx$$
$$= \frac{1}{2^k}\, a_{2^m+k} \int \psi^2(x)\, dx$$

so that

$$a_{2^m+k} = 2^k \int f(x)\, \psi(2^m x - k)\, dx \tag{7.6.5}$$

because $\int \psi^2(x)\, dx = 1$.

In view of the fact that

$$\int_{-\infty}^{\infty} \phi^2(x)\, dx = 1,$$

it follows that the coefficient a_0 is given by

$$a_0 = \int f(x)\, \phi(x)\, dx. \tag{7.6.6}$$

Usually, the limits of integration in the orthogonality conditions are from $-\infty$ to $+\infty$, but the integrand in each case is only nonzero for the finite length of the shortest wavelet or scaling function involved. The limits of integration on (7.6.5) and (7.6.6) may extend over several intervals, provided the wavelets and scaling functions are not wrapped. Since $f(x)$ is one cycle of a periodic function, which repeats itself in adjacent intervals, all contributions to the integrals from outside the unit interval $(0 \le x < 1)$ are included by integrating from $x = 0$ to $x = 1$ for the wrapped functions. Consequently, results (7.6.5) and (7.6.6) can be expressed as

$$a_{2^m+k} = 2^m \int_0^1 f(x)\, \psi(2^m x - k)\, dx \tag{7.6.7}$$

and

$$a_0 = \int_0^1 f(x)\,\phi(x)\,dx, \tag{7.6.8}$$

where $\phi(x)$ and $\psi(2^m x - k)$ involved in (7.6.7) and (7.6.8) are wrapped around the unit interval $(0 \le x < 1)$ as many times as needed to ensure that their whole length is included in $(0 \le x < 1)$.

The discrete wavelet transform (DWT) is an algorithm for computing (7.6.7) and (7.6.8) when a signal $f(x)$ is sampled at equally spaced intervals over $0 \le x < 1$. We assume that $f(x)$ is a periodic signal with period one and that the scaling and wavelet functions wrap around the interval $0 \le x < 1$. The integrals (7.6.7) and (7.6.8) can be computed to the desired accuracy by using $\phi(x)$ and $\psi(2^m x - k)$. However, a special feature of the DWT algorithm is that (7.6.7) and (7.6.8) can be computed without generating $\phi(x)$ and $\psi(2^m x - k)$ explicitly. The DWT algorithm was first introduced by Mallat (1989b) and hence is known as *Mallat's pyramid algorithm* (or *Mallat's tree algorithm*). For a detailed information on this algorithm, the reader is also referred to Newland (1993a,b).

7.7 Exercises

1. Show that the two-scale equation associated with the linear spline function

$$B_1(t) = \begin{cases} 1 - |t|, & 0 < |t| < 1 \\ 0, & \text{otherwise} \end{cases}$$

is

$$B_1(t) = \frac{1}{2}\, B_1(2t+1) + B_1(2t) + \frac{1}{2}\, B_1(2t-1).$$

Hence, show that

$$\sum_{k=-\infty}^{\infty} \left|\hat{\phi}(\omega + 2\pi k)\right|^2 = 1 - \frac{2}{3}\sin^2\frac{\omega}{2}.$$

2. Use the Fourier transform formula (7.4.43) for $\hat{\psi}(\omega)$ of the Franklin wavelet ψ to show that ψ satisfies the following properties:

(a) $\hat{\psi}(0) = \int_{-\infty}^{\infty} \psi(t)\,dt = 0,$

(b) $\int_{-\infty}^{\infty} t\,\psi(t)\,dt = 0,$

(c) ψ is symmetric with respect to $t = -\frac{1}{2}$.

3. From an expression (7.4.41) for the filter, show that

$$\hat{m}(\omega) = \frac{\left(2 + 3\cos\omega + \cos^2\omega\right)}{\left(1 + 2\cos^2\omega\right)}$$

and hence, deduce

$$\hat{\psi}(2\omega) = \exp(-i\omega)\left[\frac{2 - \cos\omega + \cos^2\omega}{1 + 2\cos^2\omega}\right]\hat{\phi}(\omega).$$

4. Using result (7.4.20), prove that

$$\frac{\hat{B}_n(\omega)}{\hat{B}_n\left(\frac{\omega}{2}\right)} = \left(\frac{1 + e^{-\frac{i\omega}{2}}}{2}\right)^n.$$

Hence, derive the following:

(a) $\hat{B}_n(\omega) = \frac{1}{2^n}\sum_{k=0}^{n}\binom{n}{k}\exp\left\{-\frac{ik\omega}{2}\right\}\hat{B}_n\left(\frac{\omega}{2}\right),$

(b) $B_n(t) = \frac{1}{2^{n-1}}\sum_{k=0}^{n}\binom{n}{k}B_n(2t - k).$

5. Obtain a solution of (7.5.22) for the following cases:

(a) $c_0 = c_1 = \frac{1}{\sqrt{2}}, \quad c_2 = c_3 = 0,$

(b) $c_0 = c_2 = \frac{1}{2\sqrt{2}}, \quad c_1 = \frac{1}{\sqrt{2}}, \quad c_3 = 0,$

(c) $c_0 = \sqrt{2}, \quad c_1 = c_2 = c_3 = 0.$

6. If the generating function is defined by (7.3.3), then show that

(a) $\sum_{n=-\infty}^{\infty} c_n = \sqrt{2},$

(b) $\sum_{n=-\infty}^{\infty} c_{2n} = \sum_{n=-\infty}^{\infty} c_{2n+1} = \frac{1}{\sqrt{2}}.$

7. Using the Strang (1989) accuracy condition that $\hat{\phi}(\omega)$ must have zeros of n when $\omega = 2\pi,\ 4\pi,\ 6\pi,\ldots$, show that

$$\sum_{k=-\infty}^{\infty} (-1)^k\, k^m\, c_k = 0, \quad m = 0,1,2,\ldots,(n-1).$$

8. Show that

(a) $\displaystyle\sum_{k=-\infty}^{\infty} c_k^2 = 1,$

(b) $\displaystyle\sum c_k\, c_{k+2m} = 0,\ \ m \neq 0,$

where c_k are coefficients of the scaling function defined by (7.3.5).

(c) Derive the result in (b) from the result in Exercise 5.

9. Given six wavelet coefficients $c_k\ (N = 6)$, write down six equations from (7.5.50a,b)–(7.5.52). Show that these six equations generate the Daubechies scaling function (7.5.50) and the Daubechies D6 wavelet (7.5.51).

10. Using the properties of $\hat{m}$ and $\hat{m}_1$, prove that

(a) $\hat{\phi}\left(\frac{\omega}{2}\right) = \left[\bar{\hat{m}}\left(\frac{\omega}{2}\right) + \bar{\hat{m}}\left(\frac{\omega}{2}+\pi\right)\right]\hat{\phi}(\omega) + \left[\bar{\hat{m}}_1\left(\frac{\omega}{2}\right) + \bar{\hat{m}}_1\left(\frac{\omega}{2}+\pi\right)\right]\hat{\psi}(\omega),$

(b) $\exp\left(-\frac{i\omega}{2}\right)\hat{\phi}\left(\frac{\omega}{2}\right)$

$$= \left[\exp\left(-\frac{i\omega}{2}\right)\bar{\hat{m}}\left(\frac{\omega}{2}\right) - \exp\left(-\frac{i\omega}{2}\right)\bar{\hat{m}}\left(\frac{\omega}{2}+\pi\right)\right]\hat{\phi}(\omega)$$

$$+\left[\exp\left(-\frac{i\omega}{2}\right)\bar{\hat{m}}_1\left(\frac{\omega}{2}\right) - \exp\left(-\frac{i\omega}{2}\right)\bar{\hat{m}}\left(\frac{\omega}{2}+\pi\right)\right]\hat{\psi}(\omega).$$

11. If $\hat{m}(\omega) = \frac{1}{2}\left(1+e^{-i\omega}\right)\left(1-e^{-i\omega}+e^{-2i\omega}\right) = e^{-\frac{3i\omega}{2}}\cos\left(\frac{3\omega}{2}\right)$, show that it satisfies the condition (7.3.4) and $\hat{m}(0) = 1$. Hence, derive the following results

(a) $\hat{\phi}(\omega) = \exp\left(-\frac{3i\omega}{2}\right)\dfrac{\sin\left(\frac{3\omega}{2}\right)}{\left(\frac{3\omega}{2}\right)},$

(b) $\sum_{k=-\infty}^{\infty} \left|\hat{\phi}(\omega + 2\pi k)\right|^2 = \frac{1}{9}(3 + 4\cos\omega + 2\cos 2\omega)$,

(c) $\phi(x) = \begin{cases} \frac{1}{3}, & 0 \le x \le 3 \\ 0, & \text{otherwise} \end{cases}$.

(d) $c_n = \int_{-\infty}^{\infty} \phi(x)\, \overline{\phi(x-n)}\, dx = \frac{1}{3}\int_0^3 \phi(x-n)\, dx = \frac{1}{3}\int_n^{n+3} \phi(x)\, dx$.

12. Show that, for any $x \in [0,1]$,

(a) $\sum_{k=-\infty}^{\infty} \phi(x-k) = 1$ and (b) $\sum_{k=-\infty}^{\infty} (c+k)\, \phi(x-k) = x$,

where $c = \frac{1}{2}\left(3 - \sqrt{3}\right)$.

Hence, using (a) and (b), show that

(c) $2\phi(x) + \phi(x+1) = x + 2 - c$,

(d) $\phi(x+1) + 2\phi(x+2) = c - x$,

(e) $\phi(x) - \phi(x+2) = x + c + \left(\sqrt{3} - 2\right)$.

13. Use (7.3.31) and (7.4.64) to show that

$$\psi(x) = -c_0\, \phi(2x) + (1 - c_0)\, \phi(2x-1) - (1 - c_3)\, \phi(2x-2) + c_0\, \phi(2x-3).$$

14. Using (7.4.64), prove that $\psi(x)$ defined in Exercise 13 satisfies the following properties:

(a) supp $\psi(x) \subset [0,3]$,

(b) $\int_{-\infty}^{\infty} \psi(x)\, \psi(x-k)dx = \begin{Bmatrix} 0, & k \ne 0 \\ 1, & k \ne 0 \end{Bmatrix}$,

(c) $\int_{-\infty}^{\infty} \psi(x-k)\, \phi(x)dx = 0$ for all $k \in \mathbb{Z}$.

Chapter 8

Newland's Harmonic Wavelets

"Wavelets are without doubt an exciting and intuitive concept. The concept brings with it a new way of thinking, which is absolutely essential and was entirely missing in previously existing algorithms."

Yves Meyer

8.1 Introduction

So far, all wavelets have been constructed from dilation equations with real coefficients. However, many wavelets cannot always be expressed in functional form. As the number of coefficients in the dilation equation increases, wavelets get increasingly longer and the Fourier transforms of wavelets become more tightly confined to an octave band of frequencies. It turns out that the spectrum of a wavelet with n coefficients becomes more boxlike as n increases. This fact led Newland (1993a,b) to introduce a new harmonic wavelet $\psi(x)$ whose spectrum is exactly like a box, so that the magnitude of its Fourier transform $\hat{\psi}(\omega)$ is zero except for an octave band of frequencies. Furthermore, he generalized the concept of the harmonic wavelet to describe a family of mixed wavelets with the simple mathematical structure. It is also shown that this family provides a complete set of orthonormal basis functions for signal analysis.

This chapter is devoted to Newland's harmonic wavelets and their basic properties.

8.2 Harmonic Wavelets

Newland (1993a) introduced a real even function $\psi_e(t)$ whose Fourier transform is defined by

$$\hat{\psi}_e(\omega) = \begin{Bmatrix} \dfrac{1}{4\pi} & \text{for} \quad -4\pi \le \omega < -2\pi, \text{ and } 2\pi \le \omega < 4\pi \\ 0, & \text{otherwise} \end{Bmatrix}, \tag{8.2.1}$$

where the Fourier transform is defined by

$$\hat{f}(\omega) = \frac{1}{2\pi} \int_{-\infty}^{\infty} f(t)\, e^{-i\omega t} dt. \tag{8.2.2}$$

The inverse Fourier transform of $\hat{\psi}_e(\omega)$ gives

$$\psi_e(t) = \int_{-\infty}^{\infty} \hat{\psi}_e(\omega) \exp(i\omega t)\, d\omega = \frac{1}{2\pi t}(\sin 4\pi t - \sin 2\pi t). \tag{8.2.3}$$

On the other hand, the Fourier transform $\hat{\psi}_0(\omega)$ of a real odd function $\psi_0(t)$ is defined by

$$\hat{\psi}_0(\omega) = \begin{Bmatrix} \dfrac{i}{4\pi} & \text{for} \quad -4\pi \le \omega < -2\pi \\ -\dfrac{i}{4\pi} & \text{for} \quad 2\pi \le \omega < 4\pi \\ 0, & \text{otherwise} \end{Bmatrix}. \tag{8.2.4}$$

Then, the inverse Fourier transform gives

$$\psi_0(t) = \int_{-\infty}^{\infty} \hat{\psi}_0(\omega) \exp(i\omega t)\, d\omega = \frac{1}{2\pi t}(\cos 2\pi t - \cos 4\pi t). \tag{8.2.5}$$

The harmonic wavelet $\psi(t)$ is then defined by combining (8.2.3) and (8.2.5) in the form

$$\begin{aligned} \psi(t) &= \psi_e(t) + i\psi_0(t) \\ &= \frac{1}{2\pi i t}\left[\exp(4\pi i t) - \exp(2\pi i t)\right]. \end{aligned} \tag{8.2.6}$$

The real and imaginary parts of $\psi(t)$ are shown in Figures 8.1(a) and 8.1(b).

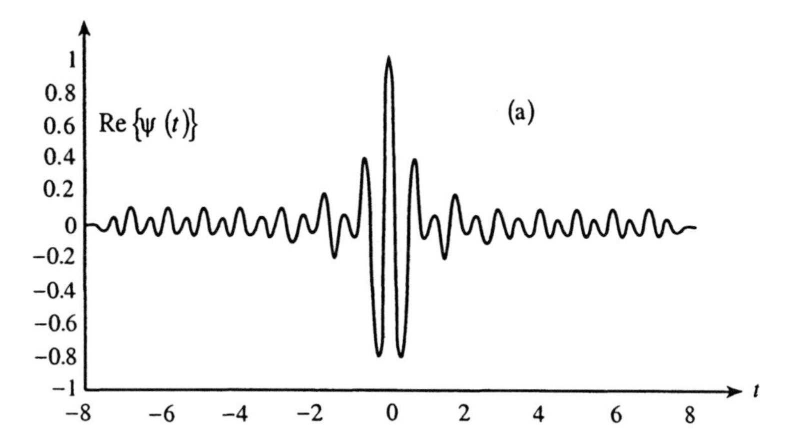

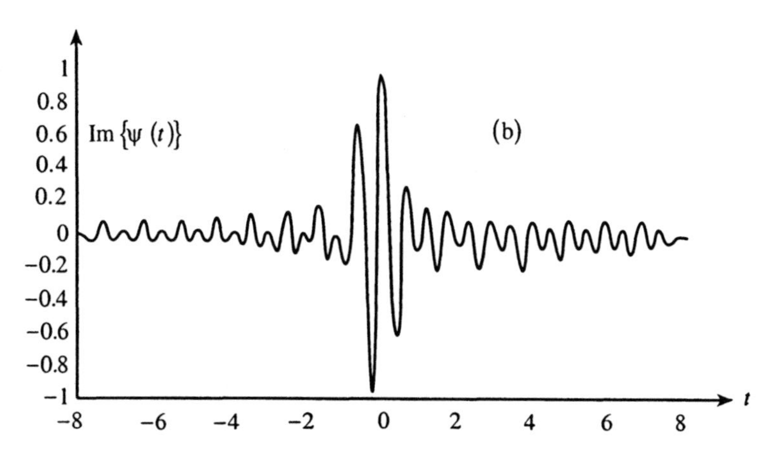

Figure 8.1. (a) Real part of $\psi(t)$; and (b) imaginary part of $\psi(t)$.

Clearly, the Fourier transform of $\psi(t)$ is given by

$$\hat{\psi}(\omega) = \hat{\psi}_e(\omega) + i\,\hat{\psi}_0(\omega) \tag{8.2.7}$$

so that, from (8.2.1) and (8.2.4), we obtain the Fourier transform of the harmonic wavelet $\psi(t)$

$$\hat{\psi}(\omega) = \begin{Bmatrix} \dfrac{1}{2\pi}, & 2\pi \le \omega < 4\pi \\ 0, & \text{otherwise} \end{Bmatrix}. \tag{8.2.8}$$

For the general harmonic wavelet $\psi(t)$ at level m and translated in k steps of size 2^{-m}, we define

$$\hat{\psi}(\omega) = \begin{Bmatrix} \frac{1}{2\pi} 2^{-m} \exp\left(-\frac{i\omega k}{2^m}\right), & 2\pi 2^m \le \omega < 4\pi 2^m \\ 0, & \text{otherwise} \end{Bmatrix}, \tag{8.2.9}$$

where m and k are integers.

The inverse Fourier transform of (8.2.9) gives

$$\psi\left(2^m t - k\right) = \frac{\exp\left\{4\pi i\left(2^m t - k\right)\right\} - \exp\left\{2\pi i\left(2^m t - k\right)\right\}}{2\pi i\left(2^m t - k\right)}, \tag{8.2.10}$$

where m is a nonnegative integer and k is an integer.

The level of the wavelet is determined by the value of m so that, at the level $(m = 0)$, the Fourier transform (8.2.9) of the wavelet occupies bandwidth 2π to 4π, as shown in (8.2.8). At level $m = -1$ with bandwidth 0 to 2π, we define

$$\hat{\psi}(\omega) = \begin{Bmatrix} \frac{1}{2\pi} e^{-i\omega k}, & 0 \le \omega < 2\pi \\ 0, & \text{otherwise} \end{Bmatrix} \tag{8.2.11}$$

so that the inverse Fourier transform gives the so-called harmonic scaling function

$$\phi(t - k) = \frac{\exp\{2\pi i(t - k)\} - 1}{2\pi i(t - k)}. \tag{8.2.12}$$

Evidently, the choice of the harmonic wavelet and the scaling function seem to be appropriate in the sense that they form an orthogonal set. If $\hat{\psi}(\omega)$ is the Fourier transform of $\psi(t)$, then the Fourier transform of $g(t) = \psi\left(2^m t - k\right)$ is

$$\hat{g}(\omega) = 2^{-m} \exp\left(-\frac{i\omega k}{2^m}\right) \psi\left(\omega 2^{-m}\right). \tag{8.2.13}$$

Clearly, the Fourier transforms of successive levels of harmonic wavelets decrease in proportion to their increasing bandwidth, as shown in Figure 8.2. For $\omega < 0$, they are always zero.

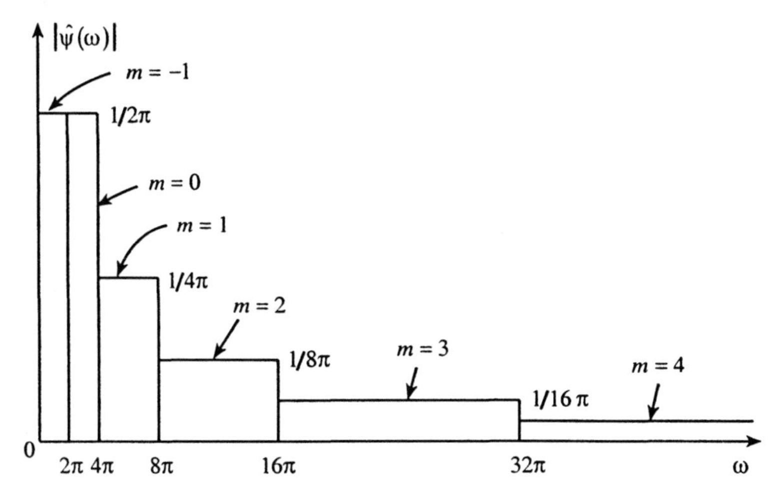

Figure 8.2. Fourier transforms of harmonic wavelets at levels $m = 0, 1, 2, 3, 4$.

In order to prove orthogonality of wavelets and scaling functions, we need the general Parseval relation (3.4.37) in the form

$$\int_{-\infty}^{\infty} f(t)\,\overline{g(t)}\,dt = 2\pi \int_{-\infty}^{\infty} \hat{f}(\omega)\,\overline{\hat{g}}(\omega)\,d\omega, \tag{8.2.14}$$

where $f, g \in L^2(\mathbb{R})$ and the factor 2π is present due to definition (8.2.2).

We also need another similar result of the form

$$\int_{-\infty}^{\infty} f(t)\,g(t)\,dt = 2\pi \int_{-\infty}^{\infty} \hat{f}(\omega)\,\hat{g}(-\omega)\,d\omega, \tag{8.2.15}$$

where $f, g \in L^2(\mathbb{R})$.

This result follows from the following formal calculation:

$$\begin{aligned}\int_{-\infty}^{\infty} f(t)\,g(t)\,dt &= \int_{-\infty}^{\infty} dt \int_{-\infty}^{\infty} \hat{f}(\omega)\,d\omega_1 \int_{-\infty}^{\infty} \hat{g}(\omega_2)\,d\omega_2 \\ &\qquad\qquad \times \exp\{i(\omega_1+\omega_2)t\} \\ &= 2\pi \int_{-\infty}^{\infty} \hat{f}(\omega_1)\,d\omega_1 \int_{-\infty}^{\infty} \hat{g}(\omega_2)\,\delta(\omega_1+\omega_2)\,d\omega_2\end{aligned}$$

$$= 2\pi \int_{-\infty}^{\infty} \hat{f}(\omega_1)\, \hat{g}(-\omega_1)\, d\omega_1$$

$$= 2\pi \int_{-\infty}^{\infty} \hat{f}(\omega)\, \hat{g}(-\omega)\, d\omega, \quad (\omega_1 = \omega).$$

Theorem 8.2.1 (Orthogonality of Harmonic Wavelets). The family of harmonic wavelets $\psi(2^m t - k)$ forms an orthogonal set.

Proof. To prove this theorem, it suffices to show orthogonality conditions:

$$\int_{-\infty}^{\infty} \psi(t)\, \psi(2^m t - k)\, dt = 0 \quad \text{for all } m, k, \tag{8.2.16}$$

$$\int_{-\infty}^{\infty} \psi(t)\, \overline{\psi}(2^m t - k)\, dt = 0 \quad \text{for } m \neq 0. \tag{8.2.17}$$

We put $g(t) = \psi(2^m t - k)$ so that its Fourier transform is

$$\hat{g}(\omega) = 2^{-m} \exp\left(-\frac{i\omega k}{2^m}\right) \hat{\psi}(2^{-m}\omega) \tag{8.2.18}$$

and then apply (8.2.15) to obtain

$$\int_{-\infty}^{\infty} \psi(t)\, g(t)\, dt = 2\pi \int_{-\infty}^{\infty} \hat{\psi}(\omega)\, \hat{g}(-\omega)\, d\omega. \tag{8.2.19}$$

If $\psi(t)$ and $g(t)$ are two harmonic wavelets, they have the one-sided Fourier transforms as shown in Figure 8.2, so that the product $\hat{\psi}(\omega)\, \hat{g}(-\omega)$ must always vanish. Thus, the right-hand side of (8.2.19) is always zero for all k and m, that is,

$$\int_{-\infty}^{\infty} \psi(t)\, \psi(2^m t - k)\, dt = 0 \quad \text{for all } m, k. \tag{8.2.20}$$

To prove (8.2.17), we apply (8.2.14) so that

$$\int_{-\infty}^{\infty} \psi(t)\, \overline{g(t)}\, dt = 2\pi \int_{-\infty}^{\infty} \hat{\psi}(\omega)\, \overline{\hat{g}}(\omega)\, d\omega. \tag{8.2.21}$$

Clearly, wavelets of different levels are always orthogonal to each other because their Fourier transforms occupy different frequency bands so that the product $\hat{\psi}(\omega)\,\overline{\hat{g}}(\omega)$ is zero for $m \neq 0$.

On the other hand, at the same level $(m = 0)$, we have

$$\hat{g}(\omega) = \exp(-i\omega k)\,\hat{\psi}(\omega). \tag{8.2.22}$$

Substituting this result in (8.2.21) and the value of $\hat{\psi}(\omega)$ from (8.2.8) gives

$$\int_{-\infty}^{\infty} \psi(k)\,\overline{\psi}(t-k)\,dt = \frac{1}{2\pi}\int_{2\pi}^{4\pi} \exp(ik\omega)\,d\omega = 0, \tag{8.2.23}$$

provided $\exp(4\pi i k) = \exp(2\pi i k)$, $k \neq 0$. This gives $\exp(2\pi i k) = 1$ for $k \neq 0$. Thus, all wavelets translated by any number of unit intervals are orthogonal to each other. Although (8.2.23) is true for $m = 0$, the same result (8.2.23) is also true for other levels except that the unit interval is now that for the wavelet level concerned. For instance, for level m, the unit interval is 2^{-m} and translation is equal to any multiple of 2^{-m}. The upshot of this analysis is that the set of wavelets defined by (8.2.10) forms an orthogonal set. Wavelets of different levels (different values of m) are always orthogonal, and wavelets at the same level are orthogonal if one is translated with respect to the other by a unit interval (different values of k).

In view of (8.2.20), it can be shown that

$$\int_{-\infty}^{\infty} \psi^2\left(2^m x - k\right)\,dx = 0. \tag{8.2.24}$$

Theorem 8.2.2 (Normalization).

$$\int_{-\infty}^{\infty} \left|\psi\left(2^m t - k\right)\right|^2 dt = 2^{-m}. \tag{8.2.25}$$

Proof. It follows from (8.2.21) that

$$\int_{-\infty}^{\infty} \psi(t)\,\overline{\psi}(t)\,dt = 2\pi\int_{-\infty}^{\infty} \hat{\psi}(\omega)\,\overline{\hat{\psi}}(\omega)\,d\omega. \tag{8.2.26}$$

Using (8.2.18) in (8.2.26) gives

$$\int_{-\infty}^{\infty} \psi\left(2^m t - k\right) \overline{\psi}\left(2^m t - k\right) dt = 2\pi 2^{-2m} \int_{-\infty}^{\infty} \hat{\psi}\left(\omega 2^{-m}\right) \overline{\hat{\psi}}\left(\omega 2^{-m}\right) d\omega. \quad (8.2.27)$$

It follows from (8.2.8) that

$$\hat{\psi}\left(\omega 2^{-m}\right) = \frac{1}{2\pi} \qquad \text{for } 2\pi 2^{-m} \le \omega < 4\pi 2^{m}$$

so that (8.2.27) becomes

$$\int_{-\infty}^{\infty} \left|\psi\left(2^m t - k\right)\right|^2 dt = 2\pi 2^{-2m} \int_{2\pi 2^m}^{4\pi 2^m} \frac{1}{(2\pi)^2} \cdot d\omega = 2^{-m}.$$

This implies the property of normality.

8.3 Properties of Harmonic Scaling Functions

We follow the harmonic wavelet terminology due to Newland (1993a) to discuss properties of harmonic scaling functions. Newland (1993a) first introduced the even Fourier transform

$$\hat{\phi}_e(\omega) = \begin{cases} \dfrac{1}{4\pi}, & -2\pi \le \omega < 2\pi \\ 0, & \text{otherwise} \end{cases} \quad (8.3.1)$$

to define an even scaling function

$$\phi_e(x) = \frac{\sin 2\pi x}{2\pi x}. \quad (8.3.2)$$

Similarly, the odd Fourier transform given by

$$\hat{\phi}_0(\omega) = \begin{cases} \dfrac{i}{4\pi}, & -2\pi \le \omega < 0 \\ -\dfrac{i}{4\pi}, & 0 \le \omega < 2\pi \\ 0, & \text{otherwise} \end{cases} \quad (8.3.3)$$

gives an odd scaling function

$$\phi_0(x) = \frac{(1 - \cos 2\pi x)}{2\pi x}. \quad (8.3.4)$$

All of these results allow us to define a complex scaling function $\phi(x)$ by

$$\phi(x) = \phi_e(x) + i\,\phi_0(x) \tag{8.3.5}$$

so that

$$\phi(x) = \frac{1}{2\pi i x}\left\{\exp(2\pi i x) - 1\right\}. \tag{8.3.6}$$

Its Fourier transform $\hat{\phi}(\omega)$ is given by

$$\hat{\phi}(\omega) = \begin{Bmatrix} \dfrac{1}{2\pi}, & 0 \le \omega < 2\pi \\ 0, & \text{otherwise} \end{Bmatrix}. \tag{8.3.7}$$

The real and imaginary parts of the harmonic scaling function (8.3.6) are shown in Figures 8.3(a) and 8.3(b).

Theorem 8.3.1 (Orthogonality of Scaling Functions). The scaling functions $\phi(x)$ and $\phi(x-k)$ are orthogonal for all integers k except $k = 0$.

Proof. We substitute Fourier transforms $\hat{\phi}(\omega)$ and $\mathscr{F}\{\phi(x-k)\} = \exp(-i\omega k)\,\hat{\phi}(\omega)$ in (8.2.15) to obtain

$$\int_{-\infty}^{\infty} \phi(x)\,\phi(x-k)\,dx = 2\pi \int_{-\infty}^{\infty} \hat{\phi}(\omega)\,\hat{\phi}(-\omega)\,\exp(i\omega k)\,d\omega = 0. \tag{8.3.8}$$

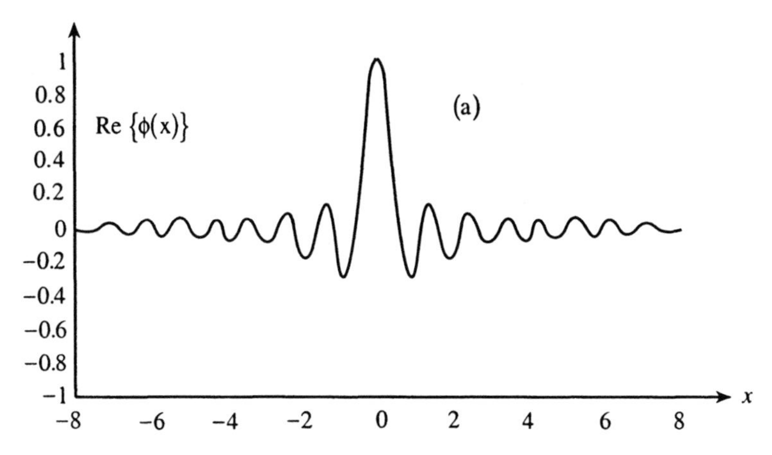

Figure 8.3. (a) Real part of the scaling function ϕ.

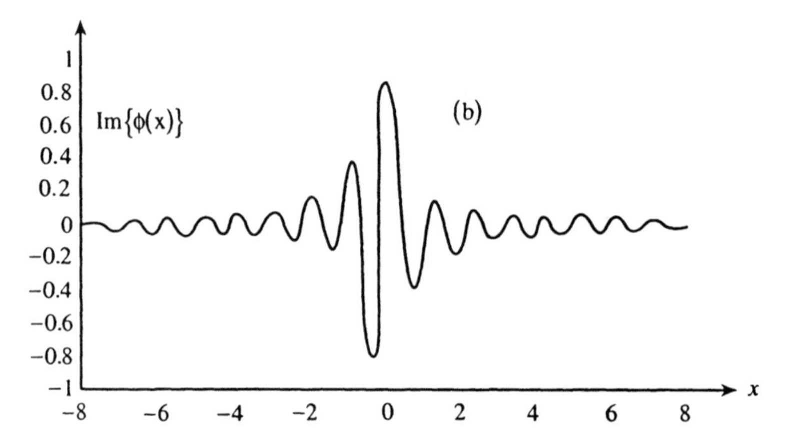

Figure 8.3. (b) Imaginary part of the scaling function ϕ.

The right-hand side is always zero for all k because $\hat{\phi}(\omega)$ is the one-sided Fourier transform given by (8.3.7).

On the other hand, we substitute $\hat{\phi}(\omega)$ and $\mathscr{F}\{\phi(x-k)\}=\exp(-ik\omega)\,\hat{\phi}(\omega)$ in (8.2.14) to obtain

$$\begin{aligned}\int_{-\infty}^{\infty}\phi(x)\,\overline{\phi}(x-k)\,dx &= 2\pi\int_{-\infty}^{\infty}\hat{\phi}(\omega)\,\overline{\hat{\phi}}(\omega)\exp(ik\omega)\,d\omega \\ &= \frac{1}{2\pi}\int_{0}^{2\pi}\exp(ik\omega)\,d\omega, \qquad \text{by (8.3.7)} \\ &= \frac{1}{2\pi i k}\left[\exp(2\pi i k)-1\right]=0 \quad \text{for } k\neq 0. \qquad (8.3.9)\end{aligned}$$

This shows that $\phi(x)$ and $\phi(x-k)$ are orthogonal for all integers k except $k=0$.

It can also be shown that

$$\int_{-\infty}^{\infty}\psi\left(2^{m}x-k\right)\phi(x-n)\,dx=0 \text{ for all } m,k,n(m\geq 0), \qquad (8.3.10)$$

$$\int_{-\infty}^{\infty}\psi\left(2^{m}x-k\right)\overline{\phi}(x-n)\,dx=0 \text{ for all } m,k,n(m\geq 0). \qquad (8.3.11)$$

Theorem 8.3.2 (Normalization).

$$\int_{-\infty}^{\infty} |\phi(x-k)|^2 = 1. \tag{8.3.12}$$

Proof. We have the identity (8.2.14) so that

$$\int_{-\infty}^{\infty} \phi(x-k)\,\bar{\phi}(x-k)\,dk = 2\pi \int_{-\infty}^{\infty} \hat{\phi}(\omega)\,\bar{\hat{\phi}}(\omega)\,d\omega .$$

Thus,

$$\int_{-\infty}^{\infty} |\phi(x-k)|^2\,dk = 2\pi \int_{0}^{2\pi} \frac{1}{(2\pi)^2}\,d\omega = 1 \qquad \text{by (8.3.7).}$$

This completes the proof.

8.4 Wavelet Expansions and Parseval's Formula

An arbitrary (real or complex) function $f(x)$ can be expanded in terms of complex harmonic wavelets in the form

$$f(x) = \sum_{m=-\infty}^{\infty} \sum_{k=-\infty}^{\infty} \left\{ a_{m,k}\,\psi\left(2^m x - k\right) + \tilde{a}_{m,k}\,\overline{\psi}\left(2^m x - k\right) \right\}, \tag{8.4.1}$$

where the complex coefficients $a_{m,k}$ and $\tilde{a}_{m,k}$ are defined by

$$a_{m,k} = 2^m \int_{-\infty}^{\infty} f(x)\,\overline{\psi}\left(2^m x - k\right)\,dx, \tag{8.4.2}$$

$$\tilde{a}_{m,k} = 2^m \int_{-\infty}^{\infty} f(x)\,\psi\left(2^m x - k\right)\,dx. \tag{8.4.3}$$

In terms of these coefficients, the contribution of a single complex wavelet to the function $f(x)$ is given by

$$a_{m,k}\,\psi\left(2^m x - k\right) + \tilde{a}_{m,k}\,\overline{\psi}\left(2^m x - k\right). \tag{8.4.4}$$

Adding all these terms gives the expansion (8.4.1).

We next give a formal proof of the Parseval formula

$$\int_{-\infty}^{\infty} |f(x)|^2\,dx = \sum_{m=-\infty}^{\infty} \sum_{k=-\infty}^{\infty} 2^{-m} \left(\left|a_{m,k}\right|^2 + \left|\tilde{a}_{m,k}\right|^2 \right). \tag{8.4.5}$$

Multiplying (8.4.1) by $\bar{f}(x)$ and then integrating the result from $-\infty$ to ∞ term by term gives

$$\int_{-\infty}^{\infty} f(x)\,\overline{f(x)}\,dx = \sum_{m=-\infty}^{\infty}\sum_{k=-\infty}^{\infty}\left[a_{m,k}\int_{-\infty}^{\infty}\overline{f(x)}\,\psi\left(2^m x-k\right)dx\right.$$

$$\left.+\tilde{a}_{m,k}\int_{-\infty}^{\infty}\bar{f}(x)\,\overline{\psi}\left(2^m x-k\right)dx\right]. \tag{8.4.6}$$

We use (8.4.2) and (8.4.3) to replace the integrals on the right-hand side of (8.4.6) by $\bar{a}_{m,k}$ and $\bar{\tilde{a}}_{m,k}$ so that (8.4.6) becomes

$$\int_{-\infty}^{\infty}|f(x)|^2\,dx = \sum_{m=-\infty}^{\infty}\sum_{k=-\infty}^{\infty}2^{-m}\left(\left|a_{m,k}\right|^2+\left|\tilde{a}_{m,k}\right|^2\right).$$

It may be noted that, for real functions $f(x)$, $\bar{a}_{k,m}=\tilde{a}_{k,m}$ so that the expansion (8.4.1) can be simplified.

Another interesting proof of the Parseval formula (8.4.5) is given by Newland (1993a) without making any assumption of the wavelet expansion (8.4.1).

We next define complex coefficients in terms of the scaling function in the form

$$a_{\phi,k} = \int_{-\infty}^{\infty} f(x)\,\bar{\phi}(x-k)\,dx, \tag{8.4.7}$$

$$\tilde{a}_{\phi,k} = \int_{-\infty}^{\infty} f(x)\,\phi(x-k)\,dx. \tag{8.4.8}$$

In view of the orthogonality and normalizaiton properties of $\psi\left(2^m x-k\right)$ and $\phi(x-k)$, it can be shown that any arbitrary function $f(x)$ can be expanded in the form

$$f(x)=\sum_{k=-\infty}^{\infty}\left\{a_{\phi,k}\,\phi(x-k)+\tilde{a}_{\phi,k}\,\bar{\phi}(x-k)\right\}$$

$$+\sum_{m=0}^{\infty}\sum_{k=-\infty}^{\infty}\left\{a_{m,k}\,\psi\left(2^m x-k\right)+\tilde{a}_{m,k}\,\overline{\psi}\left(2^m x-k\right)\right\}. \tag{8.4.9}$$

Newland (1993a) proved that this expansion (8.4.9) is equivalent to (8.4.1).

8.5 Concluding Remarks

Newland (1993b) generalized the concept of the harmonic wavelet to describe a family of mixed wavelets with the simple mathematical structure

$$\psi_{r,n}(x)=\frac{\exp(2\pi i n x)-\exp(2\pi i r x)}{2\pi i(n-r)x}. \tag{8.5.1}$$

It is proved that this family provides a complete set of orthogonal basis functions for signal analysis. When real numbers r,n are appropriately chosen, these mixed wavelets whose frequency content increases according to the musical scale can be created. These musical wavelets provide greater frequency discrimination than is possible with harmonic wavelets whose frequency band is always an octave. A major advantage for all harmonic wavelets is that they can be computed by an effective parallel algorithm rather than by the series algorithm needed for the dilation wavelet transform.

8.6 Exercises

1. Prove the following results (Newland, 1993a):

(a) $\int_{-\infty}^{\infty}\psi(2^m t-k)\,\psi(2^n t-\ell)\,dt=0$ for all $m,k,n,\ell\,(m,n\geq 0)$.

(b) $\int_{-\infty}^{\infty}\psi(2^m t-k)\,\overline{\psi}(2^n t-\ell)\,dt=0$ for all $m,k,n,\ell\,(m,n\geq 0)$.

(c) When $m=n$ and $k=\ell$, the above result 1(b) becomes

$$\int_{-\infty}^{\infty}\left|\psi(2^m t-k)\right|^2 dt=2^{-m}.$$

(d) $\int_{-\infty}^{\infty}\phi(x-m)\,\overline{\phi}(x-n)\,dx=0$ for all $m,n,\ m\neq n$.

(e) $\int_{-\infty}^{\infty}\psi(2^m x-k)\,\phi(x-\ell)\,dx=0$ for all $m,k,\ell\,(m\geq 0)$.

(f) $\int_{-\infty}^{\infty} \psi(2^m x - k)\, \bar{\phi}(x - \ell)\, dx = 0$ for all $m, k, \ell\, (m \geq 0)$.

2. Show that (Newland, 1993a)

(a) $a_{\phi,k} = 2\pi \int_{-\infty}^{\infty} \hat{f}(\omega)\, \bar{\hat{\phi}}(\omega) \exp(i\omega k)\, d\omega.$

(b) $\sum_{k=-\infty}^{\infty} a_{\phi,k}\, \phi(x - k)$

$$= 2\pi \sum_{k=-\infty}^{\infty} \int_{-\infty}^{\infty} d\omega_1 \int_{-\infty}^{\infty} \hat{f}(\omega_1)\, \bar{\hat{\phi}}(\omega_1)\, \hat{\phi}(\omega_2) \exp(i\omega_2 x) \times \exp\{i(\omega_1 - \omega_2)k\}\, d\omega_2$$

$$= \int_0^{2\pi} \hat{f}(\omega) \exp(i\omega x)\, d\omega.$$

3. Prove that (Newland, 1993a)

(a) $a_{m,k} = 2\pi \int_{-\infty}^{\infty} \hat{f}(\omega)\, \bar{\hat{\psi}}(\omega 2^{-m}) \exp(i\omega k 2^{-m})\, d\omega.$

(b) $\tilde{a}_{m,k} = 2\pi \int_{-\infty}^{\infty} \hat{f}(-\omega)\, \hat{\psi}(\omega 2^{-m}) \exp(-i\omega k 2^{-m})\, d\omega.$

4. Show that the wavelet expansion (8.4.1) is equivalent to that of (8.4.9).

5. Prove Parseval's formula (8.4.5) without making any assumption of the wavelet expansion (8.4.1).

6. If the Fourier transform of a wavelet $\psi_{r,n}(x)$ is

$$\hat{\psi}_{r,n}(\omega) = \begin{Bmatrix} \dfrac{1}{2\pi(n-r)}, & 2\pi r \leq \omega < 2\pi n \\ 0, & \text{otherwise} \end{Bmatrix},$$

show that

$$\psi_{r,n}(x) = \frac{\exp(2\pi i n x) - \exp(2\pi i r x)}{2\pi i (n-r) x}.$$

7. Introducing translation of the wavelet by $s = k(n-r)^{-1}$, generalize the result of Exercise 6 in the form

$$\psi_{r,n}(x-s)=\frac{\exp\{2\pi i n(x-s)\}-\exp\{2\pi i r(x-s)\}}{2\pi i(n-r)(x-s)},$$

where

$$\hat{\psi}_{r,n}(\omega)=\left\{\begin{array}{ll}\dfrac{\exp(-i\omega s)}{2\pi(n-r)}, & 2\pi r\le\omega<2\pi n\\ 0, & \text{otherwise}\end{array}\right\}.$$

If $r=2^m$ and $n=2^{m+1}$, show that the wavelet $\psi_{r,n}(x)$ reduces to that given by (8.2.10).

8. Prove the following results for $\psi_{r,n}(x)$ in Exercise 7 with $s_1=k_1(n-r)^{-1}$ and $s_2=k_2(n-r)^{-1}$:

(a) $\int_{-\infty}^{\infty}\psi_{r,n}(x-s_1)\,\psi_{r,n}(x-s_2)\,dx=0$ for any k_1 and k_2,

(b) $\int_{-\infty}^{\infty}\psi_{r,n}(x-s_1)\,\overline{\psi}_{r,n}(x-s_2)\,dx=0$ for $k_1\ne k_2$,

(c) $\int_{-\infty}^{\infty}|\psi_{r,n}(x-s)|^2\,dx=(n-r)^{-1}$ for $s=k(n-r)^{-1}$.

Chapter 9

Wavelet Transform Analysis of Turbulence

"The phenomenon of turbulence was discovered physically and is still largely unexplored by mathematical techniques. At the same time, it is noteworthy that the physical experimentation which leads to these and similar discoveries is a quite peculiar form of experimentation; it is very different from what is characteristic in other parts of physics. Indeed, to a great extent, experimentation in fluid dynamics is carried out under conditions where the underlying physical principles are not in doubt, where the quantities to be observed are completely determined by known equations. The purpose of the experiment is not to verify a proposed theory but to replace a computation from an unquestioned theory by direct measurements. Thus wind tunnels are, for example, used at present, at least in part, as computing devices of the so-called analog type (or, to use a less widely used, but more suggestive, expression proposed by Wiener and Caldwell: of the measurement type) to integrate the nonlinear partial differential equations of fluid dynamics."

John von Neumann

"The use of the wavelet transform for the study of turbulence owes absolutely nothing to chance or fashion but comes from a necessity stemming from the current development of our ideas about turbulence. If, under influence of the statistical approach, we had lost the need to study things in physical space, the advent of supercomputers and the associated means of visualization have revealed a zoology specific to turbulent flows, namely, the existence of coherent structures and their elementary interactions, none of which are accounted for by the statistical theory."

Marie Farge

9.1 Introduction

Considerable progress has been made over the last three decades in our understanding of turbulence through new developments of theory, experiment, and computation. More and more evidence has been accumulated for the physical description of turbulent motions in both two and three dimensions. Consequently, turbulence is now characterized by a remarkable degree of order even though turbulence is usually defined as disordered fluid flows. In spite of tremendous progress, there are still a number of open questions and unsolved problems. These include coherent structures and intermittency effects, singularities of the Navier-Stokes equations, non-Gaussian statistics of turbulent flows, perturbations to the small scale produced by nonisotropic, non-Gaussian, and inhomogeneous large-scale motions, and measurements and computations of small-scale turbulence. No complete theory is yet available for the problem of how the eddy structure of turbulence evolves both under the action of mean distortion and even during the mutual random interaction of eddies of different sizes or scales.

Most of the progress has been based on the Navier-Stokes equations combined with the Fourier transform analysis. However, there are certain major difficulties associated with the Navier-Stokes equations. First, in three dimensions, there are no general results for the Navier-Stokes equations on existence of solutions, uniqueness, regularity, and continuous dependence on the initial conditions. However, such results exist for the two-dimensional Navier-Stokes equations. Second, there are indications that solutions of three-dimensional Navier-Stokes equations can be singular at certain places and at certain times in the flow. Third, another difficulty arises from the strong nonlinear convective term in the equation. This nonlinearity leads to an infinite number of equations for all possible moments of the velocity field. This system of equations is very complicated in the sense that any subsystem is always nonclosed because it contains more unknowns than the number of equations in a given subsystem. For example, the dynamical equation for second-order moments involves third-order moments, that for third-order moments involves fourth-order moments, and so on. This is the so-called *closure problem* in the statistical theory of turbulence. This is perhaps the major difficulty of the turbulence theory. For any physical system with strong interaction, such as turbulent flows, it is not easy to guess what kind of closure is consistent with the Navier-Stokes equations. Various closure models for turbulence, including the quasi-normal model (see Monin and Yaglom, 1975) have been suggested. They are hardly consistent with physical analysis, experimental measurements, and, more recently, with direct numerical simulations of turbulence. Fourth, in the limit as $\nu \to 0$ $(R \to \infty)$, the nature of the Navier-Stokes equations changes because the nonlinear convective term dominates over the linear viscous term. Therefore, for fully developed turbulence, as $R \to \infty$, the second-order viscous term vanishes. Consequently, the second-order Navier-Stokes equations reduce to the first-order Euler equations. Thus, a slightly viscous fluid flow can lead to a *singular perturbation* of the inviscid fluid motions. Mathematically, the Navier-Stokes equations lead to a singular perturbation problem. Another major difficulty in modeling the structure and dynamics of turbulence is the wide range of length and time scales over which variations occur. However, in recent years, a broad class of self-similar dynamical processes has been developed as a possible means of characterizing turbulent flows.

Traditionally, the Fourier transform approach to turbulence has been successful due to the fact that the Fourier transform breaks up a function (or signal) into different sine waves of different amplitudes and wavenumbers (or frequencies). In fact, the classical theory of turbulent flows was developed in the Fourier transform space by introducing the Fourier energy spectrum $\hat{E}(k)$ of a function $f(x)$ in the form

$$\hat{E}(k) = \left|\hat{f}(k)\right|^2. \tag{9.1.1}$$

However, $\hat{E}(k)$ does not give any local information on turbulence. Since $\hat{f}(k)$ is a complex function of a real wavenumber k, it can be expressed in the form

$$\hat{f}(k) = \left|\hat{f}(k)\right| \exp\{i\theta(k)\}. \tag{9.1.2}$$

The phase spectrum $\hat{\theta}(k)$ is totally lost in the Fourier transform analysis of turbulent flows, and only the modulus of $\hat{f}(k)$ is utilized. This is possibly another major weakness of the Fourier energy spectrum analysis of turbulence since it cannot take into consideration any organization of the turbulent field. Also, the rate of energy dissipation is distributed very intermittently in both space and time. This is usually modeled by the breakdown of eddies, and the flux of energy is assumed to flow from larger to smaller eddies so that the turbulence is generated to small scales, where it is dissipated by viscosity. Evidently, there is a need for introducing a flux of kinetic energy which also depends on position. For a real description of real turbulence, there is a need for a representation that decomposes the flow field into contributions of different length scales, different positions, and different directions.

The idea of a hierarchy of vortices is usually employed in the study of turbulence. Combined with the theory of scales, a model of turbulence as a vortex system of different sizes with random amplitude functions leads to the statistical description of turbulence. Kolmogorov (1941a,b) used this approach to derive his famous spectral law for isotropic and homogeneous turbulence. In this idea of a hierarchy of vortices, the velocity field can be represented in terms of Fourier integral transforms. This representation seems to be unsatisfactory for the following reason. Each Fourier component in the decomposition of the velocity vector potential corresponds to a coherent vortex structure over the entire space. But the strong nonlinear interaction of the spatial temporal modes in turbulence results in the effect that periodic solutions representing coherent vortex systems are not typical structural components of the turbulent motion. The processes involving energy transfer, deformation, and vortex decomposition are described by the local conditions of the turbulent flow.

Therefore, the Fourier transform analysis does not have the ability to provide a local description of turbulent flows. In fact, the scale, position, and direction involved in the flow field are completely lost in this analysis. Moreover, the Fourier transform cannot describe the multifractal structure of fully developed turbulence. The new method of wavelet transform analysis may enable representation of quantities that depend on scale, position, and direction, and hence it has the ability to give local information about the turbulent flows.

In a series of papers, Farge and her associates (Farge, 1992; Farge and Holschneider, 1989, 1990; Farge and Rabreau, 1988, 1989; Farge et al., 1990a,b, 1992, 1996, 1999a,b) introduced new concepts and ideas to develop a new and modern approach to turbulence based on the wavelet transform analysis. They showed that the wavelet transform can be used to define local energy density, local energy spectrum, and local intermittency, to determine singularities, and to find extrema of derivatives at different positions and scales. These studies reveal that both wavelet and fractal analyses seem to be very useful and effective mathematical tools for investigating the self-similarity, coherent structures, intermittency, and local nature of the dynamics and other features of turbulent flows. Meneveau (1991, 1993) initiated wavelet transform analysis for the study of time-dependent three-dimensional computations of the velocity field in a turbulent flow. He also provided the first direct evidence that energy flows from small to large scales in some regions of turbulence. This is a remarkable new phenomenon that cannot be studied by using Fourier transform analysis.

This chapter is devoted to a brief discussion of Fourier transform analysis and the wavelet transform analysis of turbulence based on the Navier-Stokes equations. Included are fractals, multifractals, and singularities in turbulence. This is followed by Farge's and Meneveau's wavelet transform analyses of turbulence in some detail. Special attention is given to the adaptive wavelet method for computation and analysis of turbulent flows. Many references related to applications of the wavelet transform in turbulence are cited in the bibliography.

9.2 Fourier Transforms in Turbulence and the Navier-Stokes Equations

It is well-known that a Fourier transform decomposes a function or a signal $f(x)$ into different sine waves of different amplitudes and wavelengths. In general, the Fourier transform of a signal $f(x)$ can be expressed as

$$\hat{f}(k) = \left|\hat{f}(k)\right| \exp\left\{i\hat{\theta}(k)\right\}, \tag{9.2.1}$$

where $\left|\hat{f}(k)\right|$ and $\hat{\theta}(k)$ are called the *amplitude spectrum* and the *phase spectrum*, respectively.

The *energy* (or *power*) *spectrum* of a signal is defined by

$$\hat{E}(k) = \left|\hat{f}(k)\right|^2, \tag{9.2.2}$$

so that the total energy of the signal $f(x)$ is given by

$$E = \int_{-\infty}^{\infty} \hat{E}(k)\, dk = \int_{-\infty}^{\infty} \left|\hat{f}(k)\right|^2 dk. \tag{9.2.3}$$

Clearly, it follows from (9.2.2) and (9.2.3) that the energy spectrum and the total energy depend only on the amplitude and are completely independent of the phase $\hat{\theta}(k)$. In other words, the Fourier transform does not provide any local or structural information on the signal. In spite of this major weakness, the Fourier transform has been useful to analyze stationary stochastic signals. In particular, the Fourier transform is fairly successful in the theory of a homogeneous turbulent velocity field confined within a box of volume a^3. In the case of three-dimensional turbulence, the Fourier transform of the velocity field $\boldsymbol{u}(\boldsymbol{x})$ has three components, each of the form $(j = 1, 2, 3)$,

$$\hat{u}_j(\boldsymbol{k}) = \frac{1}{(2\pi)^{3/2}} \int_{-\infty}^{\infty} \int_{-\frac{a}{2}}^{a/2} u_j(\boldsymbol{x}) \exp(i\boldsymbol{k} \cdot \boldsymbol{x})\, d\boldsymbol{x}, \tag{9.2.4}$$

where $\hat{u}_j(\boldsymbol{k})$ can be expressed in terms of amplitude and phase by

$$\hat{u}_j(\boldsymbol{k}) = \left|\hat{u}_j(\boldsymbol{k})\right| \exp\{i\theta_j(\boldsymbol{k})\}. \tag{9.2.5}$$

Since $u_j(\boldsymbol{x})$ and $u_j(\boldsymbol{k})$ are random functions, it is necessary to define statistical quantities, of which the most important are the energy spectrum tensor $\Phi_{ij}(\boldsymbol{k})$ and cross correlation between components. Application of Fourier transforms shows that

$$\overline{\hat{u}_i^*(\boldsymbol{k})\, \hat{u}_j(\boldsymbol{k})} = \frac{1}{(2\pi)^{3/2}}\, a^3\, \Phi_{ij}(\boldsymbol{k}) \tag{9.2.6}$$

and

$$\overline{\hat{u}_i^*(\boldsymbol{k})\, \hat{u}_j(\boldsymbol{k}')} = \frac{1}{(2\pi)^{3/2}}\, \Phi_{ij}(\boldsymbol{k})\, \delta(\boldsymbol{k} - \boldsymbol{k}'), \tag{9.2.7}$$

where the bar represents an average over space and the asterisk denotes the complex conjugate. One of the most important properties of the turbulent flow is the correlation tensor of a homogeneous (stationary space $\boldsymbol{x}$) stochastic velocity field defined by

$$R_{ij}(\boldsymbol{r}, \boldsymbol{x}, t) = \overline{u_i(\boldsymbol{x})\, u_j(\boldsymbol{x} + \boldsymbol{r})}, \tag{9.2.8}$$

where $\boldsymbol{r}$ is the distance between simultaneous velocity fluctuations. Evidently,

$$-\rho\, R_{ij}(0, \boldsymbol{x}, t) = -\rho\, \overline{u_i\, u_j} = \tau_{ij} \tag{9.2.9}$$

is called the *Reynolds stress tensor*.

The *energy spectrum tensor* in turbulence is defined as the Fourier transform of the covariance tensor $R_{ij}(\boldsymbol{r}, \boldsymbol{x}, t)$ by

$$\Phi_{ij}(\boldsymbol{k} \cdot \boldsymbol{x}, t) = \frac{1}{(2\pi)^{3/2}} \int_{-\infty}^{\infty} \exp(-i\boldsymbol{k} \cdot \boldsymbol{r})\, R_{ij}(\boldsymbol{r}, \boldsymbol{x}, t)\, d\boldsymbol{r}, \tag{9.2.10}$$

so that the inverse Fourier transform is given by

$$R_{ij}(\boldsymbol{r}, \boldsymbol{x}, t) = \frac{1}{(2\pi)^{3/2}} \int_{-\infty}^{\infty} \exp(i\boldsymbol{k} \cdot \boldsymbol{r})\, \Phi_{ij}(\boldsymbol{k}, \boldsymbol{x}, t)\, d\boldsymbol{k}, \tag{9.2.11}$$

where the integration is over all wavenumber $\boldsymbol{k}$-space. A spectrum tensor ψ_{ij} function of the single scale variable $k = |\boldsymbol{k}|$ can be obtained by averaging over all directions of the vector argument $\boldsymbol{k}$ so that

$$\psi_{ij}(k) = \int \Phi_{ij}(\boldsymbol{k})\, dS(k), \tag{9.2.12}$$

where this integration is taken in $\boldsymbol{k}$ space over a sphere of radius k of which $dS(k)$ is an element. The *energy (power) spectrum* is then defined by

$$E(k, t) = \frac{1}{(2\pi)^{3/2}} \Psi_{ii}(k) \tag{9.2.13}$$

so that the total energy $E = \overline{\frac{1}{2} u_i^2}$ is the integral of $E(k, t)$ over all k from 0 to ∞, that is,

$$\overline{\frac{1}{2} u_i^2} = \frac{1}{2} \int \Phi_{ii}(\boldsymbol{k})\, d\boldsymbol{k} = \int_0^{\infty} E(k, t)\, dk. \tag{9.2.14}$$

Thus, it follows from (9.2.8) and (9.2.12) that

$$\begin{aligned} E(k, t) &= \frac{1}{(2\pi)^{3/2}} \Psi_{ii}(k) = \frac{1}{(2\pi)^{3/2}} \int \Phi_{ii}(\boldsymbol{k})\, dS(k) \\ &= \frac{1}{(2\pi)^{3/2}} \int |\hat{u}_i(\boldsymbol{k})|^2\, dS(k). \end{aligned} \tag{9.2.15}$$

The turbulent energy spectrum $E(k, t)$ represents the distribution of contributions to $\frac{1}{2}\overline{u_i^2}$ with respect to wavenumber (or scale) regardless of direction, and this is one of the most important characteristics of any turbulent (or three-dimensional wave) field. Thus, the study of the energy spectrum $E(k, t)$ is the central problem in the dynamics of turbulence. However, the information carried by the phase function $\theta_i(k)$ disappears completely in the definition of the energy spectrum.

One of the most common approaches to the study of turbulence is to use the Navier-Stokes equations together with the continuity equation in the Fourier transform space. In tensor notation, the Navier-Stokes equations for an unsteady motion of an incompressible viscous fluid of constant density ρ and kinematic viscosity ν and the continuity equation are

$$\frac{\partial u_i}{\partial t} + u_m \frac{\partial u_i}{\partial x_m} = -\frac{\partial p}{\partial x_i} + \nu \nabla^2 u_i + F_i, \tag{9.2.16}$$

$$\frac{\partial u_i}{\partial x_i} = 0, \tag{9.2.17}$$

where $u_i = u_i(x,t)$ is the velocity field, p is the normal pressure divided by ρ, it is often called the *kinematic pressure*, and F_i are the external body forces. The continuity equation (9.2.17) is kinematic in nature and is unaffected by the energy dissipation process in the fluid due to viscosity.

It is important to point out that the use of the Navier-Stokes equations is perhaps justified for the study of turbulence because the Mach number of incompressible turbulent flows is relatively small.

Using the continuity equation (9.2.17), the Navier-Stokes equation (9.2.16) in the absence of the external field of forces $(F_i \equiv 0)$ can be written as

$$\frac{\partial u_i}{\partial t} + \frac{\partial}{\partial x_m}(u_i\, u_m) = -\frac{\partial \rho}{\partial x_i} + \nu \nabla^2 u_i. \tag{9.2.18}$$

Taking the divergence of this equation and using (9.2.17) gives the Poisson equation

$$\nabla^2 p = -\frac{\partial^2 (u_i\, u_m)}{\partial x_i\, \partial x_m}. \tag{9.2.19}$$

Eliminating the pressure from the Navier-Stokes equations, we obtain

$$\frac{\partial u_i}{\partial t} - \nu \nabla^2 u_i = -\frac{1}{2} P_{ijm}(\nabla)(u_j\, u_m), \tag{9.2.20}$$

where

$$P_{ijm}(\nabla) = \frac{\partial}{\partial x_m} P_{ij}(\nabla) + \frac{\partial}{\partial x_j} P_{im}(\nabla), \tag{9.2.21}$$

$$\nabla = \left(\frac{\partial}{\partial x_1}, \frac{\partial}{\partial x_2}, \frac{\partial}{\partial x_3} \right), \tag{9.2.22}$$

and

$$P_{ij}(\nabla) = \delta_{ij} - \frac{1}{\nabla^2} \frac{\partial^2}{\partial x_i \, \partial x_j}. \tag{9.2.23}$$

The Fourier transform of the Navier-Stokes equations is

$$\left(\frac{\partial}{\partial t} + \nu \, k^2 \right) \hat{u}_i(\boldsymbol{k}, t) = -ik_m \, P_{ij}(\boldsymbol{k}) \int u_j(\boldsymbol{q}) \, u_m(\boldsymbol{k} - \boldsymbol{q}) \, d^3\boldsymbol{q}, \tag{9.2.24}$$

where $\hat{u}_i(\boldsymbol{k}, t)$ is the Fourier transform of $u_i(x_i, t)$, and

$$P_{ij}(\boldsymbol{k}) = \delta_{ij} - \frac{k_i \, k_j}{k^2}. \tag{9.2.25}$$

The velocity $u_i(\boldsymbol{x}, t)$ is represented as a linear combination of plane waves, each corresponding to a characteristic size $O(k_i^{-1})$ in some direction i. However, the information related to position in physical space is completely hidden, which is a major drawback when dealing with the space of intermittency of turbulent flow. It has been recognized that turbulence has a set of localized structures, often called *coherent structures*, even at a very high Reynolds number $R = (U\ell/\nu)$ or at a very low viscosity. In many practical applications in aeronautics and meteorology, R varies in between 10^6 and 10^{12}. These coherent structures are organized spatial features, which repeatedly occur and undergo a characteristic temporal life cycle. There are many examples of such structures which play a central role in the time and space intermittency of turbulence. The classical model of turbulence is based on ensemble time or space average, but this idea is of no use for the description of coherent structure. On the other hand, the Navier-Stokes equations in physical space provide no explicit information about scales of motion. This information is often useful for modeling and physical insight into turbulent flows. This difficulty requires a representation that decomposes the flow field into contributions of different positions as well as different scales.

One of the most important features of a turbulent flow is the transfer of kinetic energy from large to small scales of motion due to the nonlinear (convective) term, which acts as the source of energy transfer. Denoting the nonlinear transfer of energy to wavenumbers of magnitude k by $T(k, t)$, the three-dimensional energy spectrum $E(k,t)$ for isotropic turbulence satisfies the evolution equation

$$\frac{\partial E}{\partial t} = T(k,t) - 2\nu\, k^2\, E(k,t), \tag{9.2.26}$$

where $T(k,t)$ is formally defined in terms of triple products of fluctuating velocity and thus embodies the closure problem due to the nonlinear term in the Navier-Stokes equations. Equation (9.2.26) is made up of contributions of the inertial, nonlinear, and viscous terms of the Navier-Stokes equations. It follows from the continuity equation (9.2.17) that the pressure term does not make any contribution to (9.2.26). This implies that the net effect of the pressure field is to conserve the total energy in the wavenumber space. Only the nonlinear term in the Navier-Stokes equations is responsible for the net energy transfer from large to smaller eddies or scales – a mechanism by which large eddies decay. The total spectral flux of energy through wavenumber k to all smaller scales is given by

$$\pi(k,t) = \int_k^\infty T(k',t)\, dk' = -\int_0^k T(k',t)\, dk', \tag{9.2.27}$$

so that

$$\int_0^\infty T(k',t)\, dk' = 0, \tag{9.2.28}$$

which also follows from the conservation of energy by the nonlinear term in (9.2.16). Consequently, the evolution equation (9.2.26) leads to

$$\frac{\partial}{\partial t}\left(\frac{1}{2}\overline{u_i\, u_i}\right) = \frac{\partial}{\partial t}\int_0^\infty E(k,t)\, dk = -\varepsilon(t), \tag{9.2.29}$$

and it follows from (9.2.26) that

$$\varepsilon(t) = 2\nu \int_0^\infty k^2 E(k,t)\, dk. \tag{9.2.30}$$

This clearly represents the overall rate of energy dissipation and exhibits that small-scale (or high wavenumbers) components are dissipated more rapidly by viscosity than large-scale (or low wavenumbers) components.

Based on the usual arguments of equilibrium and stationarity, it is easy to conclude that the ensemble average of the flux must equal to the overall rate of energy dissipation, so that $\langle \pi(k,t)\rangle_{ens} = \varepsilon(t)$ in the inertial range

$\eta << k^{-1} << \ell$, where ℓ is the integral scale and η is the Kolmogorov microscale. Physically, the mechanism of energy transfer is described by simplified assumptions such as the successive breaking down of eddies or as the generation of small scales by the stretching and folding of vortices. Over the scales of motion of size k^{-1}, there is a net flux of kinetic energy to smaller scales that is equal to the time average of $\pi(k,t)$. However, $\pi(k,t)$ does not depend on *position* because the Fourier transform is used in the preceding analysis. This means that information related to position in physical space is completely absent in the theory of Kolmogorov (1941a,b), which neglects the phenomenon of intermittency.

It is also well-known that the rate of dissipation $E(x,t)$ is distributed very intermittently – a feature which increases with the Reynolds number and its moments also increase with the Reynolds number according to power-laws in the inertial range (see Kolmogorov, 1962). This allows a self-consistent statistical and geometrical representation of ε in terms of multifractals (Benzi et al., 1984; Frisch and Parisi, 1985). The power-law behavior of spatial moments of the energy dissipation can be modeled naturally within the framework of the breakdown of eddies with an additional assumption that the flux of energy to smaller scales shows spatial fluctuations. As the scales of motion become smaller, these spatial fluctuations accumulate and can then lead to very intermittent distributions of the energy dissipation. This clearly suggests that there is a need for defining a flux of kinetic energy instead of (9.2.27) which incorporates information on positions. In spite of these weaknesses of the Fourier analysis of turbulence, the upshot of the preceding description is that pressure and nonlinear inertial terms separately conserve the total energy of turbulence, whereas the linear viscous term dissipates the energy. Based on the assumption of self-similarity, Kolmogorov (1941a,b) and Oboukhov (1941) formulated a general statistical theory of turbulence, which is known as the *universal equilibrium theory*. This formulation represents a significant step in the development of the statistical theory of turbulence.

In order to study the energy spectrum function $E(k,t)$, Kolmogorov classified the spectrum into three major ranges, which are assumed to be independent. These ranges are called the large eddies $(kl \ll 1)$, the energy-containing eddies $(kl \sim 1)$, and the small eddies $(kl \gg 1)$, where l is the characteristic length scale of the energy-containing eddies or the differential length scale of the mean flow as a whole. For instance, the spectrum function $E(k,t)$ attains its maximum value at $k_l = l^{-1}$. The basic assumption of the Kolmogorov theory is that at a very high Reynolds number, the turbulent flow at the very small scales (large wavenumbers) is approximately similar to a state of statistical equilibrium and hence, this part of the spectrum is called the *equilibrium or quasi-equilibrium range*. Further, the motion of the small eddies is assumed to be statistically independent of that in the energy-containing range. The energy-containing scales of the motion may be inhomogeneous and anisotropic, but this feature is lost in the cascade so that at much smaller scales the motion is locally homogeneous and isotropic. Hence, the statistical properties of the turbulent motion in the equilibrium range must be completely determined by the physical parameters that are relevant to the dynamics of this part of the spectrum only. The motion associated with the equilibrium range $(kl \gg 1)$ is uniquely determined only by two physical parameters, ε and ν. The consequence of this assumption is that the small-scale statistical characteristics of the velocity fluctuations in different turbulent flows with high Reynolds numbers can differ only by the length scales, which depend on ε and ν. According to the theory of Kolmogorov, the turbulent motion in the equilibrium range is dissipated by viscosity at the rate ε so that $E(k)$ is a function of ε, ν, and k. The net energy supply from the small wavenumbers is transferred by the nonlinear inertial interactions to larger and larger wavenumbers until the viscous dissipation becomes significant. Clearly, the Reynolds number must be very large for the existence of a statistical range of equilibrium. A necessary condition for this is $k_l \ll k_d$, where k_d is the location of the wavenumber at which the viscous dissipation first becomes dominant. In other words, the viscous dissipation takes place predominantly at the upper part of the equilibrium range, that is, at large wavenumbers $k \gg k_d \gg k_l$. Thus, for $R \gg 1$, there exist two independent and widely separated regions $k \sim k_l$ (energy source) and near $k \sim k_d$ (energy sink), which are connected through a continuous set of wavenumbers k such that $k_l \ll k \ll k_d$. In other words, the Kolmogorov inertial range lies between the largest scale $l\left(l^{-1} = k_l\right)$, where the energy is supplied by external forces, and the smallest scales $d\left(d^{-1} = k_d\right)$, where the energy is dissipated by viscosity. This confirms the existence of an intermediate part of the energy spectrum, the so-called *inertial range*

$(k_l \ll k \ll k_d)$, where (i) the local energy transfer is significant, (ii) the properties of the statistical ensemble are independent of all features of energy input except its rate, and (iii) the viscous dissipation is insignificant. In this case, the nonlinear convection is quite significant, and the energy spectrum is therefore independent of viscosity ν so that $E(k)$ depends only on ε and k. In a state of statistical equilibrium, the rate of energy input and the rate of energy dissipation. On a simple dimensional ground, the wavenumber spectrum of kinetic energy or the energy spectrum function in the inertial range takes the form

$$E(k) = C_K \varepsilon^{2/3} k^{-5/3} F\left(\frac{k}{k_d}\right), \tag{9.2.31}$$

provided $k \gg k_l$, and where C_K is a nondimensional universal parameter, called the *Kolmogorov constant*, $k_d = \left(\varepsilon/\nu^3\right)^{1/4}$ is the characteristic dissipation wavenumber and $F(x)$ is a universal dimensionless function. In homogeneous turbulence $\varepsilon \sim \left(u^3/l\right)$ (see Batchelor, 1967), so that $k_d = \left(\frac{ul}{\nu}\right)^{3/4} k_l$. Clearly, $k_d \gg k_l$ is an essential requirement so that the Reynolds number $R = \left(\frac{ul}{\nu}\right)$ must be large. As R increases (or ν decreases), the viscous dissipation would become predominant for larger and larger wavenumbers. According to Kolmogorov's hypothesis, for sufficiently large R there exists a significant range of wavenumbers with $k_l \ll k \ll k_d$, then, in this inertial range, both energy content and energy dissipation are negligible and the spectral energy flux $\varepsilon(k) = \varepsilon$ is independent of wavenumbers k. The molecular viscosity ν then becomes insignificant, $F\left(\frac{k}{k_d}\right)$ in (9.2.31) becomes asymptotically constant for $k \ll k_d$, and then $F\left(\frac{k}{k_d}\right) \approx 1$. Consequently, the energy spectrum in the inertial range reduces to the form

$$E(k) = C_K \varepsilon^{2/3} k^{-5/3}. \tag{9.2.32}$$

This is called the *Kolmogorov-Oboukhov energy spectrum* in isotropic and homogeneous turbulence and received strong experimental support by Grant et al. (1962) in the early 1960s and later with a value of $C_k \approx 1.44 \pm 0.06$. Several experimental observations suggested that the spectrum power lies somewhere between $\frac{5}{3}$ and $\frac{7}{4}$. Even though the experimental accuracy is not very high, most experiments in oceanic and atmospheric turbulence strongly support the $-\frac{5}{3}$ spectrum law. Very recently, Métais and Lesieur (1992) proposed the structure-function model of turbulence with the spectral eddy viscosity based upon a kinetic energy spectrum in space. Their analysis gives the best agreement with the Kolmogorov $k^{-5/3}$ spectrum law and the Kolmogorov constant $C_k \approx 1.40$.

Soon after Kolmogorov's pioneering work, considerable progress was made on a detailed study of different physical mechanisms of energy transfer of turbulence. Several authors including Heisenberg (1948a,b), Lin (1948), Chandrasekhar (1949, 1956), Batchelor (1967), and Sen (1951, 1958), have investigated these problems. Of these physical energy transfer mechanisms, Heisenberg's eddy viscosity transfer was found to be more satisfactory at that time. Based on the assumption that the role of small eddies in the nonlinear transfer process is very much similar to that of molecules in viscous dissipation mechanisms, Heisenberg suggested that these small eddies act as an effective viscosity produced by the motions of the small eddies and the mean-square vorticity associated with the large eddies. He used this assumption to formulate the energy balance equation in the form

$$\frac{\partial}{\partial t}\int_0^k E(k,t)\,dt = -2\left(\nu + \frac{\eta_k}{\rho}\right)\int_0^k k^2 E(k,t)\,dk, \tag{9.2.33}$$

where η_k is the eddy viscosity defined by Heisenberg in the form

$$\eta_k = \rho\kappa\int_k^\infty \left\{\frac{E(k,t)}{k^3}\right\}^{1/2} dk, \tag{9.2.34}$$

where κ is a numerical constant.

Thus, the main problem of turbulence is to determine the spectrum function $E(k,t)$ satisfying the integrodifferential equation (9.2.33) for all subsequent time when $E(k,t)$ is given at $t=0$. For details of the problems, the reader is referred to Debnath (1978, 1998).

9.3 Fractals, Multifractals, and Singularities in Turbulence

Mandelbrot (1982) first introduced the idea of a fractal as a self-similar geometric figure that consists of an identical motif repeating itself in an ever-decreasing scale. This can be illustrated by the famous triadic Koch curve (see Figure 9.1), which can be constructed geometrically by successive iterations. The construction begins with a line segment of unit length $(L(1)=1)$, called the *initiator*. Divide it into three equal line segments. Then, replace the middle segment by an equilateral triangle without a base. This completes the first step $(n=1)$ of the construction, giving a curve of four line segments, each of length $\ell=\frac{1}{3}$, and the total length is $L\left(\frac{1}{3}\right)=\frac{4}{3}$. This new shape of the curve is called the *generator*. The second step $(n=2)$ is obtained by replacing each line segment by a scaled-down version of the generator. Thus, the second-generation curve consists of $N=4^2$ line segments, each of length $\ell=\left(\frac{1}{3}\right)^2$, with the total length of the curve $L(\ell)=\left(\frac{4}{3}\right)^2$. Continuing this iteration process successfully leads to the triadic Koch curve of total length $L(\ell)=\left(\frac{4}{3}\right)^n$, where $\ell=3^{-n}$, as shown in Figure 9.1. The name *triadic* is justified because individual line segments at each step decrease in length by a factor of 3. Obviously, the Koch curve at the end of many iterations $(n\to\infty)$ would have a wide range of scales. At any stage of the iteration process, the curve possesses several important features. First, when a part of it is expanded by a factor of 3^n, it looks similar (except for reorientation) to that obtained in n previous steps. Second, self-similarity is built into the construction process. Third, there is no way to draw a tangent at each corner leading to a tangentless (or nondifferentiable) curve. Finally, this leads to the idea that self-similar fractals are invariant to dilation.

In terms of the box-counting algorithm in fractal geometry, the minimum $N(\ell) = 4$ boxes of size $\left(\frac{1}{3}\right)$ are needed to cover the line in the Koch curve in Figure 9.1(b). Similarly, at least $N(\ell) = 4^2$ boxes of size $\ell = \left(\frac{1}{3}\right)^2$ are required

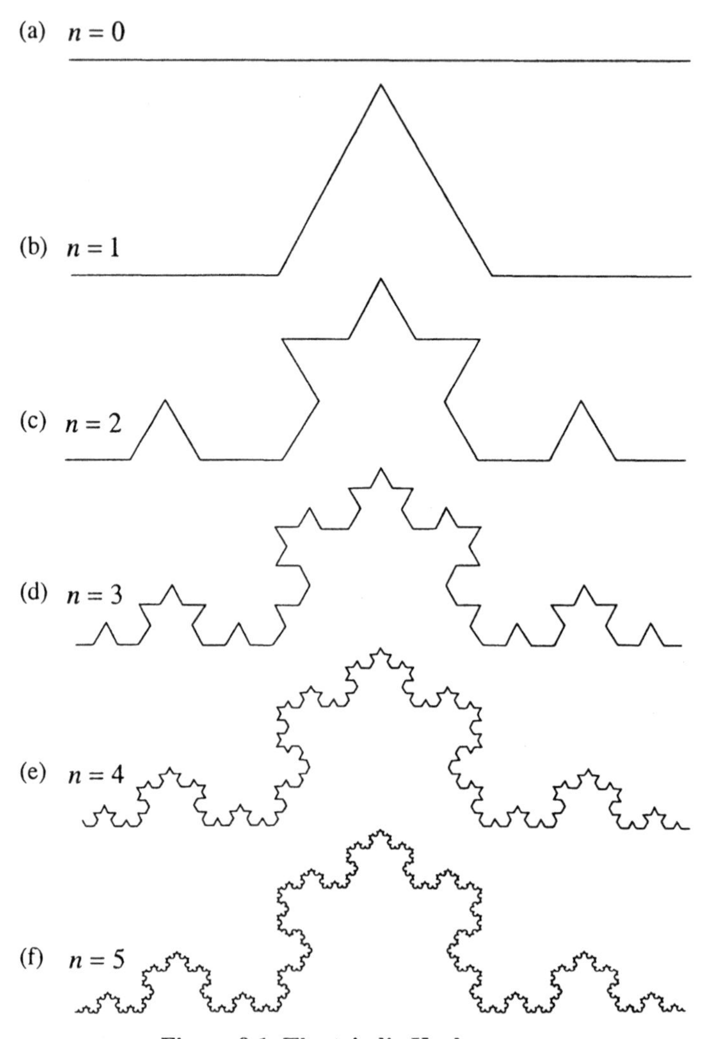

Figure 9.1. The triadic Koch curve.

to cover the line in Figure 9.1(c). In general, a minimum of $N(\ell) = 4^n$ boxes of size $\ell = \left(\frac{1}{3}\right)^n$ are needed to cover the Koch curve obtained at the nth step. On the other hand, the total length $L(3^{-n}) = \left(\frac{4}{3}\right)^n$ at the nth iteration is obtained at a finer resolution of 3^{-n}. As the resolution increases microscopically $(n \to \infty)$, the length of the Koch curve also increases without limit. This shows a striking contrast to an ordinary curve whose length remains the same for all resolutions. The intrinsic parameter that measures this property is called the *fractal Hausdorff dimension D*, which is defined by

$$D = \lim_{\ell \to 0} \frac{\log N(\ell)}{\log\left(\frac{1}{\ell}\right)} = \lim_{\ell \to 0} \frac{\log\left\{\frac{L(\ell)}{\ell}\right\}}{\log\left(\frac{1}{\ell}\right)}, \tag{9.3.1}$$

where $L(\ell) = \ell N(\ell) = \ell^{1-D}$ for small number ℓ.

For the triadic Koch curve, $N(\ell) = 4^n$ and $\ell = 3^{-n}$, so that its fractal dimension is given by

$$D = \frac{\log 4}{\log 3} \approx 1.2628 > 1 \tag{9.3.2}$$

and is noninteger and greater than one. The reason for this conclusion is due to the convolutedness of the Koch curve, which becomes more and more convoluted as the resolution becomes finer and finer. When the curve is highly convoluted, it effectively covers a two-dimensional area, that is, the one-dimensional curve fills up a space of dimension two. In general, a fractal surface has a dimension greater than two, and its dimension could become as large as three for a very highly convoluted surface, so that it can essentially cover a three-dimensional volume. This leads to a general result that the fractal Hausdorff dimension of a set is a measure of its space-filling ability.

Other famous examples include computer simulation of a diffusion-limited aggregation process, electrical discharges on insulators, which obey laws similar to diffusion-limited aggregation, and the resulting spark patterns. Computer simulations of such scale-invariant processes in three dimensions give a Hausdorff dimension between two and three. One of the most remarkable three-dimensional highly branching lightnings has the Hausdorff dimension $D \approx 2.4$ or greater.

Mandelbrot (1982) also gave a more formal definition of a fractal as a set with Hausdorff dimension strictly greater than its topological dimension. This is similar to the Euclidean dimension of ordinary objects, but the fractal dimension D is noninteger and represents the basic measure of the space-filling ability of a fractal set. The topological dimension $E=1$ for lines; for planes and surfaces, $E=2$; and for spheres and other finite volumes, $E=3$. In general, $D>E$. Mandelbrot also conjectured that fractals and the fractal Hausdorff dimension could be used effectively to model many phenomena in the real world.

While he was studying fractal geometry, Mandelbrot (1974) first recognized that the Kolmogorov statistical equilibrium theory of isotropic and homogeneous turbulence is essentially based on some basic assumptions, which include the hierarchy of self-similar eddies (or scales) of different orders and the energy cascade from larger and smaller eddies. This observation led him to believe that the structure of turbulence may be either locally or globally self-similar fractals. The problem of intermittency has also stimulated tremendous interest in the study of kinematics of turbulence using fractals and fractal dimensions (see Mandelbrot, 1974, 1975). It is believed that the slow decay described by the Kolmogorov $k^{-5/3}$ law indicates a physical situation in which vortex sheets are infinitely convoluted. Mandelbrot recognized that these surfaces are so convoluted in the limit as $\nu \to 0$ as to occupy a space of fractal Hausdorff dimension between two and three. Then, he first proposed fractal analysis of turbulent flows and predicted that multiplicative cascade models, when continued indefinitely, lead to the dissipation of energy, which is confined to a set of noninteger Hausdorff dimension. His fractal approach to turbulence received much attention after the introduction of a simple β-model by Frisch et al. (1978). They studied the β-model with special emphasis on its dynamical and fractal aspects. In addition, they explained both the geometrical and the physical significance of the fractal model of turbulent flows.

Experimental results of Anselmet et al. (1984) neither supported the β-model of Frisch nor the log-normal model of Kolmogorov. This meant that there was no uniform fractal model that could fully describe the complex structure of turbulent flows. Then, Frisch and Parisi (1985) have shown that intermittent distributions can be explained in terms of singularities of varying strength; all are located on interwoven sets of different fractal dimensions, and hence, Frisch and Parisi introduced the name *multifractals*. At the same time, Halsey et al. (1986) introduced $f(\alpha)$ for the fractal dimensions of sets of singularities characterizing multifractals. In their multifractal model of turbulence, they used the scale-invariance property, which is one of the remarkable symmetries of the Euler equations. In the meantime, the fractal facets of turbulence received considerable attention from Sreenivasan and Meneveau (1986) and Vassilicos (1992, 1993). Their analysis revealed some complicated geometric features of turbulent flows. They showed that several features of turbulence could be described approximately by fractals and that their fractal dimensions could be measured. Unfortunately, these studies can hardly prove that turbulence can be described fully by fractals. Indeed, these models now constitute a problem in themselves in the sense that properties of turbulent flows can be used to find the value of fractal dimension D. Thus, fractal models of turbulence have not yet been fully successful.

Due to several difficulties with fractal models of turbulence, multifractal models with a continuous spectrum of fractal dimension $D(h)$ have been developed by several authors, including Meneveau and Sreenivasan (1987a,b) (p-model) and Benzi et al. (1984) (random β-model). These models produced scale exponents which are in agreement with experimental results with a single free parameter. However, it is important to point out that both the multifractal model and log-normal models lack true dynamical motivation. Recently, Frisch and Vergassola (1991) developed another multifractal model which enables them to predict a new form of universality for the energy spectrum $E(k)$ in the dissipation range. This model involves a universal function $D(h)$, called *fractal dimension*, which cannot be given by phenomenological theory. This new form of universal law has received good experimental support from Gagne and Castaing (1991), but it is not consistent with Kolmogorov's similarity hypothesis. They have analyzed a wide range of turbulence data with Reynolds numbers from 10^3 to 10^7.

Finally, we close this section by adding some comments on the possible development of singularities in turbulence. Mandelbrot (1975) has remarked that "the turbulent solutions of the basic equations involve singularities or 'near-singularities' (approximate singularities valid down to local viscous length scales where the flow is regular) of an entirely new kind." He also stated that "the singularities of the solutions of the Navier-Stokes equations can only be fractals." In his authoritative review, Sreenivasan (1991) described the major influence of the fractal and multifractal formalisms in understanding certain aspects of turbulence, but he pointed out some inherent problems in these formalisms with the following comment, "However, the outlook for certain other aspects is not so optimistic, unless magical inspiration or breakthrough in analytical tools occur."

During the last decade, some progress has been made in the understanding of the implications of self-similar energy spectra of turbulence. It was shown by Thomson and Tait (1879) in their study of oscillations that when the Fourier power spectrum of a function $f(x)$ has a self-similar form

$$E(k) \sim k^{-2p}, \tag{9.3.3}$$

where p is an integer, then there exists a discontinuity in the $(p-1)$ order derivative of $f(x)$. For example, the energy spectrum of a single shock $f(x) = \operatorname{sgn} x$ is $E(k) \sim k^{-2}$ as $k \to \infty$. However, the energy spectrum such as $E(k) \sim k^{-2p}$, where p is not an integer, implies the existence of singularities that are more severe than mere discontinuities in the flow field. The singularity could be localized at one or a few points of the function such as $f(x) = \dfrac{\sin x}{x}$ (accumulating function) or could be global in the sense that $f(x)$ is singular at all or almost all x, as in the case of the Weierstrass function (see Falconer, 1990). These two very different types of functions may have identical self-similar energy spectra of the form (9.2.3) but always have *different phase spectra*. They also have a fractal (or K-fractal according to Vassilicos and Hunt's (1991) ideas) property in common; both are characterized by nontrivial Kolmogorov dimensions $D_K > 1, D_K' > 0$, where D_K is the Kolmogorov dimension of the entire function and D_K' is the Kolmogorov dimension of the intersection of a function with the x-axis, that is, the zero crossings. However, when the above two functions have the same energy spectrum similarity exponent p, they do not have the same values of D_K and D_K'. Moreover, their structure is also different in the Hausdorff sense, and the Hausdorff dimensions D_H and D_H' for the accumulating function are trivial in the sense that $D_H = 1$ and $D_H' = 0$, whereas those of the Weierstrass function are nontrivial, $D_H > 1$ and $D_H' > 0$. It has been conjectured by Mandelbrot (1982) that $D_H = D_K$ for H-fractals. Some of the major quantities involved in the statistical approach to turbulent flows are correlations and spectra. Self-similar cascades are usually associated with the power spectrum of the form

$$\Gamma(p) \sim k^{-p}. \tag{9.3.4}$$

For example, $p = 5/3$ corresponds to the Kolmogorov spectrum for small-scale turbulence, $p = 1$ characterizes the convective-inertial subrange, and $p = 5/3$ also corresponds to the Batchelor spectrum of a passive scalar in the inertial subrange. The question is whether the self-similarity leading to such spectra is *local* or *global*. Both local spectra are of the form (9.3.4) at large wavenumbers k, where p may not take integral values and p is related to the Kolmogorov dimension D_K of the interface, so that this relation can be used to derive the value of D_K in turbulence, which is in agreement with experimental findings. For a locally self-similar interface, the exponent $p = 2 - D_K'$, where D_K' is the Kolmogorov dimension of the interface with a linear cut, whereas for a globally self-similar interface, $p = 2 + E - D_H$, where E is the topological dimension and D_H is the Hausdorff dimension of the interface. Finally, it has been indicated by Vassilicos (1993) that the value of D_K may be a more accurate measure of spectra of locally self-similar interfaces than the direct measurement of the spectrum itself. Also, the value of D_K may be a more accurate criterion of high Reynolds number turbulence than the existence of self-similar spectra of the form (9.3.4). In the case of the Kolmogorov spectrum, $E(k) \sim k^{-5/3} (p = 5/6)$, which implies that the small-scale turbulence at a very high Reynolds number contains near-singularities that are either simple or nonisolated. Recent experimental findings and direct numerical simulations of turbulence have shown that the small scales of turbulent flows contain long and slender vortex tubes. Some of the vortex tubes may carry near-singularities, provided these vortex tubes are Lundgren vortices, which are asymptotic solutions of the Navier-Stokes equations in the limit as time $t \to \infty$. However, it has not yet been confirmed whether the picture of the small scales of turbulence where vortex tubes dominate the finest scales survives in the limit as $R \to \infty$.

Indeed, several theoretical works and experimental observations revealed that turbulence possesses some singularities in the velocity field or vorticity field. Sarkar's (1985) analytical treatment confirmed that finite-time cusp singularities always exist for essentially any arbitrary set of initial data and are shown to be generic. Newer experimental methods (Hunt and Vassilicos, 1991) also provide evidence of spiraling streamlines and streaklines within eddies, and thin layers of large vorticity grouped together (Schwarz, 1990); both of these features are associated with accumulation points in the velocity field. It also follows from solutions of the Navier-Stokes equations (Vincent and Meneguzzi, 1991) and She et al., 1991) that very large deviations exist in isolated eddies with complicated internal structure. These studies identify regions of intense vorticity so that streamlines form spirals. The Kolmogorov inertial energy spectrum $k^{-5/3}$ also implies that there must be singularities in the derivatives of the velocity field on scales where the rate of energy dissipation is locally very large. It has been suggested by Moffatt (1984) that the accumulation points of discontinuities associated with spiral structures could give rise to fractional power laws k^{-2p} with $1<2p<2$. The question also arises whether the self-similarity leading to the Kolmogorov spectrum is local or global. Moffatt's analysis (see Vassilicos, 1992) revealed that spiral singularities are responsible for noninteger power of self-similar spectra k^{-2p}. It is also now known that locally self-similar structures have a self-similar high wavenumber spectrum with a noninteger power $2p$. Thus, the general conclusion is that functions with the Kolmogorov spectrum have some kinds of singularities and accumulation points, unless they are fractal functions with singularities everywhere, since they are everywhere continuous but nowhere differentiable. Thus, the upshot of this discussion is that the statistical structure of the small-scale turbulent flows is determined by local regions where the velocity and any other associated scalar functions have very large derivatives or have rapid variations in their magnitude or that of their derivatives. These are regions surrounding points that are singular. It remains an open question whether the nature of this singularity is due to random fluctuations of the turbulent motions resulting from their chaotic dynamics or to the presence of localized singular structures originating from an internal organization of the turbulent flows.

9.4 Farge's Wavelet Transform Analysis of Turbulence

It has already been indicated that the dynamics of turbulent flows depends not only on different length scales but on different positions and directions. Consequently, physical quantities such as energy, vorticity, enstrophy, and pressure become highly intermittent. The Fourier transform cannot give the local description of turbulent flows, but the wavelet transform analysis has the ability to provide a wide variety of local information of the physical quantities associated with turbulence. Therefore, the wavelet transform is adopted to define the space-scale energy density by

$$\tilde{E}(\ell,x)=\frac{1}{\ell}\left|\tilde{f}(\ell,x)\right|^2, \tag{9.4.1}$$

where $\tilde{f}(\ell,x)$ is the wavelet transform of a given function (or signal) $f(x)$.

It is helpful to introduce a *local energy spectrum* $\tilde{E}(\ell,x_0)$ in the neighborhood of x_0 (see Farge, 1992) by

$$\tilde{E}_x(\ell,x_0)=\frac{1}{\ell}\int_{-\infty}^{\infty}\tilde{E}(\ell,x)\,\chi\left(\frac{x-x_0}{\ell}\right)dx, \tag{9.4.2}$$

where the function χ is considered as a filter around x_0. In particular, if χ is a Dirac delta function, then the local wavelet energy spectrum becomes

$$\tilde{E}(\ell,x_0)=\frac{1}{\ell}\left|\tilde{f}(\ell,x_0)\right|^2. \tag{9.4.3}$$

The local energy density can be defined by

$$\tilde{E}(x)=C_\psi^{-1}\int_0^{\infty}\tilde{E}(\ell,x)\,\frac{d\ell}{\ell}. \tag{9.4.4}$$

On the other hand, the *global wavelet spectrum* is given by

$$\tilde{E}(\ell)=\int_{-\infty}^{\infty}\tilde{E}(\ell,x)\,dx. \tag{9.4.5}$$

This can be expressed in terms of the Fourier energy spectrum $\hat{E}(k)=\left|\hat{f}(k)\right|^2$ so that

$$\tilde{E}(\ell)=\int_{-\infty}^{\infty}\tilde{E}(k)\left|\hat{\psi}(\ell k)\right|^2 dk, \tag{9.4.6}$$

where $\hat{\psi}(\ell k)$ is the Fourier transform of the analyzing wavelet ψ. Thus, the global wavelet energy spectrum corresponds to the Fourier energy spectrum smoothed by the wavelet spectrum at each scale.

Another significant feature of turbulence is the so-called *intermittency phenomenon*. Farge et al. (1992) used the wavelet transform to define the *local intermittency* as the ratio of the local energy density and the space averaged energy density in the form

$$I(\ell,x_0)=\frac{\left|\tilde{f}(\ell,x_0)\right|^2}{\left\langle\left\langle\left|\tilde{f}(\ell,x)\right|^2\right\rangle\right\rangle}, \tag{9.4.7}$$

where

$$\left\langle\left\langle\left|\tilde{f}(\ell,x)\right|^2\right\rangle\right\rangle = \int_{-\infty}^{\infty} \left|\tilde{f}(\ell,x)\right|^2 dx. \tag{9.4.8}$$

If $I(\ell,x_0)=1$ for all ℓ and x_0, then there is no intermittency, that is, the flow has the same energy spectrum everywhere, which then corresponds to the Fourier energy spectrum. According to Farge et al. (1990), if $I(\ell,x_0)=10$, the point at x_0 contributes ten times more than average to the Fourier energy spectrum at scale ℓ. This shows a striking contrast with the Fourier transform analysis, which can describe a signal in terms of wavenumbers only but cannot give any local information. Several authors, including Farge and Rabreau (1988), Farge (1992), and Meneveau (1991) have employed wavelets to study homogeneous turbulent flows in different configurations. They showed that during the flow evolution, beginning from a random vorticity distribution with a k^{-3} energy spectrum, the small scales of the vorticity become increasingly localized in physical space. Their analysis also revealed that the energy in the two-dimensional turbulence is highly intermittent which may be due to a condensation of the vorticity field into vortex like coherent structures. They have also found that the smallest scales of the vorticity are confined within vortex cores. According to Farge and Holschneider (1989, 1990), there exist quasisingular coherent structures in two-dimensional turbulent flows. These kinds of structures are produced by the condensation of vorticity around the quasisingularities already present in the initial data. Using the wavelet transform analysis, Meneveau (1991) first measured the local energy spectra and then carried out direct numerical simulations of turbulent shear flows. His study reveals that the mean spatial values of the turbulent shear flow agree with their corresponding results in Fourier space, but their spatial variations at each scale are found to be very large, showing non-Gaussian statistics. Moreover, the local energy flux associated with very small scales exhibits large spatial intermittency. Meneveau's computational analysis of the spatial fluctuations of $T(k,t)$ shows that the average value of $T(k,t)$ is positive for all small scales and negative for large scales, indicating the transfer of energy from large scales to small scales so that energy is eventually dissipated by viscosity. This finding agrees with the classical cascade model of three-dimensional turbulence. However, there is a striking new phenomenon that the energy cascade is reversed in the sense that energy transfer takes place from small to large scales in many places in the flow field. Perrier et al. (1995) confirmed that the mean wavelet spectrum $\tilde{E}(k)$ is given by

$$\tilde{E}(k)=\int_0^\infty \tilde{E}(x,k)\,dx. \tag{9.4.9}$$

This result gives the correct Fourier exponent for a power-law of the Fourier energy spectrum $E(k)=Ck^{-p}$, provided the associated wavelet has at least $n>2^{-1}(p-1)$ vanishing moments. This condition is in agreement with that for determining cusp singularities. Based on a recent wavelet analysis of a numerically calculated two-dimensional homogeneous turbulent flow, Benzi and Vergassola (1991) confirmed the existence of coherent structures with negative exponents. Thus, their study reveals that the wavelet transform analysis has the ability not only to give a more precise local description but also detect and characterize singularities of turbulent flows. On the other hand, Argoul et al. (1988, 1990) and Everson et al. (1990) have done considerable research on turbulent flows using wavelet analysis. They showed that the wavelet analysis has the ability to reveal Cantor-like fractal structure of the Richardson cascade of turbulent eddies.

9.5 Adaptive Wavelet Method for Analysis of Turbulent Flows

Several authors, including Farge (1992) and Schneider and Farge (1997), first introduced the adaptive wavelet method for the study of fully developed turbulence in an incompressible viscous flow at a very high Reynolds number. In a fully developed turbulence, the nonlinear convective term in the Navier-Stokes equations becomes very large by several orders of magnitude than the linear viscous term. The Reynolds number $R=(U\ell/\nu)$ represents the ratio of the nonlinear convective term and the viscous term. In other words, R is proportional to the ratio of the large excited scales and the small scales where the linear viscous term is responsible for dissipating any instabilities.

Unpredictability is a key feature of turbulent flows, that is, each flow realization is different even though statistics are reproducible as long as the flow configuration and the associated parameters remain the same. Many observations show that in each flow realization localized coherent vortices whose motions are chaotic are generated by their mutual interactions. The statistical analysis of isotropic and homogeneous turbulence is based on L^2-norm ensemble averages and hence is hardly sensitive to the presence of coherent vortices which have a very weak contribution to the L^2-norm. However, coherent vortices, are fundamental components of turbulent flows and therefore, must be taken into account in both statistical and numerical models.

Leonard (1974) developed a classical model, called the Large Eddy Simulation (LES), to compute fully developed turbulent flows. In this model, separation is introduced by means of linear filtering between large-scale active modes and small-scale passive modes. This means that the flow evolution is calculated deterministically up to cutoff scale while the influence of the subgrid scales onto the resolved scales is statistically modeled. Consequently, vortices in strong nonlinear interaction tend to smooth out, and any instabilities at subgrid scales are neglected. Thus, LES models have problems of backscatter, that is, transfer of energy from subgrid scales to resolved scales due to nonlinear instability. The LES model takes into account backscatter, but only in a locally averaged manner. Further progress in the hierarchy of turbulent models is made by using Reynolds Averaged Navier-Stokes (RANS) equations, where the time averaged mean flow is calculated and fluctuations are modeled, in this case, only steady state solutions are predicted.

During the last decade, wavelet analysis has been introduced to model, analyze, and compute fully developed turbulent flows. According to Schneider and Farge (2000), wavelet analysis has the ability to disentangle coherent vortices from incoherent background flow in turbulent flows. These components are inherently multiscale in nature and have different statistics with different correlations. Indeed, the coherent vortices lead to the non-Gaussian distribution and long-range correlations, whereas the incoherent background flow is inherently characterized by the Gaussian statistics and short-range correlations. This information suggests a new way of splitting the turbulent flow into active coherent vortex modes and passive incoherent modes. The former modes are computed by using wavelet analysis, whereas the latter modes are statistically modeled as a Gaussian random process. This new and modern approach is called the *Coherent Vortex Simulation* (CVS) and was developed by Farge et al. (1999a,b). This approach is significantly different from the classical LES which is essentially based on a linear filtering process between large and small scales without any distinction between Gaussian and non-Gaussian processes. The CVS takes advantage of a nonlinear filtering process defined in a wavelet space between Gaussian and non-Gaussian modes with different scaling laws but without any scale separation. The major advantage of the CVS treatment compared to the LES is to reduce the number of computed active modes for a given Reynolds number and control the Gaussian distribution of the passive degrees of freedom to be statistically modeled.

Turbulent flows are characterized by a fundamental quantity, called the *vorticity vector*, $\boldsymbol{\omega} = \nabla \times \boldsymbol{u}$. Physically, the vorticity field is a measure of the local rotation rate of the flow, its angular velocity.

Eliminating the pressure term from (9.2.16) by taking the curl of (9.2.16) leads to the equation for the vorticity field in the form

$$\frac{\partial \boldsymbol{\omega}}{\partial t} = (\boldsymbol{\omega} \cdot \nabla)\, \boldsymbol{u} - (\boldsymbol{u} \cdot \nabla)\, \boldsymbol{\omega} + \nu \nabla^2 \boldsymbol{\omega} + \nabla \times \boldsymbol{F}. \tag{9.5.1}$$

This is well-known as the *convection-diffusion equation* of the vorticity. The left-hand side of this equation represents the rate of change of vorticity, whereas the first two terms on the right-hand side describe the rate of change of vorticity due to stretching and twisting of vortex lines. In fact, the term $(\boldsymbol{\omega} \cdot \nabla)\, \boldsymbol{u}$ is responsible for the vortex-stretching mechanism (vortex tubes are stretched by velocity gradients) which leads to the production of vorticity. The third term on the right-hand side of (9.5.1) represents the diffusion of vorticity by molecular viscosity. In the case of two-dimensional flow, $(\boldsymbol{\omega} \cdot \nabla)\, \boldsymbol{u} = 0$, so the vorticity equation (9.5.1) without any external force can be given by

$$\frac{\partial \boldsymbol{\omega}}{\partial t} + (\boldsymbol{u} \cdot \nabla)\, \boldsymbol{\omega} = \nu \nabla^2 \boldsymbol{\omega} \tag{9.5.2}$$

so that only convection and conduction occur. This equation combined with the equation of continuity,

$$\nabla \cdot \boldsymbol{u} = 0, \tag{9.5.3}$$

constitutes a closed system which is studied by periodic boundary conditions.

In terms of a stream function ψ, the continuity equation (9.5.3) gives

$$u = \frac{\partial \psi}{\partial y} \quad \text{and} \quad v = -\frac{\partial \psi}{\partial x}, \tag{9.5.4a,b}$$

so that the vorticity $\omega = (v_x - u_y)$ satisfies the Poisson equation for the stream function ψ as

$$\nabla^2 \psi = \omega. \tag{9.5.5}$$

The total kinetic energy is defined by

$$E(t) = \frac{1}{2} \iint_D u^2 (x,t)\, dx, \tag{9.5.6}$$

and the total enstrophy is defined by

$$Z(t) = \frac{1}{2} \iint_D \omega^2 (x,t)\, dx. \tag{9.5.7}$$

We make reference to Frisch (1995) to express the enstrophy and the dissipation of energy as

$$\frac{dZ}{dt} = -2\nu P, \quad \frac{dE}{dt} = -2\nu Z, \tag{9.5.8a,b}$$

where the *palinstrophy* P is given by

$$P(t) = \frac{1}{2} \iint_D |\nabla \omega|^2\, dx. \tag{9.5.9}$$

The energy and enstrophy spectra are written in terms of the Fourier transform

$$E(\kappa) = \frac{1}{2} \sum_{\kappa - \frac{1}{2} \le |\boldsymbol{\kappa}| \le \kappa + \frac{1}{2}} |\hat{u}(\boldsymbol{\kappa})|^2, \tag{9.5.10}$$

$$Z(\kappa) = \frac{1}{2} \sum_{\kappa - \frac{1}{2} \le |\boldsymbol{\kappa}| \le \kappa} |\hat{\omega}(\boldsymbol{\kappa})|^2, \tag{9.5.11}$$

where $\boldsymbol{\kappa} = (k, \ell)$. The quantities $E(\kappa)$ and $Z(\kappa)$ measure the amount of energy or enstrophy in the band of wavenumbers between κ and $\kappa + d\kappa$. The spectral distribution of energy and enstrophy are related to the expression $\kappa^2 E(\kappa) = Z(\kappa)$.

During the last two decades, several versions of the Direct Numerical Simulation (DNS) have been suggested to describe the dynamics of turbulent flows. Using DNS, the evolution of all scales of turbulence can only be computed for moderate Reynolds numbers with the help of supercomputers. Due to severe limitations of DNS, Fröhlich and Schneider (1997) have recently developed a new method, called the *adaptive wavelet method*, for simulation of two- and three-dimensional turbulent flows at a very high Reynolds number. This new approach seems to be useful for simulating turbulence because the inherent structures involved in turbulence are localized coherent vortices evolving in multiscale nonlinear dynamics. Fröhlich and Schneider used wavelet basis functions that are localized in both physical and spectral spaces, and hence the approach is a reasonable compromise between grid-point methods and spectral methods. Thus, the space and space-adaptivity of the wavelet basis seem to be effective. The fact that the basis is adapted to the solution and follows the time evolution of coherent vortices corresponds to a combination of both Eulerian and Lagrangian methods. Subsequently, Schneider and Farge (2000) discussed several applications of the adaptive wavelet method to typical turbulent flows with computational results for temporally growing mixing layers, homogeneous turbulent flows, and for decaying and wavelet forced turbulence. They used the adaptive wavelet method for computing and analyzing two-dimensional turbulent flows. At the same time, they discussed some perspectives for computing and analyzing three-dimensional turbulent flows with new results. They also have shown that the adaptive wavelet approach provides highly accurate results at high Reynolds numbers with many fewer active modes than the classical pseudospectral method, which puts a limit on the Reynolds numbers because it does not utilize the vortical structure of high Reynolds number flows. The reader is referred to all papers cited above for more detailed information on the adaptive wavelet method for computing and analyzing turbulent flows.

9.6 Meneveau's Wavelet Analysis of Turbulence

In this section, we closely follow Meneveau's (1991) analysis of turbulence in the orthonormal wavelet representation based on the wavelet transformed Navier-Stokes equations. We first introduce the three-dimensional wavelet transform of a function $f(x)$ defined by

$$w(r,x) = W_{(r,x)}^{[f]} = \frac{r^{-3/2}}{\sqrt{C_\psi}} \int_{-\infty}^{\infty} \psi\left(\frac{\xi - x}{r}\right) f(\xi)\, d^3\xi, \tag{9.6.1}$$

where $\psi(x) = \psi(|x|)$ is the isotropic wavelet satisfying the admissibility condition

$$C_\psi = \int_{-\infty}^{\infty} |k|^{-1} |\hat{\psi}(k)|^2\, d^3k. \tag{9.6.2}$$

The inversion formula is given by

$$f(x)=\frac{1}{\sqrt{C_\psi}}\int_{-\infty}^{\infty}dr\int_{-\infty}^{\infty}r^{-3/2}\,\psi\left(\frac{x-\xi}{r}\right)\,w(r,\xi)\,\frac{d^3\xi}{r^4}. \tag{9.6.3}$$

The invariance of energy of the system can be stated as

$$\int_{-\infty}^{\infty}\{f(x)\}^2\,d^3x=C_\psi^{-1}\int_0^{\infty}dr\int_{-\infty}^{\infty}\{w(r,x)\}^2\,\frac{d^3x}{r^4}. \tag{9.6.4}$$

As in the one-dimensional case, the wavelet transform $w(r,x)$ can also be obtained from the Fourier transform $\hat{f}(k)$ of $f(x)$ so that

$$w(r,x)=\frac{1}{(2\pi)^3}\,\frac{1}{\sqrt{C_\psi}}\,r^{3/2}\int_{-\infty}^{\infty}\hat{\psi}^*(rk)\,\hat{f}(k)\,e^{i(k\cdot x)}\,d^3k. \tag{9.6.5}$$

This can also be inverted to obtain the inversion formula

$$\hat{f}(k)=\frac{1}{\sqrt{C_\psi}}\int_0^{\infty}dr\int_{-\infty}^{\infty}r^{3/2}\,\hat{\xi}(rk)\,\exp(-ik\cdot x)\,w(r,x)\,\frac{d^3x}{r^4}. \tag{9.6.6}$$

In view of the translational property, the wavelet transform commutes with differentiation in the space variables so that

$$\nabla\cdot W_{(r,x)}[f]=W_{(r,x)}[\nabla\cdot f] \tag{9.6.7}$$

and

$$\nabla W_{(r,x)}[f]=W_{(r,x)}[\nabla f]. \tag{9.6.8}$$

We now define $w_i(r,x)$ as the wavelet transform of the fluctuating part of the divergence-free velocity field $u_i(x)$. In vector notation, these quantities are denoted by $\boldsymbol{w}(r,x)$ and $\boldsymbol{u}(x)$, which depend on time t, but for notational clarity, we simply omit the time dependence. It follows from (9.6.7) that $\boldsymbol{w}(r,x)$ is divergence-free.

We apply (9.6.5) to the Fourier-transformed Navier-Stokes equations (9.2.24), where the velocities on the right-hand side have been replaced by the inverse transform of $w_i(r,x)$, so that the evolution equation for $w_i(r,x)$ is

$$\left(\frac{\partial}{\partial t}-\nu\nabla^2\right)w_i(r,x)=\frac{1}{(2\pi\,C_\psi)^{3/2}}\int_{r'}dr'\int_{r''}dr''\int\int_{x'\,x''}$$

$$w_j(r',x')\,w_k(r'',x'')\,I_{ijk}(r,x;r',r'',x',x'')\,\frac{d^3x'\,d^3x''}{r'^4\,r''^4}, \tag{9.6.9}$$

where

$$I_{ijk}(r,x;r',r'',x',x'') = (-i)(r\,r'r'')^{3/2} \int_k \int_q k_k\, P_{i_j}(k)\,\hat{\psi}^*(r\,k)\,\hat{\psi}(r'q)$$
$$\times \hat{\psi}\left\{r''(k-q)\right\} \exp\left[i\left\{k\cdot(x-x'')+q\cdot(x''-x')\right\}\right] d^3k\, d^3q. \tag{9.6.10}$$

We multiply (9.6.9) by $w_i(r,x)$ and then add over the components I to obtain the local energy equation

$$\frac{\partial}{\partial t}\, e(r,x) = t(r,x) - \varepsilon(r,x) + \nu \frac{\partial}{\partial x_j}\left[w_i\left(\frac{\partial w_i}{\partial x_j} + \frac{\partial w_j}{\partial x_i}\right)\right], \tag{9.6.11}$$

where

$$e(r,x) = \frac{1}{2}\sum_{i=1}^{3}\left[w_i(r,x)\right]^2 \tag{9.6.12}$$

represents the local density of kinetic energy at scale r and

$$t(r,x) = \int dr' \int dr'' \iint w_i(r,x)\; w_j(r',x')\; w_k(r,x'')$$
$$\times I_{ijk}(r,x;r',r'',x',x'')\, \frac{d^3x'\, d^3x''}{r'^4\, r''^4}, \tag{9.6.13}$$

is the local transfer of kinetic energy at scale r at position x. This term shows interactions among triads of scales (r,r',r'') as well as interactions among triads of positions (x,x',x''). The term $\varepsilon(r,x)$ describes the dissipation of energy at scale size r and is given by

$$\varepsilon(r,x) = \nu\, \frac{\partial w_i}{\partial x_j}\left[\frac{\partial w_i}{\partial x_j} + \frac{\partial w_j}{\partial x_i}\right]. \tag{9.6.14}$$

In view of the Parseval formula (6.3.9), the local transfer conserves energy so that

$$\int_r dr \int_x t(r,x)\, \frac{d^3x}{r^4} = 0. \tag{9.6.15}$$

The total flux of kinetic energy through scale r at position x is defined by integrating the rate of change in the local energy due to nonlinear interactions over all scales larger than r so that

$$\pi(r,x) = -\int_r^{\infty} t(r',x)\, \frac{dr'}{r'^4}, \tag{9.6.16}$$

where the negative sign shows a decrease in energy of the large scales associated with a positive flux. This total flux term is somewhat similar to (9.2.27) in the wavelet representation.

All the preceding results are not very useful in turbulence theory, but they illustrate the fact that there are complicated interactions of the wavelet transform $w_i(r,\boldsymbol{x})$ involved at different scales and different positions. These nonlocal and interscale interactions are essentially described by the complicated quantity $I_{ijk}(r,\boldsymbol{x};r',r'',\boldsymbol{x}',\boldsymbol{x}'')$. This quantity arises from the fact that, in general, the triads are not closed as they are in the Fourier representation, that is, there is no detailed energy conservation in the wavelet representation. However, it is almost impossible to make further progress on this wavelet formulation without making appropriate assumptions and approximations of I_{ijk}.

On the other hand, if the velocity field is known, quantities including $e(r,\boldsymbol{x})$, $t(r,\boldsymbol{x})$, and $\pi(r,\boldsymbol{x})$ can be computed by taking the wavelet transform of the Navier-Stokes equation combined with expressing the nonlinear terms in terms of the original velocity field. Meneveau (1991) described the discrete formulation of the evolution equation for the local kinetic energy density $e^{(m)}[\boldsymbol{i}]$ at scale r_m and position $y = 2^m(h_1 i_1, h_2 i_2, h_3 i_3)$ where $\boldsymbol{i} = (i_1, i_2, i_3)$ denotes the position of a rectangular grid with uniform mesh sizes h_1, h_2, h_3. He obtained the evolution equation for the local kinetic energy density $e^{(m)}[\boldsymbol{i}]$ given by

$$\frac{\partial}{\partial t} e^{(m)}[\boldsymbol{i}] = t^{(m)}[\boldsymbol{i}] - \upsilon^{(m)}[\boldsymbol{i}], \tag{9.6.17}$$

where $t^{(m)}[\boldsymbol{i}]$ is the nonlinear term representing the local transfer of kinetic energy at scales r_m at position $\boldsymbol{i}$, and $\upsilon^{(m)}[\boldsymbol{i}]$ is the viscous term representing dissipation and viscous transport of kinetic energy. The equation (9.6.17) is somewhat similar to (9.2.27), but it depends on the position as well.

We write the expressions for $t^{(m)}[\boldsymbol{i}]$ as

$$t^{(m)}[\boldsymbol{i}] = -\sum_{i=1}^{3}\sum_{q=1}^{7} w_i^{(m,q)}[\boldsymbol{i}] \left\{ u_j \frac{\partial u_i}{\partial x_j} + \frac{1}{\rho}\frac{\partial p}{\partial x_i} \right\}^{(m,q)} [\boldsymbol{i}], \tag{9.6.18}$$

where the pressure involved in this equation is obtained by solving the Poisson equation, and $w_i^{(m,q)}[\boldsymbol{i}]$ is the wavelet coefficient of the ith component of the velocity field.

The term $t^{(m)}[\boldsymbol{i}]$ does conserve energy on the whole so that

$$\sum_{m=1}^{M} \sum_{i_1, i_2, i_3} t^m[i] = 0, \tag{9.6.19}$$

which follows from the zero value of the volume integral of $(u \cdot \nabla u + \nabla p)$ for homogeneous turbulence, and from the condition of orthonormality of the wavelets.

The viscous term is given by

$$\upsilon^{(m)}[i] = \nu \sum_{i=1}^{3} \sum_{q=1}^{7} w_i^{(m,q)}[i] \left\{-\nabla^2 u_i\right\}^{(m,q)}[i]. \tag{9.6.20}$$

Finally, the flux of the kinetic energy term $\pi^{(m)}[i]$ in a spatial region of size r_m and position $[i]$ can be calculated by summing the density transfer of all larger scales at that position so that

$$\pi^{(m)}[i] = -\sum_{n=m}^{M} 2^{3(M-n)} t^{(n)}\left[2^{m-n} i\right]. \tag{9.6.21}$$

We next characterize the local kinetic energy at every scale in turbulence. In Fourier transform analysis, the quantity $\hat{E}(k)$ represents the power-spectral density in a band dk of wavenumbers. However, the spatial information is completely lost due to the nonlocal nature of the Fourier modes. If $u(x)$ is a one-dimensional finite energy function with mean zero and $\hat{u}(k)$ is its Fourier transform, the total energy is given by

$$\int_{-\infty}^{\infty} u^2(x)\, dx = \frac{1}{2\pi} \int_{-\infty}^{\infty} \hat{u}(k)\, \hat{u}^*(k)\, dk = \int_{0}^{\infty} E(k)\, dk, \tag{9.6.22}$$

where $E(k)$ represents the energy spectrum, and the wavenumber k is related to the distance r so that $r = 2\pi k$.

In wavelet analysis, the total energy can be written in terms of the wavelet energies in the form

$$\int_{-\infty}^{\infty} u^2(x)\, dx = \int_{0}^{\infty} E_w(k)\, dk, \tag{9.6.23}$$

where $E_w(r, x)$ is the continuous wavelet transform of $u(x)$ and

$$E_w(k) = \frac{1}{2\pi} \frac{1}{C_\psi} \int_{-\infty}^{\infty} w^2(r(k), x)\, dx. \tag{9.6.24}$$

This represents the energy density at wavenumbers k. This spectrum function is similar to the Fourier spectrum $E(k)$ but is not the same at each k because of the finite bandwidth involved in the wavelet transform.

For more detailed information on energy transfer and flux in the wavelet representation and the intermittent nature of the energy, the reader is referred to Meneveau (1991).

Answers and Hints for Selected Exercises

2.17 Exercises

1. (b) Use (2.6.8) and (2.6.9).

5. If f vanishes in some bounded interval and $f' = 0$, then $f = 0$.

6. Use the parallelogram law.

8. Use Schwarz's inequality to show that $\|x + y\| = \|x\| + \|y\|$ if and only if x and y are linearly independent.

13. Every finite dimensional normed space is complete.

14. No. Use the result that $C([a,b])$ is incomplete with respect to the norm

$$\|f\| = \left(\int_a^b |f|^2 \, dx \right)^{\frac{1}{2}}.$$

15. No. Compare with Exercise 14.

16. Yes. Use continuity of the inner product.

18. For (c) and (d), use an orthonormal sequence.

20. No.

35. $x = \sum_{n=1}^{\infty} (\alpha_n a_n)$.

47. Use an example of the set $S = \{(x, 0) : x \in \mathbb{R}\}$ and the point $x_0 = (0, 1)$.

48. $y = \sum_{n=1}^{\infty} (x, e_n) e_n$.

50. Consider an example $x_n(t) = n \exp\left[-n^6 (t - t_0)^2\right]$.

52. Define g as the composition of the orthogonal projection onto F with f.

53. The sequence $(1,0,0,\cdots), (0,1,0,0,\cdots), (0,0,1,0,0,\cdots), \cdots$ is a complete orthonormal system in the space ℓ^2.

59. (a) $f \in L^1(\mathbb{R})$, (b) $f \notin L^1(\mathbb{R})$,

(c) $f_r \in L^1(\mathbb{R})$ if $r > 1$ and $\|f_r\|_1 = \left(\frac{2r}{r-1} \right)$.

3.14 Exercises

1. (b) Hint: $f(t) = -\frac{1}{a}\frac{d}{dt}\left[\exp\left(-at^2\right)\right]$.

1. (c) Hint: $e^t = u$, $\hat{f}(\omega) = \Gamma(1-\omega)$, where $\Gamma(x)$ is the Gamma function.

1. (f) $\sqrt{\frac{\pi}{a}}\exp\left(-\frac{\omega^2}{4a} - \frac{ib\omega}{2a} + \frac{b^2}{4a}\right)$.

1. (g) $(i\omega)^n$,

1. (h) $2\Gamma(a)\cos\left(\frac{a\pi}{2}\right)|\omega|^{-a}$.

1. (i) Hint: Use (3.2.11) and then Duality Theorem 3.4.10. Draw a figure for $f(t)$ and $\hat{f}(\omega) = \Delta_{2a}(\omega)$.

1. (j) Hint: Use (3.3.5) combined with Example 3.2.3.

$$\hat{f}(\omega) = \frac{\sin(\omega - \omega_o)\tau}{(\omega - \omega_o)} + \frac{\sin(\omega + \omega_o)\tau}{(\omega + \omega_o)}.$$

Draw the graphs of $f(t)$ and $\hat{f}(\omega)$.

1. (k) $(-i)^n\sqrt{2\pi}\,P_n(\omega)\,\chi_1(\omega)$, $P_n(x)$ is the Legendre polynomial of degree n.

1. (l) $2\pi\,\delta(\omega - a)$.

11. Hint: $F'(t) = f(t)$ for almost all $t \in \mathbb{R}$, and then take the Fourier transform.

15. (a) $\gamma(t) = \left(1 - \frac{3}{2}|t| + \frac{1}{2}|t|^3\right)H(1-|t|)$. (b) $\gamma(t) = \exp(-a|t|)$.

17. Hint: $F_\lambda(\omega) = \lambda F(\lambda\omega) = \lambda\hat{\Delta}(\lambda\omega) = \hat{\Delta}_\lambda(\omega)$.

18. (b) Hint: Use the Dirichlet kernel

$$D_k(z) = \sum_{n=-k}^{k} z^k = z^{-k}\sum_{n=0}^{2k} z^n = z^{-k}\left(\frac{z^{2k+1}-1}{z-1}\right).$$ Put $z = e^{it} \neq 1$ so that

$$D_k\left(e^{it}\right) = z^{-k}\,\frac{\exp\left\{i(2k+1)\frac{t}{2}\right\}}{\exp\left(\frac{it}{2}\right)}\cdot\frac{\sin\left\{(2k+1)\frac{t}{2}\right\}}{\sin\left(\frac{t}{2}\right)} = \frac{\sin\left\{(2k+1)\frac{t}{2}\right\}}{\sin\left(\frac{t}{2}\right)}.$$

Use $F_n(z) = \dfrac{1}{n+1} \displaystyle\sum_{k=0}^{n} D_k(z).$

$$2\,F_n(z)\sin^2\frac{t}{2} = \frac{1}{n+1}\sum_{k=0}^{n} 2\sin\frac{t}{2}\sin\,(2k+1)\,\frac{t}{2}$$

$$= \frac{1}{n+1}\sum_{k=0}^{n}\left[\cos kt - \cos(k+1)t\right]$$

$$= \frac{1}{n+1}\left[1-\cos(n+1)t\right] = \left(\frac{2}{n+1}\right)\sin^2\left\{\left(\frac{n+1}{2}\right)t\right\}.$$

19. Hint:

$$\hat{f}(\omega) = \sqrt{\frac{\pi}{\alpha}}\exp\left(-\frac{\omega^2}{4\alpha}\right), \quad \int_{-\infty}^{\infty} f^2(t)dt = \sqrt{\frac{\pi}{2\alpha}}, \quad \int_{-\infty}^{\infty}\hat{f}^2(\omega)d\omega = \pi\sqrt{\frac{2\pi}{\alpha}}.$$

$$\mathscr{F}\{f^2(t)\} = \mathscr{F}\{\exp(-2\alpha t^2)\} = \sqrt{\frac{\pi}{2\alpha}}\exp\left(-\frac{\omega^2}{8\alpha}\right).$$

$$\mathscr{F}\{t^2 f^2(t)\} = (-i)^2\frac{d^2}{d\omega^2}\hat{f}_{(\omega)} = -\frac{d^2}{d\omega^2}\left[\sqrt{\frac{\pi}{2\alpha}}\exp\left(-\frac{\omega^2}{8\alpha}\right)\right].$$

$$\int_{-\infty}^{\infty} t^2\exp\left(-2\alpha t^2\right)dt = \frac{1}{4\alpha^{3/2}}\cdot\sqrt{\frac{\pi}{2}}.$$

$$\int_{-\infty}^{\infty}\omega^2\hat{f}^2(\omega)\,d\omega = \frac{\pi}{\alpha}\int_{-\infty}^{\infty}\omega^2\exp\left(-\frac{\omega^2}{2\alpha}\right)d\omega = \pi\sqrt{2\pi\alpha}\,.$$

$$\sigma_t^2 = \int_{-\infty}^{\infty} t^2 f^2(t)dt \div \int_{-\infty}^{\infty} f^2(t)dt = \sqrt{\frac{\pi}{2}}\cdot\frac{1}{4\alpha^{3/2}} \div \sqrt{\frac{\pi}{2\alpha}} = \frac{1}{4\alpha}.$$

$$\sigma_\omega^2 = \frac{\pi\sqrt{2\pi\alpha}}{\pi\sqrt{\dfrac{2\pi}{\alpha}}} = \alpha\cdot \text{ Hence, } \sigma_t^2\,\sigma_\omega^2 = \frac{1}{4} \Rightarrow \sigma_t\sigma_\omega = \frac{1}{2}.$$

20. $$\phi(t) = \frac{A_0}{2\sqrt{\pi a}}\exp\left\{-\frac{(t-t_0)^2}{4a}\right\}.$$

21. $$\hat{\phi}(\omega) = \left[a + b\cos\left(\frac{n\pi\omega}{\omega_0}\right)\right]\exp(-i\omega t_0)\,\hat{\chi}_{\omega_0}(\omega).$$

22. Hint: For (a) and (b), use results (3.3.9) and (3.3.11)

(c) $$y(x) = \frac{1}{2\pi}\int_{-\infty}^{\infty}\frac{\hat{f}(k)\,e^{ikx}dk}{\left(\sigma^2 - k^2 + 2iak\right)}.$$

(d) $$y(x) = e^{-x}\int_{-a}^{x} e^{\xi} f(\xi)\, d\xi + e^{x}\int_{x}^{a} e^{-\xi} f(\xi)\, d\xi.$$

26. Hint: Construct a differential equation similar to that in Exercise 23.

29. (b) $$\hat{f}(\omega,\sigma) = \left(\frac{4}{\omega\sigma}\right) \sin(a\omega)\, \sin(a\sigma).$$

31. $u(x) = \int_{-\infty}^{\infty} w(\xi)\, G(\xi,x)\, d\xi$, where $w(x) = W(x)/EI$, and

$$G(\xi,x) = \frac{1}{\pi}\int_{0}^{\infty} \frac{\cos k(x-\xi)\, dk}{k^4 + a^4}, \quad a^4 = \kappa/EI.$$

34. $$u(x,t) = \int_{-\infty}^{\infty} f(\xi)\, G(x,t;\, \xi,0)\, d\xi + \int_{0}^{t} d\tau \int_{-\infty}^{\infty} q(\xi,\tau)\, G(x,t;\, \xi,\tau)\, d\xi,$$

where $G(x,t;\xi,\tau) = \dfrac{1}{\sqrt{4\pi\kappa(t-\tau)}} \exp\left[-\dfrac{(x-\xi)^2}{4\kappa(t-\tau)}\right].$

39. $$u(x,t) = (4at+1)^{-\frac{1}{2}} \exp\left(-\frac{at^2}{4at+1}\right).$$

40. $G(x,t) = \dfrac{1}{(2\pi)^3} \iiint \exp\{i(\boldsymbol{\kappa}\cdot\boldsymbol{x})\} \dfrac{\sin\alpha t}{\alpha}\, d\boldsymbol{\kappa}$, where $\alpha = \left(c^2\kappa^2 + d^2\right)^{\frac{1}{2}}$.

42. $$\langle v\rangle = m\, v_m \exp\left(-\frac{1}{4}\, \pi\, v_m^2\right) + v_0,$$

and

$$B^2 = \frac{\alpha}{8\pi^2} + \frac{1}{2}\, m^2\, v_m^2 \left[\exp\left\{-\frac{1}{\alpha}\cdot 2\pi^2\, v_m^2\right\} - 1\right].$$

4.10 Exercises

1. Since $\int_{-\infty}^{\infty} g(\tau - t)\, dt = 1$, the result follows. The result implies that the set of the Gabor transforms of f with the Gaussian window decomposes the Fourier transform of f exactly.

2. (b) Hint. $\tilde{f}_g(t,\omega) = \left(f,\, \bar{g}_{t,\omega}\right)$ and use the Parseval identity of the Fourier transform.

3. Derive $\sigma_t^2 = \sqrt{a}$.

Use $\int_{-\infty}^{\infty} \exp\left(-a\,x^2\right)\,dx = \sqrt{\frac{\pi}{2}}$ and then differentiate with respect to a to find

$$\int_{-\infty}^{\infty} x^2 \exp\left(-a\,x^2\right)\,dx = \frac{1}{a}\sqrt{\frac{\pi}{4a}}.$$

Replace a by $(2a)^{-1}$ in the above result to derive

$$\|g\|_2 = \left(8\pi\,a\right)^{-\frac{1}{4}},$$

$$\sigma_t^2 = \left(8\pi\,a\right)^{\frac{1}{4}} \left\{\frac{1}{4\pi\,a} \cdot \frac{\sqrt{\pi}}{2} \cdot (2a)^{3/2}\right\}^{\frac{1}{2}} = \sqrt{a}.$$

4. For a tight frame, $A = B$.

$$\sum_{n=1}^{3} \left|\left(x, e_n\right)\right|^2 = \left|x_2\right|^2 + \left|\frac{\sqrt{3}}{2}\,x_1 + \frac{1}{2}\,x_2\right|^2 + \left|\frac{\sqrt{3}}{2}\,x_1 - \frac{1}{2}\,x_2\right|^2 = \frac{3}{2}\,\|x\|^2.$$

10. Put $\tau = 0$ in the second result, multiply the resulting expression by $\exp\left(2\pi\,i\,\omega\,t\right)$, and integrate the identity over $\omega \in [-b, b]$. The right-hand side is equal to $f(t)$ by the Fourier inversion theorem, and the left-hand side follows from the definition of the Zak transform combined with integrating the exponential.

11. Since

$$\omega^2 \left|\hat{\chi}_{[0,1]}(\omega)\right|^2 = 4\sin^2\left(\frac{\omega}{2}\right),$$

the second integral is infinite.

12. $$\hat{\chi}_{[0,1]}(\omega) = \exp\left(-\frac{i\,\omega}{2}\right) \frac{\sin\left(\frac{\omega}{2}\right)}{\left(\frac{\omega}{2}\right)}.$$

5.10 Exercises

1. (a) $$W_f(t,\omega) = \sqrt{2\pi}\,\exp\left[-2\left(\frac{t^2}{\sigma^2} + \frac{\sigma^2\omega^2}{4}\right)\right].$$

(c) $$W_f(t,\omega) = 2\exp\left[-\left\{\frac{2\pi t^2}{\sigma^2} + \frac{\sigma^2}{2\pi}(\omega - \omega_0 t)^2\right\}\right].$$

(d) $$W_f(t,\omega) = 2\exp\left[-\left\{\frac{2\pi}{\sigma^2}(t - t_0)^2 + \frac{\sigma^2}{2\pi}(\omega - \omega_0)^2\right\}\right].$$

8. $W_f(t,\omega) = 2(\omega - \omega_0)^{-1} \sin\{2(\omega - \omega_0)(T - |t|)\}\, H(T - |t|)$.

9. Hint: Use Example 5.2.6 and result (5.3.3).
$$W_f(t,\omega) = \frac{\pi}{2}|A|^2\left[\delta(\omega - \omega_0) + \delta(\omega + \omega_0) + 2\delta(\omega)\cos\{2(\omega_0 t + \theta)\}\right].$$

10. Hint: See Auslander and Tolimieri (1985).

13. Hint: Use $f_n g_n - f g = f_n g_n - f_n g + f_n g - f g$ and then apply Schwarz's inequality.

17. $W_f(n,\theta) = |A|^2 \sum_{k=-\infty}^{\infty} \delta(\theta - an - k\pi)$.

18. Hint: $f_m(n) = f(n)\, m_f(n)$, $g_m(n) = g(n)\, m_g(n)$.

22. $\psi_0(x)$ is a Gaussian signal and $\psi_0\left(\frac{x}{\rho}\right)$ is also a Gaussian signal as in Exercise 1(d) with $t_0 = 0$ and $\omega_0 = 0$.

6.6 Exercises

3. Physically, the convolution determines the wavelet transforms with dilated bandpass filters.

11. Write $\cos\omega_0 t = \frac{1}{2}\left(e^{i\omega_0 t} + e^{-i\omega_0 t}\right)$ and calculate the Fourier transform. $\hat{f}(\omega)$ has a maximum at $\omega = \pm\omega_0$, and then maximum values become more and more pronounced as σ increases.

12. $\hat{f}(\omega)$ has a maximum at the frequency $\omega = \omega_0$. Due to the jump discontinuity of $f(t)$ at time $t = \pm a$, $|\hat{f}(\omega)|$ decays slowly as $|\omega| \to \infty$. In fact, $\hat{f}(\omega) \notin L^1(\mathbb{R})$.

13. $\left(\mathcal{W}_\psi f\right)(a,b)=\dfrac{1}{\sqrt{a}}\left[\displaystyle\int_b^{b+\frac{a}{2}} f(t)\,dt-\int_{b+\frac{a}{2}}^{b+a} f(t)\,dt\right]$. Put $t=x+\dfrac{a}{2}$ in the integral $\displaystyle\int_{b+\frac{a}{2}}^{b+a} f(t)\,dt$ to get the answer.

15. Check only $\|\psi\|=1$ and that $\psi_{m,n}$ make up a tight frame with frame constant 1 (see Daubechies, 1992, page 117).

7.7 Exercises

1.
$$\hat{\phi}(\omega)=\int_0^1 (1-t)\,e^{-i\omega t}dt+\int_{-1}^0 (1+t)\,e^{-i\omega t}dt$$
$$=\int_0^1 (1-t)\,e^{-i\omega t}dt+\int_0^1 (1-t)\,e^{i\omega t}dt=\left(\frac{\sin\frac{\omega}{2}}{\frac{\omega}{2}}\right)^2.$$

$$\sum_{k=-\infty}^{\infty}\left|\hat{\phi}(\omega+2\pi k)\right|^2=16\sin^4\left(\frac{\omega}{2}\right)\sum_{k=-\infty}^{\infty}\frac{1}{(\omega+2\pi k)^4}$$
$$=\left(1-\frac{2}{3}\sin^2\frac{\omega}{2}\right)\quad\text{by (7.4.38).}$$

2. Hint: (a) follows from $\hat{\psi}(0)=0$ and (b) follows from $\left(\dfrac{d\hat{\psi}}{d\omega}\right)_{\omega=0}=0$.

(c) follows from $\hat{\psi}(\omega)=\exp\left(\dfrac{i\omega}{2}\right)\hat{f}(\omega)$ and

$$\psi(-t-1)=\frac{1}{2\pi}\int_{-\infty}^{\infty}\hat{f}(\omega)\,\exp[-i\omega(t+1)]\,d\omega$$
$$=\frac{1}{2\pi}\int_{-\infty}^{\infty}\hat{f}(\omega)\,\exp\left[-i\omega\left(t+\frac{1}{2}\right)\right]d\omega.$$

Also, (7.4.43) implies that $\hat{f}(\omega)$ is even and hence,

$$\psi(-t-1) = \frac{1}{2\pi} \int_{-\infty}^{\infty} \hat{f}(\omega) \exp\left(i\omega t + \frac{1}{2} i\omega\right) d\omega$$

$$= \frac{1}{2\pi} \int_{-\infty}^{\infty} \hat{\psi}(\omega) e^{it\omega} d\omega = \psi(t).$$

5. (a) $\phi(x) = 1, \quad 0 \le x < 1.$

   $\psi(x)$ is the Haar wavelet.

   (b) $\phi(x) = B_2(x).$

   $\psi(x) = x, \; 0 \le x < \frac{1}{2}; \; \psi(x) = 2 - 3x, \; \frac{1}{2} \le x < 1.$

   (c) $\phi(x) = \delta(x), \quad \psi(x) = \delta(x).$

6. (a) It follows from (7.3.5) that

$$\int_{-\infty}^{\infty} \phi(x)\, dx = \sqrt{2} \sum_{n=-\infty}^{\infty} c_n \int_{-\infty}^{\infty} \phi(2x-n)\, dx, \; (2x - n = t).$$

   (b) Use $\hat{m}(\pi) = \frac{1}{\sqrt{2}} \sum_{k=-\infty}^{\infty} (-1)^k c_k = 0.$

7. $\hat{\phi}(2\pi) = \hat{m}(\pi)\, \hat{m}\left(\frac{\pi}{2}\right) \hat{m}\left(\frac{\pi}{2^2}\right) \cdots$ and then $\hat{m}(\omega)$ has a zero of order n at $\omega = \pi$ if $\frac{d^m \hat{m}(\omega)}{d\omega^m} = 0$ when $\omega = \pi$ for $m = 0, 1, 2, \cdots (n-1)$. This gives the result.

8. (a) Hint: Write

$$\phi(x) = \sqrt{2} \sum_k c_k\, \phi(2x-k), \quad \phi(x) = \sqrt{2} \sum_m c_m\, \phi(2x-m)$$

$$\int_{-\infty}^{\infty} \phi^2(x)\, dx = 2 \sum_k \sum_m c_k\, c_m \int_{-\infty}^{\infty} \phi(2x-k)\, \phi(2x-m)\, dx$$

   which is zero when $k \ne m$. But, when $k = m$

   $\int_{-\infty}^{\infty} \phi^2(x)\, dx = \sum_k c_k^2 \int_{-\infty}^{\infty} \phi^2(t)\, dt$ which gives the result.

   (b) Corresponding to the scaling function ϕ defined by (7.3.5), the wavelet may be written as

$$\psi(x) = \sum_k (-1)^k\, c_k\, \phi(2x + k - N + 1).$$

Use the orthogonality condition

$$(\psi(x), \psi(x-m)) = \int_{-\infty}^{\infty} \psi(x)\, \psi(n-m)\, dn = 0 \text{ for all } m \text{ except } m = 0.$$

Substitute $\psi(x)$ in this integral so that

$$\int_{-\infty}^{\infty} \psi(x)\, \psi(n-m)\, dx = \sum_k \sum_s (-1)^{k+s} c_k\, c_s \int_{-\infty}^{\infty} \phi(2x+k-N+1)$$

$$\times \phi(2x+s-N+1-2m)\, dx.$$

The right-hand integral is zero unless $k = s - 2m$ so that

$$\int_{-\infty}^{\infty} \psi(x)\, \psi(x-m)\, dx = \sum_k (-1)^{2(k+m)} c_k\, c_{k+2m} \int_{-\infty}^{\infty} \phi^2(2t)\, dt.$$

This is always zero, except $m = 0$. This gives the result.

(c) When $m = 0$, we have

$$\sum_k (-1)^k\, c_k = 0$$

so that

$$\sum_{k=\text{even}} c_k = \sum_{k=\text{odd}} c_k = \frac{1}{\sqrt{2}} \text{ and } \left(\sum_{k=\text{even}} c_k\right)^2 + \left(\sum_{k=\text{odd}} c_k\right)^2 = 1.$$

Multiplying gives $\sum_{k=0}^{N-1} c_k^2 + 2 \sum_{k=0}^{N-1} \sum_{m=1}^{(N/2-1)} c_k\, c_{k+2m} = 1.$

This gives the result by 8(a).

9.
$$c_0 + c_1 + c_3 + c_4 + c_5 = \sqrt{2}$$
$$c_0 - c_1 + c_2 - c_3 + c_4 - c_5 = 0$$
$$-c_1 + 2c_2 - 3c_3 + 4c_4 - 5c_5 = 0$$
$$-c_1 + 4c_2 - 9c_3 + 16c_4 - 25c_5 = 0$$
$$c_0c_2 + c_1c_3 + c_2c_4 + c_3c_5 = 0$$
$$c_0c_4 + c_1c_5 = 0$$
$$c_0^2 + c_1^2 + c_2^2 + c_3^2 + c_4^2 + c_5^2 = 1.$$

11. Hint. (a) $\hat{\phi}(\omega) = \prod_{k=1}^{\infty} m\left(\frac{\omega}{2^k}\right) = \prod_{k=1}^{\infty} \exp\left(-\frac{3i\omega}{2}\cdot\frac{1}{2^k}\right)\cos\left(\frac{3\omega}{2}\cdot\frac{1}{2^k}\right)$

$$= \prod_{k=1}^{\infty} \exp\left(-\frac{ix}{2^k}\right)\cos\left(\frac{x}{2^k}\right), \left(x = \frac{3\omega}{2}\right)$$

$$= \prod_{k=1}^{\infty} \exp\left(-\frac{ix}{2^k}\right)\cdot\prod_{k=1}^{\infty}\cos\left(\frac{x}{2^k}\right)$$

$$= \exp(ix)\cdot\frac{\sin x}{x} = \exp\left(-\frac{3i\omega}{2}\right)\frac{\sin\left(\frac{3\omega}{2}\right)}{\left(\frac{3\omega}{2}\right)}.$$

(b) $\sum_{k=-\infty}^{\infty} \left|\phi(\omega+2\pi k)\right|^2 = \frac{4}{9}\sin^2\left(\frac{3\omega}{2}\right)\sum_{k=-\infty}^{\infty}\frac{1}{(\omega+2\pi k)^2}$

$$= \frac{4}{9}\sin^2\left(\frac{3\omega}{2}\right)\cdot\frac{1}{4}\operatorname{cosec}^2\left(\frac{\omega}{2}\right)$$

$$= \frac{1}{9}\left(\frac{\sin 3t}{\sin t}\right)^2 \quad \left(t = \frac{\omega}{2}\right)$$

$$= \frac{1}{9}\left(\frac{\sin(2t+t)}{\sin t}\right)^2 = \frac{1}{9}(1+2\cos\omega)^2$$

$$= \frac{1}{9}(3+4\cos\omega+2\cos 2\omega).$$

This means that condition (b) in Theorem 7.3.1 is not satisfied.

(c) follows from (a).

12. From (a) and (b), $\phi(x) = 0,\ x < 0$ or $x > 3$.

$$\phi(x)+\phi(x+1)+\phi(x+2) = 1,$$

$$c\phi(x)+(c-1)\phi(x-1)+(c-2)\phi(x+2) = x.$$

Eliminating $\phi(x+2), \phi(x)$, and $\phi(x+1)$ gives (a), (b), and (c) respectively.

8.6 Exercises

2. (a) Replace $f(x)$ and $\phi(x-k)$ by their Fourier inversion formulas in (8.4.7) and then apply

$$\int_{-\infty}^{\infty} \exp\left\{i(\omega_1 - \omega_2)\, x\right\} dx = 2\pi\, \delta(\omega_1 - \omega_2).$$

2. (b) Replace $a_{\phi,k}$ by 2(a) and $\phi(x-k)$ by its Fourier inversion formula to obtain

$$\sum_{k=-\infty}^{\infty} a_{\phi,k}\, \phi(x-k) = 2\pi \sum_{k=-\infty}^{\infty} \int_{-\infty}^{\infty} \hat{f}(\omega_1)\, d\omega_1 \int_{-\infty}^{\infty} \bar{\hat{\phi}}(\omega_1)\, \hat{\phi}(\omega_2) \exp(i\omega_2 x)$$
$$\times \exp\left\{i(\omega_1 - \omega_2)\, k\right\} d\omega_2.$$

Use the Poisson summation formula

$$\frac{1}{2\ell} \sum_{k=-\infty}^{\infty} \exp(-i x k \pi/\ell) = \sum_{m=-\infty}^{\infty} \delta(x - 2m\ell)$$

with $x = \omega_2 - \omega_1$ and $\ell = \pi$ so that

$$\sum_{k=-\infty}^{\infty} \exp\left\{i(\omega_1 - \omega_2)k\right\} = 2\pi \sum_{m=-\infty}^{\infty} \delta(\omega_2 - \omega_1 - 2\pi m).$$

The product $\bar{\hat{\phi}}(\omega_1)\, \hat{\phi}(\omega_2)$ is zero unless ω_1 and ω_2 both lie in $[0, 2\pi]$, which is the case when $m = 0$. Then, $\omega_1 = \omega_2$ and $\bar{\hat{\phi}}(\omega_1)\, \hat{\phi}(\omega_2) = (2\pi)^{-2}$ give the last integral formula.

3. Use the inverse Fourier transform

$$\overline{\psi}(2^m x - k) = 2^{-m} \int_{-\infty}^{\infty} \overline{\psi}(\omega 2^{-m}) \exp(i\omega k 2^{-m}) \exp(-i\omega x)\, d\omega$$

in (8.4.2) and then apply

$$\int_{-\infty}^{\infty} \exp\left\{i(\omega_1 - \omega_2)x\right\} dx = 2\pi\delta(\omega_1 - \omega_2)$$

to obtain 3(a).

4. (8.4.1) and (8.4.9) are identical, provided

$$\sum_{k=-\infty}^{\infty} a_{\phi,k}\, \phi(x-k) = \sum_{m=-\infty}^{-1} \sum_{k=-\infty}^{\infty} a_{m,k}\, \psi(2^m x - k).$$

Use 2(b) and show that

$$\sum_{m=-\infty}^{-1} \sum_{k=-\infty}^{\infty} a_{m,k}\, \psi\left(2^m x - k\right)$$

$$= \sum_{m=-\infty}^{-1} \sum_{k=-\infty}^{\infty} 2\pi\, 2^{-m} \int_{-\infty}^{\infty} d\omega_1 \int_{-\infty}^{\infty} \hat{f}(\omega_1)\, \hat{\psi}\left(\omega_1 2^{-m}\right) \psi\left(\omega_2 2^{-m}\right)$$

$$\times \exp\left(i\omega_2 x\right) \exp\left\{i\left(\omega_1 - \omega_2\right) k\, 2^{-m}\right\} d\omega_2.$$

Use Poisson's summation formula

$$\sum_{k=-\infty}^{\infty} \exp\left\{i\left(\omega_1 - \omega_2\right) k\, 2^{-m}\right\} = 2\pi\, 2^m \sum_{r=-\infty}^{\infty} \delta\left(\omega_2 - \omega_1 - 2\pi r\, 2^m\right)$$

and the fact that $\overline{\psi}\left(\omega_1 2^{-m}\right) \psi\left(\omega_2 2^{-m}\right)$ is zero unless $m = 0$ in the above sum and it is equal to (2π). Consequently,

$$\sum_{m=-\infty}^{-1} \sum_{k=-\infty}^{\infty} a_{m,k}\, \psi\left(2^m x - k\right)$$

$$= \sum_{m=-\infty}^{-1} \int_{2\pi 2^m}^{2\pi 2^{m-1}} \hat{f}(\omega)\, e^{i\omega x} d\omega = \int_0^{2\pi} \hat{f}(\omega)\, e^{i\omega x} d\omega.$$

5. Use results in 3(a) and 3(b) in $\left|a_{k,m}\right|^2 = a_{k,m}\, \overline{a}_{k,m}$ and $\left|\tilde{a}_{k,m}\right|^2 = \tilde{a}_{k,m}\, \overline{\tilde{a}}_{k,m}$. These lead to double integrals. Then, sum over k and m, which involves

$$\sum_{k=-\infty}^{\infty} \exp\left\{i\left(\omega_2 - \omega_1\right) k\, 2^{-m}\right\}$$

and its complex conjugate. Then, use the Poisson summation formula

$$\frac{1}{2\ell} \sum_{k=-\infty}^{\infty} \exp\left\{-\frac{i\,x\,k\,\pi}{\ell}\right\} = \sum_{r=-\infty}^{\infty} \delta\left(x - 2\,r\,\ell\right)$$

with $x = \omega_2 - \omega_1$ and $\ell = \pi\, 2^m$ to obtain

$$\sum_{k=-\infty}^{\infty} \exp\left\{i\left(\omega_1 - \omega_2\right) k\, 2^{-m}\right\} = 2\pi\, 2^m \sum_{r=-\infty}^{\infty} \delta\left(\omega_2 - \omega_1 - 2\pi r\, 2^m\right).$$

Consequently,

$$\sum_{m=-\infty}^{\infty} \sum_{k=-\infty}^{\infty} 2^{-m} \left|a_{k,m}\right|^2 = (2\pi)^3 \sum_{m=-\infty}^{\infty} \int_{-\infty}^{\infty} \hat{f}(\omega)\, \bar{\hat{f}}(\omega)\, \overline{\hat{\psi}}\left(\omega 2^{-m}\right) \hat{\psi}\left(\omega 2^{-m}\right) d\omega$$

$$= 2\pi \int_0^{\infty} \hat{f}(\omega)\, \bar{\hat{f}}(\omega)\, d\omega$$

and hence

$$\sum_{m=-\infty}^{\infty} \sum_{k=-\infty}^{\infty} 2^{-m} \left(|a_{k,m}|^2 + |\tilde{a}_{k,m}|^2 \right) = 2\pi \int_0^{\infty} \left\{ \hat{f}(\omega)\, \bar{\hat{f}}(\omega) + \hat{f}(-\omega)\, \bar{\hat{f}}(-\omega) \right\} d\omega$$

$$= 2\pi \int_{-\infty}^{\infty} \hat{f}(\omega)\, \bar{\hat{f}}(\omega)\, d\omega = \int_{-\infty}^{\infty} |f(x)|^2\, dx.$$

Bibliography

The following bibliography is not, by any means, a complete one for the subject. For the most part, it consists of books and papers to which reference is made in the text. Many other selected books and papers related to material of the subject have been included so that they may serve to stimulate new interest in future study and research.

Anselmet, F., Gagne, Y., Hopfinger, E.J., and Antonia, R.A. (1984), High-order velocity structure functions in turbulent shear flows, *J. Fluid Mech.* **140**, 63-89.

Argoul, F., Arnéodo, A., Elezgaray, J., Grasseau, G., and Murenzi, R. (1988), Wavelet transform of fractal aggregates, *Phys. Lett.* **A135**, 327-333.

Argoul, F., Arnéodo, A., Elezgaray, J., Grasseau, G., and Murenzi, R. (1990), Wavelet transform analysis of self-similarity of diffusion limited aggregrate and electro-deposition clusters, *Phys. Rev.* **A41**, 5537-5560.

Argoul, F., Arnéodo, A., Grasseau, G., Gagne, Y., Hopfinger, E.J., and Frisch, U. (1989), Wavelet analysis of turbulence reveals the multifractal nature of the Richardson cascade, *Nature* **338**, 51-53.

Aslaksen, E.W. and Klauder, J.R. (1968), Unitary representations of the affine group, *J. Math. Phys.* **9**, 206-211.

Aslaksen, E.W. and Klauder, J.R. (1969), Continuous representation theory using the affine group, *J. Math. Phys.* **10**, 2267-2275.

Auslander, L. and Tolimieri, R. (1985), Radar ambiguity functions and group theory, *SIAM J. Math. Anal.* **16**, 577-601.

Balian, R. (1981), Un principe d'incertitude fort en théoreie da signal ou en méchanique quantique, *C.R. Acad. Sci., Paris*, **292**, 1357-1362.

Batchelor, G.K. (1967), *The Theory of Homogeneous Turbulence*, Cambridge University Press, Cambridge.

Battle, G. (1987), A block spin construction of ondelettes, Part 1: Lemarié functions, *Commun. Math. Phys.* **110**, 601-615.

Benedetto, J.J. and Frazier, M.W. (ed.) (1994), *Wavelets: Mathematics and Applications*, CRC Press, Boca Raton, Florida.

Benzi, R., Paladin, G., Parisi, G., and Vulpiani, A. (1984), On the multifractal nature of fully developed turbulence and chaotic systems, *J. Phys.* **A17**, 3521-3531.

Benzi, R. and Vergassola, M. (1991), Optimal wavelet transform and its application to two-dimensional turbulence, *Fluid Dyn. Res.* **8**, 117-126.

Bertrand, J. and Bertrand, P. (1992), A class of affine Wigner functions with extended covariance properties, *J. Math. Phys.* **33**, 2515-2527.

Beylkin, G. (1992), On the representation of operators in bases of compactly supported wavelets, *SIAM J. Numer. Anal.* **29**, 1716-1740.

Beylkin, G., Coifman, R., and Rokhlin, V. (1991), Fast wavelet transforms and numerical algorithms, *Commun. Pure Appl. Math.* **44**, 141-183.

Boashash, B. (1992), *Time-Frequency Signal Analysis, Methods and Applications*, Wiley-Halsted Press, New York.

Burt, P. and Adelson, E. (1983a), The Laplacian pyramid as a compact image code, *IEEE Trans. Commun.* **31**, 482-540.

Burt, P. and Adelson, E. (1983b), A multiplication spline with application to image mosaics, *ACM Trans. Graphics* **2**, 217-236.

Carleson, L. (1966). Convergence and growth of partial sums of Fourier series, *Acta Math.* **116**, 135-157.

Chandrasekhar, S. (1949), On Heisenberg's elementary theory of turbulence, *Proc. Roy. Soc. London* **A200**, 20-33.

Chandrasekhar, S. (1956), A theory of turbulence, *Proc. Roy. Soc. London* **A210**, 1-19.

Choi, H.I. and Williams, W.J. (1989), Improved time-frequency representation of multicomponent signals using exponential kernels, *IEEE Trans. Acoust. Speech, Signal Process.* **37** (6), 862-871.

Chui, C.K. (1992a), On cardinal spline wavelets, in *Wavelets and Their Applications*, (Ed. R.B. Ruskai et al.), Jones and Bartlett, Boston, 419-438.

Chui, C.K. (1992b), *An Introduction to Wavelets*, Academic Press, New York.

Chui, C.K. (1992c), *Wavelets: A Tutorial in Theory and Applications*, Academic Press, New York.

Chui, C.K. and Shi, X. (1993), Inequalities of Littlewood-Paley type for frames and wavelets, *SIAM J. Math. Anal.* **24** (1), 263-277.

Chui, C.K. and Wang, J.Z. (1991), A cardinal spline approach to wavelets, *Proc. Amer. Math. Soc.* **113**, 785-793.

Chui, C.K. and Wang, J.Z. (1992), On compactly supported spline wavelets and a duality principle, *Trans. Amer. Math. Soc.* **330**, 903-915.

Claasen, T.A.C.M. and Mecklenbräuker, W.F.G. (1980), The Wigner distribution-A tool for time-frequency signal analysis, Part I: Continuous time signals; Part II: Discrete time signals; Part III: Relations with other time-frequency signal transformations, *Philips J. Res.* **35**, 217-250; 276-300; 372-389.

Cohen, A. (1990a), Ondelettes, analyses multirésolutions et filtres miroir en quadrature, *Ann. Inst. H. Poincaré, Anal. non linéaire* **7**, 439-459.

Cohen, A. (1990b), *Ondelettes, analyses multirésolutions et traitement numérique du signal*, Ph.D. Thesis, Université Paris, Dauphine.

Cohen, A. (1992), *Biorthogonal Wavelets, in Wavelets – A Tutorial in Theory and Applications*, Academic Press, New York, 123-152.

Cohen, A. (1995), *Wavelets and Multiscale Signal Processing*, Chapman and Hall, London.

Cohen, A., Daubechies, I., and Feauveau, J. (1992), Biorthogonal basis of compactly supported wavelets, *Commun. Pure and Appl. Math.* **45**, 485-560.

Cohen, A. and Daubechies, I. (1993), On the instability of arbitrary biorthogonal wavelet packets, *SIAM J. Math. Anal.* **24** (5), 1340-1354.

Cohen, L. (1966), Generalized phase-space distribution functions, *J. Math. Phys.* **7** (5), 781-786.

Cohen, L. (1989), Time-frequency distributions: Review, *Proc. IEEE* **77** (7), 941-981.

Cohen, L. and Posch, T. (1985), Positive-time frequency distribution functions, *IEEE Trans. Acoust. Speech Signal Process.*, **33**, 31-38.

Cohen, L. and Zaparovanny, Y. (1980), Positive quantum joint distributions, *J. Math. Phys.* **21**, 794-796.

Coifman, R.R., Jones, P., and Semees, S. (1989), Two elementary proofs on the L^2 boundedness of Cauchy integrals on Lipschitz curves, *J. Amer. Math. Soc.* **2**, 553-564.

Coifman, R.R., Meyer, Y., and Wickerhauser, M.V. (1992a), Wavelet analysis and signal processing, in *Wavelets and Their Applications* (Ed. M.B. Ruskai et al.), Jones and Bartlett, Boston, 153-178.

Coifman, R.R., Meyer, Y., and Wickerhauser, M.V. (1992b), Size properties of wavelet packets, in *Wavelets and Their Applications*, (Ed. M.B. Ruskai et al.), Jones and Bartlett, Boston, 453-470.

Daubechies, I. (1988a), Time-frequency localization operators: A geometric phase space approach, *IEEE Trans. Inform. Theory* **34**, 605-612.

Daubechies, I. (1988b), Orthogonal bases of compactly supported wavelets, *Commun. Pure Appl. Math.* **41**, 909-996.

Daubechies, I. (1990), The wavelet transform, time-frequency localization and signal analysis, *IEEE Trans. Inform. Theory* **36**, 961-1005.

Daubechies, I. (1992), *Ten Lectures on Wavelets*, NSF-CBMS Regional Conference Series in Applied Math. **61**, SIAM Publications, Philadelphia.

Daubechies, I. (1993), Different perspectives on wavelets, *Proceedings of the Symposium on Applied Mathematics* **47**, 15-37, American Mathematical Society, Providence.

Daubechies, I., Grossmann, A., and Meyer, Y. (1986), Painless nonorthogonal expansion, *J. Math. Phys.* **27** 1271-1283.

Daubechies, I. and Klauder, J. (1985), Quantum mechanical path integrals with Wiener measure for all polynomial Hamiltonians II, *J. Math. Phys.* **26**, 2239-2256.

Daubechies, I. and Lagarias, J. (1991), Two-scale difference equations I. Existence and global regularity of solutions, *SIAM J. Math. Anal.* **22**, 1388-1410.

Daubechies, I. and Lagarias, J. (1992), Two-scale difference equations II. Local regularity, infinite products of matrices and fractals, *SIAM J. Math. Anal.* **23**, 1031-1055.

De Bruijn, N.G. (1967), Uncertainty principles in Fourier analysis, In *Inequalities* (ed. O. Shisha), Academic Press, Boston, 57-71.

De Bruijn, N.G. (1973), A theory of generalized functions with applications to Wigner distribution and Weyl correspondence, *Nieuw. Arch. Voor Wiskunde* **21**, 205-280.

Debnath, L. (1978), *Oceanic Turbulence*, Memoir Series of the Calcutta Mathematical Society, Calcutta.

Debnath, L. (1995), *Integral Transforms and Their Applications*, CRC Press, Boca Raton, Florida.

Debnath, L. (1998a), Wavelet transforms, fractals, and turbulence, in *Nonlinear Instability, Chaos, and Turbulence*, Vol. I, (Ed. L. Debnath and D.N. Riahi), Computational Mechanics Publications, WIT Press, Southampton, England.

Debnath, L. (1998b), Brief introduction to history of wavelets, *Internat. J. Math. Edu. Sci. Tech.* **29**, 677-688.

Debnath, L. (1998c), Wavelet transforms and their applications, *Proc. Indian Natl. Sci. Acad.* **64A**, 685-713.

Debnath, L. (2001), *Wavelet Transforms and Time-Frequency Signal Analysis*, Birkhäuser, Boston.

Debnath, L. and Mikusinski, P. (1999), *Introduction to Hilbert Spaces with Applications*, Second Edition, Academic Press, Boston.

Dirac, P.A.M. (1958), *The Principles of Quantum Mechanics* (Fourth Edition), Oxford University Press, Oxford.

Donoho, D. (1994), Interpolating wavelet transforms, *J. Appl. Comput. Harmonic Analy.*, **2**, 1-12

Duffin, R.J. and Schaeffer, A.C. (1952), A class of nonharmonic Fourier series, *Trans. Amer. Math. Soc.* **72**, 341-366.

Duhamel, P. and Vetterli, M. (1990), Fast Fourier transforms: A tutorial review and a state of the art, *Signal Process.* **19** (4), 259-299.

Erlebacher, G., Hussani, M.Y., and Jameson, L.M. (1966), *Wavelets, Theory and Applications*, Oxford University Press, Oxford.

Everson, R., Sirovich, L., and Sreenivasan, K.R. (1990), Wavelet analysis of turbulent jet, *Phys. Lett.* **A145**, 314-319.

Falconer, K.J. (1990), *Fractal Geometry, Mathematical Foundations and Applications*, John Wiley, Chichester.

Farge, M. (1992), Wavelet transforms and their applications to turbulence, *Ann. Rev. Fluid Mech.* **24**, 395-457.

Farge, M., Goirand, E., Meyer, Y., Pascal, F., and Wickerhauser, M.V. (1992), Improved predictability of two-dimensional turbulent flows using wavelet packet compression, *Fluid Dyn. Res.* **10**, 229-242.

Farge, M., Guezenne, J., Ho, C.M., and Meneveau, C. (1990), *Continuous Wavelet Analysis of Coherent Structures*, Proceedings of the Summer Progress Centre for Turbulence Research, Stanford University-NASA Ames, 331-398.

Farge, M. and Holschneider, M. (1989), Analysis of two-dimensional turbulent flow, *Proc. Scaling, Fractals, and Nonlinear Variability in Geophysics II*, Barcelona.

Farge, M. and Holschneider, M. (1990a), Interpolation of two-dimensional turbulence spectrum in terms of singularity in the vortex cores, *Europhys. Lett.* **15**, 737-743.

Farge, M., Holschneider, M., and Colonna, J.F. (1990b), Wavelet analysis of coherent structures in two-dimensional turbulent flows, in *Topological*

Fluid Mechanics (Ed. A.K. Moffatt and A. Tsinober), Cambridge University Press, Cambridge, 765-766.

Farge, M., Kevlahan, N., Perrier, V., and Goirand, E. (1996), Wavelets and turbulence, *Proc. IEEE* **84** (4), 639-669.

Farge, M., Kevlahan, N., Perrier, V., and Schneider, K. (1999a), Turbulence analysis, modelling and computing using wavelets, *Wavelets in Physics* (Ed. J.C. van den Berg), Cambridge University Press, Cambridge, 117-200.

Farge, M., Schneider, K., and Kevlahan, N. (1999b), Non-Gaussianity and coherent vortex simulation for two-dimensional turbulence using an adaptive orthonormal wavelet basis, *Phys. Fluids* **11** (8), 2187-2201.

Farge, M. and Rabreau, G. (1988), Transformé en ondelettes pour detecter et analyser les structures cohérentes dans les ecoulements turbulents bidimensionnels, *C.R. Acad. Sci. Paris Ser. II* **307**, 1479-1486.

Farge, M. and Rabreau, G. (1989), Wavelet transform to analyze coherent structures in two-dimensional turbulent flows, *Proceedings on Scaling, Fractals and Nonlinear Variability in Geophysics I*, Paris.

Flandrin, P. (1992), Wavelet analysis and synthesis of fractional Brownian motion, *IEEE Trans. Inform. Theory* **38** (2), 910-916.

Fourier, J. (1822). *Théorie Analytique de la Chaleur*, English Translation by A. Freeman, Dover Publications, New York, 1955.

Frisch, U. and Parisi, G. (1985), On the singularity structure of fully developed turbulence, in *Turbulence and Predictability in Geophysical Fluid Dynamics and Climate Dynamics* (Ed. M. Ghil, R. Benzi, and G. Parisi), North Holland, Amsterdam, 84-125.

Frisch, U., Sulem, P.L., and Nelkin, M. (1978), A simple dynamical model of intermittent fully developed turbulence, *J. Fluid Mech.* **87**, 719-736.

Frisch, U. and Vergassola, M. (1991), A prediction of the multifractal model; the intermediate dissipation range, *Europhys. Lett.* **14**, 439-450.

Frisch, U. (1995), *Turbulence. The Legacy of A.N. Kolmogorov*, Cambridge University Press, Cambridge.

Fröhlich, J. and Schneider, K. (1997), An adaptive wavelet-vaguelette algorithm for the solution of PDEs, *J. Comput. Phys.* **130**, 174-191.

Gabor, D. (1946), Theory of communications, *J. Inst. Electr. Eng. London*, **93**, 429-457.

Gagne, Y. and Castaing, B. (1991), A universal non-globally self-similar representation of the energy spectra in fully developed turbulence, *C.R. Acad. Sci.* Paris, **312**, 414-430.

Gelfand, I.M. (1950), Eigenfunction expansions for an equation with periodic coefficients, *Dokl. Akad. Nauk. SSR* **76**, 1117-1120.

Glauber, R.J. (1964), in *Quantum Optics and Electronics* (Ed. C. de Witt, A. Blandin, and C. Cohen-Tannoudji), Gordon and Breach, New York.

Grant, H.L., Stewart, R.W., and Moilliet, A. (1962), Turbulence spectra from a tidal channel, *J. Fluid Mech.* **12**, 241-268.

Grossmann, A. and Morlet, J. (1984), Decomposition of Hardy functions into square integrable wavelets of constant shape, *SIAM J. Math. Anal.* **15**, 723-736.

Haar, A. (1910), Zur Theorie der orthogonalen funktionen-systeme, *Math. Ann.* **69**, 331-371.

Hall, M.G. (1972), Vortex breakdown, *Ann. Rev. Fluid Mech.* **4**, 195-218.

Halsey, T.C., Jensen, M.H., Kadanoff, L.P., Procaccia, I., and Shraiman, B.I. (1986), Fractal measures and their singularities: The characterization of strange sets, *Phys. Rev.* **A33**, 1141-1151.

Heil, C. and Walnut, D. (1989), Continuous and discrete wavelet transforms, *SIAM Rev.* **31**, 628-666.

Heisenberg, W. (1948a), Zur statistischen theori der turbulenz, *Z. Phys.* **124**, 628-657.

Heisenberg, W. (1948b), On the theory of statistical and isotropic turbulence, *Proc. Roy. Soc. London* **A195**, 402-406.

Hlawatsch, F. and Boudreaux-Bartels, F. (1992), Linear and quadratic time-frequency signal representations, *IEEE Signal. Process. Mag.* **9** (2), 21-67.

Hlawatsch, F. and Flandrin, P. (1992), The interference structure of the Wigner distribution and related time-frequency signal representations, in *The Wigner Distribution-Theory and Applications in Signal Processing*, (Ed. W.F.G. Mecklenbrauker), Elsevier, Amsterdam.

Holschneider, M. (1988), L'analyse d'objets fractals et leu transformaé en ondelettes, These, Université de Provence, Marseille.

Holschneider, M., Kronland-Martinet, R., Morlet, J., and Tchamitchian, P. (1989), A real-time algorithm for signal analysis with the help of the wavelet transform, in *Wavelets, Time-Frequency Methods and Phase Space*, (Ed. J.M. Combes, et al.), Springer-Verlag, Berlin, 289-297.

Holschneider, M. and Tchamitchian, P. (1991), Pointwise regularity of Riemann's "nowhere-differentiable" function, *Invent. Math.* **105**, 157-175.

Huang, Z., Kawall, J.G., and Keffer, J.F. (1997), Development of structure within the turbulent wake of a porous body. Part 2. Evolution of the three-dimensional features, *J. Fluid Mech.* **329**, 117-136.

Hunt, J.C.R. and Vassilicos, J.C. (1991), Kolmogorov's contributions to the physical and geometrical understanding of small-scale turbulence and recent developments, *Proc. Roy. Soc. London* **A434**, 183-210.

Hunt, J.C.R., Kevlahan, N.K.-R., Vassilicos, J.C., and Farge, M. (1993), Wavelets, fractals, and Fourier transforms: Detection and analysis of structure, in *Wavelets, Fractals, and Fourier Transforms*, (Ed. M. Farge, J.C.R. Hunt, and J.C. Vassilicos), Clarendon Press, Oxford.

Hussain, J.J., Schoppa, W., and Kim, J. (1997), Coherent structures near the wall in a turbulent channel flow, *J. Fluid Mech.* **332**, 185-214.

Jaffard, S. and Meyer, Y. (1996), *Wavelet Methods for Pointwise Regularity and Local Oscillations of Functions*, Volume 123, American Mathematical Society, Providence, RI.

Jähne, B. (1995), *Digital Image Processing*, Springer Verlag, New York.

Janssen, A.J.E.M. (1981a), Weighted Wigner distribution vanishing on lattices, *J. Math. Anal. Appl.* **80**, 156-167.

Janssen, A.J.E.M. (1981b), Gabor representations of generalized functions, *J. Math. Anal. Appl.* **83**, 377-394.

Janssen, A.J.E.M. (1982), On the locus and spread of pseudo-density functions in the time-frequency plane, *Philips J. Res.* **37**, 79-110.

Janssen, A.J.E.M. (1984), Gabor representation and Wigner distribution of signals, *Proc. IEEE* 41, B.2.1-41, B.24.

Janssen, A.J.E.M. (1988), The Zak transform: A signal transform for sampled time-continuous signals, *Philips J. Res.* **43**, 23-69.

Janssen, A.J.E.M. (1992), The Smith-Barnwell condition and non-negative scaling functions, *IEEE Trans. Inform. Theory* **38**, 884-885.

Jimenez, J. (1981), *The Role of Coherent Structures in Modeling Turbulence and Mixing*, Springer-Verlag, New York.

Kahane, J.-P. and Katznelson, Y. (1966). Sur les ensembles de divergence des séries trigonometriques, *Studia Math.* **26**, 305-306.

Kaiser, G. (1994), *A Friendly Guide to Wavelets*, Birkhäuser, Boston.

Kirkwood, J.G. (1933), Quantum statistics of almost classical ensembles, *Phys. Rev.* **44**, 31-37.

Klauder, J. and Skagerstam, B.S. (1985), *Coherent States*, World Scientific Publications, Singapore.

Kolmogorov, A. (1926). Une série de Fourier-Lebesgue divergent partout, *C.R. Acad. Sci. Paris* **183**, 1327-1328.

Kolmogorov, A.N. (1941a), The local structure of turbulence in an incompressible fluid with very large Reynolds numbers, *Dokl. Akad. Nauk. SSSR* **30**, 301-305.

Kolmogorov, A.N. (1941b), Dissipation of energy under locally isotropic turbulence, *Dokl. Akad. Nauk. SSSR* **32**, 16-18.

Kolmogorov, A.N. (1962), A refinement of previous hypotheses concerning the local structure of turbulence of a viscous incompressible fluid at high Reynolds numbers, *J. Fluid Mech.* **13**, 82-85.

Kovacevic, J. and Vetterli, M. (1992), Nonseparable multidimensional perfect reconstruction filter banks and wavelet bases for $\mathbb{R}^n$, *IEEE Trans. Inform. Theory* **38**, 533-555.

Landau, L. and Lifschitz, E. (1959), *Fluid Mechanics*, Addison-Wesley, New York.

Lemarié, P.G. (1988), Une nouvelle base d'ondelettes de $L^2(\mathbb{R}^n)$, *J. Math. Pure Appl.* **67**, 227-236.

Lemarié, P.G. (1989), Bases d'ondelettes sur les groupes de Lie stratifiés, *Bull. Soc. Math. France* **117**, 211-232.

Lemarié, P.G., ed., (1990), *Les Ondelettes en 1989*, Lecture Notes in Mathematics, Volume 1438, Springer Verlag, Berlin.

Lemarié, P.G. (1991), La propriété de support minimal dans les analyses multirésolution, *C. R. Acad. Sci. Paris* **312**, 773-776.

Lemarié, P.G. and Meyer, Y. (1986), Ondelettes et bases hilbertiennes, *Rev. Mat. Iberoam.* **2**, 1-18.

Leonard, A. (1974), Energy cascade in large eddy simulations of turbulent fluid flows, *Adv. Geophys.* **18A**, 237-249.

Lewis, A.S. and Knowles, G. (1992), Image compression using the 2D-wavelet transform, *IEEE Trans. Image Process.* **1**, 244-265.

Lin, C.C. (1948), Note on the law of decay of isotropic turbulence, *Proc. Nat. Acad. Sci.* U.S.A., **34**, 230-233.

Loughlin, P., Pitton, J., and Atlas, L.E. (1994), Construction of positive time-frequency distributions, *IEEE Trans. Signal Process.* **42**, 2697-2836.

Low, F. (1985), *Complete Sets of Wavepackets, A Passion for Physics – Essays in Honor of Geoffrey Chew* (Ed. C. De Tar et al.), World Scientific, Singapore, 17-22.

Lukacs, M. (1960), *Characteristic Functions*, Griffin Statistical Monographs, New York.

Mallat, S. (1988), Multiresolution representation and wavelets, Ph.D. Thesis, University of Pennsylvania.

Mallat, S. (1989a), Multiresolution approximations and wavelet orthonormal basis of $L^2(\mathbb{R})$, *Trans. Amer. Math. Soc.* **315**, 69-88.

Mallat, S. (1989b), A theory for multiresolution signal decomposition: The wavelet representation, *IEEE Trans. Patt. Recog. And Mach. Intell.* **11**, 678-693.

Mallat, S. (1989c), Multifrequency channel decompositions of images and wavelet models, *IEEE Trans. Acoust. Signal Speech Process.* **37**, 2091-2110.

Mallat, S. and Hwang, W.L. (1991), Singularity detection and processing with wavelets, *IEEE Trans. Inform. Theory* **38**, 617-643.

Malvar, H.S. (1990a), Lapped transforms for efficient transforms/subband coding, *IEEE Trans. Acoust. Speech Signal Process.* **38**, 969-978.

Malvar, H. (1990b), The design of two-dimensional filters by transformations, *Seventh Annual Princeton Conference on ISS*, Princeton University Press, Princeton, New Jersey, **247**, 251-264.

Mandelbrot, B.B. (1974), Intermittent turbulence in self-similar cascades: Divergence of high moments and dimension of the carrier, *J. Fluid Mech.* **62**, 331-358.

Mandelbrot, B.B. (1975), On the geometry of homogeneous turbulence with stress on the fractal dimension of the iso-surfaces of scalars, *J. Fluid Mech.* **72**, 401-420.

Mandelbrot, B.B. (1982), *The Fractal Geometry of Nature*, W.H. Freeman, New York.

Margenau, H. and Hill, R.N. (1961), Correlation between measurements in quantum theory, *Prog. Theor. Phys.* **26**, 722-738.

Mecklenbräuker, W. and Hlawatsch, F. (1997), *The Wigner Distribution*, Elsevier, Amsterdam.

Meneveau, C. (1991), Analysis of turbulence in the orthonormal wavelet representation, *J. Fluid Mech.* **232**, 469-520.

Meneveau, C. (1993), Wavelet analysis of turbulence: The mixed energy cascade, in *Wavelets, Fractals, and Fourier Transforms* (Ed. M. Farge, J.C.R. Hunt, and J.C. Vassilicos), Oxford Univrsity Press, Oxford, 251-264.

Meneveau, C. and Sreenivasan, K.R. (1987a), Simple multifractal cascade model for fully developed turbulence, *Phys. Rev. Lett.* **59**, 1424-1427.

Meneveau, C. and Sreenivasan, K.R. (1987b), In *Physics of Chaos and Systems Far From Equilibrium* (Ed. M.D. Van, and B. Nichols), *Nucl. Phys. B*, North Holland, Amsterdam, 2-49.

Métais, O. and Lesieur, M. (1992), Spectral large-eddy simulation of isotropic and stably stratified turbulence, *J. Fluid Mech.* **239**, 157-194.

Meyer, Y. (1990), *Ondelettes et opérateurs*, tomes I, II, and III, Herman, Paris.

Meyer, Y. (1993a), *Wavelets, Algorithms and Applications* (translated by R.D. Ryan), SIAM Publications, Philadelphia.

Meyer, Y. (1993b), *Wavelets and Operators*, Cambridge University Press, Cambridge.

Micchelli, C. (1991), Using the refinement equation for the construction of prewavelets, *Numer. Alg.* **1**, 75-116.

Moffatt, H.K. (1984), Simple topological aspects of turbulent vorticity dynamics, in *Turbulence and Chaotic Phenomena in Fluids* (Ed. T. Tatsumi), Elsevier, New York, 223-230.

Monin, A.S. and Yaglom, A.M. (1975), *Statistical Fluid Mechanics: Mechanics of Turbulence*, Vol. 2, MIT Press, Cambridge.

Morlet, J., Arens, G., Fourgeau, E., and Giard, D. (1982a), Wave propagation and sampling theory, Part I: Complex signal land scattering in multilayer media, *J. Geophys.* **47**, 203-221.

Morlet, J., Arens, G., Fourgeau, E., and Giard, D. (1982b), Wave propagation and sampling theory, Part II: Sampling theory and complex waves, *J. Geophys.* **47**, 222-236.

Mouri, H. and Kubotani, H. (1995), Real-valued harmonic wavelets, *Phys. Lett.* **A201**, 53-60.

Mouri, H., Kubotani, H., Fugitani, T., Niino, H., and Takaoka, M. (1999), Wavelet analysis of velocities in laboratory isotropic turbulence, *J. Fluid Mech.* **389**, 229-254.

Myint-U., T. and Debnath, L. (1987), *Partial Differential Equations for Scientists and Engineers* (Third Edition), Prentice Hall, New York.

Newland, D.E. (1993a), Harmonic wavelet analysis, *Proc. Roy. Soc. London* **A443**, 203-225.

Newland, D.E. (1993b), *An Introduction to Random Vibrations, Spectral and Wavelet Analysis*, (Third Edition), Longman Group Limited, London, England.

Newland, D.E. (1994), Harmonic and musical wavelets, *Proc. Roy. Soc. London* **A444**, 605-620.

Novikov, E.A. (1989), Two-particle description of turbulence, Markov property and intermittency, *Phys. Fluids* **A1**, 326-330.

Novikov, E.A. (1990), The effects of intermittency on statistical characteristics of turbulence and scale similarity of breakdown coefficients, *Phys. Fluids* **A2**, 814-820.

Oboukhov, A.M. (1941), On the distribution of energy in the spectrum of turbulent flow, *Dokl. Akad. Nauk. SSSR* **32**, 19-21.

Oboukhov, A.M. (1962), Some specific features of atmospheric turbulence, *J. Fluid Mech.* **13**, 77-81.

Onsager, L. (1945), The distribution of energy in turbulence, *Phys. Rev.* **68**, 286-292.

Onsager, L. (1949), Statistical hydrodynamics, *Nuovo Cimento Suppl.* **6**, 279-287.

Perrier, V., Philipovitch, T., and Basdevant, C. (1995), Wavelet spectra compared to Fourier spectra, *J. Math. Phys.* **36**, 1506-1519.

Ramchandran, K. and Vetterli, M. (1993), Best wavelet packet bases in a rate-distortion sense, *IEEE Trans. Image Process.* **2** (2), 160-175.

Richardson, L.F. (1926), Atmospheric diffusion shown on a distance-neighbour graph, *Proc. Roy. Soc. London* **A110**, 709-737.

Rihaczek, A.W. (1968), Signal energy distribution in time and frequency, *IEEE Trans. Inform. Theory* **14**, 369-374.

Ruskai, M.B., Beylkin, G., Coifman, R., Daubechies, I., Mallat, S., Meyer, Y., and Raphael, L. (ed.) (1992), *Wavelets and Their Applications*, Jones and Bartlett, Boston.

Saffman, P.G. (1990), Vortex dynamics and turbulence, in *Proceedings of the NATO Advanced Workshop on the Global Geometry of Turbulence: Impact of Nonlinear Dynamics*, 1-30.

Saffman, P.G. and Baker, G.R. (1979), Vortex interactions, *Ann. Rev. Fluid Mech.* **11**, 95-122.

Sarker, S.K. (1985), Generalization of singularities in nonlocal dynamics, *Phys. Rev.* **A31**, 3468-3472.

Schempp, W. (1984), Radar ambiguity functions, the Heisenberg group and holomorphic theta series, *Proc. Amer. Math. Soc.* **92**, 103-110.

Schneider, K. and Farge, M. (1997), Wavelet forcing for numerical simulation of two-dimensional turbulence, *C.R. Acad. Sci. Paris Série II* **325**, 263-270.

Schneider, K. and Farge, M. (2000), Computing and analyzing turbulent flows using wavelets, in *Wavelet Transforms and Time-Frequency Signal Analysis* (Ed. L. Debnath), Birkhauser, Boston, 181-216.

Schneider, K., Kevlahan, N.K.R., and Farge, M. (1997), Comparison of an adaptive wavelet method and nonlinearly filtered pseudo-spectral methods for two-dimensional turbulence, *Theor. Comput. Fluid Dyn.* **9**, 191-206.

Schwarz, K.W. (1990), Evidence for organized small-scale structure in fully developed turbulence, *Phys. Rev. Lett.* **64** 415-418.

Sen, N.R. (1951), On Heisenberg's spectrum of turbulence, *Bull. Cal. Math. Soc.* **43**, 1-7.

Sen, N.R. (1958), On decay of energy spectrum of isotropic turbulence, *Proc. Natl. Inst. Sci. India* **A23**, 530-533.

Shannon, C.E. (1949), Communications in the presence of noise, *Proc. IRE* **37**, 10-21.

She, Z.S., Jackson, E., and Orszag, S.A. (1991), Structure and dynamics of homogeneous turbulence, models and simulations, *Proc. Roy. Soc. London* **A434**, 101-124.

Sheehan, J.P. and Debnath, L. (1972), On the dynamic response of an infinite Bernoulli-Euler beam, *Pure Appl. Geophys.* **97**, 100-110.

Sreenivasan, K.R. (1991), Fractals and multifractals in fluid turbulence, *Ann. Rev. Fluid Mech.* **23**, 539-600.

Sreenivasan, K.R. and Meneveau, C. (1986), The fractal facets of turbulence, *J. Fluid Mech.* **173**, 357-386.

Stadler, W. and Shreeves, R.W. (1970), The transient and steady state response of the infinite Bernoulli-Euler beam with damping and an elastic foundation, *Quart. J. Mech. and Appl. Math.* **XXII**, 197-208.

Strang, G. (1989), Wavelets and dilation equations, *SIAM Review* **31**, 614-627.

Strang, G. (1993), Wavelet transforms versus Fourier transforms, *Bull. Amer. Math. Soc.* **28**, 288-305.

Strang, G. and Nguyen, T. (1996), *Wavelets and Filter Banks*, Wellesley-Cambridge Press, Boston.

Stromberg, J.O. (1982), A modified Franklin system and higher order spline systems on $\mathbb{R}^n$ as unconditional bases for Hardy spaces, in *Conference on Harmonic Analysis in Honor of Antoni Zygmund, II* (Ed. W. Beckner et al.), Wadsworth Mathematics Series, Wadsworth, Belmont, California, 475-493.

Tchamitchan, P. (1987), Biorthogonalité et théorie des opérateurs, *Rev. Mat., Iberoam.* **3**, 163-189.

Thomson, Sir William and Tait, P.G. (1879), Treatise on natural philosophy, Cambridge University Press, New edition: *Principles of Mechanics and Dynamics* (1962), Dover Publications, New York.

Townsend, A.A. (1951), On the fine scale structure of turbulence, *Proc. Roy. Soc. London* **A208**, 534-556.

Vaidyanathan, P.P. (1987), Theory and design of M-channel maximally decimated quadrature mirror filters with arbitrary M, having the perfect reconstruction property, *IEEE Trans. Acoust. Speech Signal Process.* **35**, 476-492.

Vaidyanathan, P.P. (1993), *Multirate Systems and Filter Banks*, Prentice-Hall, Englewood Cliffs, New Jersey.

Vaidyanathan, P.P. and Hoang, P.-Q. (1988), Lattice structures for optimal design and robust implementation of two-channel perfect-reconstruction QMF banks, *IEEE Trans. Acoust. Speech Signal Process.* **36**, 81-94.

Vassilicos, J.C. (1992), The multi-spiral model of turbulence and intermittency, in *Topological Aspects of the Dynamics of Fluids and Plasmas*, Kluwer, Amsterdam, 427-442.

Vassilicos, J.C. (1993), Fractals in turbulence, in *Wavelets, Fractals and Fourier Transforms: New Developments and New Applications*, (Ed. M. Farge, J.C.R. Hunt, and J.C. Vassilicos), Oxford University Press, Oxford, 325-340.

Vassilicos, J.C. and Hunt, J.C.R. (1991), Fractal dimensions and spectra of interfaces with applications to turbulence, *Proc. Roy. Soc. London* **A435**, 505-534.

Vetterli, M. (1984), Multidimensional subband coding: Some theory and algorithms, *Signal Process.* **6**, 97-112.

Vetterli, M. and Kovacevic, J. (1995), *Wavelets and Subband Coding*, Prentice-Hall, Englewood Cliffs, New Jersey.

Ville, J. (1948), Théorie et applications de la notion de signal analytique, *Cables Transm.* **2A**, 61-74.

Vincent A. and Meneguzzi, M. (1991), The spatial structure and statistical properties of homogeneous turbulence, *J. Fluid Mech.* **225**, 1-20.

Vincent, A. and Meneguzzi, M. (1994), The dynamics of vorticity tubes in homogeneous turbulence, *J. Fluid Mech.* **258**, 245-254.

Von Neumann, J. (1945), *Mathematical Foundations of Quantum Mechanics*, Princeton University Press, Princeton.

Walsh, J.L. (1923), A closet set of normal orthogonal functions, *Amer. J. Math.* **45**, 5-24.

Wickerhauser, M.V. (1994), *Adapted Wavelet Analysis from Theory to Software*, AK Peters, Wellesley, Massachusetts.

Wigner, E.P. (1932), On the quantum correction for thermodynamic equilibrium, *Phys. Rev.* **40**, 749-759.

Wilcox, C. (1960), The synthesis problem for radar ambiguity functions, MRC Technical Report 157, Mathematics Research Center, U.S. Army, University of Wisconsin, Madison.

Woodward, P.M. (1953), *Probability and Information Theory, with Applications to Radar*, Pergamon Press, London.

Wornell, G.W. (1995), *Signal Processing with Fractals: A Wavelet-Based Approach*, Prentice-Hall, Englewood Cliffs, New Jersey.

Yamada, M., Kida, S., and Ohkitani, K. (1993), Wavelet analysis of PDEs in turbulence, in *Unstable and Turbulent Motion of Fluid* (Ed. S. Kida), World Scientific Publishing, Singapore, 188-190.

Yamada, M. and Ohkitani, K. (1991), An identification of energy cascade in turbulence by orthonormal wavelet analysis, *Prog. Theor. Phys.* **86**, 799-815.

Zak, J. (1967), Finite translation in solid state physics, *Phys. Rev. Lett.* **19**, 1385-1397.

Zak, J. (1968), Dynamics of electrons in solids in external fields, *Phys. Rev.* **168**, 686-695.

Zayed, A. (1993), *Advances in Shannon's Sampling Theory*, CRC Press, Boca Raton, Florida.

Zhao, Y., Atlas, L.E., and Marks, R. (1990), The use of cone-shaped kernels for generalized time-frequency representations of nonstationary signals, *IEEE Trans. Acoust., Speech Signal Process.* **38**, 1084-1091.

Index

C

D

G

Printed in the United States
54301LVS00001B/15

9 780817 642044